QC Scona.
21.2
P67
1971

000357

LEARNING RESOURCE CENTRE
GRANT MacEWAN COMMUNITY COLLEGE

Date Due

DEC 18 1972		
FEB 7 1973		
FEB 19 1973		
OCT 10 1974		
MAR 19 1991		
Dec 13		

Scona

QC   Pollack, Herman W
21.2   Applied physics
.P67
1971

50-1

Applied Physics

PRENTICE-HALL INTERNATIONAL, INC., *London*
PRENTICE-HALL OF AUSTRALIA, PTY., LTD., *Sydney*
PRENTICE-HALL OF CANADA, LTD., *Toronto*
PRENTICE-HALL OF INDIA PRIVATE LIMITED., *New Delhi*
PRENTICE-HALL OF JAPAN, INC., *Tokyo*

APPLIED PHYSICS

Second Edition

HERMAN W. POLLACK

Chairman, Physical Sciences Division
Orange County Community College

PRENTICE-HALL, INC., ENGLEWOOD CLIFFS, NEW JERSEY

© 1971, 1964 by
PRENTICE-HALL, INC.
Englewood Cliffs, New Jersey

All rights reserved. No part of this book
may be reproduced in any way, or by any
means, without permission in writing from
the publisher.

Current printing (last digit):

10   9   8   7   6   5   4   3   2   1

13-041350-X

Library of Congress Catalog Card Number: 76-158610
Printed in the United States of America

**Dedicated to the memory of
My mother, Sarah, and my father, Max,
and to
My wife's mother, Lillian, and her father, Sydney**

Preface

The first edition of Applied Physics was intended to be used in a course designed to service technical curricula at the two-year colleges. At the time the first edition was published, it was the author's contention that Physics for the technician had to lay the theoretical base and develop in the student calculative skills for the technical courses which follow in a two-year curriculum. This text attempted to accomplish these ends in the first edition, and very little has happened since 1964 to indicate that the objective of the second edition should be any other.

The second edition requires two years of high school college preparatory mathematics. More important, a freshman college technical mathematics course should be taken concurrently with the physics. Thus, the student's mathematical skills will keep pace with the physics. This text, and the approach used, although not written with the Dr. Harris model* in mind, comes surprisingly close to the recommendations stated.

The author wishes to point out several changes which were recommended and made:

1. Approximately ninety per cent of the problems at the end of each chapter are new.
2. Approximately ninety per cent of the text has been rewritten. Some material has been deleted and some added.
3. The C.G.S. system of units has been deleted. The emphasis is upon the f.p.s. and M.K.S. systems of units.

*See "Technical Education in the Junior College/New Programs for New Jobs." Dr. Norman C. Harris, AAJC, p. 70.

4. Several chapters have been divided into two chapters in order to facilitate pacing.
5. The sequence of several of the chapters has been changed for better continuity. This is especially true in electricity.
6. The presentation of wave motion and sound now precedes that of light.
7. All tables have been moved into the Appendix. This will eliminate the need for "hunting" when a value is desired.

Answers to the even-numbered problems are included in the Appendix.

The author owes much to his colleagues and to his family who were so very helpful, patient, and encouraging throughout the preparation of the original manuscript. Having suffered through one manuscript, it is remarkable that they were able to find the courage to negotiate another.

Special recognition and thanks are extended to Mr. Paul Hlavaty and Professor Reuben Benumof and Mrs. Laura B. Wyman for review and criticisms offered. The author also wishes to thank his secretary, Mrs. Regina Westeris, for all the typing which she did beyond the call of duty. And finally, the author wishes to thank Mr. Leon J. Liguori (Prentice-Hall) for his work in the final preparation of this text and Mr. Matthew Fox (Prentice-Hall) for his faith in requesting a second edition.

HERMAN W. POLLACK

Contents

1 Some Basic Concepts 1

1.1. The Role of Technical Physics, 1; 1.2. How to Study, 2; 1.3. Scalar and Vector Quantities, 3; 1.4. Standardizing Basic Units, 3; 1.5. Operation with Units, 5; 1.6. Some Mathematics, 9; 1.7. The Slide Rule: A Start, 12.

2 Forces 14

2.1. Definition of a Force, 14; 2.2. Action and Reaction: Newton's Third Law, 15; 2.3. Vectors, 15; 2.4. Graphic Summation of Vectors, 17; 2.5. Graphical Addition of More than Two Vectors, 19; 2.6. Mathematical Addition of Vectors, 22.

3 Concurrent Coplanar Forces 26

3.1. Free-Body and Force Diagrams. Equilibrium, 26; 3.2. Concurrent Coplanar Forces, 29; 3.3. The Boom, 30; 3.4. The Toggle, 31; 3.5. The Simple Truss, 33.

4 Nonconcurrent Coplanar Forces 39

4.1. Moment of Force, 39; 4.2. Equilibrium: Parallel Forces, 41; 4.3. Equilibrium: Nonconcurrent Nonparallel Forces, 44; 4.4. Center of Gravity: Experimental Method, 46; 4.5. Center of Gravity: Regular Areas, 46; 4.6. Center of Gravity: Solids, 49.

5 Friction 59

5.1. Friction and Its Coefficient, 59; 5.2. Friction and Vectors: The Level Plane, 61; 5.3. Friction and Vectors: The Inclined Plane, 64; 5.4. Rolling Friction, 69.

6 Elasticity 74

6.1. Stress-Strain, 74; 6.2. Modulus of Elasticity, 75; 6.3. Modulus of Shear: Rigidity, 76; 6.4. Bulk Modulus: Volume Elasticity, 77; 6.5. Compressibility, 78.

7 Linear Motion 80

7.1. Frames of Reference and Motion, 80; 7.2. Special Relativity, 81; 7.3. Speed, Velocity, and Acceleration, 85; 7.4. Constant Velocity, 86; 7.5. Constant Acceleration: Variable Velocity, 87; 7.6. Gravitational Acceleration: Trajectories, 91.

8 Newton's Laws of Motion 102

8.1. The First Law of Motion, 102; 8.2. The Second Law of Motion, 104; 8.3. The Third Law of Motion, 105; 8.4. Applications of the Laws of Motion, 105; 8.5. D'Alembert's Principle, 108.

9 Angular Motion 112

9.1. Relationship: Angular-Linear, 112; 9.2. Centripetal Acceleration, 115; 9.3. Harmonic Motion, 117; 9.4. Harmonic Motion: Simple Pendulum, 118; 9.5. Harmonic Motion: The Helical Spring, 119.

10 Laws of Angular Motion 124

10.1. Moment of Inertia, 124; 10.2. Radius of Gyration, 125; 10.3. Moment of Inertia of an Area: Second Moment, 126; 10.4. Moment of Inertia About Any Axis, 126; 10.5. Harmonic Motion: Torsion Pendulum, 128; 10.6. Laws of Angular Motion, 130; 10.7. Centripetal Forces, 131.

11 Work, Energy, Power 135

11.1. Work, 135; 11.2. Energy: Potential and Kinetic, 137; 11.3. Power, 140; 11.4. The Prony Brake, 141; 11.5. Mechanical Energy: Efficiency, 142; 11.6. The Inclined Plane, 143; 11.7. The Screw Thread, 144; 11.8. The Block and Tackle, 145; 11.9. The Chain Hoist, 146; 11.10. Friction in Machines, 147.

12 Momentum, Impact 151

12.1. Momentum and Impulse, 151; 12.2. Impact: Elastic and Inelastic, 153; 12.3. Angular Momentum, 159.

13 Liquids 162

13.1. Properties of Fluids, 162; 13.2. Pressure, 164; 13.3. Transmission of Pressure, 168; 13.4. Manometers, 170; 13.5. Surface Tension: Adhesion and Cohesion, 176.

14 Buoyancy 184

14.1. Archimedes' Principle, 184; 14.2. Specific-Weight and Specific-Gravity Determinations: Solids, 186; 14.3. Specific-Weight and Specific-Gravity Determinations: Liquids, 190; 14.4. Hydrometry, 193; 14.5. Viscosity, 195.

15 Liquids: Motion 198

15.1. Bernoulli's Theorem, 198; 15.2. Quantity Flow: The Continuity Principle, 199; 15.3. Horsepower, 202; 15.4. Impact of Fluid Flow: Stationary Obstruction, 202; 15.5 *Vena Contracta* and the Coefficient of Discharge, 203; 15.6. Flow Through an Orifice, 204. 15.7 The Nozzle, 206; 15.8. The Venturi Meter, 207; 15.9. The Pitot Tube for Measuring Pressures, 208.

16 The Expansion of Materials 215

16.1. Temperature, 215; 16.2. Fahrenheit and Centigrade Conversions, 216; 16.3. Absolute-Temperature Scales, 217; 16.4. Linear Expansion of Materials, 219; 16.5. Area Expansion of Materials, 220; 16.6. Volume Expansion of Materials, 221; 16.7. Expansion of Gases, 224; 16.8. Instruments Used as Temperature Indicators, 225.

17 Specific Heat, Heat Transfer 229

17.1. Specific Heat, 229; 17.2. Equilibrium Through Mixing, 231; 17.3. Calorimetry, 232; 17.4. Heat of Combustion, 236; 17.5. Heat Energy, 236; 17.6. Heat Transfer: Convection, 240; 17.7. Heat Transfer: Conduction, 240; 17.8. Heat Transfer: Radiation, 242.

18 Gases 246

18.1. The Gas Laws: Boyle's Law and Charles' Law, 246; 18.2. The Universal Gas Law and Its Constant R, 248; 18.3. Gases and Specific Heat, 251; 18.4. The Kinetic Theory of Heat, 253.

Contentsxiii

19 P-V Diagrams 259

19.1. P-V Diagrams, 259; 19.2. Constant Volume: Isometric Change, 260; 19.3. Constant Pressure: Isobaric Change, 261; 19.4. Constant Temperature: Isothermal Change, 263; 19.5. Constant Heat: Isentropic Change—Adiabatic Expansion, 264; 19.6. The Carnot Cycle and the P-V Diagram, 267; 19.7. The Thermodynamic Temperature Scale, 269.

20 Electrostatics 272

20.1. The Electric Charge, 272; 20.2. Units, 275; 20.3. The Electric Field, 280; 20.4. Potential Energy, Potential, Potential Difference, and Equipotential Surfaces, 281; 20.5. The Electric Field for a Sphere and a Charged Wire, 287; 20.6. The Electric Field Associated with Two Parallel Plates, 289.

21 Capacitance 294

21.1. Capacitance: Parallel-Plate, 294; 21.2. Capacitance: Spherical and Tubular, 296; 21.3. Capacitors in Series and Parallel, 297; 21.4. Dielectric Strength, 300; 21.5. Energy of a Charged Capacitor, 300.

22 Electric Current: Simple Circuits 304

22.1. Conductors, Insulators, and Dielectrics, 304; 22.2. Ohm's Law, 307; 22.3. Simple Circuits, 310; 22.4. Batteries in Series and Parallel, 315; 22.5. Resistivity, 317; 22.6. Temperature Coefficient of Resistance, 319; 22.7. Power, 321.

23 Circuits and Measuring Instruments 329

23.1. Multiple-Battery Circuits, 329; 23.2. Kirchhoff's Laws, 332; 23.3. The d'Arsonval Galvanometer, 336; 23.4. The Voltmeter, 338; 23.5. The Ammeter, 339; 23.6. Current-Voltage Measurement, 340; 23.7. Resistance Measurement: Ammeter-Voltmeter Method, 341; 23.8. Resistance Measurement: The Ohmmeter, 343; 23.9. Resistance Measurement: the Slide-Wire Wheatstone Bridge, 343; 23.10. Resistance Measurement: the Box Wheatstone Bridge, 344; 23.11. Potentiometer Measurement: emf, Resistance, Current, 346.

24 Magnetism 356

24.1. Magnets, Magnetic Poles, and Forces, 356; 24.2. Coulomb's Law, 358; 24.3. Magnetic-Field Intensity, 360; 24.4. Flux Density, 362; 24.5. Torque on a Magnet, 364; 24.6. The Ampere, 365; 24.7. The Force on a Charge in a Magnetic Field, 366; 24.8. The Force on a Straight Wire Placed Into a Magnetic Field, 367; 24.9. The Force and Torque on a Coil Placed Into a Magnetic Field, 368.

25 Magnetic Fields and Circuits 372

25.1. Ampère's Theorem, 372; 25.2. A Loop of Wire, 373; 25.3. The Solenoid and the Toroid, 374; 25.4. The Long Straight Wire, 375; 25.5. The Force Between Two Current-Carrying Wires, 375; 25.6. Ferromagnetism and the Hysteresis Loop, 376; 25.7. Magnetic Circuits: mmf, 378; 25.8. The Earth's Magnetic Field, 361.

26 Induction 384

26.1. Induced Voltage in a Straight Wire, 384; 26.2. Induced Voltage in a Moving Coil, 385; 26.3. Lenz's Law, 388; 26.4. Generators and Motors, 390; 26.5. Mutual Inductance, 397; 26.6. Self-Inductance, 399; 26.7. Growth and Decay: Inductance-Resistance, 401; 26.8. Growth and Decay: Resistance-Capacitance, 403; 26.9. Energy, 405; 26.10. Eddy Currents, 405.

27 Alternating Current 408

27.1. The Sine Wave: Instantaneous emf, 408; 27.2. Average emf, 410; 27.3. Effective emf, 412; 27.4. Graphic Representation, 413; 27.5. Pure Resistance, 415; 27.6. Pure Inductance, 415; 27.7. Pure Capacitance, 417; 27.8. Resistance and Inductance, 418; 27.9. Resistance and Capacitance, 420; 27.10. Resistance, Inductance, and Capacitance, 422; 27.11. The Power in an ac Circuit, 425; 27.12. Resonance, 426.

Contents   xv

28   Wave Motion and Sound   429

28.1. Transverse and Longitudinal Waves, 429; 28.2. Particle Displacement: Velocity, 433; 28.3. Stationary Waves, 437; 28.4. Vibrating String, 438; 28.5. The Vibrating Rod, 440; 28.6. Organ Pipes, 443; 28.7. Resonance, 445; 28.8. Doppler Effect for Sound, 447.

29   Light: The Point Source   453

29.1. Development of Light Theory, 453; 29.2. The Velocity of Light, 455; 29.3. Luminous Flux and Radiant Flux, 459; 29.4. The Point Source, 461; 29.5. Photometry, 464.

30   Mirrors and Thin Lenses   468

30.1. Plane Mirrors, 468; 30.2. Fundamental Concepts and Conventions, 470; 30.3. Convex and Concave Mirrors, 472; 30.4. Refraction and the Index of Refraction, 475; 30.5. Thin Lenses, 479; 30.6. Thin Lenses in Contact, 488; 30.7. Aberrations, 490.

31   Multiple Lenses   494

31.1. The Eye, 494; 31.2. Lens Systems, 496; 31.3. The Magnifying Glass, 500; 31.4. The Compound Microscope, 502; 31.5. Telescopes, 503; 31.6. The Camera and the Projector, 507.

32   A Further Study of Refraction   511

32.1. Displacement, 511; 32.2. Deviation, 513; 32.3. Spectra, 516; 32.4. Dispersion, 518; 32.5. Direct-Vision Prisms: Angular Separation and Achromatization, 520.

33   Interference and Diffraction   525

33.1. The Double Slit, 525; 33.2. Interference in a Thin Film; the Half-Wavelength Shift, 530; 33.3. Newton's Rings, 532; 33.4. Single-Slit Diffraction, 534; 33.5. Diffraction Gratings, 537; 33.6. Crystal Diffraction, 539; 33.7. Resolving Power, 540.

34 The Polarization of Light 545

34.1. Polarization, 545; 34.2. Polarization by Reflection, 548; 34.3. Polarization by Refraction, 549; 34.4. Polarization by Absorption and Scattering, 551; 34.5. Interference and Polarization, 552.

35 Electron-Beam Acceleration and Deflection 555

35.1. Thermionic Emission, 555; 35.2. Rectifiers and the Triode, 556; 35.3. Transistors, 559; 35.4. The Thyratron, 561; 35.5. The Crookes Tube, 562; 35.6. Cathode Rays, 563; 35.7. The Cathode-Ray Tube, 564; 35.8. The Mass Spectrograph, 569; 35.9. The Cyclotron and Particle Acceleration, 571; 35.10. The Photoelectric Effect, 575; 35.11. X-Rays, 578.

36 The Atom 581

36.1. Atomic Structure, 581; 36.2. The Hydrogen Series, 583; 36.3. Bohr's Hydrogen Atom, 585; 36.4. Electron Distribution, 589; 36.5. X-Rays, 592; 36.6. Electron Spin, 593.

37 Transmutation 595

37.1. Radioactivity, 595; 37.2. The Radioactive Series, 597; 37.3. The Cloud Chamber, 599; 37.4. Artificially Induced Transmutation, 602.

38 Energy: The Elementary Particle 606

38.1. Nuclear Reactions, 606; 38.2. Binding Energy, 608; 38.3. Fusion, 610; 38.4. Fission, 611; 38.5. Elementary Particles, 614.

Appendix 619

Table A.1. Standard Prefixes—Multiple and Submultiple Units 619
Table A.2. International System of Units (SI Units) 620
Table A.3. Conversion Units 621
Table A.4. Determination of Radius of Gyration 622
Table A.5. Specific Weight 623
Table A.6. Coefficients of Friction 624
Table A.7. Elasticity 625
Table A.8. Density of Solids and Liquids 626
Table A.9. Surface Tension 627
Table A.10. Capillarity 628
Table A.11. Expansion of Solids and Liquids 629
Table A.12. Specific Heat and Latent Heat of Solids 630
Table A.13. Heat of Combustion 631
Table A.14. Thermal Conductivity K 632
Table A.15. Specific Heat of Gases 633
Table A.16. Dialectric Coefficient k 634
Table A.17. Resistivity ρ and Temperature Coefficient of Resistance α 635
Table A.18. Velocity of Sound in Various Materials 636
Table A.19. Luminous Efficiency 637
Table A.20. Index of Refraction n (for wavelength of sodium D, 5893 Å) 638
Table A.21. Fraunhofer Lines 639
Table A.22. Index of Refraction for Types of Glass 640
Table A.23. Isotopes 641
Table A.24. Shell Distribution of Electrons 642
Table A.25. The Elements 645
Table A.26. Physical Constants 647
Table A.27. Trigonometric Functions 648
Table A.28. Logarithms 649

Answers to Even-Numbered Problems 651

Index 661

1

Some Basic Concepts

1.1. The Role of Technical Physics

Physics is the science of motion and matter. It deals with energy and inanimate matter. As a quantitative science, it uses the scientific method in an effort to support its theories, hypotheses, and conjectures. This scientific method relies upon observation, logical thinking, and conclusions. An attempt is made to verify these conclusions using direct and indirect measurements.

Before the twentieth century, physics was divided into topics that included mechanics, heat, fluids, sound, electricity, and light. It was a highly departmentalized science of matter and energy. However, even though these topics were taught as separate units, it was never intended that physics be thought of as a disjointed science. The development of the relationships between the various areas of physics was left to the student. Fundamental laws always have and always will permeate all of physics. The search for a better way to teach these relationships is an important search. However, it should always be kept in mind that physics for the technician must be a supporting course for his subsequent technical courses.

The development of the future technician's cultural talents is important. Of even greater importance is the development of his technical prowess. Since the technician will be required to do the "legwork" for the professional engineer, it is important to understand that he will be selling his ability to perform, and not his ability to theorize.

1.2. How to Study

Before we plunge into the study of physics, it is important that we look at some of the dynamics of studying. Appropriate study habits can make the study of physics a rewarding experience. The student should ask himself three questions: (1) What is studying and how does it differ from reading for pleasure? (2) What are some of the the techniques of studying? (3) When is the act of studying complete?

One of the main characteristics of reading for pleasure is to get the overall meaning from a paragraph, section, or even a chapter. When reading a novel, it is seldom read so that each sentence is dissected for meaning, mechanics, etc. It is enough that the sentence contributes to the theme at hand.

In contrast, studying requires continuous comprehension of the material being read. The reading rate is variable and depends upon the extent and depth of understanding. Studying is a process of constant evaluation of the material, sifting the supporting statements from the main theme statements, and concentrating one's efforts on the main theme statements.

A recommended sequence for the successful study of technical material is the following:

1. Before a new assignment is begun, the previously assigned materials should be reviewed.
2. The new material should be "sight-read" for overall meaning. References to earlier laws, principles, derivations, illustrations, or illustrated problems should be checked, reviewed, and noted at this time.
3. The new assignment is now studied.
4. Note-taking is a technique which could be the difference between success and failure. The purpose is to record important material, yet leave the student relatively free to take part in class discussion, or to follow the lecture as it proceeds. A chronic note-taker is always behind the classroom events. If the assignment for the day is read, the student need only record the supplementary material discussed in class. In addition, the notes, therefore, become supplementary to the text material instead of duplicating it. It should include laws, principles, definitions, defining equations, supporting drawings, and the solution of homework problems.

A recommended approach to the solution of technical problems is as follows:

1. Develop a feel for the scope of the problem by rough-reading it.
2. Read the problem bit by bit, identify the known and unknown quantities, and

discard all superfluous materials. Assign symbols to all known and unknown quantities.
3. Search your mind, notes, and text for a definition, principle, law, equation, or equations containing all variables. You may need to deduce or derive your own equation or equations; or you may come up with more than one relationship.
4. If several equations are involved, it is usually better to reduce these equations to one equation. Then all numerical values, including their units, are substituted and all manipulations carried out to a completed solution.

1.3. Scalar and Vector Quantities

A *vector quantity* is one which possesses both magnitude and direction. A vector is not completely defined if one of these conditions is missing. A force is only one of the many classes of vectors which we shall study. Remember, however, that the conditions necessary for a complete definition of forces, like vectors, apply to other physical quantities, such as velocity, torque, or electric intensity. Note, too, that vectors cannot be added or subtracted by the usual methods of algebra. As we proceed with the study of forces, we shall develop methods of vector addition and subtraction that may be applied to the study of other vector phenomena.

If, on the other hand, one of the conditions necessary for a complete definition of a vector quantity is missing, the quantity is termed a *scalar*. Scalar quantities may be added by the usual algebraic methods. Quantities such as time and energy are scalar quantities. They possess magnitude but no direction. Note that such quantities as speed and torque are scalar quantities, unless the direction as well as the magnitude is given. "A car traveling east" cannot be represented as a vector quantity because one of the essential ingredients—magnitude—is missing.

1.4. Standardizing Basic Units

The metric system is used throughout all scientific work. Since this text is intended for students who will enter the world of engineering, illustrated problems will be worked in both the metric and the British systems wherever possible.

In the metric system, the decimal point is shifted to achieve multiples and submultiples. The standard prefixes as used in the "International System of Units" are shown in Appendix Table A.1. In 1866, Congress authorized the use of the metric system of units in the United States. In 1893, Congress authorized the Office of Weights and Measures to establish equivalents between the metric and British systems. The unit of metric length, the *meter*, was defined as the distance between two scratches on a platinum-iridium bar at atmospheric pressure and at the temperature at which ice freezes. The unit mass was standardized and represented by a particular *kilogram* cylinder. Both bar and cylinder are kept in the Paris archives. (Bureau des Poids et Mesures, at Serres.)

In 1960, the Eleventh General Conference of Weights and Measures redefined the unit of length so that it could be independently reproduced. The international meter is now 1,650,763.73 wavelengths in a vacuum of the red radiation from krypton-86 between the energy levels $^2p_{10}$ and $^5d_5$. Using this standard, spectral lines can be compared to uncertainty limits of within 2 parts in 10^9. Translation to the meter bar is accurate to within 1 part in 10^7.*

The present standard for the unit mass appears to be adequate, since it is capable of being compared within a few points in 10^9. However, it should be noted that since 1960–1961 the defining standard for the scale of atomic masses has been carbon-12 instead of oxygen-16.

Still another very important basic unit is the *second*. Before 1956, the second was defined as 1/86,400 of a mean solar day, using the rotating Earth as the clock. In 1956, the second was redefined as 1/31,556,925.9747 the time for the Earth to rotate about the Sun. The second as now established is equal to 9,192,631,770 vibrations of the cesium atom. However, since greater accuracy is needed, it appears that soon the second will be related to the hydrogen maser.

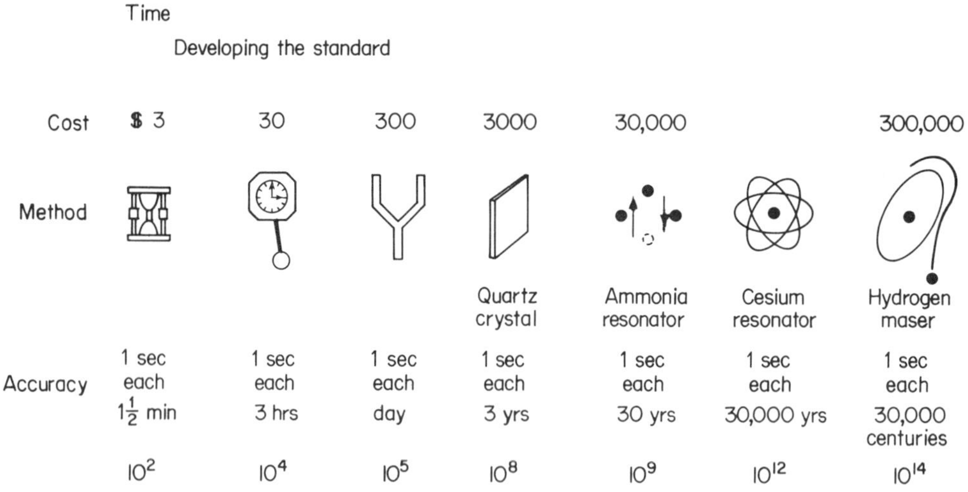

Time
Developing the standard

Cost	$ 3	30	300	3000	30,000		300,000
Method	hourglass	pendulum clock	tuning fork	Quartz crystal	Ammonia resonator	Cesium resonator	Hydrogen maser
Accuracy	1 sec each 1½ min	1 sec each 3 hrs	1 sec each day	1 sec each 3 yrs	1 sec each 30 yrs	1 sec each 30,000 yrs	1 sec each 30,000 centuries
	10^2	10^4	10^5	10^8	10^9	10^{12}	10^{14}

Fig. 1.1

* Purer wavelength sources have been developed using the atomic-beam method, and lasers exhibit purities of within 1 part in 10^{12} or 10^{14}. *Engr. Opportunities*, p46 (April 1968).

Figure 1.1* shows the development of clocks and their accuracies in an effort to define the second. It is interesting to note the cost of developing the various degrees of accuracy. The cesium resonator is accurate to within 1 second in 30,000 years; whereas it is expected that the hydrogen maser will not vary by more than 1 second every 30,000 centuries!

The ampere, the Kelvin degree, and the candela will be discussed in subsequent chapters.

The *light year* was devised by astronomers because of the vast distances encountered in the study of the universe. The light year represents the distance light travels in 1 year: 9.45×10^5 meters, or 6 million-million miles.

1.5. Operation with Units

As a result of the standards described in Sec. 1.4, several conversions from metric to British units are possible.

The British units of *length* are as follows:

$$\frac{3600}{3937} \text{ meter (m)} = 1 \text{ yard (yd)} = 3 \text{ feet (ft)}$$

$$1 \text{ ft} = \frac{1200}{3937} = 0.3048 \text{ m} = 30.48 \text{ centimeters (cm)}$$

$$1 \text{ inch (in.)} = 1/12 \text{ ft} \times 30.48 \text{ cm} = 2.540 \text{ cm}$$

The *pound* (lb) force is 0.45359243 the attraction of gravity on the kilogram at sea level and 45° latitude. Although the concept of mass and weight are covered more thoroughly in Chapter 8, some preliminary remarks are in order now.

Mass is expressed in kilograms, grams, etc. and relates to the inherent property of matter without reference to gravitational attraction. This inherent property of all matter—called *inertia*—is its ability to resist motion when at rest and to resist a change of velocity when it is in motion. Mass is proportional to inertia and gives a quantitative measure of inertial resistance.

Weight, on the other hand, may change, depending upon the gravitational attraction on a material. Thus a particular mass will have one force value (weight) at sea level and another force value (weight) at another distance from the Earth's surface.

Everyone has read about—even though they may not have experienced—weightlessness. When a mass gets far enough away from the gravitational attraction of the Earth, it loses its weight which, as stated, is directly related

---

* Figure 1.1 is modified from Fig. 2 in "Measurement Accuracy," *Engineering Opportunities*, April, 1968.

to the gravitational attraction. Its mass, or ability to resist a change of state of the mass, remains unchanged. Thus a metal object and cork object both are weightless in outer space and will float freely if released there. However, it will take different magnitudes of force to start the metal and the cork objects moving and different magnitudes of force to stop the two objects or to change their direction of motion.

Actually gravitational attraction is zero only when the mass is at an infinite distance from the Earth. Theoretically then weightlessness cannot be achieved by increasing the distance between the mass and the Earth since there will always be some interplay. However, a spaceship orbiting around the Earth is in effect always falling toward the Earth. Relative to each other, the ship and the objects inside the ship are all falling toward the Earth at the same rate. Thus the objects do not experience any movement toward the Earth. As a matter of fact any object dropped inside the spaceship will not "fall" because it is already falling at the same rate of speed as the spaceship. The objects inside the ship are said to be weightless.

It is possible to relate mass m to its weight w using the equation

$$w = mg = \text{kg} \times \text{m/sec}^2 = \text{newtons}$$

or

$$w = mg = \text{g-cm/sec}^2 = \text{dynes}$$

$$\left\{ \begin{array}{l} m = \text{mass: g or kg} \\ w = \text{force: dynes or newtons} \\ g = \text{gravitational accelaration:} \\ \quad \text{cm/sec}^2 \text{ or m/sec}^2 \end{array} \right\}$$

From the equation above, the mass in the British system of units becomes

$$m \text{ (slugs)} = \frac{w \text{ (lb)}}{g \text{ (ft/sec}^2)} \quad \left\{ \begin{array}{l} m = \text{mass: slugs} \\ w = \text{weight: lb} \\ g = \text{gravitational attraction: ft/sec}^2 \end{array} \right\}$$

In 1960, The Eleventh General Conferences on Weights and Measures denoted six fundamental units.* Some of the work of this conference has already been referred to in Sec. 1.4. These units in the metric (meter-kilogram-second-ampere) system include the *fundamental units:*

mass	kilogram (kg)
length	meter (m)
time	second (s) or (sec)
current	ampere (A)
temperature	kelvin (°K)
luminous intensity	candela (cd)

In mechanics, it should be noted that it is possible to express all quantities in terms of units of three fundamental quantities: *mass, length,* and *time.* Thus as noted above, in the metric *rationalized* system (henceforth designated as mks), the unit of mass is the *kilogram* (kg), the unit of length is the *meter* (m), and the unit of time is the *second* (sec). In another metric system of units (henceforth designated as the cgs system), the unit of mass is the *gram* (g),

---

* Amended by the executive body (1962).

the unit of length is the *centimeter* (cm), and the unit of time is the *second*. In the British system (also known as the fps system), the unit of mass is the *slug*, the unit of length is the *foot* (ft), and the unit of time is the *second*.

Table 1.1 shows the three systems of units most commonly used in physics. It is to be noted that any fundamental unit, multiplied or divided by another fundamental unit, is called a *derived unit*. Table A.2 in the Appendix shows many more derived units.

Table 1.1. Systems of Units

Dimension	Metric Absolute		British Gravitational
	cgs	mks	fps
length*	cm	m	ft
mass*	g	kg	lb-sec^2/ft *or* slugs
time*	sec	sec	sec
force	dyne	nt (newton)	lb
energy	dyne-cm *or* erg	nt-m *or* joule	ft-lb
moment of inertia	g-cm^2	kg-m^2	ft-lb-sec^2 *or* slug-ft^2
momentum	g-cm/sec	kg-m/sec	lb-sec
temp	centigrade degree	centigrade degree	Fahrenheit degree
abs temp*	Kelvin degree*	Kelvin degree	Rankine degree

* Fundamental units

In order to completely describe a physical quantity, both the name of the unit as well as its magnitude must appear in the notation. Thus "3.1416" is a pure number, but "3.1416 miles" is a physical quantity.

Any equation which relates physical quantities must be dimensionally homogeneous, i.e., both sides of the equation must be reducible to the same units. Frequently, if an equation is checked for homogeneity and the units are not the same on both sides of the equation, an omission or incorrect substitution has been made when setting up the equalities.

Most conversions from one system of units to another may be performed using the conversion constants from Table A.3 in the Appendix.

Conversions may be accomplished by using units as whole-number or fractional multipliers. It is important that the student get accustomed to referring to "1 inch equals 2.54 centimeters" as "2.54 centimeters per inch" and then writing the latter statement as "2.54 cm/in."

Example 1

Convert 18 in. to centimeters.

Solution

$$18 \,\cancel{\text{in.}} \times 2.54 \,\frac{\text{cm}}{\cancel{\text{in.}}} = 45.7 \text{ cm}$$

Example 2

A locomotive weighing 250 tons is traveling at the rate of 50 miles per hour (mph). Express (1) the weight of the locomotive in cgs and mks units; (2) the speed of the locomotive in cgs and mks units.

Solution

1. The weight of the locomotive in cgs and mks units is

$$\text{weight (cgs)} = 250 \,\cancel{\text{tons}} \times 2000 \,\frac{\cancel{\text{lb}}}{\cancel{\text{ton}}} \times 453.6 \,\frac{\text{g}}{\cancel{\text{lb}}}$$

$$= 2.50 \times 10^2 \times 2 \times 10^3 \times 4.536 \times 10^2$$

$$= 2.268 \times 10^8 \text{ g}$$

$$\text{weight (mks)} = 2.268 \times 10^8 \,\cancel{\text{g}} \times \frac{1 \text{ kg}}{10^3 \,\cancel{\text{g}}}$$

$$= 2.268 \times 10^5 \text{ kg}$$

2. The speed of the locomotive in cgs and mks units is

$$\text{speed (cgs)} = 50 \,\frac{\cancel{\text{mi}}}{\cancel{\text{hr}}} \times 5280 \,\frac{\cancel{\text{ft}}}{\cancel{\text{mi}}} \times 12 \,\frac{\cancel{\text{in.}}}{\cancel{\text{ft}}} \times 2.54 \,\frac{\text{cm}}{\cancel{\text{in.}}} \times \frac{1 \text{ hr}}{60 \,\cancel{\text{min}}} \times \frac{1 \,\cancel{\text{min}}}{60 \text{ sec}}$$

$$= 5 \times 10^1 \times 5.28 \times 10^3 \times 1.2 \times 10^1 \times 2.54 \times \frac{1}{6 \times 10^1} \times \frac{1}{6 \times 10^1}$$

$$= 2.235 \times 10^3 \text{ cm/sec} = 2235 \text{ cm/sec}$$

$$\text{speed (mks)} = 2.235 \times 10^3 \,\frac{\cancel{\text{cm}}}{\text{sec}} \times \frac{\text{m}}{10^2 \,\cancel{\text{cm}}}$$

$$= 2.235 \times 10 \text{ m/sec} = 22.35 \text{ m/sec}$$

Example 3

Convert 80 mi/hr² to m/sec².

Solution

$$\text{acceleration (mks)} = 80 \,\frac{\text{mi}}{\text{hr}^2} \times \left[\frac{\text{hr}}{60 \text{ min}}\right]^2 \times \left[\frac{\text{min}}{60 \text{ sec}}\right]^2 \times 5280 \,\frac{\text{ft}}{\text{mi}} \times \frac{12 \text{ in.}}{\text{ft}}$$

$$\times \frac{2.54 \text{ cm}}{\text{in.}} \times \frac{1 \text{ m}}{10^2 \text{ cm}}$$

$$= 80 \,\frac{\cancel{\text{mi}}}{\cancel{\text{hr}^2}} \times \frac{\cancel{\text{hr}^2}}{60^2 \,\cancel{\text{min}^2}} \times \frac{\cancel{\text{min}^2}}{60^2 \text{ sec}^2} \times 5280 \,\frac{\cancel{\text{ft}}}{\cancel{\text{mi}}} \times \frac{12 \,\cancel{\text{in.}}}{\cancel{\text{ft}}}$$

$$\times \frac{2.54 \text{ cm}}{\text{in.}} \times \frac{\text{m}}{10^2 \text{ cm}}$$

$$= \frac{80 \times 5280 \times 12 \times 2.54 \text{ m}}{60^2 \times 60^2 \times 10^2 \text{ sec}^2}$$

$$= 0.993 \text{ m/sec}^2$$

1.6. Some Mathematics

At this point, it is a good idea to review some mathematics which the student will need to use in the very next chapter. It is assumed that the student is capable of solving simple algebraic equations.

Trigonometry

The six trigonometric functions of a *right* triangle are very important and related to Fig. 1.2(a). They are:

$$\text{sine } A = \frac{\text{opposite}}{\text{hypotenuse}} = \frac{a}{c} \qquad \text{cosecant } A = \frac{\text{hypotenuse}}{\text{opposite}} = \frac{c}{a}$$

$$\text{cosine } A = \frac{\text{adjacent}}{\text{hypotenuse}} = \frac{b}{c} \qquad \text{secant } A = \frac{\text{hypotenuse}}{\text{adjacent}} = \frac{c}{b}$$

$$\text{tangent } A = \frac{\text{opposite}}{\text{adjacent}} = \frac{a}{b} \qquad \text{cotangent } A = \frac{\text{adjacent}}{\text{opposite}} = \frac{b}{a}$$

Another useful relationship for the right triangle, Fig. 1.2(a), is

$$c^2 = a^2 + b^2$$

The law of sines is related to Fig. 1.2(b) as

$$\frac{a}{\sin A} = \frac{b}{\sin B} = \frac{c}{\sin C}$$

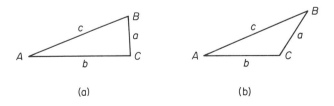

Fig. 1.2

The law of cosines is related to Fig. 1.2(b) as

$$c^2 = a^2 + b^2 - 2ab \cos C$$

The algebra, beyond the solution of simple equations, which needs mentioning here deals with exponents and the solution of the quadratic equation.

Exponents

$$a^x = a \cdot a \cdot a \cdots x \text{ factors}$$

$$a^{-x} = \frac{1}{a^x}$$

$$a^0 = 1$$

$$a^{x/y} = \sqrt[y]{a^x} = [\sqrt[y]{a}]^x$$

$$a^x \cdot a^y = a^{x+y}$$

$$\frac{a^x}{a^y} = a^{x-y}$$

$$(a^x)^y = a^{xy}$$

$$(a \cdot b \cdot c \cdots)^x = a^x \cdot b^x \cdot c^x \cdots$$

$$\left[\frac{a}{b}\right]^x = \frac{a^x}{b^x}$$

Quadratic Equations

To solve a quadratic equation using the methods of "*completing the square*," you must apply the following rules:

1. All terms in x and x^2 must be in the left-hand member of the equation; all constant terms must be in the right-hand member of the equation.
2. Divide both sides of the equation by the coefficient of x^2.
3. Square one-half the coefficient of x and add this quantity to both sides of the equation.
4. Extract the square root of both sides of the equation and apply the $\pm$ to the right-hand side.
5. Transpose the left-hand member constant to the right-hand side of the equation and combine to obtain the roots.

Example 4

Solve the equation $2x^2 - x - 6 = 0$, using the methods of completing the square.

Solution

Apply step 1:

Sec. 1.6 Some Mathematics 11

$$2x^2 - x = 6$$

Apply step 2:
$$x^2 - \frac{1}{2}x = \frac{6}{2} = 3$$

Apply step 3:
$$x^2 - \frac{1}{2}x + \frac{1}{16} = 3 + \frac{1}{16}$$

Apply step 4:
$$x - \frac{1}{4} = \pm \sqrt{3\frac{1}{16}}$$

Apply step 5:
$$x = \pm \sqrt{\frac{49}{16}} + \frac{1}{4}$$
$$= \pm \frac{7}{4} + \frac{1}{4}$$
$$= +2; \ = -\frac{3}{2}$$

To solve equations using the *quadratic formula,* proceed as follows:

1. Leave the equation in the form $ax^2 + bx + c = 0$.
2. Isolate the coefficients of x^2, x, and the constant term.
3. Insert quantities into the quadratic formula
$$x = \frac{-b \pm \sqrt{b^2 - 4ac}}{2a}$$

Example 5

Solve the equation in Example 4 using the quadratic formula.

Solution

Apply step 1:
$$2x^2 - x - 6 = 0$$

Apply step 2:
$$\text{coefficient of } x^2 = 2$$
$$\text{coefficient of } x = -1$$
$$\text{constant term} = -6$$

Apply step 3:

$$x = \frac{-b \pm \sqrt{b^2 - 4ac}}{2a}$$

$$= \frac{-(-1) \pm \sqrt{(-1)^2 - 4(2)(-6)}}{2 \times 2} = \frac{1 \pm \sqrt{1 + 48}}{4}$$

$$= \frac{1 \pm \sqrt{49}}{4} = \frac{1 \pm 7}{4}$$

$$= 2; = -\frac{3}{2}$$

1.7. The Slide Rule: A Start

Multiplication

It is desired to multiply 2 × 3 using the D and C scales. To do this, set the hairline on the 2 on the D scale. Then slide the left index (the left-hand 1 on the C scale) to the hairline. Next slide the hairline to the 3 on the C scale. You may now read the answer on the D scale. The answer is 6. To multiply larger numbers, first reduce them to the powers of 10, then proceed as before.

Division

It is desired to divide 6 by 2. Set the hairline on the 6 on the D scale. Then slide the 2 under the hairline. Next move the hairline to the left index. You may now read the answer 3 under the hairline on the D scale. (Study the remaining operations as they are needed).

Problems

1. Review Sec. 1.2, "How to Study." A full classroom discussion may uncover other aids in your quest for success in physics.
2. Take one of the problems below and analyze it according to the "recommended approach to the solution of problems" as outlined in Sec. 1.2.
3. State the conditions necessary for complete definition of a vector quantity. How does a scalar differ from a vector quantity?
4. Convert the following from the metric system to the British system of units; (a) 55 m; (b) 145 cm; (c) 2000 cm/hr^2; (d) 3×10^{10} cm/sec to mph; (e) 400 cal/g to Btu/lb.*
5. Convert the following from the British system to the mks system of units: (a) 15 mi; (b) 200 slugs; (c) 6000 Btu; (d) 3412 Btu/hr to cal/sec; (e) 45 mph.

* See Appendix Table A.3 for cal and Btu conversion scales.

6. The British units of thermal conductivity are 15 Btu × in./hr × ft² × F°. Change these units to the metric units of cal × cm/sec × cm² × C°.
7. In Fig. 1.2(a), given side a equal to 12 in. and side b equal to 20 in., calculate the length of the hypotenuse and the two angles A and B.
8. In Fig. 1.2(b), given side a equal to 12 in., side b equal to 20 in., and angle A equal to 15°, calculate the missing sides and angles.

2

Forces

2.1. Definition of a Force

In Chapter 1, *inertia* was defined as that property of matter which resists any change of state of motion or of rest. As will be seen in Chapter 7, a *force* is required to change the state of motion or of rest of a body of matter. This force may be, simply stated, a push or pull. More specifically, a force is that condition which tends to produce a change of shape of a body; or it is that something which tends to overcome the inertial resistance to the change of state of motion or of rest of a body. That is, a body in motion (constant straight-line velocity) will resist any attempt (a force) to change its motion. Any body which is at rest will resist any attempt (a force) to set it in motion. This inherent characteristic of matter is present whether the object is subjected to the Earth's gravitational field or whether it is floating around in outer space under conditions of weightlessness.

Let us examine an object in outer space out of range of the Earth's gravitational field. Inherently, the object possesses certain qualities, which keeps it stationary. To destroy this equilibrium state and thus move the object, a certain force must be applied. Notice that since the object is weightless, this force must be acting against the inertial resistance, which we designate as its mass. Notice also that once the object reaches a certain velocity, the force may be removed and as long as there is no other force operating (such as air resistance) trying to stop motion, the object's inertial resistance will be operating to keep the object in motion. Only a force operating to overcome this resistance to

Sec. 2.3 *Vectors* **15**

bringing the object to rest can change its equilibrium state and bring it to rest. It is important to note that to affect a change of direction of motion requires an external force, since the inertia of the object dictates that its velocity act along a straight line.

Gravitational forces act on every particle of a body and on all objects on the Earth's surface. Forces which act on every particle of an object are called *distributed forces*. In the solution of many problems, it is convenient to consider these distributed forces as applied to a point. A rope fastened to a box imparts a force, when pulled, to all particles in the box; these are distributed forces. The point of application of these forces is taken to be the center of gravity of the box.

2.2. Action and Reaction: Newton's Third Law*

Newton's third law states that for every *action* there is an equal and opposite *collinear reaction*. Therefore, forces act in pairs. Action forces create the reaction forces. When the action force is removed, the reaction force stops to operate. Their magnitudes are always equal and always oppose each other. Newton's first law of motion refers to an equilibrium state, which means that the action and reaction forces are equal. (See Chapter 8.)

Thus a person sets his foot on the rung of a ladder and pushes on the rung in preparation for going up the ladder. The rung must exert an equal and opposite force to the force exerted by the foot. The rung cannot create a reaction force until the person's foot has applied the action force.

Later, we shall see that if the rung of the ladder cannot support the force of the foot, the rung takes on an accelerated motion. That is, the rung must either break or the ladder must move and therefore action no longer equals reaction.

2.3. Vectors

We have stated that a *vector* must have magnitude, and that it must act in a certain direction.** Since a force possesses magnitude and direction, it is considered to be a vector.

The *magnitude* is the size of the force and is represented by such units as the pound, the dyne, the newton, etc. The magnitude must include the number

---

* Newton's second law of motion will be treated in Chapter 8.
** Note that in addition, a "bound" vector needs a point of application.

and name of the unit—for example, 5 lb. The direction in which the force acts is usually given as an angle with the *x*-axis; e.g., 15° with the *x*-axis.

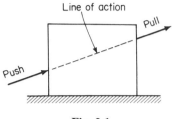

Fig. 2.1

A line on a diagram drawn in the direction of a force (indicated by an arrowhead) and through the point of application of the force is called the *action line* of the force and may be extended backward indefinitely. The *principle of transmissibility* states that the force may be considered to act anywhere along this extended line of action without changing the effect created by this force. It permits us to replace a "pull" force with a "push" force, as shown in Fig. 2.1.

Graphically, an arrow may be drawn to represent a force. If this arrow has the proper length and is pointed in the proper direction, the force is said to be *defined*. If the arrow is applied at the proper place, the force is said to be *completely defined*.

The magnitude may be represented by any convenient unit of length, provided that this unit of length is maintained throughout the solution; e.g., if 1 cm = 2 dynes, a 6-dyne force would be represented by an arrow 3 cm long. If 1 cm = 2 lb, an arrow 4 cm long would be read as representing an 8-lb force. The choice of scales is a matter of convenience.

In many instances, it is much easier to describe a force with a graphic representation—even if the diagram is drawn freehand—than to attempt to describe the force with words. Figure 2.2(a) shows a *space diagram* and Fig. 2.2(b) show the *vector diagram*. The vector diagram is invaluable to the solution of physics problems, since it yields information about the problem at a glance.

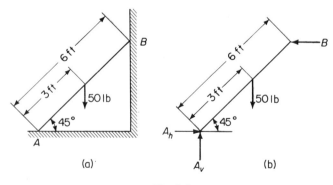

Fig. 2.2

2.4. Graphic Summation of Vectors

When the two vectors A and B pull on an object, Fig. 2.3(a), point O will move along a straight line in a direction along OR. Thus the direction taken by O may be achieved by replacing A and B by *one* vector R acting along OR. The magnitude of force R—called the *resultant*—may be determined by the methods of this chapter. The resultant force R replaces two or more forces and produces the same net effect as the forces which it replaces. This process is referred to as *vector addition*. The resultant is a vector quantity: it has magnitude and direction.

Parallelogram Method

Two vectors may be added vectorially using the *parallelogram method*. If, at a point, a vector is drawn to represent a force and if from the same point another vector is drawn representing a second force, then the diagonal of a parallelogram, applied at the same point, will be the resultant vector of the two forces (see Fig. 2.3a). It should be pointed out that the other diagonal connecting points A and B (not shown) will not cause point O to move in the direction of vector R.

Example 1

As shown in Fig. 2.3(a), assume two boys pulling on two ropes fastened to point O and pulling in the directions shown. Force A is 80 lb and force B is 60 lb. The

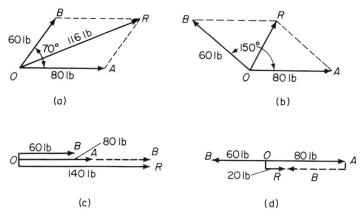

Fig. 2.3

angle between them is 70°. With what magnitude and in what direction will O move?

Solution

For convenience, we have decided to use 1 in. = 10 lb. Therefore, arrow A will be 8 in. long and arrow B will be 6 in. long. The tails of the arrows will be anchored at O and drawn with an angle of 70° between them as shown in Fig. 2.3(a).

Now draw a dotted line from point A parallel to OB and another dotted line from point B parallel to OA. These four lines—OB, OA, and the two dotted lines—form a parallelogram. The diagonal from O to R is the resultant. Note that its tail is anchored at O and its head is at R.

Using the same scale (1 in. = 10 lb), the length of vector R is measured and its magnitude determined. A protractor will determine the angle at which the resultant acts.

If the vectors and angle are drawn to length, the resultant will measure 11.6 in., which makes the magnitude equal to 11.6 in. × 10 lb/in. = 116 lb. A protractor will reveal that R is acting at an angle of 30° with OA.

Example 2

Assume both boys to be pulling as shown in Fig. 2.3(b) at an angle of 150°. Determine the resultant R.

Solution

Assume 1 in. = 10 lb; OA will be anchored at O and will be 8 in. long; OB will be 6 in. long and anchored at O. The parallelogram is formed as before and the diagonal OR drawn as shown.

The length of OR is measured and found to be 4.1 in. long, which yields 4.1 in. × 10 lb/in. = 41 lb. A protractor shows OR acting at 47° with OA.

Example 3

Assume the magnitudes of the two forces to be the same as in Examples 1 and 2. (1) If the vectors are as shown in Fig. 2.3(c), what is the resultant? (2) If the vectors are as shown in Fig. 2.3(d), what is the resultant?

Solution

1. Using the same scale, 1 in. = 10 lb, draw vectors A and B as shown in Fig. 2.3(c). Move vector B so that it is in the position shown in the figure. Since both vectors pull on O in the same direction, the resultant is their algebraic sum. Thus, if OB is 6 in. long and OA is 8 in. long, R is 14 in. long. Using our scale of 1 in. = 10 lb, R equals 14 in. × 10 lb/in. = 140 lb and acts toward the right.

2. In Fig. 2.3(d), since force A pulls on O to the right and since force B pulls on O to the left along the same line of action, the two vectors must subtract. The resultant direction will be that of the larger vector. Thus, if OA is 8 in. long and OB is 6 in. long, then R is $8 - 6 = 2$ in. Using the same scale, 2 in. $\times$ 10 lb/in. $=$ 20 lb. Since 80 lb is greater than 60 lb and acts toward the right, the resultant will act toward the right.

The Triangle Method

The two vectors A and B, Fig. 2.4(a), may be added using the *triangle method*. The procedure is to move vector B to the position shown in Fig. 2.4(b), or to

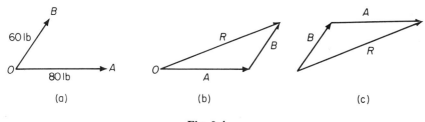

Fig. 2.4

move vector A to the position shown in Fig. 2.4(c). The triangles in both figures are completed. The hypotenuse in each triangle is the resultant R.

2.5. Graphic Addition of More than Two Vectors

When two or more vectors are to be combined and replaced by a single vector, any two vectors may be selected and replaced by a resultant vector. This resultant vector is then combined with another vector into another resultant. The process is repeated until only a single resultant remains.

Example 4

Using the parallelogram method, find the resultant of the three forces A, B, and C shown in Fig. 2.5(a).

Solution

The forces to be considered are A, B, and C. Select forces A and B (any other two forces could have been selected) and form a parallelogram as shown in Fig. 2.5(b).

20   Forces   Chap. 2

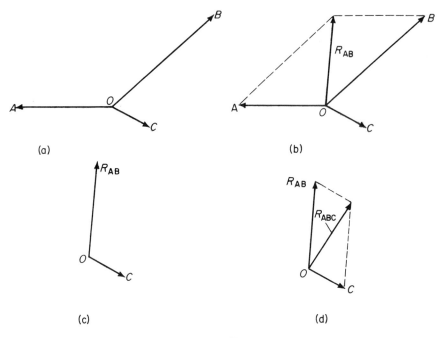

Fig. 2.5

The diagonal R_{AB} of the parallelogram is drawn. This diagonal is the resultant of the two forces A and B, and replaces them. The problem reduces to a system of two forces R_{AB} and C as shown in Fig. 2.5(c). Using the resultant R_{AB} and C, a new parallelogram is formed as shown in Fig. 2.5(d). The diagonal resultant R_{ABC} is the resultant of the original three forces.

It is to be noted that as each force from the original system is used, we treat the new set of conditions as a completely new problem. It is also important to note that each force is used only once.

Example 5

Given the same three forces shown in Fig. 2.6(a), find the resultant using the triangle method.

Solution

Select two forces OC and OB (Fig. 2.6a). Displace OB to CR so that the tail of arrow CR is anchored at C (the head of the arrow OC). Next draw the resultant R_{BC} so that its tail is anchored at O and its head is at point R. Since this resultant R_{BC} is a vector force which may be used in place of forces OB and OC, the system has degenerated to a system of two forces; R_{CB} and OA. The process is

Sec. 2.5 Graphical Addition of More than Two Vectors

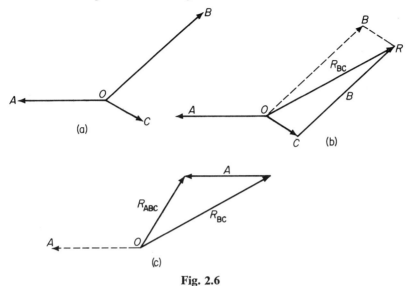

Fig. 2.6

repeated as shown in Fig. 2.6(c) using the resultant R_{BC} and OA to obtain the resultant for the system R_{ABC}.

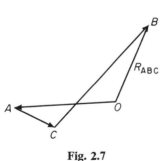

Fig. 2.7

In Fig. 2.7, the same problem is solved using the *polygon method*. Resultant R_{ABC} is arrived at by starting with any one of the forces and displacing each of the others, one at a time, until all vectors have been used once. The gap between the tail of the first arrow used and the head of the last arrow used yields the resultant. Note that the tail of the resultant R_{ABC} is anchored at O and that the head of R_{ABC} touches the head of arrow CB.

Example 6

Using the system of forces shown in Fig. 2.6(a), find the resultant by the polygon method.

Solution

Draw vector OA, shown in Fig. 2.7, in the same direction and to the same length as OA in Fig. 2.6(a). Displace vector OC to AC, and OB to CB. All vectors in Fig. 2.7 must be drawn to the same length and in the same direction as shown in Fig. 2.7. The resultant R_{ABC} may now be drawn as shown in Fig. 2.7.

2.6. Mathematical Addition of Vectors

In preceeding sections, it was pointed out that two vectors may be replaced by one vector which would have the same net effect upon the point of application as the two (or more) vectors had on the point of application. It is also true that a single vector may be replaced by two (or more) vectors. This process of replacing a single vector by two vectors is called *resolution of forces* into *components*. Any two components may be used to replace a resultant. However, if the two components selected to replace a resultant coincide with

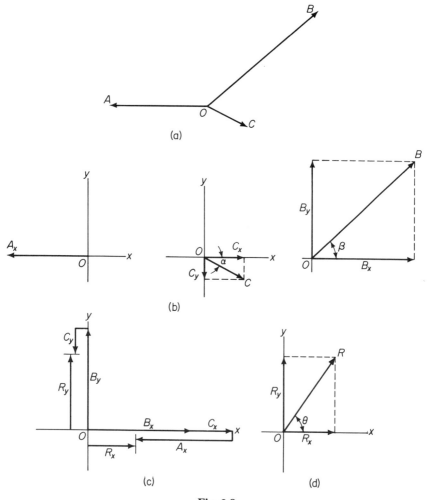

Fig. 2.8

Sec. 2.6   Mathematical Addition of Vectors   23

the x-axis and the y-axis, it is then relatively easy to calculate their value using the sine and cosine functions.

If it is desired to find the resultant of several vector forces mathematically, the method most widely used is the following:

1. Resolve each of the forces, Fig. 2.8(a), into its own components along the x-axis and the y-axis, Fig. 2.8(b).
2. Add the x-components, Fig. 2.8(c).
3. Add the y-components, Fig. 2.8(c).
4. Construct a parallelogram and compute the resultant, Fig. 2.8(d), using the Pythagorean theorem*:

$$OR = \sqrt{(\sum R_y)^2 + (\sum R_x)^2}$$

5. Having computed R_y, R_x, and R, the angle the resultant R makes with the x-axis may be computed from the $\sin \theta$, $\cos \theta$, or $\tan \theta$ (see Fig. 2.8d).

It is very important that the student understand the preceding concept: R, the resultant, is always taken as plus. The sign of the component is taken as plus if it points to the right or up. It is taken as minus if it points to the left, or down. The angle θ is taken as the angle the resultant makes with the x-axis.

Example 7

Calculate the resultant of the forces in Fig. 2.9(a).

Solution

Resolve each force into its own x- and y-components as shown in Fig. 2.9(b). Add all components along the x-axis in Fig. 2.9(b):

$$\sum R_x = -56 + 20.8 + 61.3 = 26.1 \text{ lb}$$

Add all components along the y-axis in Fig. 2.9(b):

$$\sum R_y = 0 - 12 + 51.4 = 39.4 \text{ lb}$$

From Fig. 2.9(c), it can be seen that

$$R = \sqrt{(39.4)^2 + (26.1)^2} = 47.3 \text{ lb}$$

To calculate the angle θ,

$$\tan \theta = \frac{R_y}{R_x} = \frac{39.4}{26.1} = 1.510$$

$$\theta = 56°30'$$

The complete solution is shown in Fig. 2.9(c).

* The symbol $\sum$ in the following equation is the Greek letter *sigma* and means "the sum of."

24   Forces   Chap. 2

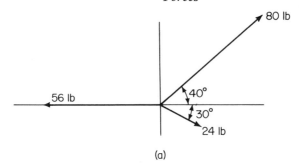

(a)

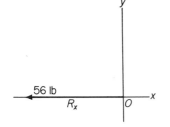 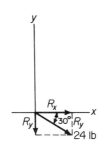 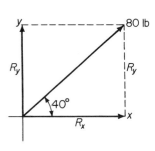

$R_x = -56 \cos 0° = -56$ lb. $R_x = 24 \cos 30° = 20.8$ lb. $R_x = 80 \cos 40° = 61.3$ lb.
$R_y = -56 \sin 0° = 0$ lb $R_y = -24 \sin 30° = -12.0$ lb. $R_y = 80 \sin 40° = 51.4$ lb.

(b)

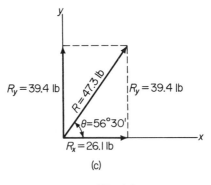

(c)

Fig. 2.9

Problems

1. In your own words, describe the concept of force in terms of the inertia of a body of material. Give several examples of natural and generated forces.
2. Find the resultant of the following forces, using the parallelogram method (use a

scale of $\frac{1}{2}$ in. = 10 lb): (a) 80 lb acting east and 50 lb acting at 45° with the x-axis; (b) 100 lb acting 30° with the x-axis in the first quadrant, and 50 lb acting at 60° with the x-axis in the second quadrant.

3. Using the parallelogram method, combine the following forces and solve graphically:

F_a = 200 lb at 90°
F_b = 100 lb at 30°
F_c = 150 lb at 315°

4. Using the parallelogram method, find the resultant of the following forces:

A = 5 lb at 30°
B = 8 lb at 135°
C = 4 lb at 240°
D = 6 lb at 0°

5. Solve Prob. 3 using the polygon method.
6. Solve Prob. 4 using the polygon method.
7. Mathematically resolve the following forces into their horizontal and vertical components: (a) 50 lb at 45°; (b) 100 lb at 60°; (c) 90 lb at 240°; (d) 40 lb at 120°; (e) 120 lb at 300°.
8. Mathematically resolve the following forces into their horizontal and vertical components: (a) 5 lb at 20°; (b) 8 lb at 70°; (c) 6 lb at 130°; (d) 12 lb at 260°; (e) 15 lb at 335°.
9. Solve Prob. 3 mathematically.
10. Solve Prob. 4 mathematically.
11. Draw a vector diagram of the following forces, measuring all angles in a counterclockwise direction from the positive x-axis; solve mathematically for the resultant: 30 lb at 60°; 45 lb at 90°; 20 lb at 135°; 10 lb at 180°; 50 lb at 240°; 25 lb at 330°.
12. Using the data in Prob. 4, rotate the axes counterclockwise through 30° holding the four forces stationary. Drop perpendiculars to the new set of axes. Solve mathematically for the resultant. Does the rotation of the axes have any effect on the final results?

3

Concurrent Coplanar Forces

3.1. Free-Body and Force Diagrams. Equilibrium

Figure 3.1(a) shows a *free-body* diagram. The free-body diagram is a sketch of the conditions of the problem. It need not be to scale, but should show all the conditions of the problem. The force diagram, Fig. 3.1(b), is a vector representation of the problem.

Two types of forces may act on a body: *contact forces* and *field forces*. Contact forces are those which come into contact with a body. Field forces are those which act on a body but do not come into contact with it. Magnetic, electrostatic, and gravitational forces are a few examples of field forces. All

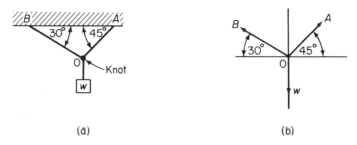

Fig. 3.1

Sec. 3.1   *Free-Body and Force Diagrams. Equilibrium*   27

contact and field forces shown on a free-body diagram (Fig. 3.1a) should be shown on the force diagram (Fig. 3.1b) Note that A and B are contact forces; w is a gravitational force and as such is a field force.

From the student's standpoint, when drawing a vector diagram, he should assume the "viewpoint of the body." That is, ask yourself, "What is the effect of a push or pull *on* the body?" In Fig. 3.1(b), the student should ask himself, "What effect does w have on the knot?" The answer is, "w is pulling down on the knot." Therefore, the vector representing the magnitude of w should be shown pulling down on the knot. Again, "What is the effect of A on the knot? What is the effect of B on the knot?" The answers to these questions are shown in Fig. 3.1(b).

There are several free-body diagrams which appear frequently in problems. They are shown in Fig. 3.2.

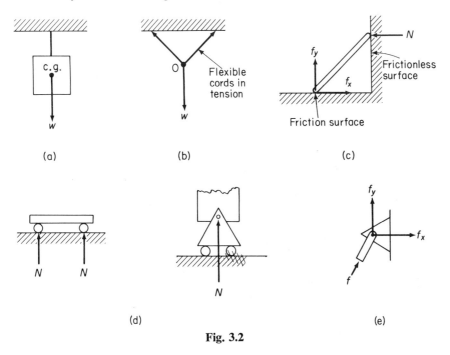

Fig. 3.2

In Fig. 3.2(a), the weight of the body is a measure of the gravitational attraction of the Earth on the body. This gravitational attraction is taken as acting down and through the center of gravity of the body.

Flexible members, such as cords, cables, or ropes, are shown as tensile forces, acting along the member as shown in Fig. 3.2(b). Flexible members cannot support compression forces. (See Chapter 6.) They will collapse.

If a smooth frictionless surface is in contact with a second surface, the force is shown as acting perpendicular (*normal*) to the frictionless surface. This is shown at the top of Fig. 3.2(c). At the bottom of the free-body diagram, friction is present. It will be seen in Chapter 4 that the normal force F_y and the friction force F_x add, vectorially, to produce a resultant force at the point of contact between the two surfaces.

All rollers are represented by a force acting perpendicular to the surface with which the roller is in contact as shown in Fig. 3.2(d).

A hinge pin, Fig. 3.2(e), may exert a force in any direction. Therefore, two component forces may be selected and, when their numerical values are determined, combined into a resultant which shows its magnitude and the direction in which it acts.

The student should recall that a resultant was defined as a force which could replace a system of forces and produce the same net effect as the system. That is, if the tendency of several forces is to produce motion, then the resultant will also produce this tendency.

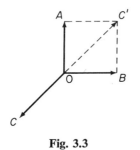

Fig. 3.3

Note that in Fig. 3.3 the forces A and B may be replaced by C'. The force C' will pull on O with the same force and in the same direction as A and B before they were replaced by C'. In Fig. 3.3, a force C, acting in the opposite direction and with the same magnitude as C', will balance C' exactly. We say C and C' are in *equilibrium*. Notice that C will also balance A and B. Thus C' is called the *resultant* of A and B, and C is called the *equilibrant* of A and B.

To summarize, forces which are not in equilibrium will have a resultant. These forces may be put into equilibrium by replacing the resultant with a force equal but acting in the opposite direction to the resultant.

Figure 3.4(a) shows three forces A, B, and C. Using the polygon method of Chapter 2, the three forces are combined graphically in Fig. 3.4(b). The resultant R is drawn with its tail at the tail of arrow A and its head at the head of arrow C. Note that if the resultant is turned around, it will close the polygon

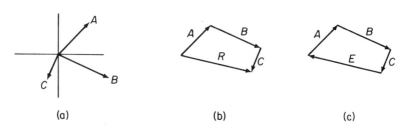

Fig. 3.4

Sec. 3.2   Concurrent Coplanar Forces   29

and it (E) will put the original system of three forces in equilibrium, as shown in Fig. 3.4(c).

It should also be noted that there is no resultant in a system of forces which is in equilibrium.

3.2. Concurrent Coplanar Forces

A system of forces which acts in one plane is said to be *coplanar*. If these forces intersect at a point, they are said to be *concurrent*. If the system of forces is in equilibrium—i.e., if the resultant is zero—then the vector sum of the forces is equal to zero. It is also true that, if the summation of these forces is zero, the *components* of these forces must also be algebraically equal to zero. The conditions necessary to prove equilibrium are

$$\sum F_x = 0 \qquad \sum F_y = 0$$

Therefore, the procedure is to resolve all forces into their components along the x- and y-axes. To have equilibrium, the sum of all the components along the x-axis must equal zero and the sum of all the components along the y-axis must equal zero.

Example 1

The system of forces in Fig. 3.5(a) are in equilibrium. Calculate the value of the vector E. Note that since E is a vector, a complete solution must include its magnitude and the direction in which it is acting.

Solution

1. Resolve each of the forces into its components as shown in Fig. 3.5(b).
2. The equations for equilibrium are

$$\sum F_x = -50 + 40 \cos 30° + E \cos \theta = 0$$
$$\sum F_y = 0 - 40 \sin 30° + E \sin \theta = 0$$

3. Solve these equations for $E \cos \theta$ and $E \sin \theta$:

$$E \cos \theta = 50 - 40(0.866) = 15.4 \text{ lb}$$
$$E \sin \theta = 40(0.500) = 20.0 \text{ lb}$$

4. Substitute these values into Fig. 3.5(b) as shown in Fig. 3.5(c) and solve for E:

$$E = \sqrt{(20.0)^2 + (15.4)^2}$$
$$= 25.2 \text{ lb}$$

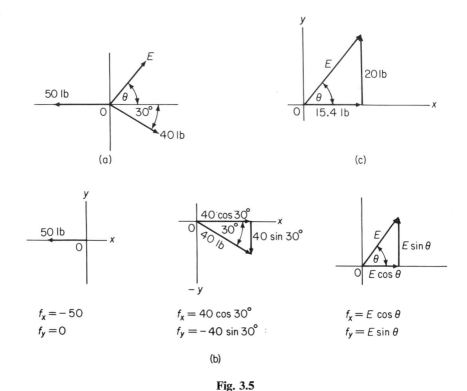

Fig. 3.5

5. The angle at which E acts is

$$\tan \theta = \frac{20}{15.4} = 1.299$$

$$\theta = 52°25' \quad \text{(in the first quadrant)}$$

3.3. The Boom

Example 2

Calculate the tension in the cord T and the force of the boom on point C. Neglect the weight of the boom (see Fig. 3.6a).

Solution

1. Draw the vector diagram as shown in Fig. 3.6(b). Note that point C is in equilibrium; otherwise, it would be in motion.

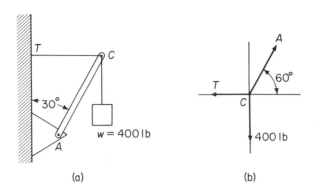

Fig. 3.6

2. Resolve all forces into x- and y-components and add using the equilibrium equations:

$$\sum F_x = A \cos 60° - T = 0$$
$$\sum F_y = A \sin 60° - 400 = 0$$

3. Solve for A and T:

From $\sum F_y$,

$$A = \frac{400}{\sin 60°} = \frac{400}{0.866} = 462 \text{ lb}$$

From $\sum F_x$,

$$T = A \cos 60° = 462(0.500) = 231 \text{ lb}$$

3.4. The Toggle

A toggle, Fig. 3.7(a), is a device which uses a force f to balance a force P.

Example 3

In Fig. 3.7(a), calculate the forces at P when point D is 2 in. above the line of action of the toggle.

Solution

1. The angles a and β may be determined from Fig. 3.7(b):

$$\sin a = \tfrac{2}{4} = 0.500$$
$$a = 30°$$

32   Concurrent Coplanar Forces   Chap. 3

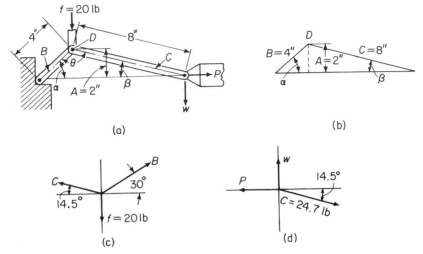

Fig. 3.7

$$\sin \beta = \tfrac{2}{8} = 0.250$$
$$\beta = 14.5°$$

2. Apply the laws of equilibrium at D in the vector diagram, Fig. 3.7(c):

$$\sum F_x = B \cos 30° - C \cos 14.5° = 0$$
$$\sum F_y = B \sin 30° + C \sin 14.5° - 20 = 0$$

3. Solve the two simultaneous equations:

$$0.866B - 0.968 C = 0$$
$$0.500B + 0.250 C = 20$$
$$B = 27.6 \text{ lb}$$
$$C = 24.7 \text{ lb}$$

4. Draw the vector diagram at P as shown in Fig. 3.7(d) and apply the laws of equilibrium:

$$\sum F_x = C \cos 14.5° - P = 0$$
$$\sum F_y = -C \sin 14.5° + w = 0$$

5. Substitute for C and solve for P and w:

$$\sum F_x = 24.7 \cos 14.5° - P = 0$$
$$P = 24.7 \cos 14.5° = 24.7(0.968) = 23.9 \text{ lb}$$

Sec. 3.5 The Simple Truss 33

$$\sum F_y = -24.7 \sin 14.5° + w = 0$$
$$w = 24.7 \sin 14.5° = 24.7(0.250) = 6.17 \text{ lb}$$

3.5. The Simple Truss

A *truss* is a frame structure in which the members are arranged so that they are subjected to either tensile or compressive stresses from loads which they carry. *Live loads* are external loads such as the weight of an automobile. *Dead loads* are caused by the weight of the members themselves, or any fixed loads such as the roof of a house which the truss supports.

Forces in the members may be calculated by selecting a point where all but two (or one) of the forces are known. After finding the unknown forces at this point, proceed to an adjacent junction, where one or two of the forces are unknown. The laws of equilibrium are applied at each junction.

Example 4

Calculate the forces in the members of the truss shown in Fig. 3.8(a), neglecting the weight of the member.

Solution

1. Figure 3.8(b) shows an analysis of the forces at each junction. Consider junction A: junction B pulls on A, junction D pushes on A, and force f_1

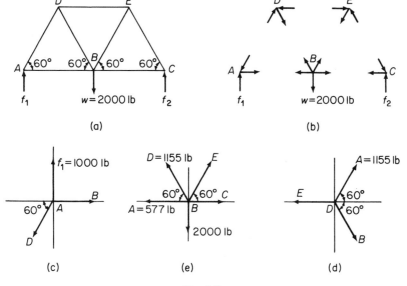

Fig. 3.8

pushes on A. It should be noted that if junction B is pulling on junction A, then junction A must be pulling on junction B with the same force. Beam AB is in tension. Also, if junction D is pushing on junction A, then junction A must be pushing on junction D with the same force. Beam AD is in compression. The student should analyze the action-reaction forces at each junction point in Fig. 3.8(b).

2. By symmetry, loads f_1 and f_2 are equal and each support one-half of w:

$$f_1 = 1000 \text{ lb}$$
$$f_2 = 1000 \text{ lb}$$

3. Select point A (point C could have been selected) and draw the vector diagram (Fig. 3.8c) of the three forces AD, AB, and f_1 acting on junction A. Compare the the forces acting on point A in Fig. 3.8(c) with the forces acting on point A in Fig. 3.8(b). Apply the equations of equilibrium at point A, Fig. 3.8(c):

$$\sum F_x = AB - AD \cos 60° = 0$$
$$\sum F_y = 1000 - AD \sin 60° = 0$$

From $\sum F_y = 0$,

$$AD = \frac{1000}{\sin 60°} = \frac{1000}{0.866} = 1155 \text{ lb}$$

From $\sum F_x = 0$,

$$AB = AD \cos 60° = 1155(0.500) = 577 \text{ lb}$$

4. Next select junction D. In Fig. 3.8(b), the three forces acting at point D are DB, DE, and DA. We know that DA is pushing on A and, therefore, AD must be pushing on D. Also, the weight w (2000 lb) is pulling on B and through member BD is attempting to pull D down. Thus Fig. 3.8(b) shows BD pulling on D. If ED is considered, it can be seen that ED is pushing on an attempt to support D. Figure 3.8(d) shows the force diagram at D. Applying the equation of equilibrium to point D, Fig. 3.8(d),

$$\sum F_x = 1155 \cos 60° + DB \cos 60° - DE = 0$$
$$\sum F_y = 1155 \sin 60° - DB \sin 60° = 0$$

From $\sum F_y$, solve for DB:

$$1155(0.866) - 0.866 DB = 0$$

$$DB = \frac{1155(0.866)}{0.866} = 1155 \text{ lb}$$

From $\sum F_x$, solve for DE:

$$1155(0.500) + 1155(0.500) - DE = 0$$
$$DE = 1155 \text{ lb}$$

5. Now draw the vector diagram for all the forces acting at junction B (Fig. 3.8e) and apply the equations of equilibrium:

$$\sum F_x = BC + BE \cos 60° - 1155 \cos 60° - 577 = 0$$
$$\sum F_y = BE \sin 60° + 1155 \sin 60° - 2000 = 0$$

From $\sum F_y$, solve for BE:

$$BC = \frac{2000 - 1155(0.866)}{0.866} = 1155 \text{ lb}$$

From $\sum F_x$, solve for BC:

$$BE = 1155 \cos 60° + 577 - 1155 \cos 60° = 577 \text{ lb}$$

Problems

1. Explain the meaning of the terms "concurrent" and "coplanar."
2. Draw the vector diagrams for the space diagrams, Fig. 3.9(a), (b), (c), (d) and (e).
3. Calculate forces F_1 and F_2 given that F_1 acts at 30° in the first quadrant, F_2 acts at 60° in the second quadrant, and a force of 1000 lb acts along the negative y-axis. All angles are with the x-axis.

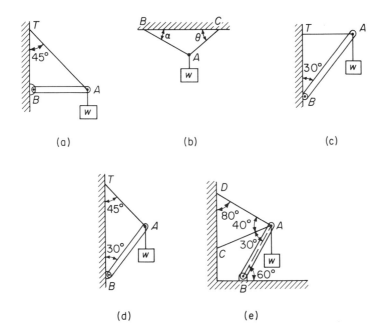

Fig. 3.9

4. Solve for A and B given the following forces: first quadrant, A at 30°; second quadrant, B at 45°; third quadrant, 30 lb at 50°; fourth quadrant, 50 lb at 60° (all angles with the x-axis).
5. Solve for P and F when a 75-lb force acts east; a 100-lb force acts north; a 50-lb force acts south of east at 30°; P acts west of north at 45°; and F acts south of west at 60°. The system is in equilibrium.
6. Calculate the value of θ and a for the following three forces: 25 lb along the negative x-axis; 10 lb at an angle of θ in the first quadrant; and 20 lb at an angle a in the fourth quadrant. All angles are with the x-axis.
7. In Fig. 3.9(a), calculate the tension T in the cord and the compression on the beam AB if $w = 50$ lb. Neglect the weight of the beam.
8. In Fig. 3.9(b), angle $a = \theta = 15°$ and $w = 80$ lb. What is the tension in B and C?
9. In Fig. 3.9(b), angle $a = 30°$, $\theta = 45°$, and $w = 100$ lb. Calculate the tension in B and C.
10. In Fig. 3.9(c), calculate the tension in cord TA and the compression in the beam AB if $w = 200$ lb. Neglect the weight of the beam.
11. In Fig. 3.9(d), calculate the tension in the cord TA and the compression in the beam AB if $w = 500$ lb. Neglect the weight of the beam.
12. In Fig. 3.9(e), if the tension in the cord AC is 400 lb, calculate the tension in the cord AD and the compression in the beam AB if w equals 1000 lb.
13. In Fig. 3.10, calculate the forces P and f if the ball weighs 30 lb.

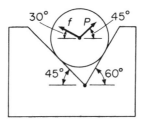

Fig. 3.10

14. Calculate the forces at points A, B, and C on the derrick in Fig. 3.11.

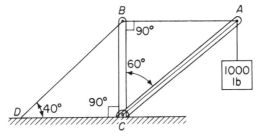

Fig. 3.11

Problems

15. Calculate the tension in the cords A, B, C, and D in Fig. 3.12.

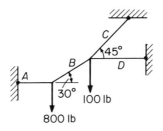

Fig. 3.12

16. Calculate the forces in each member of the truss in Fig. 3.13 and the reactions R_1 and R_2.

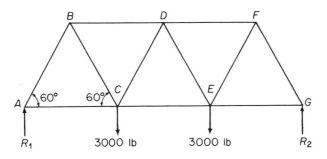

Fig. 3.13

17. Repeat Prob. 16 for Fig. 3.13 when a 2000-lb force is placed at D and acts down.
18. Repeat Prob. 16 for Fig. 3.14.

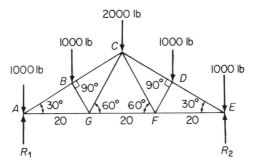

Fig. 3.14

19. Calculate the forces in each member of the truss in Fig. 3.15 and the reactions at the anchors A and C.

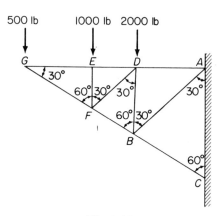

Fig. 3.15

20. In Fig. 3.7(a), calculate the forces in the members B and C and the force f when $P = 1000$ lb. The length of member $B = 3$ ft, $C = 7$ ft, and $A = 2$ ft.

4

Nonconcurrent Coplanar Forces

4.1. Moment of Force

Systems of forces may be in equilibrium without their lines of acting intersecting at a point. Such systems of forces are called *nonconcurrent*. They may be either parallel or nonparallel to each other. This section will deal with nonconcurrent coplanar forces.

Nonconcurrent forces produce a tendency to rotation, called *moment of force*, or *torque*. The moment of force is the product of the force and the perpendicular distance from the line of action of the force to the *center of rotation*. Symbolically, $M = f(d)$. The distance d must be the *perpendicular* distance to the *line of action* of the force. This perpendicular distance is referred to as the *moment arm*. Thus in Fig. 4.1(a) and (b), the forces f and the moment arms d tend to produce a clockwise rotation about the center of rotation O. In Fig. 4.1(c), the force f and the moment arm d tend to produce a counterclockwise rotation about the center of rotation O. For purposes of uniformity, this text will use a clockwise rotation to represent positive (+) rotation, and a counterclockwise rotation to represent negative (−) rotation.*

The units of torque are the units of the product of force and distance. Thus

---

* It should be noted that the reverse convention could have been used.

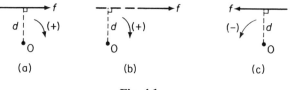

Fig. 4.1

$$f \text{ (lb)} \times d \text{ (ft)} = M \text{ (lb-ft)}$$
$$f \text{ (lb)} \times d \text{ (in.)} = M \text{ (lb-in.)}$$
$$f \text{ (g)} \times d \text{ (cm)} = M \text{ (g-cm)}$$

Example 1

Given the lever shown in Fig. 4.2, calculate the four moments for the four forces shown.

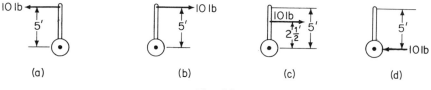

Fig. 4.2

Solution

For Fig. 4.2(a),
$$M = f \times d = -10 \text{ lb} \times 5 \text{ ft} = -50 \text{ lb-ft}$$

For Fig. 4.2(b),
$$M = f \times d = 10 \text{ lb} \times 5 \text{ ft} = 50 \text{ lb-ft}$$

For Fig. 4.2(c),
$$M = f \times d = 10 \text{ lb} \times 2.5 \text{ ft} = 25 \text{ lb-ft}$$

For Fig. 4.2(d),
$$M = f \times d = 10 \text{ lb} \times 0 \text{ ft} = 0$$

In the last example, Fig. 4.2(d), the torque is equal to zero because there is no moment arm. The 10-lb force passes through the center of rotation of the lever.

A *couple* is a pair of equal nonconcurrent forces acting in opposite directions, as shown in Fig. 4.3. Since the forces are equal and opposite, there can be no linear motion. Because both forces tend to create rotary motion about the center of rotation in the same direction, there is always an unbalance.

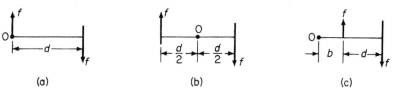

Fig. 4.3

To achieve rotary equilibrium in a couple, an equal and opposite couple is needed.

It can be shown that a couple may be replaced by the mathematical equivalent of *one* of the forces times the perpendicular distance between the two forces. The center of rotation O may be anywhere along the axis of the couple.

Example 2

In Fig. 4.3, show that for a couple, $\sum M_0 = f \times d$.

Solution

For Fig. 4.3(a),
$$\sum M_0 = (f \times d) + (f \times 0) = fd$$

For Fig. 4.3(b),
$$\sum M_0 = \frac{f \times d}{2} + \frac{f \times d}{2} = fd$$

For Fig. 4.3(c),
$$\sum M_0 = -(f \times d) + f(d + b) = -\cancel{fb} + fd + \cancel{fb} = fd$$

4.2. Equilibrium: Parallel Forces

Torque is a vector quantity and as such has magnitude, direction, and a point of application. As vectors, torques may be added vectorially. If the sum of the clockwise moments does *not* equal the sum of the counterclockwise moments, there is a net torque remaining, called the *resultant torque*. If, however, the sum of the clockwise moments *equal* the sum of the counterclockwise moments, the system is said to be in *rotational equilibrium*.

In Chapter 2, it was stated that if the forces up equal the forces down, equilibrium exists along the *y*-axis. Also, if the forces to the right equal the forces to the left, equilibrium exists along the *x*-axis.

The preceding two paragraphs may be stated as

Nonconcurrent Coplanar Forces

$$\sum M_o = 0 \quad \sum F_y = 0 \quad \sum F_x = 0$$

These are the three equations of equilibrium used for coplanar systems.

Example 3

Calculate the forces R_1 and R_2 in Fig. 4.4(a).

Solution

1. Draw the vector diagram as shown in Fig. 4.4(b). The weight of the beam is concentrated at the center of gravity of the beam. Thus

$$20 \text{ lb/ft} \times 20 \text{ ft} = 400 \text{ lb}$$

2. Note that there are no forces in the x-direction. Therefore, $\sum F_x = 0$ does not apply here.

3. From Fig. 4.4(b), use R_1 as the pivot to calculate R_2:

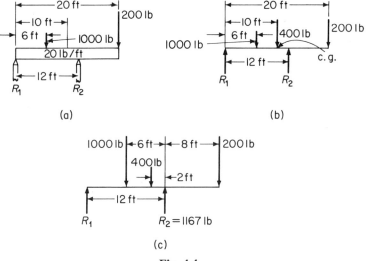

Fig. 4.4

$$\sum M_{R_2} = (1000 \times 6) + (400 \times 10) - R_2(12) + (200 \times 20) = 0$$

$$R_2 = \frac{14{,}000}{12} = 1167 \text{ lb}$$

4. From Fig. 4.4(c), use R_2 as the pivot to calculate R_1. Note that the vector diagram has been redimensioned

$$\sum M_{R_2} = (200 \times 8) - (400 \times 2) - (1000 \times 6) + 12R_1 = 0$$

$$R_1 = 433 \text{ lb}$$

Sec. 4.2   Equilibrium: Parallel Forces   43

5. From $\Sigma F_y = 0$, as a check,

$$R_1 - 1000 - 400 - 200 + 1167 = 0$$
$$R_1 = 433 \text{ lb} \quad \text{(check)}$$

Another very interesting example dealing with parallel nonconcurrent forces is the type where it is necessary to calculate the magnitude and point of application of the equilibrant.

Example 4

Solve for E and d in Fig. 4.5(a). Neglect the weight of the bar.

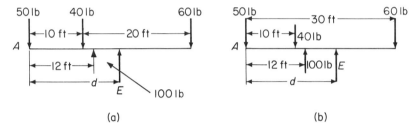

Fig. 4.5

Solution

1. Assume E to be acting at a distance d as shown in Fig. 4.5(a). If the solution shows E to be negative, the direction was assumed incorrectly.
2. Draw and dimension the vector diagram as shown in Fig. 4.5(b).
3. From Fig. 4.5(b),

$$\sum F_y = E + 100 - 50 - 40 - 60 = 0$$
$$E - 50 \text{ lb}$$
$$\sum M_A = (40 \times 10) - (100 \times 12) - Ed + (60 \times 30) = 0$$
$$d = \frac{1000}{50} = 20 \text{ ft}$$

4. As a check, moments may be taken about the 40-lb force. Thus

$$\sum M_{40} = -(50 \times 10) - (100 \times 2) - (50 \times d_1) + (60 \times 20) = 0$$
$$d_1 = \frac{500}{50} = 10 \text{ ft}$$

But
$$d = d_1 + 10 \text{ ft} = 10 + 10 = 20 \text{ ft} \quad (check)$$

4.3. Equilibrium: Nonconcurrent Nonparallel Forces

In the general case of nonparallel forces, the equations of equilibrium are
$$\sum F_x = 0, \quad \sum F_y = 0, \quad \sum M_0 = 0$$
With these three equations, it is possible to solve problems which have three unknowns. At this time it is important that the student review Fig. 3.2.

Example 5

Figure 4.6(a) shows a 20-ft ladder weighing 50 lb leaning against a smooth wall and making an angle of 60° with the ground. If a man, weighing 150 lb, stands

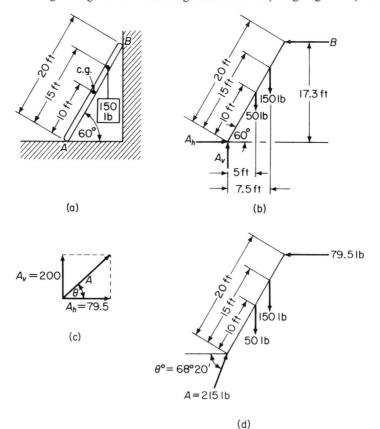

Fig. 4.6

Sec. 4.3   Equilibrium: Nonconcurrent Nonparallel Forces

three-quarters the distance up the ladder from the ground, what are the reaction forces at the wall and at the ground?

Solution

1. Using the force diagram Fig. 4.6(b), two equations for equilibrium may be written:

$$\sum F_x = A_h - B = 0$$
$$\sum F_y = A_v - 50 - 150 = 0$$

2. Using the force diagram, Fig. 4.6(b), moments are taken about A at the bottom of the ladder. This will eliminate A from the equation (no moment arm) and yield one equation in one unknown. It should also be noted that the perpendicular distances must be calculated:

$$\sum M_A = (50 \times 5) + (150 \times 7.5) - (B \times 17.3) = 0$$
$$B = \frac{1375}{17.3} = 79.5 \text{ lb}$$

3. From the equation

$$\sum F_x = A_h - B = 0,$$
$$A_h = 79.5 \text{ lb}$$

Notice that A_h is the only force opposing B in Fig. 4.6(b).

4. From the equation

$$\sum F_y = A_v - 50 - 150 = 0,$$
$$A_v = 200 \text{ lb}$$

5. It is now possible to calculate the value of A and the direction $0°$ in which it acts. From Fig. 4.6(c),

$$\tan \theta° = \frac{200}{79.5} = 2.516$$
$$\theta° = 68°20'$$
$$\sin 68°20' = \frac{200}{A}$$
$$A = \frac{200}{\sin 68°20'} = \frac{200}{0.929}$$
$$= 215 \text{ lb}$$

6. The complete force diagram is shown in Fig. 4.6(d).

4.4. Center of Gravity: Experimental Method

Gravitational attraction is the force between two bodies, or particles of matter. It is the force of attraction exerted on each of the separate particles which when added, make up the force of attraction that the Earth, as a particle, exerts. The *resultant* force of all these separate forces constitute the *weight* of the body. There is a point to which the resultant force may be applied so that it will create the same effect on the object as will all the small forces taken together. This point is called the *center of gravity* (cg) of the object. It is the pivot point at which the clockwise moments equal the counterclockwise forces for rotational equilibrium of the object.

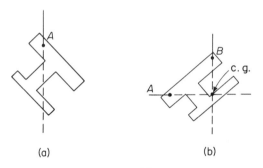

Fig. 4.7

The center of gravity of a body may be found experimentally by suspending it from two different points on the object, as shown in Fig. 4.7. The center of gravity will lie on a line directly under the point of suspension and at the point where these two lines cross. This is shown in Fig. 4.7(b).

4.5. Center of Gravity: Regular Areas

If the material from which an object is made is homogeneous and if the object is of uniform thickness, it is possible to neglect both the thickness and the density of the object when calculating the center of gravity of the object. Under these conditions the weight is taken as proportional to the area and areas are used instead of forces in the equations of equilibrium. Centers of gravity of lines, areas, and volumes are called *centroids*.

Appendix Table A.4 shows some of the more common locations of center of gravity. Using these standard forms, it is possible to calculate the center of gravity of regular objects. The general procedure is as follows:

Sec. 4.5 Center of Gravity: Regular Areas 47

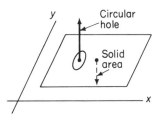

Fig. 4.8

1. Divide the body into simple areas of known or easily found centers of gravity.
2. Consider these simple areas as forces such that sectors representing "solid" areas act *down* because of gravitational attraction. Areas which are "cutouts" (holes) act *up*. Figure 4.8 shows both conditions.
3. Because the *net* area of the object will always act down for purposes of calculations, an area A_e acting at a distance X_e must be included to create equilibrium conditions.
4. Apply the equations of equilibrium and calculate the magnitude of the net area A and the distance from the center of rotation to the center of gravity, X_e.
5. Revolve the area through 90°, select a new center of rotation, and repeat steps 1 through 4. This will locate the distance Y_e.
6. The point of intersection of X_e and Y_e determine the center of gravity for the area.

Example 6

Locate the center of gravity for the area in Fig. 4.9(a).

Solution

1. From Fig. 4.9(b), the magnitudes of the areas are

$$A_1 = \text{total area of } ABCD = bh = 12 \times 10 = 120 \text{ in.}^2$$
$$A_2 = \text{area of the circle} = \pi r^2 = \pi \times 2^2 = 12.6 \text{ in.}^2$$
$$A_e = \text{net area}$$

Area A_1 acts down

Areas A_2 and A_e act up

2. From Fig. 4.9(c), these areas act at a perpendicular-moment-arm distance X from the y-axis. Thus

A_e acts at a distance X_e from the y-axis

A_1 acts at a distance of 6 in. from the y-axis

A_2 acts at a distance of 3 in. from the y-axis

3. Applying the equations of equilibrium to Fig. 4.9(c), taking moments about corner D,

$$\sum F_y = A_y = 0$$
$$A_e + A_2 - A_1 = 0$$
$$A_e = A_1 - A_2 = 120 - 12.6 = 107.4 \text{ in.}^2$$
$$\sum M_D = 0 = -(A_e \times X_e) - (A_2 \times X_2) + (A_1 \times X_1) = 0$$

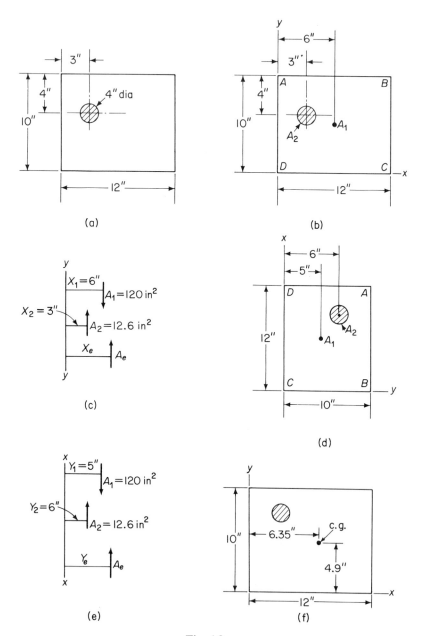

Fig. 4.9

Sec. 4.6   Center of Gravity: Solids   49

$$= -(107.4\, X_e) - (12.6 \times 3) + (120 \times 6) = 0$$

$$X_e = \frac{720.0 - 37.8}{107.4} = \frac{682.2}{107.4} = 6.35 \text{ in.}$$

4. Rotate Fig. 4.9(b) clock wise to the position shown in Fig. 4.9(d) so that the x-axis is vertical and moments may be taken about point C:

$$\sum M_C = -(A_e \times Y_e) - (A_2 \times Y_2) + (A_1 \times Y_1) = 0$$
$$= -(107.4 Y_e) - (12.6 \times 6) + (120 \times 5) = 0$$

$$Y_e = \frac{600 - 75.6}{107.4} = \frac{524.4}{107.4} = 4.9 \text{ in.}$$

5. See Fig. 4.9(f) for the location of the center of gravity.

4.6. Center of Gravity: Solids

We have seen that if solids have uniform thickness and uniform density, we may substitute *area* for force and apply the equations of equilibrium to calculate the center of gravity. Since the thickness is uniform, the center of gravity will be halfway between the faces of the material.

If the density of the material is uniform, but the thickness is *not* uniform, small regular *volumes* may be substituted for the areas. Since the thickness is nonuniform, moments must be taken about the x-, y-, and z-axes.

If the thickness is uniform but the density varies, *weight* may be used instead of area or volume. If the densities are symmetrical about a centerline (see Fig. 4.10a), the center of gravity will be on that centerline. Only two equations of equilibrium are needed.

Example 7

Figure 4.10(a) shows a composite structure about the centerline. Calculate the center of gravity. (For density values, see Appendix Table A.5.)

Solution

1. The center of gravity will be on the x-axis, Fig. 4.10(a), because the object is symmetrical about this centerline.
2. Find the center of gravity of the aluminum, *disregarding* the steel plug. From Appendix Tables A.4 and A.5, the center of gravity and weight of the aluminum may be calculated. The weight of the aluminum is

$$\text{solid aluminum} = 3 \times 4 \times 10 \times \frac{168.5}{1728} = 11.70 \text{ lb}$$

$$\text{rectangular aluminum removed} = 3 \times 2 \times 4 \times \frac{168.5}{1728} = 2.34 \text{ lb}$$

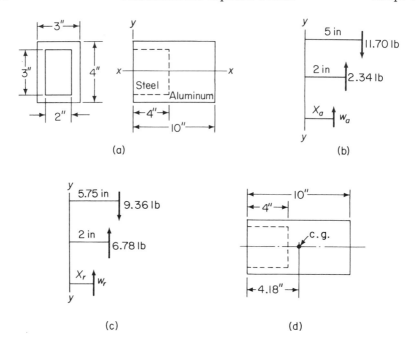

Fig. 4.10

From Fig. 4.10(b),

$$\sum w_a = 0$$
$$= w_a + 2.34 - 11.70 = 0$$
$$w_a = 9.36 \text{ lb}$$
$$\sum M = 0$$
$$= (11.70 \times 5) - (2.34 \times 2) - (w_a \times X_a) = 0$$
$$= 58.50 - 4.68 - 9.36 X_a = 0$$
$$X_a = \frac{53.82}{9.36} = 5.75 \text{ in.}$$

3. The steel plug is now inserted and the center of gravity of the composite structure calculated.
   (a) The weights of the materials are

$$w_a = 9.36 \text{ lb}$$
$$w_s = 3 \times 2 \times 4 \times \frac{488}{1728} = 6.78 \text{ lb}$$

   (b) From Fig. 4.10(a), from the left end of the composite object, the distances to the centers of gravity of each regular segment are

$$X_a = 5.75 \text{ in.}$$
$$X_s = 2 \text{ in.}$$

(c) The force diagram is shown in Fig. 4.10(c).

(d) Calculate the resultant weight from Fig. 4.10(c):

$$\sum w = w_r - 6.78 - 9.36 = 0$$
$$w_r = 16.14 \text{ lb}$$

(e) Take moments about the left end of the object:

$$\sum M_y = (6.78 \times 2) + (9.36 \times 5.75) - w_r X_r = 0$$
$$X_r = \frac{67.38}{16.14} = 4.18 \text{ in.}$$

(f) Therefore, the net weight of 16.14 lb may be concentrated along the central axis 4.18 in. from the left end of the composite object, as shown in Fig. 4.10(d).

Problems

1. A horizontal beam which weighs 100 lb has one end embedded in a concrete wall. If the beam projects beyond the wall a distance of 8 ft, what is the moment acting to break the beam at the wall?
2. A wheel has a rope wrapped around its circumference. What is the torque on the center of the shaft, if the wheel is 4 ft in diameter and a force of 80 lb is applied to the rope?
3. In Fig. 4.11, calculate the resultant torque.

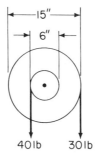

Fig. 4.11

4. The distance from the axle of a wheelbarrow to the place where a man grips the handles is 4.8 ft. Assume the center of gravity of a 150-lb load is 2 ft from the axle. (a) What force is exerted by the man? (b) What force is exerted on the axle?

5. A pair of pliers is used to grip a pin. The pliers is 10 in. long and the hinge is 3 in. from jaw end of the pliers. Assume the pliers is gripped 2 in. from the end with a force of 18 lb. (a) What is the holding force applied to the pin which is gripped? (b) What is the force on the hinge?
6. Calculate the resultant torque in Fig. 4.12 and verify your answer using the concept of couples.

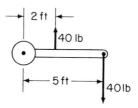

Fig. 4.12

7. Using the concept of couples, calculate the resultant torque in Fig. 4.13.

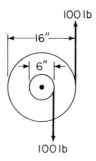

Fig. 4.13

8. A uniform bar is 4 ft long. The left end of the bar supports a 60-lb load; the right end of the bar supports a 28-lb load. (a) At what point along the bar must a pivot support be tied to establish equilibrium? (b) What weight will this pivot be required to support?
9. A uniform bar is 9 ft long and weighs 75 lb. Starting from the left end of the bar, weights are suspended every $1\frac{1}{2}$ ft, as shown in Fig. 4.14. Where must a scale be attached to the bar to create equilibrium?

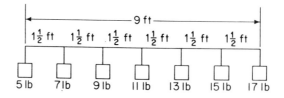

Fig. 4.14

Problems

10. Calculate R_1 and R_2 for Fig. 4.15.

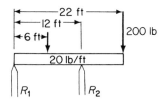

Fig. 4.15

11. Calculate the reactions at the rollers, Fig. 4.16.

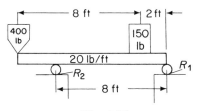

Fig. 4.16

12. A 20-ft beam weighs 400 lb and supports concentrated loads from left to right as follows: f at the extreme left end, and 250 lb at the extreme right end. A 150-lb force and a force P are concentrated at 4 ft and 16 ft, respectively, from the left end of the beam; P and f are supporting points. Calculate P and f.

13. Neglecting the weight of the bar, calculate the equilibrant E and the distance d in Fig. 4.17.

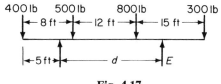

Fig. 4.17

14. Two men are at opposite ends of a 12-ft beam carrying it on their shoulders. The beam weighs 25 lb. If a 200-lb weight is hung 4 ft from the forward man, how much weight is each supporting?

15. In Prob. 14, how far forward must the man at the rear move in order to equalize the load?

16. Calculate the forces R_1 and R_2 in Fig. 4.18. Triangle A weighs 250 lb and triangle B weighs 300 lb.

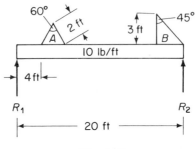

Fig. 4.18

17. Calculate f_1, f_2, and f_3 in Fig. 4.19.

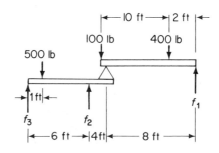

Fig. 4.19

18. In Fig. 4.20, calculate the force f required to raise a weight w of 1000 lb.

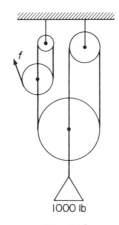

Fig. 4.20

Problems

19. In Fig. 4.21, calculate the value of T.

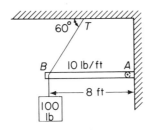

Fig. 4.21

20. A ladder leaning against a smooth wall makes an angle of 80° with the ground. The ladder weighs 75 lb and is 20-ft long. What is the reaction at the wall and ground?
21. Calculate the reaction at the ground and at the wall when a 180-lb man is $\frac{3}{4}$ the distance up the ladder in Prob. 20.
22. A 50-lb ladder, Fig. 4.22, overhangs the wall. A weight of 10 lb is fastened at the top of the ladder. A 200-lb man climbs $\frac{2}{3}$ the distance up the ladder. What is the reaction at the ground and at the wall?

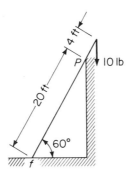

Fig. 4.22

23. A horizontal beam is hinged at a wall as shown in Fig. 4.23. Calculate the tension in the cable and the reaction at the wall if $Tf = 6$ ft.
24. Assume the angle θ in Fig. 4.23 to be $60°$ and the weight to be unknown. If T is equal to 2000 lb, calculate the reaction at the wall f and the weight w.

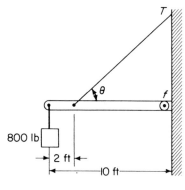

Fig. 4.23

25. Calculate T and f in Fig. 4.24.

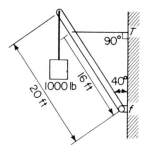

Fig. 2.24

26. Calculate T and f in Fig. 4.25.

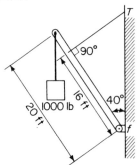

Fig. 4.25

Problems

27. Calculate the center of gravity in Fig. 4.26.

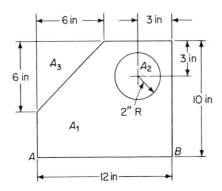

Fig. 4.26

28. Calculate the center of gravity in Fig. 4.27.

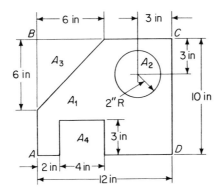

Fig. 4.27

29. Calculate the center of gravity in Fig. 4.28.

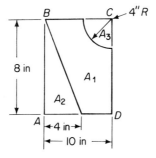

Fig. 4.28

30. Calculate the center of gravity in Fig. 4.29.

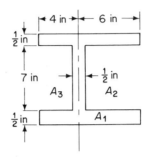

Fig. 4.29

31. In Fig. 4.30, calculate the center of gravity of the steel structure.
32. Calculate the center of gravity in Fig. 4.30 if the hole has an aluminum plug.

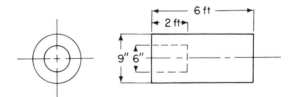

Fig. 4.30

33. Calculate the center of gravity in Fig. 4.31.

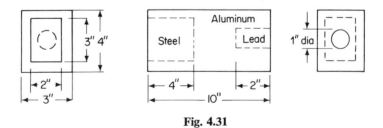

Fig. 4.31

5

Friction

5.1. Friction and its Coefficient

If a book is placed on a table and a pulling force exerted upon it, the book will not move until a sufficient force is exerted to overcome the interlocking effect between the two surfaces. This interlocking effect is called *friction*. Friction forces are *resisting* forces and as such are represented acting in an opposite direction from the applied force trying to create motion. This is shown in Fig. 5.1(a).

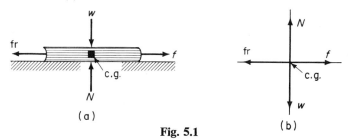

Fig. 5.1

If, starting from zero, the applied force f is gradually increased, the resisting force fr will increase and the book will remain motionless. This type of friction force is called *static* or *limiting friction*. As the magnitude of the applied force is increased, a force will be reached which will be greater than the resisting force. The equilibrium state between f and fr will be distributed such that a resultant, or unbalance, is created in favor of f and motion will take place. Note that this unbalance creates an acceleration in the motion of the book.

Another phenomenon occurs at the moment motion takes place. The force applied to the book at the instant motion just starts is greater than the force needed to keep the book moving with uniform motion. If this unbalance is maintained, an acceleration of the book takes place. However, uniform motion is an equilibrium state. There are no unbalanced forces operating. This means that the applied force f must be reduced if uniform motion equilibrium is desired. Subsequently, if the applied force f is reduced, the resisting force fr will also become smaller to maintain equilibrium. This resisting force to uniform motion is called *dynamic* or *sliding friction*.

The principles just stated are important enough to be repeated; they are as follows:

1. Friction always acts *opposite* to the direction of intended or actual motion.
2. Until motion takes place, the applied force equals the friction force.
3. The force applied to the book (in the above example) at the time motion just starts is greater than the force needed to keep the book moving with uniform motion.

Two forces other than the applied force and the friction force act on the book. One of these forces is the *gravitational force*. The other force is created by the surface on which the book is sliding, and is pushing up on the book. The gravitational force is represented by the *weight* of the book; whereas the force pushing up on the book acts normal, or at right angles, to the surface on which the book is sliding. This force is called the *normal force and should not be confused with the weight of the object.*

These four forces acting on the book may be represented by a vector diagram, inasmuch as each force has magnitude, direction, and a point of application, and therefore is a vector quantity (see Fig. 5.1).

Notice that N is normal to the surface S, fr is parallel to surface S and perpendicular to N, and w always acts toward the center of the earth, or straight down. The applied force f, however, may act in any direction. Note that the direction in which f acts will determine the direction of motion of the book, and consequently the direction of action of friction, fr. All forces may be represented as concentrated at the center of gravity of the object.

Since the book is either stationary or in uniform motion, the forces, Fig. 5.1(b), must be in equilibrium. It is therefore possible to apply the equations of equilibrium:

$$\sum F_x = f - fr = 0$$
$$\sum F_y = N - w = 0$$

Sec. 5.2 Friction and Vectors: The Level Plane

From these equations, it can be seen that

$$f = fr \quad \text{and} \quad N = w$$

The ratio μ (*mu*), is a ratio between the friction force fr and the normal force N. The ratio μ is called the *coefficient of friction*.

We may also say that, since the friction is dependent upon the normal force exerted on the book, the normal force N and fr are proportional to each other:

$$fr \propto N$$

and that μ is the proportionality constant such that

$$fr = \mu N$$

Therefore, the coefficient of friction is

$$\mu = \frac{fr}{N}$$

If fr is the force observed just as the book starts to slide, μ is called the *coefficient of static friction*. If fr is the force observed for uniform motion of the book, μ is called the *coefficient of sliding friction*. Static friction is higher than sliding friction.

Appendix Table A.6 shows some representative values for sliding, static, and rolling friction when one surface is in contact with another surface.

5.2. Friction and Vectors: The Level Plane

Case 1. FORCE PARALLEL TO THE PLANE

Example 1

A wood block is pulled along a wood surface, Fig. 5.2(a). If the block weighs 50 lb and the coefficient of friction is 0.3, what is the force required to maintain uniform motion?

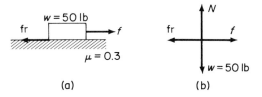

Fig. 5.2

Solution

Figure 5.2(b) shows the vector diagram of the forces on the block. The equations of equilibrium may be written as follows:

From $\sum F_y$,

$$N - 50 = 0$$
$$N = 50 \text{ lb}$$

From $fr = \mu N$,

$$fr = 0.3 \times 50 = 15 \text{ lb}$$

From $\sum F_x$,

$$f - fr = 0$$
$$f = fr = 15 \text{ lb}$$

Note that N and w have the same values in this problem.

Case 2. Force Not Parallel to the Plane

In Fig. 5.3(a), the conditions are the same as in Case 1, except that f is pulling at an angle β with the horizontal plane. Therefore, f is resolved into its x- and y-components and the equations of equilibrium are applied. From Fig. 5.3(b), it can be seen that the y-component, $f \sin \beta$, assists the normal force N in its effort to raise the block off the plane. The x-component, $f \cos \beta$, is the force used to move the block along the plane.

Example 2

The wood block, Fig. 5.3(a), weighs 50 lb and the force f is pulling the block at an angle of 20°. Calculate f, fr, and N if the coefficient of friction is 0.3.

Solution

The equations of equilibrium are applied to the vector diagram, Fig. 5.3(b).

From $\sum F_y$,

$$N + f \sin 20° - 50 = 0$$
$$N = 50 - f \sin 20°$$
$$= 50 - 0.34f$$

Sec. 5.2   Friction and Vectors: The Level Plane   63

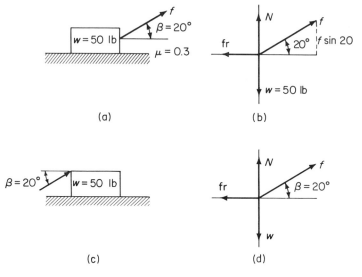

Fig. 5.3

From fr,

$$fr = \mu N = 0.3(50 - 0.34f)$$
$$= 15 - 0.102f$$

From $\sum F_x$,

$$f \cos 20° - fr = 0.94f - (15 - 0.102f) = 0$$
$$f = 14.40 \text{ lb}$$

From the friction equation,

$$fr = 15 - 0.102f = 15 - 0.102(14.4)$$
$$= 13.53 \text{ lb}$$

From $\sum F_y$,

$$N = 50 - 0.34f = 50 - 0.34(14.4)$$
$$N = 45.10 \text{ lb}$$

Note that N does not have the same value as w.

It is important to note in all cases that it makes no difference whether the block is being pushed or pulled. Figure 5.3(c) shows the block being pushed and Fig. 5.3(d) shows the vector diagram. The student should compare Fig. 5.3(d) and Fig. 5.3(b).

5.3. Friction and Vectors: The Inclined Plane

Case 3. SLIDING UNDER FORCE OF WEIGHT ONLY

Now, instead of pulling the block along the horizontal surface, the block is caused to slide down an inclined plane. The only force which could cause motion down the plane is the weight of the block; more specifically, the x-component of the weight of the block (see Fig. 5.4a and b).

In Fig. 5.4 (a), only three forces act:

1. the weight of the block, straight down;
2. the normal, perpendicular to the inclined surface;
3. the friction force, opposite to the direction of motion; since motion is down the plane, the friction vector acts up the plane.

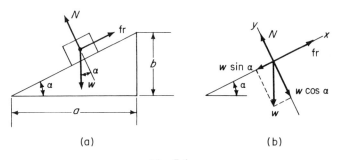

Fig. 5.4

Figure 5.4(b) shows the w-force resolved into its x- and y-components. Thus there are now four forces operating: fr is the friction vector opposing $w \sin \alpha$, which is acting to create motion down the plane; N is the normal force in opposition to the y-component of w, or $w \cos \alpha$.

Place the block on a horizontal plane. Gradually raise one end of the plane until the block slides down the plane with a *uniform motion*. The angle α which the plane makes with the horizontal is called the *angle of repose*. If the angle α is greater than the angle of repose, the block will accelerate down the plane. If the angle α is less than the angle of repose, there will be no motion. It should be recalled that it takes a greater force down the plane to start motion then to maintain uniform motion down the plane. Therefore, α must be greater than the angle of repose to start the block moving. The tangent of this angle is the coefficient of *static friction*. The tangent of the angle required to maintain uniform motion the coefficient of *sliding friction*.

In Fig. 5.4(a), it is necessary only to measure a and b. The quotient of b divided by a will yield the coefficient of friction μ. If the ratio is taken at the

Sec. 5.3   Friction and Vectors: The Inclined Plane   65

instant motion starts, then μ is the coefficient of static friction. If the ratio is taken as the block moves down the inclined plane with uniform motion, then μ is the coefficient of sliding friction.

From Fig. 5.4(b),

$$\sum F_x = fr - w \sin \alpha = 0$$
$$\sum F_y = N - w \cos \alpha = 0$$

Dividing $\sum F_x$ by $\sum F_y$:

$$\frac{fr}{N} = \frac{\cancel{w} \sin \alpha}{\cancel{w} \cos \alpha} = \tan \alpha$$

But

$$\frac{fr}{N} = \tan \alpha = \mu \quad \text{(coefficient of friction)}$$

Note that this angle α is the angle of impending or uniform motion. Thus α is the angle of repose and μ is the coefficient of friction. It is important to distinguish between the inclined angle α and the arctan μ (coefficient of friction.) The possible combinations are

1. $\alpha <$ arctan μ; there will be *no* motion down the plane;
2. $\alpha >$ arctan μ; there will be accelerated motion down the plane;
3. $\alpha =$ arctan μ; angle of repose, uniform motion down the plane.

Example 3

The weight of the block in Fig. 5.4(a) is 50 lb. Calculate (1) the angle at which the block will move with uniform motion down the plane if the friction force opposing uniform motion is 28 lb; (2) the normal force of the plane pushing against the block. (3) Analyze the angle α in terms of its effect on the motion of the block.

Solution

1. From $\sum F_x$,

$$fr - w \sin \alpha = 0$$
$$50 \sin \alpha = 28$$
$$\sin \alpha = \frac{28}{50} = 0.56 \qquad \left\{ \begin{array}{l} w = 50 \text{ lb} \\ fr = 28 \text{ lb} \end{array} \right\}$$
$$= 34°4'$$

2. From $\sum F_y$,

$$N - w \cos \alpha = 0$$
$$N = 50 \cos 34°4'$$
$$= 41.4 \text{ lb}$$

Also,

$$\tan \alpha = \frac{fr}{N} = \frac{28}{41.4} = 0.67$$
$$\alpha = 34°4'$$

3. The analysis of α and μ follows from

$$\tan^{-1} 0.67 = 34°4'$$

Thus

if $\alpha < 34°4'$, no motion
if $\alpha > 34°4'$, accelerated motion
if $\alpha = 34°4'$, uniform motion

Case 4. FORCE PARALLEL TO THE INCLINE

Suppose that the block is being pulled up the inclined plane as shown in Fig. 5.5(a). The weight w acts toward the center of the earth and the normal N acts perpendicular to the plane; f acts parallel and up the inclined plane, imparting motion to the block up the plane. Since motion is up the plane, friction fr acts to oppose motion; fr therefore acts down the inclined plane.

It should be noted in Fig. 5.5(b) that f must balance both fr and $w \sin \alpha$, the component of the weight w, to insure equilibrium for uniform motion. Also, note that the normal force N is equal to the other component of w, or $w \cos \alpha$.

Example 4

Refer to Fig. 5.5(b). Given $\mu = 0.3$, $w = 50$ lb, and the angle of the inclined plane $= 30°$, calculate the friction force and the force required to move the block up the incline.

Solution

From $\sum F_y$,

$$N - w \cos \alpha = 0$$
$$N = 50 \cos 30°$$
$$= 43.3 \text{ lb}$$

$$\left\{ \begin{array}{l} \mu = 0.3 \\ w = 50 \text{ lb} \\ \alpha = 30° \end{array} \right\}$$

Sec. 5.3 Friction and Vectors: The Inclined Plane

From the friction equation,
$$fr = \mu N = 0.3(43.3)$$
$$= 13 \text{ lb}$$

From $\sum F_x$,
$$f - fr - w \sin \alpha = 0$$
$$f = fr + w \sin \alpha = 13 + 50 \sin 30° = 0$$
$$= 38 \text{ lb}$$

It is left to the student to draw the vector diagram if the block in Fig. 5.5(a) is being pushed up the inclined plane by a force parallel to the incline. The student should compare this diagram with Fig. 5.5(b).

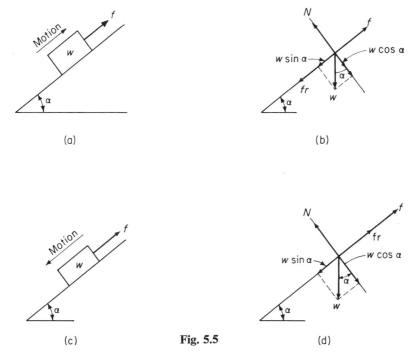

Fig. 5.5

It is important to note that in Fig. 5.5(c), the block is being pushed up the incline with a force which is not great enough to keep the block from sliding down the incline. Figure 5.5(d) shows the vector diagram. Note that $w \sin \alpha$ must be greater than $f + fr$ to have motion down the plane.

Case 5. Force Not Parallel to the Incline

Suppose the block is being pulled up an inclined plane by a force as shown in Fig. 5.6(a). The student should note that the y-component of f, $f \sin \alpha$, is aiding N, and that the x-component of w, $w \sin \alpha$, is aiding fr.

Example 5

Refer to Fig. 5.6(b). Given $\mu = 0.3$, $w = 50$ lb, the angle of incline of the plane $= 30°$, and the angle the force that f makes with the inclined plane $= 20°$, calculate f, N, and fr.

Solution

From $\sum F_y$,

$$N + f \sin \beta - w \cos \alpha = 0$$
$$N = 50 \cos 30° - f \sin 20°$$
$$= 43.3 - 0.34 f$$

$$\begin{Bmatrix} w = 50 \text{ lb} \\ \mu = 0.3 \\ \alpha = 30° \\ \beta = 20° \end{Bmatrix}$$

From the friction equation,

$$fr = \mu N = 0.3(43.3 - 0.34 f)$$
$$= 13 - 0.102 f$$

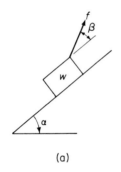

(a)

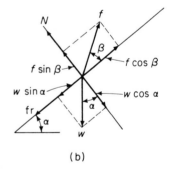

(b)

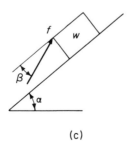

(c)

Fig. 5.6

Sec. 5.4   Rolling Friction   69

From $\sum F_x$,

$$f \cos \beta - w \sin \alpha - fr = 0$$
$$f \cos 20° - 50 \sin 30° - (13 - 0.102 f) = 0$$
$$0.94 f - 25 - 13 + 0.102 f = 0$$
$$f = \frac{38}{1.042}$$
$$= 36.47 \text{ lb}$$

From $\sum F_y$,

$$N = 43.3 - 0.34 f = 43.3 - 0.34(36.47)$$
$$= 30.9 \text{ lb}$$

From fr,

$$fr = 13 - 0.102 f = 13 - 0.102(36.47)$$
$$= 9.28 \text{ lb}$$

The student should draw the vector diagram for Fig. 5.6(c) and compare it to Fig. 5.6(b).

5.4. Rolling Friction

Figure 5.7 shows the exaggerated cross-sectional view of a roller being rolled over a smooth surface. Since the roller has weight, it will cause the surface S to be depressed as though it were made of rubber. Once the roller has passed, the surface S will return to its original shape.

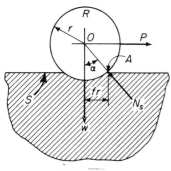

Fig. 5.7

The forces, P, w, and N_s act on the roller: the weight w acts toward the center of the earth at center of gravity of the roller, O; P is the force pulling or pushing the roller and acting parallel to the surface S; and N_s is the force exerted on the roller by the surface S. Actually, N_s is comparable to the normal force for sliding friction, and fr is the coefficient of rolling friction created by the small angle α.

If α is taken to be very small, the perpendicular distance between A and OP can be taken as very nearly equal to r. Now, taking moments about A,

$$\sum M_a = (P \times r) - (w \times fr) = 0$$
$$fr = \frac{P \times r}{w}$$

When we considered the coefficient of sliding friction μ, we noted that μ was the ratio of the resistance to sliding to the weight acting normal to the surface. This coefficient is dimensionless:

$$\mu \text{ (no units)} = \frac{fr}{N} = \frac{\cancel{lb}}{\cancel{lb}}$$

Using the same reasoning, from $fr = (p \times r)/w$, fr is the torque per unit weight and is the force required to overcome resistance to rolling friction:

$$fr = \frac{P \times r}{w} = \frac{\cancel{lb} \times \text{in.}}{\cancel{lb}} = \text{in.}$$

The discussion of fluid friction, or viscosity, will be postponed until Chapter 13.

Problems

1. A force f is applied to a 100–lb wood box resting on a wood floor. What is the least force required (a) to start the box moving? (b) to sustain uniform motion once started? Refer to Appendix Table A.6.
2. A box is pulled with a force of 30 lb over a level surface. If the box weighs 90 lb, calculate the coefficient of friction.
3. A box weighing 150 lb is pulled over a level floor. The coefficient of static friction between the two surfaces is 0.5. Calculate the least force required to start the box moving.
4. A 30–lb metal block is to be moved over a level wood surface. (a) The block is stationary and a force of 15 lb is applied in an effort to move it. Will the block move? Why? (b) Assume the same conditions as in part a, except that an *initial* booster force of 3 lb is added to aid the 15–lb force. Once motion takes place, the 3–lb booster force is removed. Will the block continue to move? Why? (c) Does the block move with uniform motion or accelerated motion? Explain.
5. A 30–lb block is pulled along a level plane with a rope which makes an angle of 30° with the horizontal, as shown in Fig. 5.3 (a). If the coefficient of friction is 0.35, calculate N, fr, and f.
6. Assume the same conditions as in Prob. 5, except that the box is pulled as shown in Fig. 5.8.

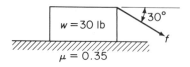

Fig. 5.8

Problems

7. Refer to Fig. 5.9. (a) What weight w will cause impending motion in the 250-lb block? The pulleys are frictionless. (b) Calculate the normal force. (c) Calculate the friction force.

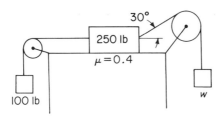

Fig. 5.9

8. In Fig. 5.10, calculate (a) the weight w required for impending motion; (b) the normal force between the surfaces; (c) the friction force.

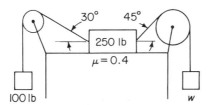

Fig. 5.10

9. In Fig. 5.11, calculate (a) the weight of the box, w, for impending motion; (b) the normal forces at A and B; (c) and the friction forces at A and B.

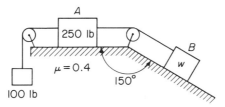

Fig. 5.11

10. Calculate the force necessary to just hold a 500-lb block on a 35° inclined plane if the coefficient of friction between the two surfaces is 0.2. The force acts parallel to the plane.
11. Repeat Prob. 10 if the force f acts parallel to the ground.
12. A 400-lb block is being pulled up a 25° incline by a force of 280 lb. Calculate the coefficient of friction between the two surfaces if the force is acting parallel to the incline.
13. Two weights, 50 lb and 125 lb, are connected by a rope. A second rope is at-

tached to the larger weight so that it creates an angle of 30° with a level plane. Calculate the tension in the latter rope when the blocks are pulled along the level plane. The coefficient of friction is 0.3.

14. A 120–lb block 4 in. × 4 in. × 12 in. high is placed on a 25° inclined plane. The coefficient of friction is 0.45. Will the block topple, slide, or stand still?

15. In Fig. 5.12, will the block topple, slide, or remain stationary? Prove.

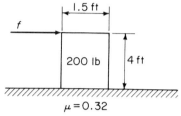

Fig. 5.12

16. In Fig. 5.13, will the block slide, topple, or remain stationary? Prove.

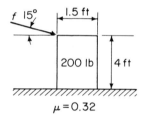

Fig. 5.13

17. In Fig. 5.14, will the block slide, topple, or remain stationary? Prove.

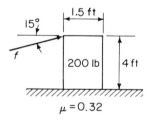

Fig. 5.14

18. A ladder is 20 ft long and weighs 60 lb. It extends 4 ft over the top of a smooth vertical wall. If the base of the ladder makes an angle of 70° with the floor, what is the reaction at the wall and the coefficient of friction at the floor?

19. Calculate the force f in Fig. 5.15 necessary to cause the block B to move up the wall.

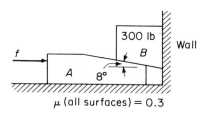

Fig. 5.15

20. Calculate the force f necessary to hold both block stationary in Fig. 5.16.

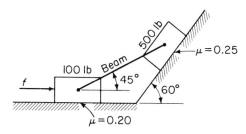

Fig. 5.16

6

Elasticity

6.1. Stress–Strain

A 3-in. piece of rubber, having x number of atoms, is stretched to a length of 4 in. Upon release of the applied force, it returns to its original 3-in. length. The important observation to be made here is not that the piece of rubber is 3-in. or 4-in. long, but that the total number of atoms remains the same. The important phenomenon is the separation of atoms from each other so that the interatomic spacing became greater. It is also important to recognize that the interatomic forces were strong enough to support this additional applied force, thus enabling the piece of rubber to return to its original length. If the force causing the atoms to separate is greater than the interatomic force holding the atoms together, the piece will deform or rupture. Every material has a characteristic limit called the *elastic limit;* if this limit is exceeded, the material will take a permanent set.

For every increase in the applied force, there is a corresponding increase in the length of the workpiece. When the workpiece is in tension, the increase in length of the work must be in the direction of the applied force. The increase in length is referred to as *tensile strain*. If the workpiece is in compression, the workpiece must decrease in length. This decrease in length is referred to as *compressive strain*. In both tensile and compressive strain, the structure of the workpiece is resisting the deformation. The area resisting these applied forces is perpendicular to the applied force f, as seen in Fig. 6.1(a). If the workpiece is in *shear*, as shown in Fig. 6.1(b), the area being deformed is *parallel* to the applied force f. This type of deformation is referred to as *shear strain*.

Sec. 6.2 Modulus of Elasticity

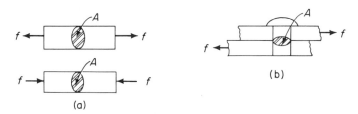

Fig. 6.1

The strain is a ratio between the original length of the workpiece and the change in length. Note that strain has no units:

$$\text{strain} = \frac{\text{change in length}}{\text{original length}} = \frac{\Delta l}{l} \quad \left\{ \begin{array}{l} \Delta l = \text{in.} \\ l = \text{in.} \\ \text{strain} = \text{no units} \end{array} \right\}$$

Further, the greater the cross-sectional area of the material, the more atoms that are in contact with each other; therefore, the greater the internal resisting force holding the material together. The internal resisting force balances the stretching force, which is proportional to the area. This force per unit area is called the *stress:*

$$\text{stress} = \frac{\text{force}}{\text{unit area}} = \frac{f}{A} \quad \left\{ \begin{array}{l} f = \text{lb} \\ A = \text{ft}^2 \\ \text{stress} = \text{lb/ft}^2 \text{ (psf)} \end{array} \right\}$$

If the force is a stretching force, it puts the material in tension and is called *tensile stress.* If the force is compressive, it is called *compressive stress.* A shearing force is called *shear stress.*

6.2. Modulus of Elasticity

Weights (forces) may be added to a material in the form of a wire, which will cause the wire to stretch. If the weights are then removed, the wire returns to its original length (perhaps not immediately, since there is some hysteresis* present). Eventually, a loading is reached for which the wire will *not* return to its original length. Robert Hooks stated that below the aforementioned loading, the ratio of stress to strain is equal to a constant. This constant is called *Young's Modulus.* It is different for different materials and appears in Appendix Table A.7 as "modulus of elasticity." The equation is

$$\text{modulus of elasticity} = \frac{\text{stress}}{\text{strain}} = \frac{f/A}{\Delta l/l}$$

$$= \frac{fl}{A \Delta l} = \frac{\text{lb-in.}}{\text{in.}^2 \text{-in.}} = \text{lb/in.}^2 \text{ (psi)}$$

* The failure of a stressed body to return immediately to equilibrium state once the loading is removed.

The modulus of elasticity may have the same value for tension and compression. The *shear modulus* need not be the same as the modulus of elasticity. To distinguish the modulus of tension from the modulus of shear, the modulus of shear is called the *modulus of rigidity*.

Example 1

An aluminum rod, 48 in. long and with an area of $\frac{1}{2}$ in.², is held vertically and firm at one end while a load of 2100 lb is suspended from the other end. If the rod stretches 0.0192 in., calculate the stress, strain, and modulus of elasticity.

Solution

$$\text{strain} = \frac{\Delta l}{l} = \frac{0.0192 \text{ in.}}{48 \text{ in.}}$$

$$= 0.0004$$

$$\text{stress} = \frac{f}{A} = \frac{2100 \text{ lb}}{\frac{1}{2} \text{ in.}^2}$$

$$= 4200 \text{ psi}$$

$$\left\{ \begin{array}{l} f = 2100 \text{ lb} \\ A = \frac{1}{2} \text{ in.}^2 \\ \Delta l = 0.0192 \text{ in.} \\ l = 48 \text{ in.} \end{array} \right\}$$

$$\text{modulus of elasticity} = \frac{\text{stress}}{\text{strain}} = \frac{f/A}{\Delta l/l} = \frac{4200 \text{ lb/in.}^2}{0.0004}$$

$$= 10.5 \times 10^6 \text{ psi}$$

6.3. Modulus of Shear: Rigidity

The modulus of shear—called the *modulus of rigidity*—is related to the change of shape of an object. There is actually no dimensional change, but only a distortion of shape, that is, as the force is applied, each successive layer of atoms is caused to slip sideways as shown in Fig. 6.2.

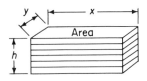
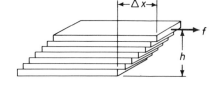

Fig. 6.2

If the force is released, the successive layers recover their original positions because of the interatomic forces at the interfaces. The modulus of rigidity is

$$\text{modulus of rigidity} = \frac{f/A}{\Delta x/h}$$

where $A = xy$.

Sec. 6.3   *Modulus of Shear: Rigidity*   77

Example 2

In Fig. 6.2, a force of 6000 lb is applied to the top of the block while the bottom of the block is held stationary. The dimensions of the block are $x = 6$ in. and $y = 4$ in. Calculate the modulus of rigidity if the block is distorted 2.4×10^{-4} in. and $h = 3$ in.

Solution

$$\text{modulus of rigidity} = \frac{f/A}{\Delta x/h} = \frac{fh}{\Delta x A}$$

$$= \frac{6000 \text{ lb} \times 3 \text{ in.}}{2.4 \times 10^{-4} \text{ in.} (6 \times 4) \text{ in.}^2}$$

$$= 3.12 \times 10^6 \text{ psi}$$

$\left\{ \begin{array}{l} f = 6000 \text{ lb} \\ \Delta x = 2.4 \times 10^{-4} \text{ in.} \\ y = 4 \text{ in.} \\ x = 6 \text{ in.} \\ h = 3 \text{ in.} \end{array} \right\}$

Another illustration of the principle of shear is illustrated in Example 3, below. Note that the area sheared is parallel to the direction of motion of the punch.

Example 3

A punch and die, Fig. 6.3, is to be used to cut a $\frac{1}{2}$-in. hole in an $\frac{1}{8}$-in. steel plate, shear strength (shear stress) of 50,000 psi. What force is required to punch the hole?

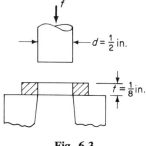

Fig. 6.3

Solution

The surface area of the bore is

$$A = \pi dt = \pi(\tfrac{1}{2})(\tfrac{1}{8})$$
$$= 0.196 \text{ in.}^2$$
$$f = A(\text{stress}) = 0.196(50,000)$$
$$= 9800 \text{ lb}$$

6.4. Bulk Modulus: Volume Elasticity

Liquids may be subjected to compression. When this is the case, the stress is called *pressure*. It will be seen later that, when applied to a liquid, this pressure is transmitted equally to all parts of it.

$$\text{stress} = \frac{\Delta f}{A} \text{ psi} = P \quad (pressure)$$

When dealing with liquids, our interest must lie in volume change rather than change in length. Therefore, we substitute volume for length, and strain becomes

$$\text{strain} = \frac{\Delta V}{V} \quad (no\ units)$$

Since shear modulus of volume elasticity is the ratio of stress to strain,

$$\text{bulk modulus} = \frac{\text{stress}}{\text{strain}} = \frac{\Delta f/A}{\Delta V/V} = \frac{P}{\Delta V/V}$$

Example 4

A pressure is applied to 1000 in.³ of water in a container. If the initial pressure is 14.7 psi and is changed to 147 psi, find (1) the stress, (2) the strain, and (3) the change in volume. The bulk modulus of water at 68°F is 0.3×10^6 psi.

Solution

1. $\quad \text{stress} = 147 - 14.7 = 132.3 \text{ psi} \quad (pressure)$
2. $\quad \text{strain} = \frac{\Delta V}{V} = \frac{P}{B} = \frac{132.3}{0.3 \times 10^6} = 441 \times 10^{-6}$
3. $\quad \text{change in volume} = 441 \times 10^{-6} \times 1000 = 0.441 \text{ in.}^3$

6.5. Compressibility

It is difficult to place a liquid in tension. It is more common to place it in compression. This *compressibility* of a substance is the reciprocal of the bulk modulus (bulk mod):

$$\text{compressibility} = \frac{\Delta V/V}{f/A} = \frac{\Delta V/V}{P}$$

Example 5

What is the compressibility of mercury if its bulk mod is 3.6×10^6 psi?

Solution

$$\text{compressibility} = \frac{1}{\text{bulk mod}} = \frac{1}{3.6 \times 10^6} = 27.8 \times 10^{-8} \text{ psi}$$

Problems

1. State and explain Hooke's law.
2. (a) If tensile or compressive forces are applied to a workpiece, is the resisting force to deformation generated by the cross-sectional area parallel or perpendicular to the direction of the applied force? (b) If the applied force creates a condition of shear, is the resisting force generated by the cross-sectional area parallel or perpendicular to the direction of the applied force?
3. The body of a bolt has a 1-in. diameter. It has a 4-ton force applied as the nut is tightened. What is the stress applied to the bolt?
4. A helical spring is 8 in. long. It is stretched to 10.5 in. What is the strain?
5. A brass rod 10 in. long and $1\frac{1}{4}$ in. in diameter elongates when a 50-ton tensile force is applied. What is the elongation?
6. A steel cable is 60 ft long. The length of the cable increases by 0.060 in. when a load of 900 lb is applied. Calculate the diameter of the cable.
7. A man who weighs 250 lb puts all his weight on one foot. Assume the average area of his leg bone to be 0.480 in.^{2} and the length of the leg bone to be 14 in. Calculate (a) the amount the bone shortens if Young's modulus is 3×10^6 psi; (b) the stress; (c) the strain.
8. The dimensions of a steel block, Fig. 6.2, are $y = 4$ in. $\times$ $x = 8$ in. $\times$ $h = 6$ in. How much does the block distort if a horizontal force of 8 tons is applied at the top of the block and the bottom is held stationary?
9. Mercury is subjected to a pressure of 751 atmospheres (atm). Determine the percentage decrease in volume.
10. What is the bulk modulus of glycerin if there is a decrease in volume of 2.9×10^{-3} in.^{3} when under a pressure of 20 atm, if the original volume was 64 in.^{3}?
11. Assume a round punch and die is to cut a $\frac{5}{8}$-in.-diameter hole into a $\frac{1}{2}$-in.-thick brass plate. The shear stress of brass is 30,000 psi. Calculate the force required to cut this hole.
12. A rectangular punch, 2 in. $\times$ 3 in., is used to cut a hole into a $\frac{1}{16}$-in.-thick sheet of phosphor bronze. The shear stress of the material is 16,000 psi. Calculate the force required to cut this hole.

7

Linear Motion

7.1. Frames of Reference and Motion

In a true sense, there is no such thing as absolute rest. All bodies move relative to other bodies, or to a reference system. This reference system is called the *frame of reference*, and the relative movement of one body in a frame of reference is called *motion*. True, two bodies traveling in the same direction at the same speed may be considered to be at rest in relation to each other. But both may be in motion relative to each other when they are each compared to the same frame of reference.

A good illustration of the above is to consider a man on a large raft floating downstream, as shown in Fig. 7.1. The distance the raft moves downstream per unit time is s. This motion is relative to the shore. If the man A does not move relative to the raft, his motion *relative to the shore* will be the same as the motion of the raft *relative to the shore*. If the motion of the man *relative to the*

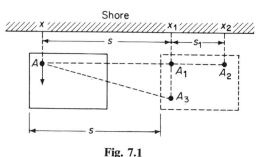

Fig. 7.1

raft is considered, the man will have no motion. In Fig. 7.1, A will be in the same place on the raft at A_1, but will have moved a distance s relative to point x on the shore.

Now assume that the man starts to walk to the right in Fig. 7.1. Note that at the end of time t, the man has moved from A_1 a distance s_1 to A_2, when the raft is used as the frame of reference. When the shore is used as a frame of reference, the man has moved a distance $s + s_1$.

If the man starts to walk perpendicular to the shore, in the direction of the arrow in Fig. 7.1, he will have moved from A_1 to A_3 with reference to the raft. With reference to the shore, he will have moved from A to A_3, along the hypotenuse of the triangle shown.

It is important to realize that motion can only be described if it is described in a frame of reference. When motion of objects are mentioned, it is implied that the Earth, or an object on the Earth, is used as the frame of reference.

Example 1

An airplane, moving in a northerly direction relative to the air at a rate of 120 mph, is being pushed toward the west by a 30-mph wind. What is the true velocity and direction of the plane relative to the Earth?

Solution

Draw a vector B along the y-axis to represent the 120-mph north velocity. Draw a vector A along the negative x-axis to represent the 30-mph west velocity. Complete the parallelogram to yield the true direction and magnitude of the plane. The resultant R is the resultant of the two velocities affecting the fight of the plane with the Earth as a reference. In other words, the plane is helped by the wind (note Fig. 7.2):

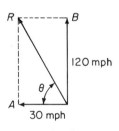

Fig. 7.2

$$R = \sqrt{120^2 + 30^2} = 123.7 \text{ mph}$$

$$\theta = \tan^{-1} \frac{120}{30} = 76°$$

7.2. Special Relativity

We can now concern ourselves with the special theory of relativity which deals with frames of reference and constant velocity. The general theory of relativity deals with frames of reference and acceleration which is beyond the scope of this text. Conservation of energy dictates that in any isolated, or closed system, the total amount of all energies in the system remains constant. Einstein included the equivalence of mass and energy. Thus in the same way that 1 Btu of heat energy is equivalent to 778 ft-lb of kinetic energy; 9×10^{16} joules of kinetic energy is equivalent to 1 kg of mass. The Einstein equation which relates energy to mass is

$$E = (\Delta m)c^2 \quad \left\{ \begin{array}{l} c = 3 \times 10^8 \text{m/sec} \\ m = \text{kg} \\ E = \text{joules} \end{array} \right\}$$

Einstein stated two postulates which explained the relative motions of frames of references. The first postulate of special relativity deals with frames of reference and constant velocity. It states that the laws of physics are the same for all frames of reference moving relative to each other at constant velocity. This means that if a person is locked in a box which is moving with a constant velocity, that person would have no way of determining his relative motion to other frames of reference no matter how many or what kind of experiments he conducted inside his frame of reference. The postulate in effect states that there is no universal frame of reference. Measurements of time and space depend upon the relative motion of that which is being observed and the observer. Two observers may be observing a phenomenon, obtain totally different results, and both be correct.

The second postulate deals with the invariance of the speed of light in free space. Thus the speed of light has the same value for all observers and it makes no difference what their state of motion. Assume a space ship is traveling at one-half the speed of light, and light is turned on at the front of the spaceship. Will the speed of the light, relative to the earth, be greater than 3×10^8 meters per second (m/sec)? Einstein's second postulate says it cannot be. It will still be traveling at 3×10^8 m/sec.

Let us now develop the relationships of two objects in various frames of reference. Assume in Fig. 7.3(a) that we are viewing P from K. K thinks he is stationary and we will consider him as being correct. A beam of light emanates from K and moves a distance s as P moves with a velocity v to P'. If, according to Einstein's postulate, the speed of light is invariant, then the distance s is proportional to the time t it takes the light to travel from P to P' and

$$s = ct$$

Solving this equation for t,

$$t = \frac{s}{c}$$

Now in Fig. 7.3(b) let us consider the viewpoint of P of the light coming from K at the speed of light c. P thinks he is stationary and also thinks K is moving with a velocity v to the left toward K'. The velocity v' now is related to the distance s and the velocity v as

Sec. 7.2 Special Relativity

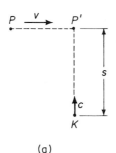
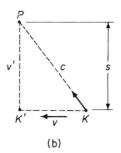

Fig. 7.3

$$v' = \sqrt{c^2 - v^2}$$

It should be noted that in both Fig. 7.3(a) and (b), the velocities are the same when viewed from the stationary frame of reference.

The distance (s) equation is therefore

$$s = v't'$$

and the time t' is

$$t' = \frac{s}{v'} = \frac{s}{\sqrt{c^2 - v^2}}$$

The ratio of the two times is

$$\frac{t}{t'} = \frac{s/c}{s/\sqrt{c^2 - v^2}} = \frac{\sqrt{c^2 - v^2}}{c} = \sqrt{1 - \frac{v^2}{c^2}}$$

Note that if the source of light K, Fig. 7.3(a), is taken as the frame of reference and is stationary, t has a particular value $t = s/c$. If, however, the source of light K is observed from a moving frame of reference P, Fig. 7.3(b), then from his viewpoint, P thinks he is stationary and that K is moving toward the left. To him the time t' for the light to get to him appears to be greater than t and equal to $t' = s/\sqrt{c^2 - v^2}$.

If P were the light source, the same arguments would apply.

The phenomenon just discussed is known as "time dilation." The equation is

$$t_0 = t\sqrt{1 - \frac{v^2}{c^2}}$$

From the discussion above, a "clock" carried by a very fast moving particle will register a time t according to the fast particle. Whereas a "clock" placed on an object in space and disassociated from the frame of reference of the fast moving particle, will register a much longer period of time t' for the particle according to the disassociated frame.

Similar arguments when applied to rest mass m_0 as seen by a stationary observer and m as the mass of the object moving relative to the observer yields:

$$m_0 = m\sqrt{1 - \frac{v^2}{c^2}}$$

Thus the rest mass m_0 of an object increases to m as its speed approaches the speed of light.

Example 2

A quantity of material, rest mass 2 kg, is accelerated to 80% the speed of light. Calculate its mass at this speed.

Solution

The mass at 80% the speed of light is:

$$m = \frac{m_0}{\sqrt{1 - \frac{v^2}{c^2}}} = \frac{2}{\sqrt{1 - \frac{(0.8 \times 3 \times 10^8)^2}{(3 \times 10^8)^2}}} \quad \left\{\begin{array}{l} m_0 = 2 \text{ kg} \\ c = 3 \times 10^8 \text{m/sec} \\ v = (0.80)(3 \times 10^8) \text{ m/sec} \end{array}\right\}$$

$$= 3\tfrac{1}{3} \text{ kg}$$

Time dilation produces other relativistic effects upon other physical relationships. The dilation effect of time in a moving frame of reference has the effect of contracting lengths. This is known as the *Lorentz-Fitzgerald contraction*. The equation is:

$$l = l_0 \sqrt{1 - \frac{v^2}{c^2}}$$

Thus the length of an object, when measured by an observer at rest when the object is in motion in another frame of reference, is measured as being shorter than if measured when not in motion.

Example 3

Calculate the relativistic length of a stick which is moving with a velocity of 85% the speed of light if the stick measures 3 ft when stationary.

Solution

The new length is

$$l = l_0 \sqrt{1 - \frac{v^2}{c^2}} = 3\sqrt{1 - \frac{(0.85 \times 186{,}000)^2}{(186{,}000)^2}} \quad \left\{\begin{array}{l} l_0 = 3 \text{ ft} \\ c = 186{,}000 \text{ mi/sec} \end{array}\right\}$$

$$= 1.58 \text{ ft}$$

7.3. Speed, Velocity, and Acceleration

A man makes a trip between two cities at a constant speed of 30 mph following road x in Fig. 7.4. He may be able to maintain a constant speed, but if he follows the road x, his direction of motion is always changing. If his direction of travel is changing, the motion is referred to as a *speed* and the 30 mph *cannot* be represented as a vector: it is a *scalar* quantity.

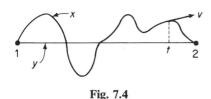

Fig. 7.4

Movement along a straight line is called a *displacement*. Therefore, a 30-mph movement between the two cities, in Fig. 7.4, along road y has both magnitude and direction and may be represented as a vector; the latter motion is called a *velocity*.

Also in Fig. 7.4, at a particular time t the man will have an instantaneous speed of 30 mph in the direction indicated by the arrow. At that instant, the man is said to have an *instantaneous velocity* which can be represented by a vector arrow.

Another way to look at Fig. 7.4 is to assume that the length of the twisting road x is 60 mi. Although the straight-line distance (road y) is 30 mi. If he chooses to use road x, he must travel at a *speed* of 60 mph to travel the 60 erratic miles in 1 hour. Thus he must travel a distance of 60 mi to be *displaced* 30 mi. His *speed* is 60 mph. His *velocity*, however, is only 30 mph.

Now, suppose the man travels between the two cities at 30 mph along road y. After 1 hour he would have traveled, or have been displaced, 30 mi. If he travels the same 30 mi by increasing and decreasing his speed, and if it

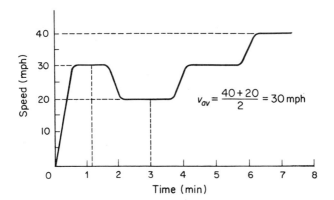

Fig. 7.5

takes him the same 1 hour, he must be averaging 30 mph. We say his *average velocity* is 30 mph. Thus if he reduces his speed to 20 mph for one time interval, he must at some point increase his speed to 40 mph for a comparable time interval. Figure 7.5 is a plot showing this average speed at 30 mph.

In the preceding discussion, it was seen that velocity is a vector quantity which has the requisite components of displacement and magnitude. Now put the man in a car on a circular racetrack. If it takes him time t to go around the track and return to the starting point, his displacement is zero. His movement around the track is a speed and not a velocity. At specific instants, this car would have specific velocities of 30 mph east, 30 mph west, etc., or *instantaneous velocities*.

If, instead of the constant velocity just discussed, the velocity changes for every increment of time, we say that the body has an *acceleration*. If the *change* of velocity *per unit time* is constant, the body is said to possess *uniform acceleration*. This presumes that the *direction* of motion does not change, but that the velocity does change uniformly *in the direction of motion*. Since acceleration is a vector quantity, it may also change at right angles to the direction of motion. An acceleration vector at right angles to the direction of motion causes a change in direction only. An acceleration vector parallel to the velocity vector causes a change in magnitude only.

Therefore, the various possibilities are as follows:

1. rest—zero velocity;
2. constant velocity—zero acceleration;
3. constantly changing velocity per unit time—constant acceleration;
4. velocity changing a variable amount per unit time—variable acceleration.

7.4. Constant Velocity

Constant velocity is the movement of a body a given, or fixed, distance for a fixed interval of time. Also, *rest* may be defined as a zero velocity. We can easily see that a body at rest is in equilibrium with the forces at play on it. A body traveling at constant velocity also must have all the forces acting on it in equilibrium. Since any unbalance will destroy equilibrium, the constant velocity of the body would be destroyed. We shall examine this more closely later in this chapter.

If a body travels a fixed distance in a given length of time, then

$$s = vt \qquad \left\{ \begin{array}{l} s = \text{distance traveled: ft } or \text{ cm)} \\ t = \text{velocity: ft/sec } or \text{ cm/sec} \\ v = \text{time: sec} \end{array} \right\}$$

If time is given in seconds and velocity is given in feet per second, the units of distance will be feet. If time is given in seconds, and velocity is given in centimeters per second, the units of distance will be centimeters:

Sec. 7.5   *Constant Acceleration: Variable Velocity*   87

$$s = v \frac{ft}{\cancel{sec}} \times t \cancel{sec} = ft$$

$$= v \frac{cm}{\cancel{sec}} \times t \cancel{sec} = cm$$

Other units are possible: mph, m/sec, ft/min, etc. The important thing to remember is that the units must be consistent throughout a particular problem.

Example 4

An automobile is traveling at a rate of 45 mph. How far in feet will the automobile have traveled in 5 hr?

Solution

$$v = 45 \frac{\cancel{mi}}{\cancel{hr}} \times 5280 \frac{ft}{\cancel{mi}} \times \frac{1 \cancel{hr}}{60 \cancel{min}} \times \frac{1 \cancel{min}}{60 \sec}$$

$$= 66 \text{ ft/sec}$$

$$t = 5 \cancel{hr} \times 60 \frac{\cancel{min}}{\cancel{hr}} \times 60 \frac{\sec}{\cancel{min}}$$

$$= 18{,}000 \text{ sec}$$

$$s = vt = 66 \frac{ft}{\cancel{sec}} \times 18{,}000 \cancel{sec}$$

$$= 1.19 \times 10^6 \text{ ft}$$

7.5. Constant Acceleration: Variable Velocity

Acceleration may be considered as a *change in velocity* for fixed units of time. Since velocity itself is a change in position for every fixed unit of time, it should be evident that an acceleration is a change rate of a change rate. Therefore, the units for acceleration are ft/sec² (which is read "feet per second-squared"); or cm/sec²; etc. If the *change* in velocity is constant, the acceleration is constant. If the velocity is increasing uniformly, the acceleration is positive; if decreasing uniformly, the velocity is called *negative acceleration*, or *deceleration*.

Note that speed in a circular path may be constant; or it may appear that we are dealing with a constant velocity. Remember that velocity is a vector quantity, and as such, both speed and direction of motion must be constant to have constant velocity. Since the direction is constantly *changing*, we are dealing with a constant acceleration. This will be studied much more thoroughly in Chapter 9.

The following rules hold true for uniformly accelerated motion:

1. initial velocity ± a change in velocity = final velocity;
2. acceleration × time = change in velocity;
3. average velocity × time = distance traveled;
4. (initial velocity + final velocity) ÷ by 2 = average velocity.

The following equations have been derived or taken directly from the preceding four statements:

(a) Since acceleration deals with a constantly increasing (or decreasing) velocity, the midpoint velocity is used to indicate the results of an acceleration. This pivot velocity is called *average velocity*. For every velocity less than average velocity, there must be a corresponding velocity greater than the average velocity. This is shown in Fig. 7.6.

From statement 4 above, the average velocity is

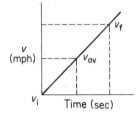

Fig. 7.6

$$v_{av} = \frac{v_f + v_i}{2}$$

v_i = initial velocity
v_f = final velocity
v_{av} = average velocity: ft/sec or cm/sec
t = time: sec
s = distance: ft or cm
a = acceleration: ft/sec² or cm/sec²

(b) From statement 3, the distance s in terms of v and t is

$$s = vt$$

(c) The distance in terms of v, t, and a is

$$s = vt + \tfrac{1}{2}at^2$$

Substituting in the average-velocity statement,

$$s = vt = \left(\frac{v_f + v_i}{2}\right)t$$

From statement 1, solving for v_f,*

$$v_f = v_i + \Delta v$$

From statement 2, which states $\Delta v = at$,

$$v_f = v_i + at$$

Therefore,

$$s = vt = \left[\frac{(v_i + at) + v_i}{2}\right]t$$

$$= \frac{2v_i t + at^2}{2}$$

$$= v_i t + \tfrac{1}{2}at^2$$

(d) To derive the distance in terms of v and a from

---

* The Greek letter Δ (*delta*) is the symbol used for "change"; thus Δv means "change in velocity."

Sec. 7.5   Constant Acceleration: Variable Velocity   89

$$v_f = v_i + at \quad \text{and} \quad s = \left(\frac{v_f + v_i}{2}\right)t$$

solve $v_f = v_i + at$ for t:

$$t = \frac{v_f - v_i}{a}$$

and substitute into

$$s = \left(\frac{v_f + v_i}{2}\right)t$$
$$= \left(\frac{v_f + v_i}{2}\right)\left(\frac{v_f - v_i}{a}\right)$$
$$= \frac{v_f^2 - v_i^2}{2a}$$

A summary of the useful equations follows:

$$v_{av} = \frac{v_f + v_i}{2} \quad \text{(acceleration } a \text{ and time } t \text{ not needed)}$$

$$at = \Delta v = v_f - v_i \quad \text{(distance } s \text{ not needed)}$$

$$2as = v_f^2 - v_i^2 \quad \text{(time } t \text{ not needed)}$$

$$\left.\begin{array}{l} s = vt \\ s = v_i t + \tfrac{1}{2}at^2 \end{array}\right\} \text{(} v \text{ here is average velocity)}$$

It is suggested that the student set up his problem solutions as shown in the illustrated problems which follow. This does not mean that these equations should be memorized. An attempt to approach these problems from a commonsense viewpoint will be very helpful.

Example 5

A train traveling in a straight line starts from rest and accelerates at a constant rate until it has reached 90 mph. Calculate the average velocity in fundamental units.

Solution

$$v = \frac{v_f + v_i}{2} = \frac{90 + 0}{2}$$
$$= 45 \text{ mph}$$

$$\left\{\begin{array}{l} v_i = 0 \text{ mph} \\ v_f = 90 \text{ mph} \\ v = ? \text{ ft/sec} \end{array}\right\}$$

$$v = 45 \frac{\cancel{mi}}{\cancel{hr}} \times 5280 \frac{\text{ft}}{\cancel{mi}} \times \frac{1 \cancel{hr}}{3600 \text{ sec}}$$
$$= 66 \text{ ft/sec}$$

Example 6

How far will the train in Example 5 have traveled in 10 min?

Solution

$$t = 10 \ \cancel{\text{min}} \times \frac{60 \text{ sec}}{\cancel{\text{min}}}$$
$$= 600 \text{ sec}$$

$\left\{ \begin{array}{l} t = 10 \text{ min} \\ s = ? \text{ ft} \end{array} \right\}$

$$s = vt = 66 \text{ ft/}\cancel{\text{sec}} \times 600 \ \cancel{\text{sec}}$$
$$= 39{,}600 \text{ ft}$$

Example 7

Assume the train travels in a straight line and accelerates from a velocity of 20 ft/sec to 50 ft/sec in 12 sec. Calculate its acceleration.

Solution

$$2as = v_f^2 - v_i^2 \quad (\text{no } t \text{ or } s)$$
$$s = v_i t = \tfrac{1}{2}at^2 \quad (\text{no } s \text{ given})$$
$$v_f = v_i + at \quad (\text{all ingredients present})$$

$\left\{ \begin{array}{l} v_i = 20 \text{ ft/sec} \\ v_f = 50 \text{ ft/sec} \\ t = 12 \text{ sec} \\ a = ? \text{ ft/sec}^2 \end{array} \right\}$

Solve the latter equation for acceleration:

$$a = \frac{v_f - v_i}{t} = \frac{50 \text{ ft/sec} - 20 \text{ ft/sec}}{12 \text{ sec}}$$
$$= 2.5 \text{ ft/sec}^2$$

Example 8

Given the same set of conditions as set forth in Example 7, calculate the distance traveled by the train.

Solution

$$s = v_i t + \tfrac{1}{2}at^2$$
$$= 20 \ \frac{\text{ft}}{\cancel{\text{sec}}} \times 12 \ \cancel{\text{sec}} + \frac{1}{2} \times 2.5 \ \frac{\text{ft}}{\cancel{\text{sec}^2}} \times (12 \ \cancel{\text{sec}})^2$$
$$= 420 \text{ ft}$$

$\left\{ \begin{array}{l} v_i = 20 \text{ ft/sec} \\ v_f = 50 \text{ ft/sec} \\ t = 12 \text{ sec} \\ a = 2.5 \text{ ft/sec}^2 \\ s = ? \text{ ft} \end{array} \right\}$

Alternate Solution 1

$$s = vt = \left(\frac{v_f + v_i}{2}\right) t = \left(\frac{50 \text{ ft/}\cancel{\text{sec}} + 20 \text{ ft/}\cancel{\text{sec}}}{2}\right) 12 \ \cancel{\text{sec}}$$
$$= 420 \text{ ft}$$

Alternate Solution 2

$$2as = v_f^2 - v_i^2$$

$$s = \frac{v_f^2 - v_i^2}{2a} = \frac{(50 \text{ ft/sec})^2 - (20 \text{ ft/sec})^2}{2(2.5 \text{ ft/sec}^2)}$$

$$= \frac{(2500 - 400) \text{ ft}^2/\text{sec}^2}{5 \text{ ft/sec}^2}$$

$$= 420 \text{ ft}$$

7.6. Gravitational Acceleration: Trajectories

Any object in free fall is acted upon by the force of gravity and will increase its velocity uniformly. If its velocity is increased uniformly, its "rate of change of velocity per unit time" will be constant. Since *rate of change of velocity* is defined as acceleration, the acceleration of a freely falling body is said to be constant. Gravitational acceleration varies for different places on the Earth's surface. The International Committee on Weights and Measures accepts the value of 32.174 ft/sec² or 9.80665 m/sec² as a standard. For the purposes of this text, rounded-off values of gravitational acceleration g will be used. Thus g will equal 32 ft/sec², or 9.80 m/sec². Also, increasing velocities will cause the value of g to be positive; decreasing velocities will cause the value of g to be negative.

Example 9

A ball is dropped over the edge of a tall building. Calculate and tabulate the velocity, distance, and acceleration of the ball each second if the ball is in free fall for 4 sec.

Solution

The velocity at the moment that the ball is released is 0 ft/sec and, obviously, it it has traveled 0 distance (see Table 7.1). The gravitational acceleration is + 32 ft/sec², since the velocity is increasing.

Table 7.1

Time (sec)	Velocity (ft/sec)	Average Velocity (ft/sec)	Distance (ft)	Acceleration (ft/sec²)
0	0	0	0	32
1	32	16	16	32
2	64	32	64	32
3	96	48	144	32
4	128	64	256	32

The change in velocity which takes place over 4 sec is

$$\Delta v = gt = 32 \text{ ft/sec}^2 \times 4 \text{ sec}$$
$$= 128 \text{ ft/sec}$$

The average velocity is

$$v_{av} = \frac{v_f + v_i}{2} = \frac{128 + 0}{2}$$
$$= 64 \text{ ft/sec}$$

$$\left\{\begin{array}{l} t = 4 \text{ sec} \\ g = +32 \text{ ft/sec}^2 \\ v_i = 0 \text{ ft/sec} \\ s = ? \text{ ft} \end{array}\right\}$$

The distance traveled in 4 sec is

$$s = vt = 64 \text{ ft/sec} \times 4 \text{ sec}$$
$$= 256 \text{ ft}$$

Alternate solution

$$s = v_i t + \tfrac{1}{2} gt^2$$
$$= 0(4) + \tfrac{1}{2}(32)(4)^2$$
$$= 256 \text{ ft} \quad (check)$$

Careful analysis of $s = v_i t + \tfrac{1}{2} at^2$ shows that this formula is written as the sum of two distances:

1. The distance $v_i t$ is a result of any initial velocity imparted to the ball and is totally unrelated to the force of gravity. If there are no opposing forces tending to stop the ball, this velocity will add distance at a *constant* rate for as long as the problem dictates.
2. The second set of circumstances found in the formula adds distance to the flight of the ball and is the *variable velocity* for a given length of time. This velocity is not a constant value. The velocity is *changing* at a fixed rate; therefore, the velocity at any two instants is never the same. This term in the formula is dependent upon gravitational acceleration, which applies an unbalanced force to the ball. This unbalance is always constant at 32 ft/sec² or 9.80 m/sec².

These two distances added together give the final position of the ball. Notice that, if there is no initial velocity, the distance $s = v_i t$ drops out of the preceding distance formula, which then reads $s = \tfrac{1}{2} gt^2$.

Example 10

The ball of Example 9 is thrown vertically upward with an initial velocity of 4 ft/sec. Calculate the maximum height reached by the ball.

Sec. 7.6 Gravitational Acceleration: Trajectories

Solution

At the top of the trajectory, the ball will have a velocity of 0 ft/sec. Since the initial velocity was 40 ft/sec, the average velocity will be

$$v = \frac{v_f + v_i}{2} = \frac{(0 + 40) \text{ ft/sec}}{2}$$
$$= 20 \text{ ft/sec}$$

It will take

$$t = \frac{v_f - v_i}{g} = \frac{(0 - 40) \text{ ft/sec}}{-32 \text{ ft/sec}^2} \quad \left\{ \begin{array}{l} v_i = 40 \text{ ft/sec} \\ v_f = 0 \text{ ft/sec} \\ g = -32 \text{ ft/sec}^2 \\ s = ? \text{ ft} \end{array} \right\}$$
$$= 1.25 \text{ sec}$$

The maximum height will be

$$s = vt = 20 \times 1.25$$
$$= 25 \text{ ft}$$

Alternate Solution

$$s = v_i t + \tfrac{1}{2} g t^2 = 40(1.25) + \tfrac{1}{2}(-32)(1.25)^2$$
$$= 25 \text{ ft} \quad (check)$$

Example 11

The ball in Example 10 will now start to come down. Assume the ball just misses the edge of the roof of the building, Fig. 7.7. Where will the ball be (1) after 2.5 sec? (2) after 4 sec?

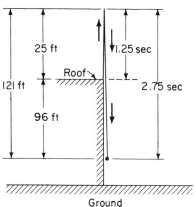

Fig. 7.7

Solution

1. The solutions reached in Example 9 and 10 are valid here. Therefore, if it took 1.25 sec to reach the maximum height, it should take the same time interval to *come back down* to the edge of the roof. This is shown in Fig. 7.7. Since g is always acting down and no other resistance to motion is assumed to operate, the velocity at the roof

edge will be 40 ft/sec. Thus for the distance traveled on the way down,

$$s = v_i t + \tfrac{1}{2}gt^2$$
$$= 0(1.25) + \tfrac{1}{2}(32)(1.25)^2$$
$$= 25 \text{ ft}$$

$$\begin{pmatrix} v_i = 0 \text{ ft/sec} \\ g = +32 \text{ ft/sec}^2 \\ t = 2.5 - 1.25 \\ = 1.25 \text{ sec} \end{pmatrix}$$

2. To determine the position of the ball 4 sec after it was thrown,

$$s = v_i t + \tfrac{1}{2}gt^2$$
$$= 0(2.75) + \tfrac{1}{2}(32)(2.75)^2$$
$$= 121 \text{ ft} \quad \textit{(from maximum position)}$$

$$\begin{pmatrix} t = 4 - 1.25 \\ = 2.75 \text{ sec} \\ v_i = 0 \text{ ft/sec} \\ g = +32 \text{ ft/sec}^2 \\ s = ? \text{ ft} \end{pmatrix}$$

Therefore, the ball is $121 - 25 = 96$ ft below the edge of the roof.

Alternate Solution

It is possible to calculate the lowest distance below the roof that the ball will reach by subtracting and disregarding the time it takes for the ball to travel from the roof-to-maximum-to-roof from the total time allowed in the problem. At the instant that the ball is exactly even with the roof, 2.5 sec of the 4 sec have been used. Therefore, 1.5 sec are left for travel below the roof. Also, the velocity of the ball after these 2.5 sec is the same as when the ball was initially thrown upward, or 40 ft/sec:

$$s = v_i t + \tfrac{1}{2}gt^2$$
$$= 40(1.5) + \tfrac{1}{2}(32)(1.5)^2$$
$$= 96 \text{ ft} \quad \textit{(check)}$$

$$\begin{pmatrix} v_i = 40 \text{ ft/sec} \\ t = 1.5 \text{ sec} \\ g = +32 \text{ ft/sec}^2 \\ s = ? \text{ ft} \end{pmatrix}$$

Now suppose the same ball from Example 11 is thrown from the top of the roof with a horizontal velocity, as shown in Fig. 7.8(a).

It can be shown experimentally that two balls, one projected horizontally and one dropped vertically from the same height, will both reach the ground at the same time. This indicates that the vertical velocity of both balls is affected by gravitational attraction, 32 ft/sec². Thus the *vertical descent* is exactly the same for both balls. They will both reach the ground at the same time, assuming the ground is level.

The horizontal velocity, neglecting air friction, *remains constant throughout the entire flight of both balls.* The horizontal velocity of the ball dropped straight down is zero. The horizontal velocity of the ball thrown horizontally is v_h and is shown in Fig. 7.8(a).

From this discussion, it is evident that at any instant two vectors may be used to describe the motion of an object in flight. These two vectors are components which may be added vectorially into a resultant vector, tangent to the flight path of the object, as shown in Fig. 7.8(a).

Sec. 7.6   *Gravitational Acceleration: Trajectories*   95

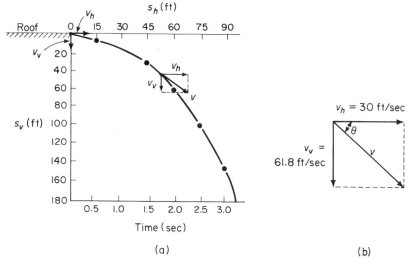

Fig. 7.8

Example 12

A ball is thrown horizontally with a velocity of 30 ft/sec. Plot the path of the ball for 3 sec.

Table 7.2

t (sec)	s_h (ft)	s_v (ft)	v_v (ft/sec)	v_{av} (ft/sec)	v_h (ft/sec)	a (ft/sec^2)
0	0	0	0	0	30	32
0.5	15	4	16	8	30	32
1.0	30	16	32	16	30	32
1.5	45	36	48	24	30	32
2.0	60	64	64	32	30	32
2.5	75	100	80	40	30	32
3.0	90	144	96	48	30	32

Solution

At point 0 (Fig. 7.8a), all the motion is along the horizontal velocity, $v_h = 30$ ft/sec (see Table 7.2). The vertical velocity is $v_v = 0$ ft/sec.
After 0.5 sec:
Horizontal distance s_h is

$$s_h = v_i t = 30 \text{ ft/sec} \times 0.5 \text{ sec} = 15 \text{ ft}$$

Vertical velocity v_v is

$$v_v = v_i + gt = 0 + 32(0.5) = 16 \text{ ft/sec} \qquad \{v_i = 0 \text{ ft/sec}\}$$

Average vertical velocity v_{av} is

$$v_{av} = \frac{v_i + v_v}{2} = \frac{0 + 16}{2} = 8 \text{ ft/sec}$$

Vertical distance traveled s_v is

$$s_v = v_{av} t = 8 \times 0.5 = 4 \text{ ft}$$

Example 13

Calculate, in Example 12, (1) the time t it takes the ball to fall 60 ft; (2) the vertical velocity at time t; (3) the magnitude and the direction of the resultant vector at time t.

Solution

1. To calculate the time t, using the equation

$$s = v_i t + \tfrac{1}{2} g t^2$$
$$60 = 0(t) + \tfrac{1}{2}(32 \text{ ft/sec}^2) \times t^2 \text{ sec}^2 \qquad \left\{ \begin{array}{l} g = +32 \text{ ft/sec}^2 \\ s = 60 \text{ ft} \\ v_i = 0 \text{ ft/sec} \end{array} \right\}$$

solve for t

$$t = \sqrt{\frac{2 \times 60}{32}}$$
$$= 1.93 \text{ sec}$$

Checking the curve, Fig. 7.8(a), when $s_v = 60$ ft, the time is approximately 1.9 sec.

2. The vertical velocity at time t is

$$v_v = v_i + gt = 0 + 32(1.93)$$
$$= 61.8 \text{ ft/sec}$$

3. The resultant vector is (Fig. 7.8b)

$$v = \sqrt{v_h^2 + v_v^2} = \sqrt{30.0^2 + 61.8^2}$$
$$= 68.7 \text{ ft/sec}$$
$$\tan \theta = \frac{61.8}{30} = 2.06$$
$$\theta = 64°6'$$

Example 14

A projectile is fired at an angle of 30° to the ground and with an initial velocity of 1000 ft/sec. Calculate the maximum height and the range of the projectile if the ground is level.

Sec. 7.6 Gravitational Acceleration: Trajectories 97

Solution

1. Resolve the 1000 ft/sec into its vertical v_v and horizontal v_h components as shown in Fig. 7.9(a):

$$v_v = 1000 \sin 30° = 500 \text{ ft/sec}$$
$$v_h = 1000 \cos 30° = 866 \text{ ft/sec}$$

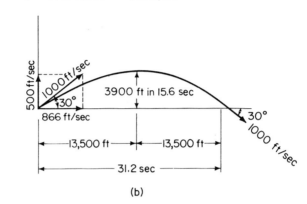

Fig. 7.9

2. The vertical velocity is affected only by the gravitational acceleration of 32 ft/sec². The horizontal velocity is not affected (neglecting air resistance) by gravitational acceleration and will cease only after an elapsed time determined by the time required for the projectile to reach its maximum height and fall back to Earth.

The *time* required to reach maximum height using equation $gt = v_f - v_i$ is

$$t = \frac{v_f - v_i}{g} = \frac{(0 - 500) \text{ ft/sec}}{-32 \text{ ft/sec}^2} \quad \left\{ \begin{array}{l} v_i = 500 \text{ ft/sec} \\ v_f = 0 \text{ ft/sec} \\ g = -32 \text{ ft/sec}^2 \end{array} \right\}$$
$$= 15.6 \text{ sec}$$

The maximum height is

$$s = \left(\frac{v_f + v_i}{2}\right) t$$
$$= \left(\frac{0 + 500}{2}\right) 15.6 \quad \left\{ \begin{array}{l} v_i = 500 \text{ ft/sec} \\ v_f = 0 \text{ ft/sec} \\ t = 15.6 \text{ sec} \end{array} \right\}$$
$$= 3900 \text{ ft}$$

3. It should be noted that the distance traveled by the projectile up is the same as the distance traveled down. Therefore, the time required to make the round trip is

$$2t = 2(15.6) = 31.2 \text{ sec}$$

The range is

$$s_R = v_h(2t) = 866(31.2)$$
$$= 27{,}000 \text{ ft}$$

It should be noted in Fig. 7.9(b) that the projectile strikes the ground at the same angle and with the same velocity with which it left the ground.

Example 15

Assume the same conditions as Example 14. Calculate (1) the vertical velocity after 10 sec; (2) the vertical distance traveled; (3) the horizontal distance traveled.

Solution

1. The vertical velocity after 10 sec is

$$v_f = gt + v_i = (-32)10 + 500 \qquad \left\{ \begin{array}{l} g = -32 \text{ ft/sec}^2 \\ t = 10 \text{ sec} \\ v_i = 500 \text{ ft/sec} \end{array} \right\}$$
$$= 180 \text{ ft/sec}$$

2. The vertical distance after 10 sec is

$$s_v = \left(\frac{v_f + v_i}{2}\right)t = \left(\frac{180 + 500}{2}\right)10$$
$$= 3400 \text{ ft}$$

Or

$$s_v = v_i t + \tfrac{1}{2}gt^2 = 500(10) + \tfrac{1}{2}(-32)(10)^2$$
$$= 3400 \text{ ft}$$

3. The horizontal distance traveled is

$$s_h = v_h t = 866 \times 10$$
$$= 8600 \text{ ft}$$

Problems

1. Convert 50 mph to (a) ft/sec; (b) cm/sec; (c) m/sec.
2. A train is moving at the rate of 90 mph. How far will it go in 6 min?
3. Explain the differences, and give at least one illustration, between the following: (a) distance and displacement; (b) speed and velocity; (c) average velocity and instantaneous velocity.

4. An airplane A passes another airplane B in flight. Airplane A has a speed of 650 mph and airplane B has a speed of 400 mph. Both speeds are relative to earth. What is the speed of A with reference to B (a) if they are flying in opposite directions? (b) if they are flying in the same direction?
5. An airplane is flying with a velocity of 650 mph east relative to still air. A 90-mph wind is blowing from the northeast. Calculate the velocity of the airplane relative to earth.
6. A man rows a boat across a stream which is 1500 ft wide. The stream is flowing at the rate of 5 mph. If he could row the boat at the rate of 8 mph in still water, how far downstream will the boat drift?
7. A boat heads northeast at a speed of 30 mph relative to the shore. If the current is 5 mph south, how far will the boat travel in 2 hr? What is the direction of the boat?
8. Two freight trains are traveling in opposite directions with velocities of 40 mph and 60 mph, respectively. If the trains are 1200 ft long, how long will it take them to pass each other?
9. Assume the trains in Prob. 8 are traveling in the same direction. How long will it take for the faster train to pass the slower train?
10. Assume a 10-ft rod is capable of achieving a velocity in the direction of its length of 60% the speed of light. What is its length at the new velocity as viewed by a stationary observer?
11. Assume a meter stick is moving in the direction of its length. How fast must it be traveling to appear to be three-fourths its true length to a stationary observer?
12. Assume a muon to be traveling at 98% the speed of light in its frame of reference. In its own frame of reference it takes 2×10^{-6} sec to decay while traveling 500 m. As viewed from Earth and using the relativistic equations, calculate (a) the distance traveled by the muon, (b) the life time of the muon.
13. Calculate the mass of an alpha particle traveling at three-fourths the speed of light. The mass of the alpha particle is essentially the mass of its two protons and two neutrons. Its rest mass is 6.696×10^{-27} kg.
14. Calculate the mass of an electron traveling at 95% the speed of light.
15. The rest mass of a helium particle is 6.65×10^{-27} kg. If its mass at velocity v is 7×10^{-27} kg, calculate the velocity of the helium particle.
16. An object is thrown into the air with a velocity of 40 mph at an angle of elevation with the ground of 60°. Calculate its horizontal velocity and vertical velocity at the instant of release?
17. A car is traveling with a speed of 65 mph. How long will it take the car to travel 15 mi?
18. An airplane traveling at the rate of 600 mph takes 2.4 hr to reach its destination. How many feet did it fly?
19. An airplane starts from rest and reaches 150 mph in 24 sec. (a) Calculate its acceleration. (b) How far will it have traveled?
20. A train, traveling in a straight line from rest, accelerates at a constant rate until it reaches 50 mph. (a) Calculate the average velocity in ft/sec. (b) How far will the train travel in 30 min?
21. An automobile, starting from rest and traveling in a straight line, accelerates

from a velocity of 15 ft/sec to 55 ft/sec in 4 sec. (a) Calculate its acceleration. (b) How far will the automobile travel?
22. An airplane has a velocity of 200 mph at a point in its flight. (a) If it has an acceleration of 2 ft/sec^2, calculate its velocity 4 minutes later. (b) How far did it travel?
23. A car is accelerated from 15 mph to 60 mph in 12 sec. Assume a constant acceleration. Calculate (a) the acceleration; (b) the time required to bring the car from rest to 15 mph; (c) the time required to bring the car from rest to 60 mph; (d) the distance traveled in accelerating from 15 mph to 60 mph.
24. A car passes a point with a velocity of 26 ft/sec; 6 seconds later, it passes the same point with the same velocity, but in the opposite direction. (a) How far did the car go if the deceleration was uniform? (b) What is the deceleration?
25. Assume a body starts from rest and that at the end of 15 sec its velocity is 90 ft/sec. How far will it travel (a) in 10 sec? (b) in 5 sec? (c) in 15 sec?
26. What is the acceleration of a bullet during its passage through a 6-in. piece of wood if it enters the wood at a speed of 1800 ft/sec and leaves with a velocity of 1400 ft/sec?
27. A ball is dropped from a 60-story building. Each floor is 12-ft high. Neglecting air resistance, calculate (a) the velocity of the ball when it reaches the ground; (b) its velocity when it passes the 20th floor; (c) the time required to reach the ground; (d) the time required to reach the 20th floor.
28. A ball is thrown straight down from the top of a 100-ft cliff with a velocity of 30 ft/sec. Calculate (a) the final velocity when it strikes the ground; (b) the time required to reach the ground.
29. Assume a ball is thrown up with an initial velocity of 40 ft/sec. Calculate (a) the maximum height reached; (b) the time required to reach maximum height; (c) the time to reach 20 ft on the way up; (d) the time to reach 20 ft on the way down.
30. A stone is dropped into a well; 3 seconds later, the splash of the stone entering the water is heard. If sound travels at the rate of 1080 ft/sec, what is the distance from the top of the well to the top of the water?
31. A ball is thrown in a horizontal direction from the top of a tall building 400 ft high. If the velocity of the ball is 30 ft/sec, calculate its range and plot its path.
32. Calculate, in Prob. 31, (a) the vertical velocity and horizontal velocity of the ball 2.5 sec after it was thrown; (b) the vertical distance and horizontal distance traveled by the ball.
33. A projectile is fired at an angle of elevation of 30° with the horizontal and with a velocity of 250 ft/sec. Calculate (a) the maximum height reached by the projectile; (b) the range of the projectile; (c) the vertical-velocity and horizontal-velocity vectors after 2 sec; (d) the resultant velocity after 2 sec; (e) the velocity when it strikes the ground.
34. A projectile is in flight for 12 sec for a distance of 3000 ft. Calculate (a) the initial velocity; (b) the angle of projection.
35. A man throws a ball at an angle of 45° with the horizontal. The ball reaches a height of 50 ft. (a) What is its velocity when it strikes the ground? (b) What is its velocity when it is 10 ft above the ground?
36. A projectile is fired with an initial velocity of 300 ft/sec at an elevation angle of

20°. If the gun is fired across level ground, find its velocity and position 3 sec after firing.

37. Figure 7.10 shows a ball being thrown at 60° and with an initial velocity of 60 ft/sec so that it misses the edge of a cliff. Calculate (a) the distance from where the ball was thrown to the edge of the cliff; (b) the velocity and the angle at which the ball strikes the ground below the cliff; (c) the total range; (d) the time required for the entire flight.

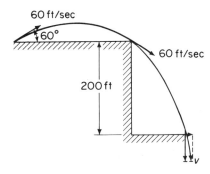

Fig. 7.10

8

Newton's Laws of Motion

8.1. The First Law of Motion

In his *Principia* (1686), Sir Isaac Newton stated the mathematical relationships now known as *the three laws of motion*.

The first law states: "A body or system of bodies in equilibrium will remain in equilibrium unless acted upon by an external unbalanced force." This says that a force is a *tendency* to produce a motion, or a change in motion. A *steady state* means that there is no motion at all, or that *uniform* motion exists. This is an equilibrium state and therefore no unbalance is present. The *status quo* will be maintained as long as there is no unbalance. As soon as a restraining force is applied (or a force is added), an unbalance is created. As long as this unbalance exists, the motion is decelerated (or accelerated) and remains so until the unbalance is removed, when equilibrium is again established.

The first law of motion indicates that matter offers resistance to acceleration. To overcome this resistance, a force must be applied. As indicated, if the applied force balances this resistance, equilibrium exists. If the applied force is greater than the resistance to motion, the object will speed up. If this "greater" force is removed, the body will decelerate.

Resistance to linear acceleration is called *inertia*. Resistance to angular motion is called *moment of inertia*. This inertia depends upon mass, not

Sec. 8.1 The First Law of Motion 103

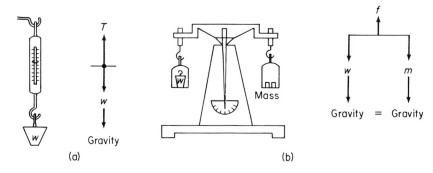

Fig. 8.1

upon the volume or the weight of the body. Mass is dependent on the quantity of matter packed into an atom, irrespective of volume.

This leads to a discussion of the difference between weight and mass. If a spring balance is used, as shown in Fig. 8.1(a), the *weight* of the object is recorded. If the same object is balanced against a standard mass, as shown in Fig. 8.1(b), w is recorded as the *mass* of the object.

When an object is weighed on a spring balance, it reaches a state of equilibrium with the force exerted by the spring. The balancing force required from the spring is the same as the gravitational force. This force changes according to the location of the object on the surface of the Earth.

The beam balance is a comparison measure against a *standard mass*. If the object has greater mass than the comparison mass, an unbalance is created, and the body will be subjected to an acceleration. The resistance to acceleration (inertia) is a measure of the mass, not of the weight. Remember that the weight changes as one proceeds farther and farther away from the surface of the Earth. But so does the attraction of gravity. As a matter of fact, the weight changes because the attraction changes. Nevertheless, if the weight is divided by the gravitational attraction on the body, the ratio, *for a given material*, will always be the same no matter where the weighing is done. This ratio, $w/g = m$, is called the *mass*, where w/g is a constant because as g changes so does w.

Mass, then, depends on inertia, whereas weight depends upon gravitational attraction. It is particularly interesting to note that, in Fig. 8.1(a), the weight vector is the gravitational vector. In Fig. 8.1(b), the gravitational vector pulling on the weight balances the gravitational vector pulling on the standard mass. Therefore, we are actually measuring the mass of our object, since we are comparing masses.

The gravitational law states that the attraction between two fixed masses is dependent on the distance between them. The masses of the two bodies remain constant anywhere on earth. If the masses remain constant, the attraction must depend on the distance between the two bodies only. Therefore, the unit of force for fixed masses is defined as the attraction of gravity at a specific location, where K is a proportionality constant:

$$f = K\frac{m_1 m_2}{d^2}$$

8.2. The Second Law of Motion

The second law of motion states: "When an unbalanced force acts upon a body, the acceleration which results will be in the direction of the force, directly proportional to the unbalanced force, and inversely proportional to the mass of the body."

From the preceding statement,

$$a \propto f \quad \text{and} \quad a \propto \frac{1}{m}$$

Therefore,

$$a \propto \frac{f}{m} \quad \text{and} \quad f \propto ma$$

To convert the foregoing to an equation,

$$f = Kma$$

The units of m and a are then chosen so that K equals 1.

In the centimeter–gram–second (cgs) system, if m is chosen as 1 g, a as 1 cm/sec^2, and f as the unit force, then K will be equal to 1. The unit of force is the *dyne*, which will accelerate a 1-g mass 1 cm/sec^2:

$$f = ma = \frac{\text{g-cm}}{\text{sec}^2} = \text{dyne}$$

In the meter–kilogram–second (mks) system, the acceleration unit is the m/sec^2. The unit of force is the newton (nt):

$$\text{nt} = \frac{\text{kg-m}}{\text{sec}^2}$$

In the foot–pound–second (fps) or "British" system, the pound is the unit of force, the unit of acceleration is ft/sec^2, and the unit of mass is the slug:

$$\text{slug} = \frac{\text{lb-sec}^2}{\text{ft}}$$

In free fall,

$$w = mg \quad \left\{ \begin{array}{l} w = \text{weight: lb} \\ m = \text{mass: slugs} \\ g = 32 \text{ ft/sec}^2 \end{array} \right\}$$

Therefore,

$$m = \frac{w}{g}$$

The British system of units will be used because it is the most convenient for engineering computations. Thus substituting w/g for m,

$$f = ma = \frac{wa}{g} \qquad \left\{\begin{array}{l} w = \text{weight: lb}^2 \\ g = 32 \text{ ft/sec} \\ a = \text{ft/sec}^2 \\ f = \text{force: lb} \end{array}\right\}$$

Therefore, the acceleration g is 32 ft/sec²; w is the weight of the body in pounds; a is the acceleration produced by the unbalance force f acting on the body. Body f and w must be in the same units (pounds) and a and g must be in the same units (feet per second per second).

8.3. The Third Law of Motion

The third law of motion states: " . . . for every force of action there is a simultaneous equal reaction force opposite to the action force." If a man pushes down on the top of a table with his finger, the table pushes back on his finger. If the finger is the action force, the table is the reaction force. The force exerted by the finger must be of the same magnitude as the force exerted by the table. If the two forces were not equal (action–reaction), an unbalance (an acceleration) would exist. The gravitational pull of the earth on the moon is another example of action–reaction.

8.4. Applications of the Laws of Motion

Example 1

A force of 100 lb is applied to a 250-lb object. Calculate the acceleration of the object.

Solution

$$f = ma = \frac{w}{g} a$$

Solve for the acceleration:

$$a = \frac{fg}{w} = \frac{100 \text{ lb} \times 32 \text{ ft/sec}^2}{250 \text{ lb}} \qquad \left\{\begin{array}{l} f = 100 \text{ lb} \\ w = 250 \text{ lb} \\ g = 32 \text{ ft/sec}^2 \end{array}\right\}$$

$$= 12.8 \text{ ft/sec}^2$$

Example 2

A 6000-lb elevator is accelerating during its ascent at the rate of 2 ft/sec². Calculate the tension in the cable (see Fig. 8.2).

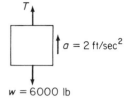

Solution

w = 6000 lb

The term w is the action force and T is the reaction force. If $T = w$, the elevator will either remain stationary or move up at a constant velocity. Since there is an unbalance,

Fig. 8.2

$$f = \frac{w}{g}a = \frac{6000 \text{ lb}}{32 \text{ ft/sec}^2} \times 2 \text{ ft/sec}^2 \quad \left\{ \begin{array}{l} w = 6000 \text{ lb} \\ g = 32 \text{ ft/sec}^2 \\ a = 2 \text{ ft/sec}^2 \end{array} \right\}$$
$$= 375 \text{ lb}$$

To produce an upward acceleration of 2 ft/sec², the 375 lb of unbalance must be added to the tension in the cable. Therefore,

$$T = w + f = 6000 + 375$$
$$= 6375 \text{ lb}$$

Example 3

Assume the same conditions as in Example 2, except that the elevator is ascending with a deceleration (negative acceleration) at the rate of 2 ft/sec². Calculate the tension T.

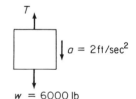

Solution

w = 6000 lb

Since the elevator is slowing down (decelerating), the acceleration is indicated as acting down. The unbalance is now in favor of w over T, and a acts down, aiding w as shown in Fig. 8.3.

Fig. 8.3

$$f = \frac{w}{g}a = \frac{6000 \text{ lb}}{32 \text{ ft/sec}^2} \times 2 \text{ ft/sec}^2$$
$$= 375 \text{ lb} \quad \left\{ \begin{array}{l} w = 6000 \text{ lb} \\ g = 32 \text{ ft/sec}^2 \\ a = 2 \text{ ft/sec}^2 \end{array} \right\}$$
$$T = w - f = 6000 - 375$$
$$= 5625 \text{ lb}$$

Example 4

A 6-lb object and a 4-lb object are hung over a pulley as shown in Fig. 8.4(a). The pulley is frictionless, and the weights of the cord and pulley are neglected.

Sec. 8.4   *Application of the Laws of Motion*   107

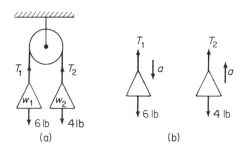

Fig. 8.4

After 3 sec, calculate (1) the acceleration of the system; (2) the velocity; (3) the distance moved by each object; (4) the tension T_E for the left-hand object; (5) the tension T for the right-hand object.

Solution

1. The unbalanced force which accelerates w is 6 lb $-$ 4 lb $=$ 2 lb. The total weight w moved is 6 lb $+$ 4 lb $=$ 10 lb. The acceleration from $f = (w/g)a$ is

$$a = \frac{fg}{w} = \frac{2 \times 32}{10} \qquad \left\{ \begin{array}{l} f = 2 \text{ lb} \\ g = 32 \text{ ft/sec}^2 \\ w = 10 \text{ lb} \end{array} \right\}$$
$$= 6.4 \text{ ft/sec}^2$$

2. In 3 sec, the velocity will be

$$v = 6.4 \text{ ft/sec}^2 \times 3 \text{ sec}$$
$$= 19.2 \text{ ft/sec}$$

The average velocity will be

$$v_{av} = \frac{19.2 + 0}{2} = 9.6 \text{ ft/sec}$$

3. The distance moved by each object will be

$$s = vt = 9.6 \times 3 = 28.8 \text{ ft}$$

The 6-lb object moves down and the 4-lb object moves up, as shown in Fig. 8.4(b).

4. To calculate the tension in the left-hand member, the reasoning is as follows: the tension T must be between the 6-lb and the 4-lb objects for motion to take place in the direction of the 6-ib object. In the left-hand member, Fig. 8.4(a), the tension T must be less than 6-lb. Therefore, $6 - T$ is the unbalance favoring the 6-lb object. Also favoring the 6-lb object is the unbalance which results from $f = ma$. Equating,

$$6 - T = \frac{6}{32} \times 6.4$$
$$T = 4.8 \text{ lb}$$

5. In the right-hand member, the tension T must be greater than 4 lb. Therefore, the unbalance is in favor of the tension T, so that $T - 4$ is equated with $f = ma$:

$$T - 4 = \frac{4}{32} \times 6.4$$
$$T = 4.8 \text{ lb} \quad (check)$$

8.5. D'Alembert's Principle

In problems where an unbalance creates an acceleration, Jean d'Alembert (1717–1783) realized that if he could add a force to oppose this unbalance, he would have a system of forces in equilibrium. This force which opposes acceleration is called the *inertia reaction* and is assumed to be equal to wa/g and acting in opposition to the acceleration. The force assumed equal to the inertia reaction force acts at the center of gravity of the body.

Example 5

The 6000-lb elevator in Example 2 is accelerated during its ascent at the rate of 2 ft/sec². Calculate the tension in the cable using d'Alembert's principle.

Solution

Referring to Fig. 8.5(a), T acts upward, pulling the elevator up. T is opposed by w, but since T is greater than w, the elevator has an acceleration upward. This creates an unbalanced condition. According to d'Alembert's principle, adding the inertia force, wa/g, in opposition to a, the whole system is placed into equilibrium and the method of vectors may be applied to the three forces T, w, and wa/g.

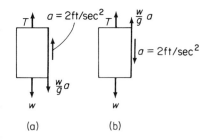

Fig. 8.5

In Fig. 8.5(a), $\sum F_y = 0$:

$$T - w - \frac{w}{g}a = 0$$

Therefore,

$$T = w + \frac{w}{g}a = 6000 + \frac{6000}{32}(2)$$
$$= 6375 \text{ lb} \quad (acting\ up)$$

Example 6

The elevator in Example 3 is ascending, but is decelerating at the rate of 2 ft/sec². Calculate the tension T in the cable.

Solution

Since the elevator is decelerating, a is shown acting down in Fig. 8.5(b). Therefore, the inertia reaction force must act up in opposition to this negative acceleration. Since $\sum F_y = 0$ in Fig. 8.5(b),

$$T + \frac{w}{g}a = w$$

Therefore,

$$T = w - \frac{w}{g}a = 6000 - \frac{6000}{32}(2)$$
$$= 5625 \text{ lb} \quad (acting\ up)$$

Problems

1. A 2500-lb automobile is traveling at a speed of 60 mph. Calculate the force needed to stop the automobile in 8 sec.
2. Calculate the force required to give a 2500-lb automobile a velocity of 60 mph (a) in 12 sec; (b) in 24 sec.
3. Calculate the force required to accelerate a 300-ton train uniformly at a rate of 2 ft/sec² on a level track (a) if friction is neglected; (b) if the friction is 800 lb.
4. A 300-ton train is moving on a level track at a speed of 50 mph. Calculate the force required to stop the train within 300 ft.
5. Neglecting friction, calculate the acceleration of the system in Fig. 8.6.

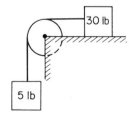

Fig. 8.6

110   *Newton's Laws of Motion*   Chap. 8

6. Calculate the acceleration and tension in the cords in Fig. 8.4(a) if $w_1 = 6$ lb and $w_2 = 10$ lb.
7. Calculate the tension in the cord in Fig. 8.4(a) if $w_1 = 2000$ lb and $w_2 = 2400$ lb.
8. Calculate the acceleration and the tension in the cord in Fig. 8.7. Neglect friction.

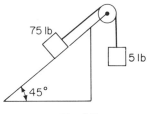

Fig. 8.7

9. Calculate the tension in the cord and the acceleration in Fig. 8.8. Neglect friction.

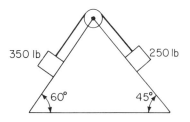

Fig. 8.8

10. Calculate the acceleration and tension in the cord in Fig. 8.7 if the coefficient of friction for the 75-lb block is 0.4.
11. Calculate the acceleration and tension in the cord in Fig. 8.8 if the coefficients of friction are 0.2 and 0.3 for the 350-lb and 250-lb blocks, respectively.
12. A 5000-lb elevator is accelerating at the rate of 4 ft/sec^2 as it ascends. Calculate the tension in the cable.
13. Solve Prob. 12 if the elevator is ascending and decelerating.
14. Solve Prob. 12 if the elevator is descending and accelerating.
15. Solve Prob. 12 if the elevator is descending and decelerating.
16. A 3000-lb elevator carrying a total load of 200 lb is ascending with a uniform acceleration. If the acceleration is 2 ft/sec^2, calculate (a) the force created on the platform by the 200-lb load; (b) the total tension in the cable; (c) the total tension on the cable if the elevator is decelerating.
17. Solve Prob. 16 if the elevator is descending.
18. An elevator which has a 200-lb load is descending and accelerating. Calculate the acceleration which creates weightlessness.
19. How long will it take a 250-lb block to slide down a 30° inclined plane if the

friction is 30 lb? (The block is 15 ft above the bottom of the plane, vertical distance.)

20. A man lowers himself from the top of a house to the ground with a rope. The rope can support a force of 150 lb. What is the least velocity with which he can reach the ground, if he weighs 175 lb and the ground is 40 ft below the top of the house?

21. A bobsled is started from rest at the top of a 20° incline. The incline is $\frac{1}{2}$-mi long and covered with snow. The coefficient of kinetic friction between the sled and the snow is 0.05. How far will the sled slide on the level ground at the bottom at the hill?

22. Assume the bobsled in Prob. 21 slides on ice on the hill and on snow on the level plain. The coefficient of friction of ice is 0.02 and of the snow 0.05. How far will the sled travel on the plain?

9

Angular Motion

9.1. Relationship: Angular–Linear

The first part of this chapter treats the kinematics of angular motion. The last part deals with the kinematics of angular motion.

Three units may be used to measure plane angles: the degree (°), the radian, and the revolution. The *radian* (rad) is defined as the angle subtended at the center of the circle by an arc equal in length to the radius of the circle. Since there are 2π rad in the circumference of a circle, there are 2π rad in 1 *revolution* (rev) of a circle.

Thus 1 circumference equals

$$2\pi \text{ rad} = 360° = 1 \text{ rev}$$

Dividing by 2π,

$$1 \text{ rad} = \frac{360}{2\pi} = 57.3°$$

Also,

$$\pi \text{ rad} = 180°$$

Dividing by 180°,

$$1° = \frac{\pi}{180} = 0.01745 \text{ rad}$$

Thus to convert degrees (θ) to radians, the following conversion equation may be used:

112

Sec. 9.1   *Relationship: Angular–Linear*   113

$$\text{rad} = \frac{\pi\theta}{180}$$

Example 1

Convert 150° to radians.

Solution

$$\text{rad} = \frac{\pi\theta}{180} = \frac{\pi\,150°}{180} \quad \{\theta = 150°\}$$

$$= \frac{5}{6}\pi \text{ rad}$$

To convert radians to degrees, the reciprocal of the above constant is used. Thus

$$\text{degrees} = \frac{180}{\pi} \times \text{rad}$$

Example 2

Convert $3\pi/4$ rad to degrees.

Solution

$$\text{degrees} = \frac{180}{\pi} \times \text{rad} = \frac{180}{\pi}\left(\frac{3}{4}\pi\right)$$

$$= 135°$$

When a wheel turns through 1 rev, or 2π rad, a point on the rim passes through a distance of $2\pi r$, or 1 circumference. If r is the radius of the wheel, this means that the linear distance $2\pi r$ is equal to 2π (the angular units) times the radius. Thus r becomes the proportionality constant between the linear units s, v, and a and the angular units θ (*theta*), ω (*omega*), and α (*alpha*). The conversion equations are

$$s = \theta r$$
$$v = \omega r$$
$$a = \alpha r$$

$\left\{\begin{array}{ll} s = \text{displacement: ft} & \theta = \text{rad} \\ v = \text{velocity: ft/sec} & \omega = \text{rad/sec} \\ a = \text{acceleration: ft/sec}^2 & \alpha = \text{rad/sec}^2 \end{array}\right\}$

Example 3

A flywheel starting from zero velocity accelerates up to 800 rpm in 20 sec. Express the velocity in rad/sec and the acceleration in rad/sec^2.

Solution

$$800 \frac{\text{rev}}{\text{min}} \times \frac{1 \text{ min}}{60 \text{ sec}} = \frac{40 \text{ rev}}{3 \text{ sec}}$$

$$\frac{40 \text{ rev}}{3 \text{ sec}} \times 2\pi \frac{\text{rad}}{\text{rev}} = \frac{80\pi \text{ rad}}{3 \text{ sec}} \quad (\textit{velocity})$$

$$\frac{80\pi \text{ rad}}{3 \text{ sec}} \times \frac{1}{20 \text{ sec}} = \frac{4\pi \text{ rad}}{3 \text{ sec}^2} \quad (\textit{acceleration})$$

Example 4

From Example 3, if the flywheel has an 18-in. diameter, what is the linear acceleration and the final velocity of a point on the rim?

Solution

From Example 3, the acceleration is

$$a = \alpha r = \frac{4\pi}{3} \times \frac{3}{4} \qquad \left\{ \begin{array}{l} \alpha = \frac{4\pi}{3} \text{ rad/sec}^2 \\ r = \frac{1.5}{2} \text{ ft} \\ \omega = \frac{80\pi}{3} \text{ rad/sec} \end{array} \right.$$
$$= \pi \text{ ft/sec}^2$$

The final velocity is

$$v = \omega r = \frac{80\pi}{3} \times \frac{3}{4}$$
$$= 20\pi \text{ ft/sec}$$

The student who is familiar with the linear equations of motion should have little trouble remembering the angular equations. Simply substitute θ for s, ω for v, and α for a (see Table 9.1).

Table 9.1

Linear Equations		Angular Equations	
$s = vt$		$\theta = \omega t$	
$s = v_i t + \frac{1}{2} a t^2$	$s = \text{ft}$	$\theta = \omega_i t + \frac{1}{2} \alpha t^2$	$\theta = \text{rad}$
$v_f - v_i = at$	$v_i = \text{initial velocity: ft/sec}$	$\omega_f - \omega_i = \alpha t$	$\omega = \text{rad/sec}$
$v_f^2 - v_i^2 = 2as$	$v_f = \text{final velocity: ft/sec}$	$\omega_f^2 - \omega_i^2 = 2\alpha\theta$	$\alpha = \text{rad/sec}^2$
	$a = \text{acceleration: ft/sec}^2$		$t = \text{sec}$
	$t = \text{time: sec}$		

Example 5

Referring to Examples 2 and 3, calculate the revolutions and the radians generated during the acceleration of the flywheel.

Sec. 9.2 Centripetal Acceleration 115

Solution

Revolutions during acceleration:

$$s = v_{av}t = \frac{1}{2} \times \frac{40 \text{ rev}}{3 \text{ sec}} \times 20 \text{ sec}$$
$$= 134 \text{ rev}$$
$$\theta = 2\pi \times \text{rev} = 2\pi \times 134$$
$$= 268\pi \text{ rad} = 841 \text{ rad}$$

Alternate Solution

Revolutions during acceleration:

$$s = v_i t + \tfrac{1}{2}at^2 = 0 + \tfrac{1}{2} \times 3.14(20)^2$$
$$= 628 \text{ ft}$$
$$\theta = \omega_i t + \tfrac{1}{2}\alpha t^2 = 0 + \tfrac{1}{2} \times 4.18(20)^2$$
$$= 838 \quad (check)$$

9.2. Centripetal Acceleration

Figure 9.1 represents a portion of a rotating object at the end of a string, length r, at the beginning and end of a time interval; A represents the initial position and B the position of the particle a short time (Δt) later. This interval of time, Δt, corresponds to the short chordal interval Δs between A and B; r

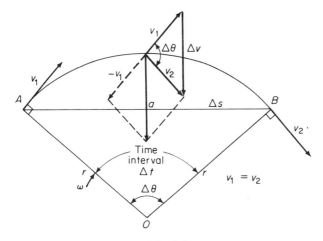

Fig. 9.1

is the radius from the center of rotation O to the path of the particle; and $\Delta\theta$ corresponds to Δs for the time interval Δt.

According to the definition of acceleration, a change in either the *magnitude* or the *direction* of the velocity will satisfy the conditions of acceleration. In Fig. 9.1, if there were no force holding the particle in its circular path, the particle at point A would move off in a straight line tangent to the circular path and perpendicular to r. Since, however, the particle is held in the circular path by r, it will, at point B, move off in a direction v_2 if r should break at the end of time interval Δt. Since the rotating object revolves about point O with no change of speed, the *magnitude* of the velocity vector will not change. Its direction of rotation will constantly be pulled toward the center O. This causes the particle continually to shift its direction toward the center of rotation O, so that its direction of movement is always perpendicular to the radius (string) r. This change of direction of the velocity of the particle is constant and therefore fulfills the conditions of acceleration. Note that if the direction of change, at the point of tangency of the vector with the radius, were not acting along r, the particle would not fall on a true circular path, since the particle, at the point of tangency, is always being pulled toward the center of rotation. This action toward the center of rotation is called *centripetal acceleration*.

Also, if the v_1 direction at A is transposed to the point B, the difference in direction may be represented by a vector Δv acting toward the center:
From Fig. 9.1,

$$\Delta s = v \Delta t$$

$$v = \frac{\Delta s}{\Delta t}$$

$\left\{ \begin{array}{l} \Delta s = \text{change in distance} \\ \Delta t = \text{change in time} \\ v = \text{velocity} \\ a = \text{acceleration} \\ r = \text{radius of rotation} \end{array} \right\}$

Also,

$$a_r = \frac{\Delta v}{\Delta t}$$

Therefore, from the similar triangles containing Δs and Δv,

$$\frac{\Delta v}{v} = \frac{\Delta s}{r}$$

Solving for Δv,

$$\Delta v = \frac{v \, \Delta s}{r}$$

Substituting,

$$a_r = \frac{\Delta v}{\Delta t} = \frac{v \, \Delta s}{r} \times \frac{1}{\Delta t} = \frac{v \, \Delta s}{r \, \Delta t}$$

Sec. 9.3   *Harmonic Motion*   117

For Δs, substitute $v\,\Delta t$:

$$a_r = \frac{v \times v\,\Delta t}{r\,\Delta t} = \frac{v^2}{r}$$

For angular acceleration, if $v = \omega r$,

$$a_r = \frac{(\omega r)^2}{r} = \omega^2 r \quad \textit{(centrifugal acceleration)}$$

Example 6

A weight at the end of a 4-ft rope is rotated parallel to the ground at a uniform rate of 20 rpm (revolutions per minute). What is the centripetal acceleration?

Solution

$$20\,\frac{\text{rev}}{\text{min}} \times \frac{1\,\text{min}}{60\,\text{sec}} = \frac{1}{3}\,\text{rps} \quad \textit{(revolutions per second)}$$

$$\frac{60\,\text{sec/min}}{20\,\text{rev/min}} = 3\,\text{sec/rev}$$

Tangential velocity:

$$v = \frac{2\pi r}{t} = \frac{2\pi\,4\,\text{ft}}{3\,\text{sec}} = \frac{8\pi}{3} = 8.37\,\text{ft/sec}$$

Acceleration:

$$a_r = \frac{v^2}{r} = \frac{8.37^2\,\text{ft}^2}{4\,\text{ft-sec}^2} = 17.5\,\text{ft/sec}^2$$

Checking,

$$\omega = \frac{v}{r} = \frac{8.37}{4} = 2.09\,\text{rad/sec}$$

$$a_r = \omega^2 r = 2.09^2 \times 4 = 17.5\,\text{ft/sec}^2$$

9.3. Harmonic Motion

Linear harmonic motion may be illustrated by projecting the shadow of point A, Fig. 9.2, upon the diameter of the circular path of A as it revolves about the center of rotation O. Various positions of point A and the corresponding shadow A_1 are shown in Fig. 9.2. It should be noted that although the movement of A about the circumference of the circle is constant for a time interval t, the movement of the shadow A_s for the same time interval is not constant. It moves along the diameter of the circle with an increasing and decreasing velocity. This motion is called *simple harmonic motion* (SHM).

The displacement of the shadow of point A is opposite and proportional to the acceleration of the particle.

The time for 1 complete revolution (360°) of point A is called the *period:*

$$T = \frac{2\pi}{\omega} \quad \left\{ \begin{array}{l} T = \text{sec} \\ \omega = \text{rad/sec} \end{array} \right\}$$

Since the displacement of the shadow is opposite to the acceleration the sign is negative in the following equation:
If

$$a = -\omega^2 s \quad \{ s = \text{displacement} \}$$

Then

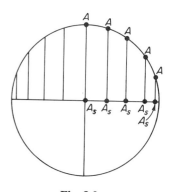

Fig. 9.2

$$\omega = \sqrt{\frac{-a}{s}}$$

Substituting into the equation for the period,

$$T = \frac{2\pi}{\omega} = \frac{2\pi}{\sqrt{-a/s}} = 2\pi \sqrt{\frac{-s}{a}}$$

Since *frequency* is the reciprocal of the period,

$$F = \frac{1}{T} = \frac{1}{2\pi}\sqrt{-a/s}$$

9.4. Harmonic Motion: Simple Pendulum

A weight w is attached to one end of a weightless string in Fig. 9.3. The other end is anchored at Q. If the weight is displaced as shown in Fig. 9.3(a), an unbalance is created and

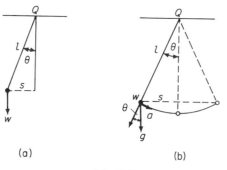

(a)   (b)

Fig. 9.3

Sec. 9.5 *Harmonic Motion: The Helical Spring* 119

$$s = l \sin \theta$$

Since the acceleration vector a is proportional to the gravitational force g, the vector component of g is the acceleration as shown in Fig. 9.3(b):

$$a = -g \sin \theta \quad (opposed\ to\ the\ direction\ of\ s)$$

If the angle θ is kept below 10°, the vectors a and s may be considered equal.*

Using the equation for period,

$$T = 2\pi \sqrt{-s/a} = 2\pi \sqrt{\frac{-l\sin\theta}{-g\sin\theta}} = 2\pi \sqrt{l/g} \quad \left\{\begin{array}{l} l = \text{length of string: ft} \\ S = \text{horizontal displacement} \\ a = \text{acceleration} \\ t = \text{period: sec/vibration} \end{array}\right\}$$

If g is a constant value, the period (or frequency) depends only upon the length l of the pendulum.

Example 7

A pendulum which is 36 in. long has a frequency of 31.5 vib/min (vibrations per minute). Calculate g.

Solution

The frequency is

$$F = 31.5 \frac{\text{vib}}{\text{min}} \times \frac{1}{60} \frac{\text{min}}{\text{sec}} = 0.525 \text{ vib/sec} \quad \left\{\begin{array}{l} l = \frac{36}{12} = 3 \text{ ft} \\ F = 31.5 \text{ vib/min} \end{array}\right\}$$

The period is

$$T = \frac{1}{F} = \frac{1}{0.525} = 1.9 \text{ sec/vib}$$

Solve $T = 2\pi \sqrt{l/g}$ for g:

$$g = \frac{4\pi^2 l}{T^2} = \frac{4\pi^2(3)}{1.9^2}$$

$$= 32.8 \text{ ft/sec}^2$$

9.5. Harmonic Motion: The Helical Spring

In Fig. 9.4, the neutral position of the object suspended from a spring is shown at B. That is, the gravitational force pulling on the object is equal to the force exerted on the object by the spring. The system is in equilibrium.

If the object is pulled down to position C and held, the force necessary

---

* The error is very small for angles θ less than 10°.

'to hold it at C will be added to the gravitational force. The sum of the gravitational force and the force necessary to hold the object at C is balanced by the upward pull of the spring.

If the force necessary to hold the object at C is removed, an unbalanced condition is created in favor of the spring, exactly equal to the force necessary to hold the object at C. The object will accelerate (up) from C to B because the upward pull of the spring is greater than the gravitational force pulling down.

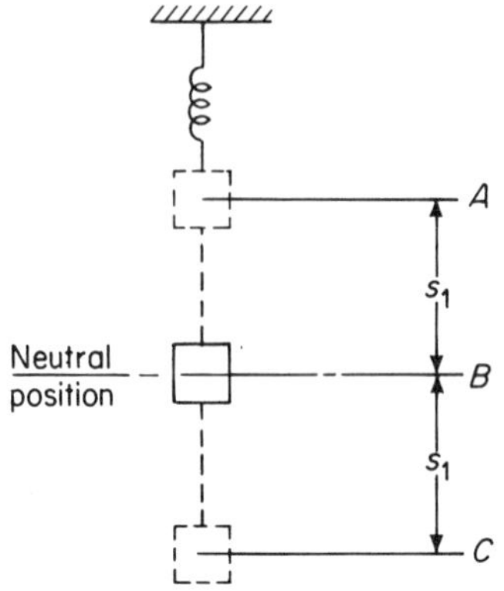

At B, the neutral position, the acceleration changes to a deceleration because the gravitational force down is greater than the upward pull of the spring. Note that this condition exists from B to A.

Fig. 9.4

The *period of oscillation* is the time required for the object to move from C to A and then back to C. If air friction and internal friction in the spring are neglected, s_1 is equal to s.

The force required to hold the block at position C is dependent upon the restoring force pulling the block up. The force is also proportional to the displacement. Thus

$$f \propto -s$$

If a constant k for a particular spring is introduced,

$$f = -ks$$

From $f = ma$,

$$ma = -ks$$

$\left\{ \begin{array}{l} A = \text{amplitude} \\ s = \text{displacement} \\ F = \text{frequency} \\ v' = \text{harmonic velocity} \end{array} \right\}$

The acceleration is

$$a = -\frac{ks}{m} = -ks\frac{g}{w}$$

The spring constant k is

$$k = -\frac{wa}{gs} = -\frac{f}{s}$$

The velocity of a particular oscillating with harmonic motion is given by the equation

$$v' = 2\pi F \sqrt{A^2 - s^2}$$

Example 8

A spring is hung from the ceiling, as in Fig. 9.4. The free length of the spring is 8 in. When a $\frac{1}{2}$-pound object is hung on the spring and equilibrium is estab-

Sec. 9.5   Harmonic Motion: The Helical Spring   121

lished, the length of the spring is 12 in. The object is then pulled down so that the spring stretches to 14 in. and released. Calculate (1) the spring constant; (2) acceleration at a point 2 in. above B; (3) the force acting 3 in. below B; (4) the acceleration at 14 in.; (5) the period; (6) the velocity at the midpoint B; (7) the velocity at 14 in.

Solution

1. The spring constant is determined as follows:

$$w = ks$$
$$k = \frac{w}{s} = \frac{\frac{1}{3}}{\frac{1}{3}} = \frac{3 \text{ lb}}{2 \text{ ft}} \qquad \left\{ \begin{array}{l} w = \frac{1}{3} \text{ lb} \\ s = 12 - 8 = 4 \text{ in.} = \frac{1}{3} \text{ ft} \end{array} \right\}$$

2. The acceleration 2 in. above B is

$$a = -\frac{kgs}{w} = -\frac{\frac{3}{2} \times 32 \times \frac{1}{6}}{\frac{1}{3}} \qquad \left\{ \begin{array}{l} k = \frac{3 \text{ lb}}{2 \text{ ft}} \\ g = 32 \text{ ft/sec}^2 \\ w = \frac{1}{3} \text{ lb} \\ s = \frac{2}{12} = \frac{1}{6} \text{ ft} \end{array} \right\}$$
$$= -16 \text{ ft/sec}^2$$

3. The force acting 3 in. below B is

$$a = -\frac{3 \times 32 \times 1}{2 \times \frac{1}{3} \times 4}$$
$$= -24 \text{ ft/sec}^2 \qquad \{ s = \frac{3}{12} = \frac{1}{4} \text{ ft} \}$$
$$f = -ks = -\frac{3}{2} \times \frac{1}{4} = -\frac{3}{8} \text{ lb}$$

4. The acceleration at 20 in. is

$$a = -\frac{kgs}{w} = -\frac{3 \times 32 \times 1}{2 \times \frac{1}{3} \times 6}$$
$$= -16 \text{ ft/sec}^2 \qquad \{ s = 14 - 12 = 2 \text{ in.} = \frac{1}{6} \text{ ft} \}$$

5. The period is

$$T = 2\pi \sqrt{s/a} = 2\pi \sqrt{\frac{1}{6 \times 16}} \qquad \left\{ \begin{array}{l} s = \frac{1}{6} \text{ ft} \\ a = 16 \text{ ft/sec}^2 \end{array} \right\}$$
$$= 0.641 \text{ vib/sec}$$

6. The velocity at point B is

$$v' = \frac{2\pi}{T} \sqrt{A^2 - s^2} = \frac{2\pi}{0.641} \sqrt{\left(\frac{1}{6}\right)^2 - 0}$$
$$= 1.63 \text{ ft/sec}$$

7. The velocity at 14 in. is

$$v' = \frac{2\pi}{T}\sqrt{A^2 - s^2} = \frac{2\pi}{0.641}\sqrt{\left(\frac{1}{6}\right)^2 - \left(\frac{1}{6}\right)^2}$$

$$= 0 \text{ ft/sec}$$

Problems

1. Convert the following degrees to radians: (a) 50°; (b) 45°; (c) 135°; (d) 245°; (e) 450°.
2. Convert the following radians to degrees: (a) $^4/_5\pi$; (b) $^7/_8\pi$; (c) $^9/_7$; (d) 6π; (e) $^7/_4\pi$.
3. Convert the following to rad/sec: (a) 400 rpm; (b) 300 rph.
4. Convert the following to rpm: (a) 8 rad/sec; (b) 350 rad/min.
5. Convert the following to rad/sec²: (a) 1000 ft/sec²; where the radius of the path is 30 in.; (b) 200 mi/hr² on a ½-mi-diameter track.
6. An automobile tire, 30 in. in diameter, is turning at the rate of 8000 rev every 10 minutes. Calculate (a) the speed of the automobile; (b) the angular velocity of the wheel.
7. A flywheel rotates at 300 rpm. How many seconds are required for the flywheel to turn 5000 rad?
8. A 24-in.-diameter pulley is driven with a belt from an 18-in.-diameter motor pulley. The belt speed is 900 ft/min. Calculate the angular velocity of each pulley in (a) rpm; (b) rad/sec.
9. A 30-in.-diameter wheel is rotating at the rate of 120 rpm. Find the linear velocity in ft/sec and the angular velocity in rad/sec of the wheel (a) at the center; (b) at the farthest point from the center; (c) at a point midway between the rim and the center of the axle.
10. The angular velocity of a point on a wheel changes from 10 rps to 240 rpm. Calculate the acceleration if the velocity change takes place in 2 min.
11. Calculate the time required for an object to change its velocity from 30 rpm to 6 rps if the acceleration is 3 rps every second.
12. A motor shaft starting from rest increases in speed uniformly every second by 270°/sec. It takes the motor 20 sec to attain full speed. What is the rpm of the motor shaft?
13. An object increases its angular velocity from 4 rad/sec to 10 rad/sec. Calculate the number of revolutions the object makes in 12 sec.
14. A crank, having an 18-in. radius, is rotated with a constant angular velocity of 360 rpm. (a) What is the tangential velocity of a point on the end of the crank? (b) What is its linear acceleration?
15. A boy holds one end of a string which has an object attached to the other end. The boy is whirling the object in a circle, radius 5 ft, with an acceleration toward the center of 50 ft/sec². What is the velocity of the object?
16. If the object in Prob. 15 makes 2 rev in 5 sec, calculate the velocity and centripetal acceleration.

Problems

17. Calculate the centripetal acceleration of a car going around a 600-ft-radius curve traveling at a rate of 50 mph.
18. A 15-in.-radius crank rotates at a uniform angular velocity of 240 rpm. Calculate (a) the tangential and radial velocities of the crank; (b) the tangential and radial accelerations of a point at the end of the crank.
19. A car rounds a curve at a speed of 45 mph. Calculate the radius of curvature of the road which will produce a radial acceleration equal to gravitational acceleration.
20. Given a pendulum which makes 80 vibrations every 60 second, if the pendulum is 14 cm long, what is the acceleration of gravity in this locality?
21. Find the length of a pendulum which is adjusted to keep correct time where $g = 32.5$ ft/sec^2 if the pendulum makes 1 cps (cycle per second).
22. A pendulum is pulled to the side so that it makes an angle of 10° with the vertical centerline. The pendulum is 30 in. long. Calculate (a) the acceleration at this position; (b) the horizontal displacement; (c) the period; (d) the frequency; (e) the tension in the string.
23. A spring is hung from a support as shown in Fig. 9.4. The free length of the spring is 6 in. When a 2-lb weight is hung at the free end, equilibrium is established at a spring of length 14 in. The spring is then stretched to a length of 18 in. and released. Find (a) the spring constant; (b) acceleration at a point 3 in. above the equilibrium position; (c) the force at a point 2 in. below the equilibrium position; (d) the acceleration at 18 in.; (e) the period; (f) the frequency of oscillation; (g) the velocity and acceleration at the equilibrium position.

10

Laws of Angular Motion

10.1. Moment of Inertia

We have defined *inertia* as the resistance of matter to acceleration. *Moment of inertia* may be defined as the resistance of matter to *angular* acceleration. It can be shown that the resistance of matter to angular acceleration also depends upon the square of the distance from the mass to the axis of rotation. Therefore, the moment of inertia (I) of a body or particle about an axis is the sum of all the products of the square of the distances from the axis of rotation to the center of gravity (c.g.) of the mass, times the magnitude of this mass. This may be written

$$I_{o-o} = m_1 s_1^2 + m_2 s_2^2 + m_3 s_3^2 + \cdots$$
$$= \sum Ms^2 = \sum \frac{w}{g} s^2$$

$\left\{\begin{array}{l} m = \text{mass} \\ s = \text{distance from center of mass to c.g. of mass} \\ I_{o-o} = \text{M of I through c.g.} \\ w = \text{weight} \\ g = 32 \text{ ft/sec}^2 \end{array}\right\}$

Example 1

A rod, of negligible weight, has two 8-lb objects at its ends as shown in Fig. 10.1. The distance from the center of gravity of each object to the center of rotation is 1.5 ft. Calculate the moment of inertia of the system.

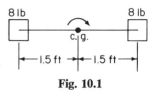

Fig. 10.1

Sec. 10.2   Radius of Gyration

Solution

The moment of inertia is

$$I_{o-o} = m_1 s_1^2 + m_2 s_2^2$$

$$= \frac{w_1}{g} s_1^2 + \frac{w_2}{g} s_2^2$$

$$= \frac{8}{32}(1.5)^2 + \frac{8}{32}(1.5)^2$$

$$= 1.125 \text{ ft-lb-sec}^2$$

$\begin{Bmatrix} s_1 = 1.5 \text{ ft} \\ s_2 = 1.5 \text{ ft} \\ w_1 = 8 \text{ lb} \\ w_2 = 8 \text{ lb} \end{Bmatrix}$

10.2. Radius of Gyration

From the moment-of-inertia equation,

$$I_{o=o} = m_1 s_1^2 + m_2 s_2^2 + m_3 s_3^2 + \cdots = mK^2$$

Each of the mass values $(m_1, m_2, m_3, \ldots)$ is taken as equal to the other. The s-values vary because they represent values from the axis of rotation of the entire mass m to the center of gravity of each small mass. If all the s-values are squared, averaged, and then the square root taken, the value K obtained is called the *radius of gyration*. It is the distance from the axis of rotation to a point where the entire mass could be considered as concentrated without changing the effect on the resistance to angular acceleration.

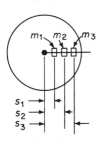

Fig. 10.2

Thus in Fig. 10.2, let us consider a series of small masses along the radius of a sphere. From the Appendix Table A.4, the radius of gyration is taken as

$$K = R\sqrt{\frac{2}{5}}$$

The effect of the entire mass may be concentrated at a point $0.632R$ from the center of rotation of the sphere.

If the object is homogeneous, the radius of gyration is independent of the density of the material. Thus, if an object is made of steel or wood, the radius of gyration is the same if the shapes of both objects are the same.

If the object is homogeneous and of uniform thickness, the area determines the value of K.

Example 2

Given a 12-lb dish which has a diameter of 8 in., calculate the moment of inertia of the dish about its central axis P–P.

Solution

From Table A.4 in the Appendix, the radius of gyration about p–p is $K = R/\sqrt{2}$. Therefore,

$$I_{p-p} = mK^2 = \frac{w}{g}\left(\frac{R}{\sqrt{2}}\right)^2$$

$$= \frac{12}{32}\left(\frac{\frac{1}{3}}{\sqrt{2}}\right)^2 = \frac{12}{32}\left(\frac{1}{18}\right)$$

$$= 0.0208 \text{ ft-lb-sec}^2$$

$$\left\{\begin{array}{l} w = 12 \text{ lb} \\ g = 32 \text{ ft/sec}^2 \\ R = \frac{4}{12} = \frac{1}{3} \text{ ft} \\ K_{p-p} = \frac{R}{\sqrt{2}} \end{array}\right\}$$

10.3. Moment of Inertia of an Area: Second Moment

The *second moment*, or the moment of inertia of an area, is derived from

$$\sum As^2$$

The second moment of an area is therefore

$$I = AK^2$$

Example 3

Calculate the moment of inertia of a circular disk about an axis parallel to the plane of the disk. The diameter of the disk is 8 in.

Solution

$$I_{o-o} = AK^2 = \pi R^2 \left(\frac{R}{2}\right)^2 = \frac{\pi R^4}{4} \text{ ft}^4$$

$$= \frac{\pi(\frac{1}{3})^4}{4}$$

$$= 0.0097 \text{ ft}^4$$

$$\left\{\begin{array}{l} R = \frac{4}{12} = \frac{1}{3} \text{ ft} \\ K_{o-o} = \frac{R}{2} \end{array}\right\}$$

10.4. Moment of Inertia About Any Axis

Appendix Table A.4 gives the radii of gyration for axes perpendicular K_p, and parallel K_o, to the plane of the objects.

Rotation about a displaced axis may be obtained from the mass equation:

$$I_{x-x} = I_{o-o} + md^2$$

Sec. 10.4   Moment of Inertia About Any Axis

If $I_{o-o} = mK^2$, then

$$I_{x-x} = mK^2 + md^2 = m(K^2 + d^2)$$
$$= \frac{W}{g}(K^2 + d^2)$$

$\begin{Bmatrix} I_{x-x} = \text{displaced } CG \text{ above} \\ \text{or below abscissa} \\ I_{o-o} = \text{axis through } CG \\ K = \text{radius of gyration} \\ d = \text{displaced distance} \end{Bmatrix}$

If the area is to be used, then

$$I_{x-x} = AK^2 + Ad^2$$
$$= A(K^2 + d^2)$$

Example 4

Calculate the moment of inertia of a rectangle 12 in. × 9 in. in the position shown in Fig. 10.3 about an axis 10 in. from the center of gravity.

Solution

The moment of inertia about the central axis is

$$I_{o-o} = AK^2 = bh\left(\frac{h}{\sqrt{12}}\right)^2$$
$$= \frac{bh^3}{12} = \frac{\frac{3}{4}(1)^3}{12}$$
$$= 0.06 \text{ ft}^4$$

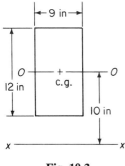

Fig. 10.3

The moment of inertia about the displaced axis is

$$I_{x-x} = AK^2 + Ad^2 = AK^2 + bhd^2$$
$$= 0.06 + \frac{3}{4}(1)(\frac{5}{6})^2$$
$$= 0.58 \text{ ft}^4$$

$\begin{Bmatrix} h = 12 \text{ in.} = 1 \text{ ft} \\ b = 9 \text{ in.} = \frac{3}{4} \text{ ft} \\ d = \frac{10}{12} = \frac{5}{6} \text{ ft} \\ k = \frac{h}{\sqrt{12}} \end{Bmatrix}$

Example 5

In Example 4, cut out a 6-in.-diameter circle so that the center of gravity of the circle and rectangle coincide as shown in Fig. 10.4.

Solution

The hole cut into the rectangle may be thought of as yielding a negative moment of inertia. Therefore, the moment of inertia of the circle is calculated about the x-x axis and subtracted from the moment of inertia about the o-o axis.
The moment of inertia of the *hole* is

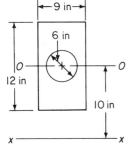

Fig. 10.4

128   Laws of Angular Motion   Chap. 10

$$I_{x-x} = AK^2 + Ad^2 = \pi R^2 \left(\frac{R}{2}\right)^2 + \pi R^2 d^2$$

$$= \frac{\pi R^4}{4} + \pi R^2 d^2 = \frac{\pi(\frac{1}{4})^4}{4} + \pi \left(\frac{1}{4}\right)^2 \left(\frac{5}{6}\right)^2$$

$$\left\{ \begin{array}{l} R = \frac{3}{12} = \frac{1}{4} \text{ ft} \\ d = \frac{10}{12} = \frac{5}{6} \text{ ft} \\ K_{o-o} = \frac{R}{2} \end{array} \right\}$$

$$= 0.14 \text{ ft}^4$$

The moment of inertia of the object with the hole is

$$I_{x-x} = 0.58 - 0.14$$
$$= 0.44 \text{ ft}^4$$

10.5. Harmonic Motion: Torsion Pendulum

In the *torsion pendulum*, the torque causing angular motion sets up a restoring force in the twisted wire. This restoring force is proportional to the angular displacement. Note from our former discussion that if the displacement is clockwise, the restoring force is counterclockwise.

From the general equation for period (from the preceding chapter),

$$T = 2\pi \sqrt{\frac{s}{a}} \quad \text{and} \quad f = ma$$

$$T = 2\pi \sqrt{\frac{s}{f/m}} = 2\pi \sqrt{\frac{m}{f/s}}$$

$$\left\{ \begin{array}{l} T = \text{period: sec/vib} \\ s = \text{displacement: ft} \\ a = \text{acceleration: ft/sec}^2 \\ m = \text{mass: slugs} \\ f = \text{force: lb} \end{array} \right\}$$

Since we are dealing with angular motion, the linear equation may be transformed into the angular equation by replacing the symbols m with I, f with L, and S with θ. Thus

$$T = 2\pi \sqrt{\frac{I}{L/\theta}}$$

$$\left\{ \begin{array}{l} T = \text{period: sec/vib} \\ I = \text{M of I: ft-lb-sec}^2 \\ L = \text{torque: ft-lb} \\ \theta = \text{displacement: rad} \end{array} \right\}$$

The torque constant of the wire is written

$$\frac{L}{\theta} = \frac{\pi n r^4}{2l}$$

$$\left\{ \begin{array}{l} n = \text{modulus of rigidity: lb-ft/in.}^2 \\ \quad \text{(Table A.7)} \\ r = \text{radius of wire: in.} \\ l = \text{length of wire: in.} \\ \theta = \text{angular displacement: rad} \end{array} \right\}$$

Using the preceding equations:
For a disk,

$$I_1 = \tfrac{1}{2}mR^2$$

Sec. 10.5   Harmonic Motion: Torsion Pendulum   129

For a ring,
$$I_2 = \tfrac{1}{2}m(R_1^2 + R_2^2)$$

If the ring is attached to a disk so that its center of gravity coincides with the center of gravity of the disk, the equation becomes

$$T = 2\pi \sqrt{\frac{I_1 + I_2}{L/\theta}} \qquad \left\{ \begin{array}{l} I_1 = \text{moment of disk} \\ I_2 = \text{moment of ring} \end{array} \right\}$$

Example 6

A steel rod 90 cm long and 4 mm in diameter is clamped at one end. If the loose end of the rod is twisted through an angle of 40°, calculate the torque on the rod.

Solution

$$L = \frac{\pi n r^4 \theta}{2 l}$$

$$= \frac{\pi (7.7 \times 10^{11})(0.2)^4 (40\pi/180)}{2 \times 90}$$

$$= 1.5 \times 10^7 \text{ dyne-cm}$$

$\left\{ \begin{array}{l} L = \text{torque: dyne-cm} \\ n = \text{coef of rigidity} \\ r = \text{radius of rod:cm} \\ \theta = \text{angle of twist: rad} \\ l = \text{length of rod: cm} \end{array} \right\}$

Example 7

A 32-lb disk has a diameter of 12 in. Calculate its moment of inertia.

Solution

The moment of inertia is

$$I = mK^2 = m\left(\frac{R}{\sqrt{2}}\right)^2 = \tfrac{1}{2}mR^2 = \tfrac{1}{2}\frac{w}{g}R^2$$

$$= \tfrac{1}{2}\left(\tfrac{32}{32}\right)\left(\tfrac{1}{2}\right)^2 = \tfrac{1}{8} \text{ ft-lb-sec}^2$$

$\left\{ \begin{array}{l} w = 32 \text{ lb} \\ R = \tfrac{6}{12} = \tfrac{1}{2} \text{ ft} \\ g = 32 \text{ ft/sec}^2 \\ K = \dfrac{R}{\sqrt{2}} \end{array} \right\}$

Example 8

The disk in Example 7 is suspended by a $\tfrac{1}{2}$-in.-diameter rod, 3 ft long. Calculate the constant of torque of the rod. The coefficient of rigidity of the rod is 11×10^6 ft-lb/in.²

Solution

$$\frac{L}{\theta} = \frac{\pi n r^4}{2l} = \frac{3.14(11 \times 10^6)(\frac{1}{4})^4}{2 \times 36} \quad \left\{ \begin{array}{l} r = \text{radius of rod} = \frac{1}{4} \text{ in.} \\ l = 36 \text{ in.} \end{array} \right\}$$

$$= 1874 \text{ lb-ft/rad}$$

Example 9

Calculate the period and frequency of the disk in Examples 7 and 8 when released from the angle of twist.

Solution

$$T = 2\pi \sqrt{\frac{I}{L/\theta}} = 2\pi \sqrt{\frac{\frac{1}{8}}{1874}}$$

$$= 0.051 \text{ sec/vib}$$

$$F = \frac{1}{T} = \frac{1}{0.051}$$

$$= 20 \text{ vib/sec}$$

10.6. Laws of Angular Motion

The first and third laws of motion are laws dealing with equilibrium conditions and as such apply to angular as well as to linear motion. The second law of motion must be restated in order for it to apply to angular motion: Thus the second low—which states that "when an unbalanced force acts upon a body, the acceleration which results will be directly proportional to the unbalanced force and inversely proportional to the mass"—might be rephrased to read: "An applied effort will yield a result directly proportional to the effect and inversely proportional to any resistance." If *applied effort* is taken to mean torque, *result* to mean angular acceleration, and *resistance* to mean the moment of inertia, the second law of motion will also apply to angular motion.

Under these conditions, the analogies between f and L, between m and I, and between a and α become evident. Therefore, $f = ma$ becomes

$$L = I\alpha \quad \left\{ \begin{array}{l} L = \text{torque } or \text{ moment of force: ft-lb} \\ I = \text{moment of inertia: ft-lb-sec}^2 \text{ (slug-ft}^2) \\ \alpha = \text{angular acceleration: rad/sec}^2 \end{array} \right\}$$

Example 10

Calculate the torque required to produce an angular acceleration of 3 rad/sec² to an object whose moment of inertia is 300 lb-ft-sec².

Sec. 10.6   Centripetal Forces

Solution

$$L = I\alpha = 300 \times 3$$
$$= 900 \text{ ft-lb}$$

$$\left\{ \begin{array}{l} \alpha = 3 \text{ rad/sec}^2 \\ I = 300 \text{ lb-ft-sec}^2 \\ L = ? \text{ ft-lb} \end{array} \right\}$$

Example 11

A 1200-lb uniform circular disk has a 6-ft diameter. Starting from rest, the disk is accelerated about a central axis by a force of 200 lb acting at its circumference. Calculate the rpm of the disk after 4 sec.

Solution

Moment of inertia:

$$I = mK^2 = \frac{w}{g}\left(\frac{R}{\sqrt{2}}\right)^2 = \frac{w}{g}\frac{R^2}{2}$$
$$= \frac{1200 \times 3^2}{32 \times 2} \frac{\text{lb ft}^2}{\text{ft/sec}^2}$$
$$= 168.7 \text{ ft-lb-sec}^2$$

$$\left\{ \begin{array}{l} w = 1200 \text{ lb} \\ g = 32 \text{ ft/sec}^2 \\ R = \frac{6}{2} = 3 \text{ ft} \\ K = \frac{R}{\sqrt{2}} \\ f = 200 \text{ lb} \end{array} \right\}$$

The torque is the force times the perpendicular distance:

$$L = fR = 200 \times 3 = 600 \text{ ft-lb}$$

The angular acceleration from $L = I\alpha$ is

$$\alpha = \frac{L}{I} = \frac{600 \text{ ft-lb}}{168.7 \text{ ft-lb-sec}^2}$$
$$= 3.56 \text{ rad/sec}^2$$

At the end of 4 sec, the average angular velocity is

$$\omega = 4 \text{ sec} \times 3.56 \text{ rad/sec}^2$$
$$= 14.24 \text{ rad/sec}$$

The rpm is

$$\frac{14.24 \text{ rad/sec} \times 60 \text{ sec/min}}{2\pi \text{ rad/rev}} = 136 \text{ rpm}$$

10.7. Centripetal Forces

Centripetal force is based on the concepts discussed in Chapter 9. That chapter pointed out that if a body moves in a circular path with uniform motion, it must possess an acceleration toward the center of rotation. This acceleration toward the center of rotation times the mass of the rotating object is called the *centripetal force* (f_{cp}). The inertia reaction force, acting outward and equal to the centripetal force, is called the *centrifugal force* (f_{cf}).

From $f = ma$ and $a = v^2/r$,

$$f_{cp} = ma = \frac{w}{g}\frac{v^2}{r}$$

and from $v = \omega r$,

$\begin{aligned} f_{cp} &= \text{centripetal force: lb} \\ f_{cf} &= \text{centrifugal force: lb} \\ w &= \text{weight: lb} \\ g &= 32 \text{ ft/sec}^2 \\ v &= \text{velocity: ft/sec} \\ r &= \text{radius: ft} \\ \omega &= \text{angular velocity: rad/sec} \end{aligned}$

$$f_{cp} = \frac{w}{g}\frac{v^2}{r} = \frac{w}{g}\frac{\omega^2 r^2}{r}$$

$$= \frac{w}{g}\omega^2 r$$

Example 12

A boy swings a 12-lb stone attached to the end of a 6-ft rope in a circular arc. If the stone has a tangential speed of 8 ft/sec, calculate the tension in the cord.

Solution

Note that in Fig. 10.5 the tension T is created by the centrifugal force f_{cf} acting away from the center of rotation. The centripetal force f_{cp} opposes this outward force by causing an acceleration toward the center of rotation:

$$\sum F_x = f_{cp} - T = 0$$
$$f_{cp} = T$$

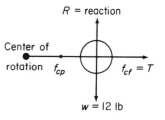

Fig. 10.5

Also,

$$\sum F_y = R - 12 = 0$$
$$R = 12 \text{ lb}$$

From

$$f_{cp} = \frac{w}{g}\frac{v^2}{r} = \frac{12}{32} \times \frac{8^2}{6}$$

therefore,

$$T = f_{cp} = 4 \text{ lb}$$

Problems

1. Illustrate and describe the concept of radius of gyration.
2. A 20-lb steel ball is whirled at the end of a 10-ft cord. Calculate the moment of inertia of the system.
3. How much does the cord in Prob. 2 have to be shortened or lengthened to produce a moment of inertia of 100 ft-lb-sec^2?
4. Calculate the moment of inertia of a 5-ton flywheel. The average radius at which the mass is concentrated is 6 ft.
5. (a) Calculate the moment of inertia of a 30-lb disk through its central axis if its diameter is 16 in. (b) Calculate the moment of inertia if the axis is parallel to the disk.
6. Two steel objects are placed respectively 8 ft and 4 ft from the center of rotation of a uniform steel bar. If the objects weigh 6 lb and 5 lb, respectively, calculate the change in moment of inertia of the system.
7. Calculate the moment of inertia of the areas shown in Fig. 10.6(a) through (d) about the central axis o–o through the center of gravity.

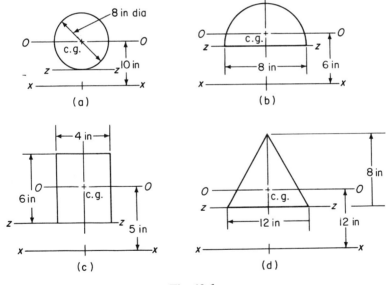

Fig. 10.6

8. Calculate the moment of inertia about an axis perpendicular (p–p) to the area for (a) Fig. 10.6(a); (b) 10.6(c).
9. Calculate the moment of inertia of the areas in Fig. 10.6(a) through (d) about the displaced axis x–x.
10. Calculate the moment of inertia of the areas in Fig. 10.6(a) through (d) if the displaced axis is at z–z.

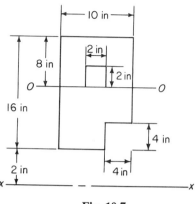
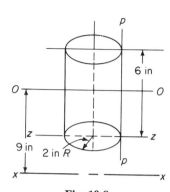

Fig. 10.7   Fig. 10.8

11. Calculate the moment of inertia about the x–x axis in Fig. 10.7.
12. In Fig. 10.8, calculate the moment of inertia about (a) the o–o axis; (b) the x–x axis; (c) the z–z axis; (d) the p–p axis.
13. A 50-lb steel plate is supported through its center of gravity by a torsion rod 36 in. long and $\frac{1}{2}$ in. in diameter. The period of the pendulum when twisted and released is 2 sec. (a) Calculate the torque constant if the modulus of rigidity is 11.4×10^6 lb-ft/in.^{2}. (b) Calculate the moment of inertia of the plate. (c) If the angle of twist is 30°, calculate the torque on the rod.
14. Calculate the torque required to produce an angular acceleration of 1.5 rad/sec^2 if the moment of inertia is 100 ft-lb-sec^2.
15. Given a 2500-lb flywheel which has a 24-in. diameter and a radius of gyration of 3 ft, calculate the torque required to increase its speed from 60 rpm to 120 rpm in 10 sec.
16. A 1600-lb uniform circular disk has a diameter of 5 ft. It is accelerated about its central axis from rest by a 150-lb force acting at the largest radius. Calculate (a) the moment of inertia; (b) the torque; (c) the angular acceleration; (d) the angular velocity after 45 sec; (e) the rpm after 5 sec.
17. A 1600-lb drum has a diameter of 2 ft. A rope is wrapped around its circumference with an object fastened to the loose end of the rope. As the drum revolves, the rope unwinds 3200 ft in 8 min. Calculate (a) the final linear velocity and acceleration of the object; (b) the force applied by the object to the rope.
18. A rope 5 ft long is tied to a 10-lb weight. If the rope cannot support a force of more than 25 lb, what is the tangential velocity of the stone when the rope breaks?
19. A 3200-lb car is traveling at 45 mph around a curve which has a radius of 200 ft. (a) What is the centripetal force holding the car on the road? (b) What should the lateral angle of the road be to keep the car from sliding off the road?
20. In Prob. 19, what must be the coefficient of friction holding the car on a flat road?

11

Work, Energy, Power

11.1. Work

Work is a concept which has a special and specific meaning in engineering technology. The popular notion of work is associated with physical effort and fatigue. The scientific concept of work is completely disassociated from fatigue. It is simply the product of force multiplied by the displacement (distance) through which the effective force acts—and the action must be in the direction of the displacement force. Remember that torque was defined as force times the perpendicular distance to the line of action of the force. Although motion must be present to have work, it is not necessary for motion to be present in order to have torque (see Fig. 11.1).

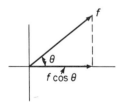

Fig. 11.1

The general equation for work is

$$\text{wk} = fs \cos \theta$$

$$\begin{cases} \text{wk} = \text{work} \\ f = \text{force} \\ s = \text{distance} \\ \theta = \text{angle between the distance moved and the action line of force} \end{cases}$$

If the force unit is the pound and the distance unit the foot, the unit of work is the *foot-pound* (ft-lb). If the force unit is the dyne and the distance

136   Work, Energy, Power   Chap. 11

unit the centimeter, the unit of work is the dyne-centimeter, or *erg*. In the mks system, if the force unit is the newton (nt) and the distance is the meter, the work unit is the newton-meter (nt-m), or *joule*.

Example 1

A box is pushed with a horizontal force of 50 lb along a floor for a distance of 20 ft. Calculate the work done while pushing the box.

Solution

$$\text{wk} = f \times s = 50 \text{ lb} \times 20 \text{ ft} \qquad \left\{ \begin{array}{l} f = 50 \text{ lb} \\ s = 20 \text{ ft} \end{array} \right\}$$
$$= 1000 \text{ ft-lb}$$

Example 2

The box in Example 1 moves along the floor while being pulled at an angle of 30° with the floor. Calculate the work done on the box.

Solution

$$\text{wk} = fs \cos \theta = 50 \text{ lb} \times 20 \text{ ft} \times \cos 30° \qquad \left\{ \begin{array}{l} f = 50 \text{ lb} \\ s = 20 \text{ ft} \\ \theta = 30° \end{array} \right\}$$
$$= 866 \text{ ft-lb}$$

Example 3

Calculate the work done on the box in Example 1 if the 50-lb force pulls straight up ($\theta = 90°$) but there is no movement.

Solution

Note that since there is no movement of the box, there can be no work. Thus

$$\text{wk} = fs \cos \theta = 50 \text{ lb} \times 0 \text{ lb} \cos 90° \qquad \left\{ \begin{array}{l} f = 50 \text{ lb} \\ s = 0 \text{ ft} \\ \theta = 90° \end{array} \right\}$$
$$= 0 \text{ ft-lb}$$

Example 4

A 24-in.-diameter handwheel is rotated by exerting a force of 30 lb at the outer rim. (1) Calculate the work done when the wheel is turned through $\frac{1}{2}$ revolution. (2) Calculate the torque exerted.

Solution

1. One revolution of the handwheel yields a distance of

$$\pi D = \pi \times 2 = 6.28 \text{ ft} \qquad \left\{ \begin{array}{l} d = \frac{24}{12} = 2 \text{ ft} \\ f = 30 \text{ lb} \\ s = \frac{1}{2} \text{ rev} \end{array} \right\}$$

The distance moved in $\frac{1}{2}$ rev is

$$s = \frac{6.28}{2} = 3.14 \text{ ft}$$

The work expended is the force times the displacement in the direction of f:

$$\text{wk} = fs = 30 \text{ lb} \times 3.14 \text{ ft}$$
$$= 94.2 \text{ ft-lb}$$

2. The torque is the force times the perpendicular distance:

$$L = fr = 30 \times 1 = 30 \text{ ft-lb} \quad \text{where} \quad r = \frac{12}{12} = 1 \text{ ft}$$

The angular units for work may be deduced by substituting L for f and θ for s.

$$\text{wk} = fs \quad \text{becomes} \quad \text{wk} = L\theta \qquad \left\{ \begin{array}{l} \text{wk} = \text{ft-lb} \\ L = \text{torque: ft-lb} \\ \theta = \text{rad} \end{array} \right\}$$

Example 5

In Example 4, the torque was found to be 30 ft-lb. In angular units,

$$\text{wk} = L\theta = 30 \times \pi \qquad \left\{ \begin{array}{l} L = 30 \text{ ft-lb} \\ \theta = \frac{1}{2} \text{ rev} = \pi \text{ rad} \end{array} \right\}$$
$$= 94.2 \text{ ft-lb} \quad (\textit{see Example } 4\text{a})$$

11.2. Energy: Potential and Kinetic

Energy may be defined as the capacity to do work. It may be found in many forms or converted from one form into another. The total amount of energy before conversion must always equal the total amount after conversion. Sometimes it is not easy to collect all the energy after conversion. Some of the possible conversions follow: a motor converts electrical energy into mechanical energy; the steam engine converts heat energy into mechanical energy; a battery converts chemical energy into electrical energy; a waterwheel converts mechanical energy into electrical energy; a windmill converts one form of mechanical energy into another form of mechanical energy.

The total amount of energy is always the same. The law of conservation of energy states that energy can be neither created nor destroyed. In the nuclear reaction, there is an interchange of mass and energy. Thus the law may be stated: "The sum total of energy and *mass* can neither be created nor destroyed."

Nevertheless, we do distinguish between two types of mechanical energy. The energy of position, we call *potential energy* (PE); the energy of motion, we call *kinetic energy* (KE).

In Fig. 11.2, if the ball, position *a*, is held a distance h' over the table, it has the potential to strike the table with a force due to its position and mass. This potential exists even though the ball may never be dropped. If the same ball be moved to position *b*, so that its height is now *h*, the ball will have a potential striking power with reference to the floor. This energy may be considered as stored energy, due entirely to its position with respect to some reference plane. (The mass remains constant throughout this entire discussion.) Notice that the ball if dropped has the *capacity* to do work, since the weight (force) will travel a distance *h* when released.

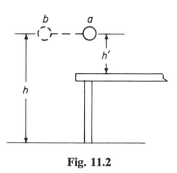

Fig. 11.2

$$PE = wh \qquad \left\{ \begin{array}{l} PE = \text{ft-lb} \\ w = \text{weight: lb} \\ h = \text{height: ft} \end{array} \right\}$$

Once the ball is released, the position, distance *h*, keeps changing; i.e., it gets less and less. Since the mass is constant, and potential energy $= fs = wh$, the potential energy must be getting less and less. Since the law of conservation of energy dictates that the sum total of energy cannot change and since the potential energy is decreasing as the distance between the reference plane and the ball is decressing, the potential energy "lost" must be present in some other form. By far the largest amount of this lost potential energy must be because of the motion of the ball. The static energy of position has been converted into the dynamic energy of motion, i.e., kinetic energy.

The work required to bring the body to rest depends upon the velocity acquired by the object. The greater the velocity, the more work expended in stopping this velocity (and mass). Or, we may consider the work required to accelerate a body from rest to its final velocity. In either case, the work required to stop or accelerate an object may be considered the kinetic energy of the body. Therefore, the equations below follow from the preceding discussion:

1. Kinetic energy,

$$KE = wk = fs$$

2. From Newton's second law,

$$f = ma$$

Sec. 11.2   *Energy: Potential and Kinetic*   139

3. Accelerating the object from rest, or stopping an object in motion,

$$v_2^2 - v_1^2 = 2as$$

$$v_2^2 = 2as \quad (\text{where } v_1 = 0)$$

$$s = \frac{v^2}{2a} \quad \begin{array}{l}(\textit{subscripts may be dropped; one of the velocities,} \\ \textit{in either case, must be zero})\end{array}$$

4. Therefore,

$$KE = fs = ma \times \frac{v^2}{2a}$$

$$= \frac{1}{2}mv^2 = \frac{1}{2}\frac{w}{g}v^2$$

$\left\{\begin{array}{l} KE = \text{kinetic energy: ft-lb} \\ f = \text{force exerted in accelerating} \\ \quad \text{a body to } v \text{ or stopping it} \\ \quad \text{after it acquires a velocity } v \\ s = \text{distance traveled} \\ wk = \text{work as result of KE} \\ a = \text{acceleration} \\ g = 32 \text{ ft/sec}^2 \\ w = \text{weight of moving object} \\ m = \text{mass of object} \\ v = \text{velocity} \end{array}\right\}$

Example 6

A 3200-lb automobile has a velocity of 30 mph. Calculate the force required to stop the automobile in a 400-ft distance.

Solution

1. Velocity:

$$v = 30\frac{\text{mi}}{\text{hr}} \times 5280 \frac{\text{ft}}{\text{mi}} \times \frac{1 \text{ hr}}{3600 \text{ sec}} \quad \left\{\begin{array}{l} v = 30 \text{ mph} \\ w = 3200 \text{ lb} \\ s = 400 \text{ ft} \end{array}\right\}$$

$$= 44 \text{ ft/sec}$$

2. Kinetic energy:

$$KE = \frac{1}{2}\frac{w}{g}v^2 = \frac{1}{2} \times \frac{3200}{32} \times 44^2$$

$$= 96{,}800 \text{ ft-lb}$$

3. Force required to stop the car:

$$KE = fs$$

$$f = \frac{KE}{s} = \frac{96{,}800}{400}$$

$$= 242 \text{ lb}$$

Consider next an object in the form of a pendulum. In Fig. 11.3, with the pendulum in position *a*, all the energy is of position and is potential energy. Upon releasing the object, the potential energy changes to kinetic energy, because the distance *s* decreases from a maximum to zero, while the velocity is increasing from zero to a maximum. Then, as the object starts to climb from position *b* to position *c*, the distance *s* increases from zero to a maximum as the velocity decreases from a maximum to zero at *c*.

Therefore, at *a*, the potential energy is a maximum and the kinetic energy is a minimum (zero). At *b*, the potential energy is zero, and the kinetic energy is a maximum. At *c*, the potential energy is again a maximum, and the kinetic energy is zero.

Similarly, the angular, or rotational kinetic energy equals

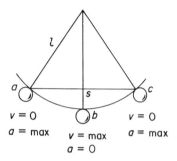

Fig. 11.3

$$\mathrm{KE} = \frac{I\omega^2}{2} \text{ ft-lb} \quad \begin{Bmatrix} I = \text{moment of inertia:} \\ \text{ft-lb-sec}^2 \\ \omega = \text{angular velocity:} \\ \text{rad/sec} \end{Bmatrix}$$

An object rolling down an incline has both kinetic energy of rotation and kinetic energy of translation. It is also true that when the object is at the top of the incline, its kinetic energy is zero and its potential energy is a maximum. At the bottom of the incline, its potential energy is zero and its kinetic energy (of translation and rotation) is a maximum (see Fig. 11.4).

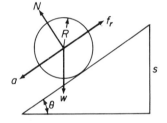

Fig. 11.4

$$\mathrm{KE} = \tfrac{1}{2}I\omega^2 + \tfrac{1}{2}mv^2$$

11.3. Power

Power is the time-rate at which energy is delivered or performed. When delivered at the rate of 550 ft-lb/sec, or 33,000 ft-lb/min, the power is equal to 1 *horsepower* (hp). Also, 0.737 ft-lb/sec equals 1 *watt*. Notice that the horsepower delivered may be controlled in two ways; if the work is held constant, the horsepower delivered may be increased by decreasing the time factor; if the time is held constant, the work delivered must be increased to increase the horsepower.

Example 7

What horsepower is needed to raise an object 50 ft in 30 sec if the object weighs 1100 lb?

Sec. 11.3   Power   141

Solution

1. Work:
$$wh = 1100 \times 50 = 55{,}000 \text{ ft-lb}$$

2. Power:
$$P = \frac{wh}{t} = \frac{55{,}000 \text{ ft-lb}}{30 \text{ sec}} = 1833 \text{ ft-lb/sec}$$

3. Horsepower:
$$\text{hp} = \frac{1833}{550} = 3.3 \text{ hp}$$

11.4. The Prony Brake

The *prony brake* is a device for measuring the power delivered by a pulley or shaft. In Fig. 11.5(a), f' is the friction force opposing the force f applied to the lever. Since R does not move, all the energy is converted into heat energy as a result of the friction. In Fig. 11.5(a):

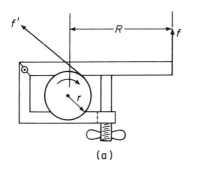
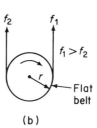

Fig. 11.5

1. The circumference of the pulley $= 2\pi r$;
2. The work necessary to overcome the friction for 1 rev of the pulley is
$$\text{wk} = fs = f' \times 2\pi r$$
3. The work per minute is
$$\text{wk} = 2\pi r f' N$$
4. The horsepower is
$$\text{hp} = \frac{2\pi r f' N}{33{,}000}$$

$\left\{\begin{array}{l} f' = \text{friction force: lb} \\ f = \text{force on lever: lb} \\ r = \text{radius of pulley: ft} \\ R = \text{length of lever: ft} \\ \text{wk} = \text{work: ft-lb} \\ N = \text{span of pulley} \end{array}\right\}$

142　　　　　　　　　　　Work, Energy, Power　　　　　　　　　　　Chap. 11

5. If the torque is fR, and if $fR = f'r$, then

$$hp = \frac{2\pi R f N}{33,000}$$

Sometimes a flat belt is wrapped around a shaft or pulley as shown in Fig. 11.5(b). If the shaft turns clockwise, f_1 will be greater than f_2. The difference between f_1 and f_2 is the friction force. This friction force f' times the circumference of the pulley will yield the work done by the machine in overcoming friction. Therefore, $f_1 - f_2$ may be substituted for f':

$$hp = \frac{2\pi r f' N}{33,000} = \frac{2\pi r \dot{N}(f_1 - f_2)}{33,000}$$

Example 8

A pulley, 30 in. in diameter and operating at 200 rpm, drives a flat belt. If the net pull by the pulley is 400 lb, what horsepower is delivered to the belt?

Solution

1. Circumference of the pulley:

$$\pi D = \pi \times 2.5 \text{ ft} = 7.85 \text{ ft}$$

2. Work required for every revolution of the pulley:

$$7.85 \times 400 = 3140 \text{ ft-lb}$$

3. Work for every minute of rotation:

$$3140 \times 200 = 628,000 \text{ ft-lb/min}$$

4. Horsepower:

$$hp = \frac{628,000}{33,000} = 19$$

11.5. Mechanical Energy: Efficiency

A *machine* is any device which takes in energy from some outside source and delivers it to some other point to do work.

The *mechanical advantage* (MA) of a machine is the ratio between the force exerted by the machine and the force put into the machine. For instance, if an automobile is being raised by a jack, the force applied to the jack handle is the input force; the force exerted by the jack on the automobile is the output force. Therefore the mechanical advantage of the jack, as a machine, is the ratio of output force to input force.

$$MA = \frac{\text{force input}}{\text{force output}} = \frac{f_o}{f_i}$$

The ratio of the distance moved by the input force during a time interval to the distance moved by the output force during the same time interval is called the *displacement ratio* (DR). The displacement ratio may also be obtained from the ratio of velocity input to velocity output.

$$\text{DR} = \frac{\text{distance input}}{\text{distance output}} = \frac{s_i}{s_o}$$

$$\text{DR} = \frac{\text{velocity input}}{\text{velocity output}} = \frac{v_i}{v_o}$$

It should be made very clear that because of losses (friction, etc.), the mechanical advantage of the system will be affected, so that any losses will be reflected in the MA ratio. These losses, however, are not reflected in the ratio of distances (DR) because, no matter how many losses there are, the distances will always remain the same. For this reason, many texts prefer to call the displacement ratio the *ideal mechanical advantage*, and the ratio of forces the *actual mechanical advantage*.

The *efficiency* (eff) of a machine is the ratio of energy output to energy input:

$$\text{eff} = \frac{\text{energy out}}{\text{energy in}} = \frac{\text{force} \times \text{distance out}}{\text{force} \times \text{distance in}} = \frac{f_o s_o}{f_i s_i}$$

Since

$$\frac{f_o}{f_i} = \text{MA} \quad \text{and} \quad \frac{s_o}{s_i} = \frac{1}{\text{DR}}$$

therefore,

$$\text{eff} = \text{MA} \times \frac{1}{\text{DR}}$$

The law of conservation of energy dictates that efficiency of any machine can never be greater than 100%. This is so because the MA ratio reflects losses and will always be less than the DR ratio. Therefore, in the foregoing efficiency equation, the MA will always be less than the DR and the efficiency must be less than 100%.

11.6. The Inclined Plane

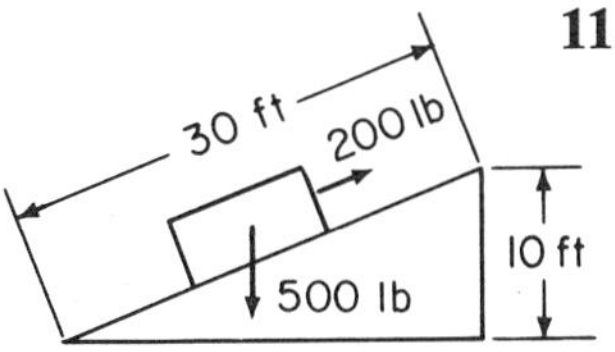

Fig. 11.6

Example 9

A conveyor belt 30 ft long is used to raise boxes into a second-story window of a warehouse. The window is 10 ft above the ground. The force required to raise a 500-lb box is 200 lb, as shown in Fig. 11.6. Calculate (1) the work input; (2) the work output; (3) mechanical advantage; (4) displacement ratio; (5) efficiency.

Solution

1. Work input:

$$\text{wk}_i = f_i s_i = 200 \times 30 = 6000 \text{ ft-lb}$$

2. Work output:

$$\text{wk}_o = f_o s_o = 500 \times 10 = 5000 \text{ ft-lb}$$

$\left\{ \begin{array}{l} s_i = 30 \text{ ft} \\ s_o = 10 \text{ ft} \\ f_i = 200 \text{ lb} \\ f_o = 500 \text{ lb} \end{array} \right\}$

3. Mechanical advantage:

$$\text{MA} = \frac{f_o}{f_i} = \frac{500}{200} = 2.5$$

4. Displacement ratio:

$$\text{DR} = \frac{s_i}{s_o} = \frac{30}{10} = 3$$

5. Efficiency:

$$\text{eff} = \frac{\text{MA}}{\text{DR}} = \frac{2.5}{3} \times 100 = 83.3\%$$

11.7. The Screw Thread

A *screw thread* is actually an inclined plane wrapped around a cylinder, as shown in Fig. 11.7(a).

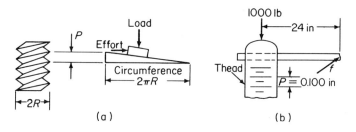

Fig. 11.7

The thread, as a machine, raises the load vertically and parallel to the centerline of the thread, a distance equal to the pitch P for every revolution. The radius is taken at the pitchline and is therefore equal to half the average pitch diameter. The turning effect is applied at right angles to the direction of the load displacement.

Sec. 11.8   *The Block and Tackle*   **145**

Example 10

A lead screw, Fig. 11.7(b), has 10 single threads per inch. A lever arm is fastened to one end of the screw. If the efficiency is 30%, find (1) the displacement ratio; (2) the MA; (3) the effort required to raise a 1000-lb load.

Solution

1. The displacement ratio is

$$\text{DR} = \frac{s_i}{s_o} = \frac{2\pi(24)}{0.100} \qquad \left\{ \begin{array}{l} s_i = 2\pi\,(24)\text{ in.} \\ s_o = 0.100\text{ in.} \end{array} \right\}$$

$$= 1507$$

2. The mechanical advantage is

$$\text{MA} = \text{DR} \times \text{eff} = 1507 \times 0.30$$
$$= 452$$

3. From $\text{MA} = f_o/f_i$, the force input is

$$f_i = \frac{f_o}{\text{MA}} = \frac{1000}{452} = 2.2\text{ lb}$$

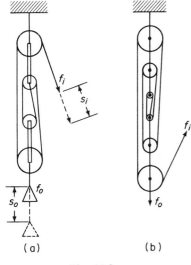

Fig. 11.8

11.8. The Block and Tackle

A *block and tackle* (Fig. 11.8a and b) is a machine which permits moving loads short distances by applying light loads moving longer distances. If load f_o moves $s_o = 1$ in., each of the strands supporting f_o must move 1 in. Since the system in Fig. 11.8(a) has four supporting strands, each of the strands must shorten (if load f_o moves up) by 1 in. Thus the total movement s_i must be 4 in. if all of the slack is to be removed. If the work output equals the work input ($f_o s_o = f_i s_i$) and s_i moves 4 times s_o, then the force f_i must be $\frac{1}{4}$ the force output (neglecting losses).

Example 11

Given the block and tackle shown in Fig. 11.8(b), if it takes 6 lb to raise a 39-lb object, calculate (1) the MA; (2) the DR; (3) the efficiency.

Solution

1. The mechanical advantage is

$$\mathrm{MA} = \frac{f_o}{f_i} = \frac{39}{6} \qquad \left\{ \begin{array}{l} f_o = 39\ \mathrm{lb} \\ f_i = 6\ \mathrm{lb} \end{array} \right\}$$

$$= 6.5$$

2. The displacement ratio is

$$\mathrm{DR} = 7 \quad (\textit{seven supporting cords})$$

3. The efficiency is

$$\mathrm{eff} = \frac{\mathrm{MA}}{\mathrm{DR}} = \frac{6.5}{7} \times 100$$

$$= 92.85\%$$

11.9. The Chain Hoist

The *chain hoist* (Fig. 11.9) differs from the block and tackle in that in the former the chain is continuous. The two upper pulleys are of different diameters but keyed (fastened) to the same shaft. So, even though the diameters are different, the pulleys revolve at the same rpm. When force f_i is applied, chain length a winds up on the large pulley, while chain length b unwinds from the small pulley. The resulting effect is to raise the load with a large MA.

Assume that f_i moves in the direction indicated. If the two upper pulleys rotate 1 rev, chain b moves (unwinds) from the small pulley a distance

$$2\pi r = \text{unwound length}$$

At the same time, side a is winding up on the large pulley a distance

$$2\pi R = \text{wound up length}$$

Therefore, the resulting shortening of the loop ab is

$$2\pi R - 2\pi r$$

But load f_o will move 1 in. while the *two supporting sides a* and b shorten 1 in. each, or a total of 2 in. Therefore, the distance the load moves is

$$s_o = \tfrac{1}{2}(2\pi R - 2\pi r)$$

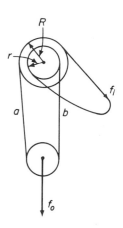

Fig. 11.9

Sec. 11.10   *Friction in Machines*   **147**

The distance s_i, moved by the effort force f_i, is

$$s_i = 2\pi R$$

The displacement ratio is

$$\text{DR} = \frac{s_i}{s_o} = \frac{2\pi R}{\frac{1}{2}(2\pi R - 2\pi r)} = \frac{2R}{R - r} \quad \left\{ \begin{array}{l} R = \text{radius—large pulley} \\ r = \text{radius—small pulley} \end{array} \right\}$$

The mechanical advantage is

$$\text{MA} = \frac{f_o}{f_i}$$

The efficiency is

$$\text{eff} = \frac{\text{MA}}{\text{DR}} = \frac{f_o(R - r)}{f_i(2R)}$$

11.10. Friction in Machines

A *reversible machine* in one which will reverse its direction of displacement as soon as the applied force causing the displacement is released. A *non-reversible machine* is one which will *not* reverse its direction of displacement when the load is removed. This is owing to the holding power of the friction present. Let

f_1 = force necessary to *raise* a load with uniform motion

f_2 = force necessary to *lower* a load with uniform motion

fr = force necessary to overcome friction

f = balancing load

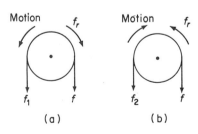

Fig. 11.10

1. To overcome the load and friction, Fig. 11.10(a),

$$f_1 = f + fr$$

2. When friction aids the input force as shown in Fig. 11.10(b),

$$f_2 = f - fr$$

3. Subtracting,

$$f_1 - f_2 = (f + fr) - (f - fr) = 2fr$$

4. Reversible machines:

$$fr = \frac{f_1 - f_2}{2}$$

5. Nonreversible machines:

$$fr = \frac{f_1 - (-f_2)}{2} = \frac{f_1 + f_2}{2}$$

Problems

1. A 3-ton beam is raised a distance of 300 ft. Calculate the work done on the beam.
2. Calculate the work required to move a 500-lb object along a level plane a distance of 40 ft if the cable tied to the object makes an angle of 20° with the plane.
3. Calculate the work required to move the 500-lb object in Prob. 2 a distance of 40 ft up a 20° plane. The cable is parallel to the plane.
4. A tank if filled from the top by a pump outside and at the base of the tank. The tank is 140 ft high and holds 600 ft^3 of water. The weight of 1 ft^3 of water is 62.4 lb. How much work is required to fill the tank?
5. A tank holds fuel oil (sp wt* = 60.14 lb/ft^3) and is 25 ft high. The tank is 16 ft in diameter. Calculate the work required to fill the tank from the top.
6. A locomotive increases its velocity from 15 mph to 45 mph on a level stretch of track. If the work required to increase its velocity is 5×10^6 ft-lb, calculate the weight of the locomotive.
7. An automobile weighs 3600 lb and can carry five passengers. (a) If the automobile is traveling at the rate of 20 mph and the average weight of each passenger is 180 lb, what is the kinetic energy of the system? (b) What is the kinetic energy of the system if the automobile increases its velocity to 60 mph?
8. A large forging press drops its ram of 15 tons a height of 5 ft once every 15 sec. If it takes $6 = 10^6$ ft-lb of kinetic energy to complete one forging, how long will it take to complete this forging?
9. A 2000-lb object is being slid up a 60° incline by a cable over a pulley. A 500-lb weight is fastened to the other end of the cable which hangs vertically down. If the 500-lb weight moves 30 ft, calculate the change in potential energy for the system.
10. A 20-lb object is hung on a 4-ft cord as a pendulum. If the object is raised 10° from the center position and released, calculate its change potential energy for a half-cycle.
11. A steel ball, 2 ft in diameter, is perched at the top of a 30° inclined plane. The length of the inclined plane is 60 ft long. Calculate (a) the potential energy of the ball; (b) the linear velocity of the ball when it reaches the bottom of the incline; (c) the total kinetic energy at the bottom of the slope. (The specific weight of steel is 488 lb/ft^3.)
12. Calculate the KE of the ball in Prob. 11 if it rolls over level ground with a velocity of 3 fps.
13. A 400-lb object is raised 50 ft in 3 min. Calculate the horsepower requirement.
14. Calculate the hp required to lift 15,000 gallons of water a height of 150 ft in 2 hr. There are 8.28 lb per gallon of water.

---

* Specific weight represents the weight per unit volume of a substance. See Chapter 13.

Problems

15. A constant drawbar pull of 60,000 lb is exerted by a locomotive on a train when increasing its speed from 20 mph to 60 mph. Calculate the hp (a) at 60 mph; (b) at 20 mph.
16. A train and all its cars weighs 900 tons. The force exerted on the train is 4000 lb. (a) Calculate the horsepower needed for the locomotive to maintain a constant speed of 30 mph. (b) Calculate the velocity of the train when the hp requirements reach 500 hp.
17. A motor, a pulley, and a clamping device are hooked up as a prony brake. The lever arm used to balance the pull of the motor is 3 ft long. A spring balance attached to the end of the lever arm shows an average pull of 15 lb at the same time that a tachometer registers 1040 rpm. What hp does the motor deliver?
18. A hand punch is used to punch holes in a $\frac{1}{4}$-in. metal plate. An operator moves the handle 18 in. and exerts a force of 15 lb on the handle. The average force required at the punch is 1000 lb. Calculate (a) the output energy; (b) the input energy; (c) the displacement ratio; (d) the mechanical advantage; (e) the efficiency.
19. A winch is used to load pianos into a truck by pulling them up an inclined plane. The back of the truck is 4 ft off the ground, the conveyor is 12 ft long, and the pianos weigh 600 lb each. If the winch exerts a 220-lb pull, find (a) work input; (b) work output; (c) the mechanical advantage; (d) the displacement ratio; (e) the efficiency.
20. (a) Calculate the hp input, in Prob. 19, to load 5 pianos if it takes 30 min to load one piano. (b) Calculate the hp output.
21. A jackscrew used to lift 1600 lb has 8 threads to the inch. The lever arm used to apply the necessary operational forces is 15 in. long. If the efficiency of the system is 40%, what force should be applied perpendicular to the lever arm?
22. Given a system of pulleys as shown in Fig. 11.11. (a) What is the displacement ratio of the system? (b) Neglecting friction, find the force required to raise 3600 lb.
23. To raise a 50-lb object with an applied force of 8 lb using the two pulley systems in Fig. 11.12 (a) and (b), calculate: (a) the displacement ratio; (b) the mechanical advantage; (c) the efficiency.

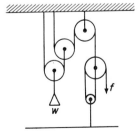

Fig. 11.11

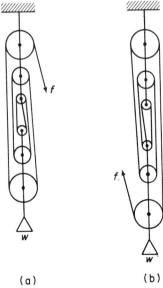

Fig. 11.12

24. Figure 11.9 shows a chain hoist which has two pulleys ($R = 12$ in.; $r = 8$ in.), as indicated and keyed to each other. If it takes a 60-lb force to raise a 350-lb object, calculate (a) the input distance moved by the input force per revolution; (b) the output distance moved by the output force per revolution; (c) the displacement ratio; (d) the mechanical advantage; (e) the efficiency of the system.
25. A lead screw of a vise has a $\frac{1}{4}$-in. pitch screw. Through one end of the lead screw is a lever which is 9 in. long. The other end of the lead screw actuates the jaws. If a 25-lb force is applied to the lever, calculate the force at the jaws (eff = 30%).

12

Momentum, Impact

12.1. Momentum and Impulse

Momentum is a concept involving mass and energy. Every object which has motion must also possess momentum. Anyone attempting to stop an object moving with a velocity v soon discovers that the object possesses momentum. This momentum results from an acceleration of the object, and may be written as

$$\text{momentum} = mv = \frac{w}{g}v \qquad \left\{ \begin{array}{l} m = \text{mass: slugs} \\ w = \text{weight: lb} \\ g = 32 \text{ ft/sec}^2 \\ v = \text{velocity: ft/sec} \end{array} \right\}$$

This equation should not be confused with the kinetic-energy equation, which is $KE = \frac{1}{2} mv^2$.

Example 1

Calculate the kinetic energy and the momentum of a 4-lb object which is thrown with a velocity of 30 mph just prior to impact.

Solution

1. The kinetic energy is

$$KE = \frac{1}{2}\left(\frac{w}{g}\right)v^2 = \frac{1}{2}\left(\frac{4}{32}\right)44^2 \qquad \left\{ \begin{array}{l} w = 4 \text{ lb} \\ v = 30 \text{ mph} \\ = 49 \text{ ft/sec} \end{array} \right\}$$
$$= 121 \text{ ft-lb}$$

2. The momentum is

$$\text{momentum} = \frac{w}{g}v = \frac{4 \text{ lb}}{32 \text{ ft/sec}^2} \times 44 \text{ ft/sec}$$

$$= 5.5 \text{ lb-sec}$$

According to Newton's second law, "the force may be considered as the time rate of change of momentum":

$$f = \frac{\Delta(mv)}{\Delta t}$$

If the mass m is a constant, the foregoing equation becomes

$$f = m\frac{\Delta v}{\Delta t}$$

But $\Delta v/\Delta t$ is acceleration, so

$$f = m\frac{\Delta v}{\Delta t} = ma$$

Note that a force acting through an interval of time—called the *impulse of a force*—will produce a change in momentum such that from

$$f = \frac{\Delta(mv)}{\Delta t}$$

impulse is equal to

$$f \Delta t = \Delta (mv) = mv_f - mv_i$$

$$\left\{ \begin{array}{l} m = \text{mass} \\ v_f = \text{final velocity} \\ v_i = \text{initial velocity} \\ t = \text{change in time} \\ f = \text{accelerating force} \end{array} \right\}$$

Example 2

Calculate the impulse which the object in Example 1 can exert in coming to rest.

Solution

The change in momentum $\Delta(mv)$ would represent 5.5 lb-sec at 44 ft/sec to 0 lb-sec at rest. Therefore, the impulse is

$$f \Delta t = \Delta(mv) = 5.5 \text{ lb-sec}$$

Example 3

If the object in Example 2 comes to rest in 18 ft, calculate the stopping force which was exerted on the object.

Solution

1. The average velocity is

$$v_{av} = \frac{v_f + v_i}{2} = \frac{0 + 44}{2} = 22 \text{ ft/sec} \qquad \left\{ \begin{array}{l} s = 18 \text{ ft} \\ v_i = 44 \text{ ft/sec} \\ v_f = 0 \text{ ft/sec} \end{array} \right\}$$

Sec. 12.2   Impact: Elastic and Inelastic

2. The time to decelerate to rest is

$$t = \frac{s}{v_{av}} = \frac{18}{22} = 0.82 \text{ sec}$$

3. The stopping force is

$$f = \frac{mv}{t} = \frac{w}{g}\left(\frac{v}{t}\right) = \frac{4}{32}\left(\frac{22}{0.82}\right)$$
$$= 3.35 \text{ lb}$$

12.2. Impact: Elastic and Inelastic

Since velocity is a vector quantity, momentum is a vector quantity, and as such, momentum has magnitude acting through a given displacement. The direction of the change in momentum is in the direction of the force creating the momentum, or in the direction of the impulse.

Newton's third law of motion deals with action and reaction forces being equal and opposite in direction. Thus

$$f_1 = f_2$$

If f_1 and f_2 act during the same time interval,

$$f_1 \Delta t = f_2 \Delta t$$

Since $f \Delta t$ is impulse,

$$\Delta(mv)_1 = \Delta(mv)_2$$

The law of conservation of momentum states: "In a closed system, the total momentum of the system remains unchanged by anything which takes place within the system." Therefore, in the equation above, any loss by $\Delta(mv)_1$ must be compensated for with a corresponding gain by $\Delta(mv)_2$.

Example 4

A gun fires a 2-oz bullet with a speed of 2400 ft/sec. If the gun weighs 6 lb, calculate the recoil velocity.

Solution

Momentum:

$$m_1 v_1 = m_2 v_2$$
$$\frac{w_1}{g} v_1 = \frac{w_2}{g} v_2$$
$$w_1 v_1 = w_2 v_2$$

$\left\{\begin{array}{l} m_1 = \text{mass—bullet} \\ m_2 = \text{mass—gun} \\ v_1 = \text{velocity—bullet} \\ v_2 = \text{velocity—gun} \\ w_1 = \frac{2}{16} \text{ lb} \\ w_2 = 6 \text{ lb} \end{array}\right\}$

Solving for v_2,

$$v_2 = \frac{w_1}{w_2} v_1 = \frac{2 \times 2400}{16 \times 6}$$

$$= 50 \text{ ft/sec}$$

In all cases of impact, the concept of the conservation of momentum applies. That is, the total momentum of a system remains unchanged. The coefficient of restitution e is a proportionality constant which relates momentum before with momentum after impact between two objects. This proportionality constant may vary between two ideal values. If the coefficient of restitution is equal to one ($e = 1$), the impact between the two bodies is said to be *perfectly elastic*. If the coefficient of restitution is equal to zero ($e = 0$), the impact is said to be *perfectly inelastic*.

Since kinetic energy is *not* conserved for *inelastic impact*, the relative velocities before and after impact are proportional to each other. *All sign conventions apply to the velocities.* Thus in Fig. 12.1(a),

$$v_1 - v_2 \propto V_2 - V_1$$

and

solve for

$$e(v_1 - v_2) = V_2 - V_1$$

$$e = \frac{V_2 - V_1}{v_1 - v_2}$$

$$\left\{\begin{array}{l} \text{velocities before impact } (v): \\ v_1 = \text{velocity of } m_1 \text{: ft/sec} \\ v_2 = \text{velocity of } m_2 \text{:ft/sec} \\ \text{velocities after impact } (V): \\ V_1 = \text{velocity of } m_1 \text{: ft/sec} \\ V_2 = \text{velocity of } m_2 \text{: ft/sec} \end{array}\right\}$$

If the two bodies m_1 and m_2 are perfectly elastic, the standard sign convention may be applied, and the sum of the momentums before impact equals the sum of the momentums after impact, Fig. 12.1. The general equation is

$$m_1 v_1 + m_2 v_2 = m_1 V_1 + m_2 V_2$$

In Fig. 12.1(a), m_1 is moving to the right and v_1 carries a plus (+) sign. But m_2 acts toward the left and therefore v_2 carries a minus (−) sign. *After*

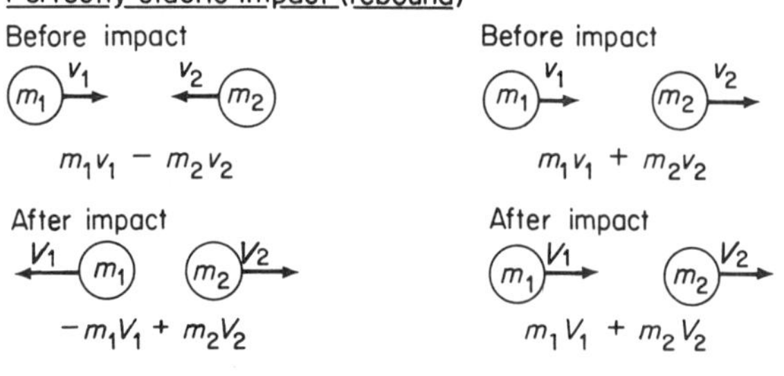

Perfectly elastic impact (rebound)

Before impact

$m_1 v_1 - m_2 v_2$

After impact

$-m_1 V_1 + m_2 V_2$

(a)

Before impact

$m_1 v_1 + m_2 v_2$

After impact

$m_1 V_1 + m_2 V_2$

(b)

Fig. 12.1

Sec. 12.2 Impact: Elastic and Inelastic

impact m_1 and m_2 reverse their directions of motion: m_1 is moving from right to left, and the sign of V_1 is minus ($-$); m_2 acts from left to right, and the sign of V_2 is plus ($+$). The preceding equation, therefore, becomes

$$m_1 v_1 + m_2(-v_2) = m_1(-V_1) + m_2 V_2$$
$$m_1 v_1 - m_2 v_2 = -m_1 V_1 + m_2 V_2$$

It is interesting to note that, if $m_1 = m_2$, the equation becomes

$$v_1 - v_2 = V_2 - V_1$$

Thus $v_1 - v_2$ is the relative velocity before impact and $V_2 - V_1$ is the relative velocity after impact. Since these two quantities are equal, the ratio equals one (1) and indicates perfect elasticity.*

For *elastic impact*, the sum of the kinetic energies is constant, since no mechanical energy is lost to the system. True, kinetic energy may be momentarily converted to potential energy, but it is immediately converted to kinetic energy again with the possibility of an interchange of energy between the two masses.

For *inelastic impact*, some of the kinetic energy may be converted to some other form of energy (heat, light, etc.). Therefore, it is possible that the total kinetic energy before impact will not equal the total kinetic energy after impact. But the sum of the momentums before impact always equals the sum of the momentums after impact, whether the impact is elastic or inelastic.

Elastic impact is shown in Fig. 12.1(a) with the two masses moving toward each other before impact and away from each other after impact.

Figure 12.1(b) shows the two masses moving in the same direction before impact and the same two bodies moving in the same direction after impact. Since both drawings show motion to the right, the sign of all the values for velocity is plus.

For perfectly inelastic impact, the direction before impact will be as shown

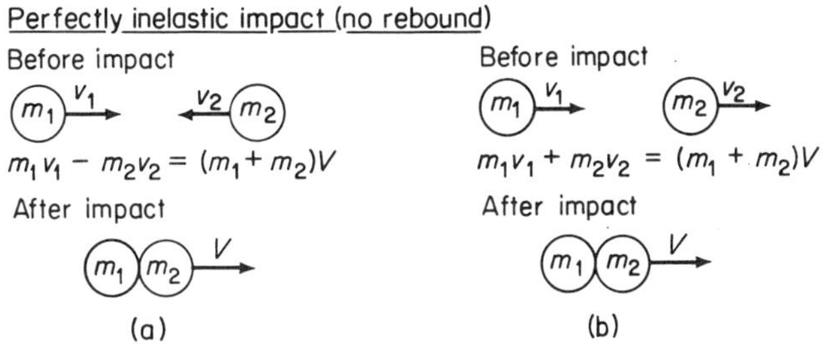

Perfectly inelastic impact (no rebound)

Before impact

$(m_1) \xrightarrow{v_1} \quad \xleftarrow{v_2} (m_2)$

$m_1 v_1 - m_2 v_2 = (m_1 + m_2) V$

After impact

$(m_1)(m_2) \xrightarrow{V}$

(a)

Before impact

$(m_1) \xrightarrow{v_1} \quad (m_2) \xrightarrow{v_2}$

$m_1 v_1 + m_2 v_2 = (m_1 + m_2) V$

After impact

$(m_1)(m_2) \xrightarrow{V}$

(b)

Fig. 12.2

---

* A more rigorous proof would be to start with the sum of the kinetic energies before impact equal to the sum of the kinetic energies after impact.

in Fig. 12.2. Note that, after impact, the directions of motion of both masses will be the same. Also note that if $e = 0$ for perfectly inelastic impact, the value of

$$e = \frac{V_2 - V_1}{v_1 - v_2}$$

becomes

$$0(v_1 - v_2) = V_2 - V_1 \quad \text{or} \quad V_2 = V_1 = V \quad \text{(see Fig. 12.2)}$$

Thus we find ourselves in the enviable position of having three equations which may be used in the solution of the problems in this part of the chapter:

1. The momentum equation:

$$m_1 v_1 + m_2 v_2 = m_1 V_1 + m_2 V_2$$

2. For perfectly elastic bodies,

$$\tfrac{1}{2} m_1 v_1^2 + \tfrac{1}{2} m_2 v_2^2 = \tfrac{1}{2} m_1 V_1^2 + \tfrac{1}{2} m_2 V_2^2$$

3. For any impact less than for perfectly elastic bodies,

$$e = \frac{V_2 - V_1}{v_1 - v_2}$$

Example 5

Two objects, 128 lb and 80 lb, have velocities of 10 ft/sec and 18 ft/sec, respectively. They are approaching each other when they collide with perfectly elastic impact. If you assume that they are moving away from each other after impact, calculate their velocities after impact (see Fig. 12.3).

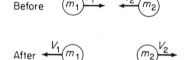

Fig. 12.3

Solution

1. The momentum equation is

$$m_1 v_1 + m_2 v_2 = m_1 V_1 + m_2 V_2$$

$$\frac{128}{32}(+10) + \frac{80}{32}(-18) = \frac{128}{32}(-V_1) + \frac{80}{32}(V_2)$$

$$\left\{ \begin{array}{l} m_1 = 128 \text{ lb} \\ m_2 = 80 \text{ lb} \\ v_1 = 10 \text{ ft/sec} \\ v_2 = 18 \text{ ft/sec} \end{array} \right\}$$

2. Solve and collect terms:

$$-5 = -4V_1 + 2.5 V_2$$

3. Since the collision is perfectly elastic, the kinetic energy equation may be used:

Sec. 12.2   *Impact: Elastic and Inelastic*   157

$$\tfrac{1}{2}m_1v_1^2 + \tfrac{1}{2}m_2v_2^2 = \tfrac{1}{2}m_1V_1^2 + \tfrac{1}{2}m_2V_2^2$$

$$\tfrac{1}{2}\left(\tfrac{128}{32}\right)(10)^2 + \tfrac{1}{2}\left(\tfrac{80}{32}\right)(18)^2 = \tfrac{1}{2}\left(\tfrac{128}{32}\right)V_1^2 + \tfrac{1}{2}\left(\tfrac{80}{32}\right)V_2^2$$

4. Solve and collect terms:

$$2420 = 8V_1^2 + 5V_2^2$$

5. Solve the former equation for V_1:

$$V_1 = \frac{5 + 2.5V_2}{4}$$

6. Substitute into the quadratic equation:

$$8\left(\frac{5 + 2.5V_2}{4}\right)^2 + 5V_2^2 = 2420$$

7. The solution to this equation leads to two roots. Only the positive root may be used:

$$V_2 = 16.5 \text{ ft/sec}$$

8. Substituting $V_2 = 16.3$ ft/sec into the linear equation and solving for V_1,

$$V_1 = \frac{5 + 2.5V_2}{4} = \frac{5 + 2.5(16.5)}{4}$$

$$= 11.6 \text{ ft/sec}$$

Note that the direction of v_1 and v_2 must have been assumed correctly, since V_1 and V_2 are both plus.

Example 6

Assume that the two masses in Example 5 are not perfectly elastic. Calculate the velocities after impact, if the coefficient of restitution is $e = 0.6$.

Solution

Here the kinetic energy equations cannot be used because the impact is not perfectly elastic and some of the kinetic energy will be lost at impact. However,

$$e = \frac{V_2 - V_1}{v_1 - v_2}$$

or

$$0.6 = \frac{V_2 - (-V_1)}{10 - (-18)} = \frac{V_2 + V_1}{28}$$

Therefore,

$$V_2 + V_1 = 16.8$$

From Example 5, the linear equation (momentum equation) is

$$2.5V_2 - 4V_1 = -5$$

Solving both equations simultaneously yields

$$V_1 = 7.2 \text{ ft/sec}$$
$$V_2 = 9.6 \text{ ft/sec}$$

Example 7

A bullet is fired into a stationary 8-lb wood block resting on a frictionless surface. The bullet weighs 2 oz and has a velocity of 1800 ft/sec. Calculate its velocity after impact (see Fig. 12.4).

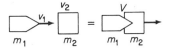

Fig. 12.4

Solution

Momentum before impact = momentum after impact:

$$m_1v_1 + m_2v_2 = m_1V_1 + m_2V_2 \qquad \begin{cases} m_1 = \frac{2}{16} = \frac{1}{8} \text{ lb} \\ m_2 = 8 \text{ lb} \\ v_1 = 1800 \text{ ft/sec} \\ v_2 = 0 \text{ ft/sec} \end{cases}$$

Note that the block has no velocity before impact and that the bullet and the block have the same velocity after impact, so that

$$V_1 = V_2 = V$$

Thus

$$\frac{\frac{1}{8}}{32}(1800) + \frac{8}{32}(0) = \frac{\frac{1}{8}}{32}V + \frac{8}{32}V$$

Solve for V:

$$V = 27.7 \text{ ft/sec}$$

It is interesting to compare the kinetic energies and note that in this instance—inelastic impact—they are *not* equal. Thus:

Kinetic energy *before* impact is

$$\text{KE} = \frac{1}{2}m_1v_1^2 + \frac{1}{2}m_2v_2^2 = \frac{1}{2}\left(\frac{\frac{1}{8}}{32}\right)(1800)^2 + \frac{1}{2}\left(\frac{8}{32}\right)(0)^2$$
$$= 6330 \text{ ft-lb}$$

Kinetic energy *after* impact is

$$\text{KE} = \frac{1}{2}m_1V_1^2 + \frac{1}{2}m_2V_2^2 = \frac{1}{2}\left(\frac{\frac{1}{8}}{32}\right)(27.7)^2 + \frac{1}{2}\left(\frac{8}{32}\right)(27.7)^2$$
$$= 97.4 \text{ ft-lb}$$

12.3. Angular Momentum

Angular momentum is the rotational equivalent of linear momentum. It has already been stated that moment of inertia I may replace mass m and that angular velocity ω may replace linear velocity v in the linear equation. The equation for angular momentum therefore becomes

$$\text{angular momentum} = I\omega \quad \begin{cases} I = \text{moment of inertia: ft-lb-sec}^2 \\ \omega = \text{angular velocity: rad/sec} \\ \text{angular momentum} = \text{ft-lb-sec} \end{cases}$$

Example 8

A 390-lb disk is 2 ft in diameter. What is its angular momentum when rotating at a speed of 1800 rpm?

Solution

Angular velocity:

$$\omega = \frac{1800}{60} \times 2\pi = 60\pi \text{ rad/sec}$$

Moment of inertia:

$$I = \frac{1}{2}\left(\frac{w}{g}\right)R^2 = \frac{1}{2}\left(\frac{390}{32}\right)(1)^2$$
$$= 6.09 \text{ ft-lb-sec}^2$$

Angular momentum:

$$I\omega = 6.09 \times 60\pi = 1147.4 \text{ ft-lb-sec}$$

Note that the angular momentum of a body will not change unless acted on by some external force. This is actually a restatement of the law of linear momentum. Therefore, the moment of inertia of a body will resist any change in rotary motion whether that body is trying to speed up or slow down. Any shift in the distribution of the the moment of inertia will change the angular velocity. It is also true that the axis of rotation will remain fixed and resist any attempt to change its orientation.

The moment of impulse may also be written by substituting the angular notations for the linear notations in the impulse equation $f \, \Delta t = \Delta(mv)$:

$$T \, \Delta t = \Delta(I\omega) \quad \begin{cases} T = \text{torque: ft-lb} \\ t = \text{time: sec} \\ I = \text{M of I: ft-lb-sec}^2 \\ \omega = \text{angular velocity: rad/sec} \\ \text{angular impulse} = \text{ft-lb-sec} \end{cases}$$

Problems

1. Explain the concept of coefficient of restitution as it relates to inelastic or partially elastic impact. Why is it permissible to use the kinetic-energy equation in the case of perfectly elastic impact? Explain.
2. Calculate, for a 500-lb projectile which has a speed of 800 mph, (a) the momentum; (b) the kinetic energy; (c) What force is required to stop the projectile in 0.5 sec?
3. A 4000-lb car is moving with a speed of 50 mph. Calculate (a) the kinetic energy of the car; (b) the momentum of the car; (c) the force required to stop the car in 0.06 sec.
4. For the car in Prob. 3, calculate (a) the impulse; (b) the time required to stop the car in 120 ft; (c) the force required to stop the car.
5. A machine gun fires 2-oz bullets at the rate of 8 bullets per second. If the velocity of the bullets is 2500 ft/sec, what is the average reaction force exerted by the gun against the support holding the gun?
6. A baseball can be thrown by a pitcher so that it is traveling 90 mph when it reaches the bat. The contact time between the bat and the ball is 0.032 sec. After leaving the bat, the ball travels in the opposite direction from the pitch with a velocity of 110 mph. The weight of the ball is $\frac{3}{8}$ lb. (a) What are the momentums of the ball before and after striking the bat? (b) What average force does the bat impart to the ball?
7. A large steel block measures 3 ft × 2 ft × 6 ft and rests on a trailer truck. The truck is traveling along a level stretch of road with a speed of 60 mph when it applies its brakes. It is able to stop at a distance of 600 ft. What is the stopping reaction force between the truck and the steel block?
8. A 200-lb man jumps (horizontally), with a speed of 8 mph, from a 70-lb rowboat. Calculate the final velocity of the boat, if it starts from rest and the friction of the water is neglected.
9. A ball is dropped from a height of 6 ft to a floor. The coefficient of restitution between the floor and the ball is 0.8. Calculate the height to which the ball will bounce.
10. A 6-lb ball (speed 60 mph) approaches a 6-lb ball (speed 80 mph). If the collision is perfectly elastic, calculate their velocities after impact. Assume they are going in opposite directions after impact. (See Fig. 12.5.)

Fig. 12.5

11. Assume the collision in Prob. 10 to be perfectly inelastic. Calculate the velocities after impact.
12. Solve Prob. 10 if the coefficient of restitution is 0.7.
13. A 6-lb ball and a 4-lb ball, traveling in the same direction, collide with perfectly elastic impact. If the 6-lb ball, before impact, is traveling at the rate of 66 ft/sec,

and the 4-lb ball is traveling at the rate of 44 ft/sec, find the velocities after impact. (See Fig. 12.6.)

Fig. 12.6

14. Assume the same conditions as in Prob. 13, except that the collision is perfectly inelastic. (See Fig. 12.6.)
15. Solve Prob. 13 when the coefficient of restitution is 0.8.
16. A 50-lb flywheel is rotating with a speed of 200 rpm. Calculate the angular momentum of the wheel if its mean radius is 2 ft.

13

Liquids

13.1. Properties of Fluids

Materials are generally classified as solids, liquids, or gases. Solids are bodies which have definite shape and size. The term *fluids*, as used in the title of this section, applies to both liquids and gases: *liquids* have definite size but take the shape of their container; *gases* have neither shape nor size and will fill a container of any size ot shape. Materials may exist, under certain conditions, in any one of the three forms cited.

Of course, the most common illustration of this is water, which may exist as a solid (ice) or as a gas (steam). Water freezes at 32°F (0°C) and boils at 212°F (100°C) at sea level. It is very slightly compressible and at 39.2°F (4°C) it reaches its minimum volume. The velocity of sound in water is 4700 ft/sec. The specific weight of water (weight per unit volume) is 62.4 lb per cubic *foot*. If 62.4 lb/ft^3 is divided by 1728 in.^{3}/ft^3, the specific weight may also be 0.0361 lb per cubic *inch*. Ice, the solid state of water, has a specific weight of 57.2 lb/ft^3, or 0.0331 lb/in.^{3}.

The *specific weight**** (sp wt) of a substance may be obtained by dividing the weight of the substance by its volume:

$$W = \frac{w}{V} \quad \left\{ \begin{array}{l} w = \text{weight: lb} \\ W = \text{sp wt: lb-ft}^3 \\ \quad (or) \text{ lb/in.}^3 \\ V = \text{volume: in.}^3 \end{array} \right\}$$

* Some physics texts use the concept of *weight density* instead of specific weight. The author prefers the term *specific weight* because this is exactly what weight per unit volume signifies. Most engineering texts prefer this notation.

Sec. 13.1   *Properties of Fluids*   **163**

Another concept which is useful is that of *density*, which is defined as *mass per unit volume:*

$$D = \frac{m}{V} \qquad \left\{ \begin{array}{l} m = \text{mass: slugs} \\ V = \text{volume: ft}^3 \\ D = \text{density: slugs/ft}^3 \end{array} \right\}$$

From the foregoing equation, if the volume is held constant, different materials have their own characteristic masses and weights. The student should visualize a cube 1 ft on each side and then realize that, even though density dictates that the volume be held to unity, the mass of the various materials is variable. Appendix Table A.8 is based on the volume of a unit cube of material.

Still another extremely important concept is that of *specific gravity* (sp gr), which is the ratio between the density of a substance and the density of water. Since density is a mass concept, it may be defined as specific weight per 32 ft/sec². The ratio of the density (sp wt) of a substance to the density (sp wt) of water may be equated and the following relationship evolved:

$$\frac{W_s}{W_w} = \text{specific gravity} \quad (a\ pure\ number) \qquad \left\{ \begin{array}{l} W_s = \text{sp wt: substance} \\ W_w = \text{sp wt: water} \end{array} \right\}$$

This means that mercury, which has a specific gravity of 13.6, has a weight of 13.6 times more than an *equal volume* of water.

Example 1

Five cubic feet of iron weighs 2250 lb. Calculate (1) the specific weight of iron; (2) the specific gravity of iron.

Solution

1. The specific weight is

$$W = \frac{w}{V} = \frac{2250}{5} \qquad \left\{ \begin{array}{l} w = 2250\text{lb} \\ V = 5\ \text{ft}^3 \\ W_w = 62.4\ \text{lb/ft}^3 \end{array} \right\}$$
$$= 450\ \text{lb/ft}^3$$

2. The specific gravity is

$$\text{sp gr} = \frac{W_i}{W_w} = \frac{450}{62.4}$$
$$= 7.2$$

In the cgs system, 1 g, or a cubic centimeter (cc, cm³) of water, is defined as the mass of pure water at 4°C. The concept of density is therefore based on the density of water as being unity. Thus

$$\text{sp gr} = \frac{\text{sp wt of substance}}{\text{sp wt of water}}$$

164 *Liquids* Chap. 13

But we have just defined the specific weight of water in the cgs system as equal to 1. It follows that the specific gravity and the specific weight are equal in the metric system. Note that in Appendix Table A.8, the column marked "Density" is also the specific-gravity value in any system of units.

13.2. Pressure

Pressure may be defined as *force per unit area*. The concept of *force* applies to an entire surface, whereas the concept of pressure applies only to the unit area. One prerequisite is that the force must act at right angles to the surface area.

It is therefore important to understand that a 10-lb force, applied to an area of 2 in.^{2}, yields a pressure of 5 psi (pounds per square inch). So that for every square inch of area, a pressure of 5 psi is applied. Thus increasing the area to 5 in.^{2} yields an *applied force* of 25 lb. The force applied has increased, but the pressure remains the same:

$$P = \frac{\text{force}}{\text{area}} = \frac{f}{A} = \frac{\text{lb}}{\text{in.}^2} = \text{psi} \qquad \left\{ \begin{array}{l} f = \text{force: lb} \\ P = \text{pressure: psi} \\ A = \text{area: in.}^2 \end{array} \right\}$$

The pressure exerted on each square inch of all objects at sea level is 14.7 psia (pound per square inch absolute). This quantity is called *one atmosphere* (atm). Two atmospheres would be 29.4 psi, and indicates another method used to express pressure.

The values for atmospheric pressure at sea level are

$$h = 14.7 \text{ psi}$$
$$= 29.92 \text{ in. of mercury}$$
$$= 34 \text{ ft of water}$$
$$= 76 \text{ cm of mercury}$$

It is also possible to relate pressure to a column of liquid. If a column of a liquid exerts a pressure at the bottom of that column, the pressure may be stated in terms of the height of the column producing that pressure. It is possible to indicate pressure in terms of inches of mercury, inches of water, feet of water, etc.

The volume of a container is

$$V = Ah$$

The weight of the container is

$$w = VW = AhW \qquad \left\{ \begin{array}{l} A = \text{area of container: in.}^2 \\ h = \text{height of container: in.} \\ V = \text{volume: in.}^3 \\ W = \text{sp wt: lb/in.}^3 \\ P = \text{pressure: psi} \end{array} \right\}$$

Thus

Sec. 13.2   *Pressure*   165

$$P = \frac{\text{force}}{\text{area}} = \frac{\text{weight}}{\text{area}} = \frac{\cancel{A}hW}{\cancel{A}} = hW$$

From the equation $P = f/A$, it can be seen that pressure is a unit-area concept.

Example 2

Calculate the pressure at the bottom of a tank if the tank is 15 in. high, 4 in. in diameter.

Solution

1. The volume of the tank is

$$V = Ah \frac{\pi d^2}{4}(h) = \frac{\pi 4^2}{4}(15) \qquad \left\{ \begin{array}{l} h = 15 \text{ in.} \\ d = 4 \text{ in.} \\ W_w = 0.0361 \text{ lb/in.}^3 \end{array} \right\}$$
$$= 188.4 \text{ in.}^3$$

2. The total weight of the water is

$$w_w = VW_w = 188.4 \, \cancel{\text{in.}^3} \times 0.0361 \text{ lb/}\cancel{\text{in.}^3}$$
$$= 6.8 \text{ lb}$$

3. The area at the bottom of the tank is

$$A = \frac{\pi d^2}{4} = \frac{\pi(4)^2}{4} = 12.6 \text{ in.}^3$$

4. The pressure is

$$P = \frac{w}{A} = \frac{6.8}{12.6}$$
$$= 0.54 \text{ psi}$$

or

$$P = hW_w = 15 \times 0.0361$$
$$= 0.54 \text{ psi}$$

Example 3

Convert 0.54 psi into (1) pounds per square foot; (2) inches of mercury; (3) feet of water; (4) centimeters of mercury; (5) atmospheres.

Solution

1. Pounds per square foot (psf):

$$P = 0.54 \frac{\text{lb}}{\cancel{\text{in.}^2}} \times 144 \frac{\cancel{\text{in.}^2}}{\text{ft}^2} \qquad \{ P = 0.54 \text{ psi} \}$$
$$= 77.8 \text{ psf}$$

166   *Liquids*   Chap. 13

2. Inches of mercury:

Specific weight of mercury (Hg):

$$W_m = W_w \times \text{sp gr of Hg} = 0.0361 \times 13.6$$
$$= 0.491 \text{ lb/in.}^3$$

Pressure:

$$P = hW_m$$

Solving for h,

$$h = \frac{P}{W_m} = \frac{0.54 \text{ lb/in.}^2}{0.491 \text{ lb/in.}^3}$$
$$= 1.1 \text{ in. of Hg}$$

3. Feet of water:

$$h = \frac{P}{W_w} = \frac{0.54 \text{ lb/in.}^2}{0.0361 \text{ lb/in.}^3 \times 12 \text{ in./ft}}$$
$$= 1.25 \text{ ft}$$

4. Centimeters of mercury:

$$h = 1.1 \text{ in.} \times 2.54 \frac{\text{cm}}{\text{in.}} = 2.8 \text{ cm of Hg}$$

5. Atmospheres:

$$\text{atm} = \frac{0.54 \text{ lb/in.}^2}{14.7 \text{ lb/in.}^2} = 0.0367 \text{ atm}$$

There are gages made which register pressure relative to atmospheric pressure. That is, one end of the gage is connected to a pressure chamber, and the other is open to the atmosphere. Thus the pressure inside the chamber is compared to the pressure outside the chamber by such a gage. These relative pressures are called *gage pressures* and are designated as *pounds per square inch gage* (psig).

In Fig. 13.1, a pressure which is 10 pounds per square inch above atmospheric pressure is designated as 10 psig.

Absolute pressure is the pressure relative to a perfect vacuum. Thus our 10 psig may be converted to read *pounds per square inch absolute* (psia), as shown in Fig. 13.1. Thus

$$14.7 + 10 = 24.7 \text{ psia}$$

A reading of 10 psia may also be read as in *pounds per square inch vacuum* (psiv) when referenced to atmospheric pressure, as shown in Fig. 13.1:

$$14.7 - 10 = 4.7 \text{ psiv}$$

Sec. 13.3   *Transmission of Pressure*   167

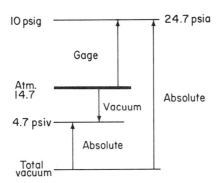

Fig. 13.1

Example 4

A pressure gage reads 20 psi. Determine the absolute pressure if atmospheric pressure is 14.4 psia.

Solution

The absolute pressure is

$$14.4 + 20 = 34.4 \text{ psia}$$

Example 5

The absolute pressure is 12 psia. (1) Convert this pressure to read vacuum. (2) Convert the vacuum pressure to inches of mercury. (3) Convert the vacuum pressure to inches of water absolute (in. of water abs).

Solution

1. Vacuum pressure:

$$14.7 - 12 = 2.7 \text{ psiv}$$

2. Inches of mercury:

$$h = \frac{P}{W_m} = \frac{2.7}{0.491} = 5.5 \text{ in. of Hg vac}$$

3. Inches of water:

$$h = \frac{P}{W_w} = \frac{12}{0.0361} = 332.4 \text{ in. of water abs}$$

13.3. Transmission of Pressure

We have seen that a pressure, when exerted on the bottom of a container filled with liquid, is due to the height of the column of liquid above the bottom. Up to this point in this chapter the force created by the liquid column has been considered without reference to any applied forces from some external source.

Now we shall consider *Pascal's principle*, which states that "a pressure applied to the surface of a fluid is transmitted undiminished in all directions throughout that fluid." These pressures always act at right angles to the area considered, whether the pressure is applied *to* the area or *by* the area.

In Fig. 13.2(a), a long pipe is attached to a wooden barrel. The column of water inside the pipe determines the pressure inside the barrel. If the pressure which the barrel can withstand is exceeded, it will break. Notice that a small-diameter pipe is capable of exerting large *forces* upon the wall of the barrel.

In Fig. 13.2(b), the external pressure is created by a piston. Assume that the face of the piston has an area of 1 in.². If a force of 1 lb is applied to the

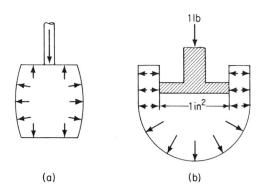

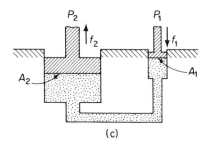

Fig. 13.2

Sec. 13.3   Transmission of Pressure   169

piston, a pressure of 1 psi is applied to the rest of the container—neglecting the pressure exerted by the height of the column inside the container. Thus every square inch of surface of the container has applied to it a pressure of 1 psi. Notice that the pressure of 1 lb/in.^{2} and the force of 1 lb are equal, because the area of the face of the piston is 1 in.^{2}. The pressure and the exerted force on the wall of the container are not equal unless the area of the wall is 1 in.^{2}.

In Fig. 13.2(c), the small piston P_1 exerts a pressure through the liquid to the bottom of the large piston. The pressure on the small piston is the same as the pressure on the large piston. The *force* exerted on the large piston, however, increases in direct ratio to the area increase. Thus, if the small piston has an area of 1 in.^{2} and exerts a pressure of 5 psi, and the large piston has an area of 4 in.^{2}, the large piston will exert a force of 5 psi $\times$ 4 in.^{2} = 20 lb. The large piston will exert a force which is four times that of the small piston.

If the small piston moves 4 in., the large piston will move $\frac{1}{4}$ the distance, or 1 in. The ratio of area-to-distance-moved is an inverse ratio. The ratio of distance-moved to the square of the diameters is also an inverse ratio. Several of the ratios discussed are now shown in equation form:

$$P_1 = \frac{f_1}{A_1}, \quad P_2 = \frac{f_2}{A_2}$$

But $P_1 = P_2$. Therefore,

$$\frac{f_1}{A_1} = \frac{f_2}{A_2} \quad \text{or} \quad f_2 = \frac{f_1 A_2}{A_1}$$

P_1 = pressure—small piston
P_2 = pressure—large piston
f_1 = force—small piston
f_2 = force—large piston
A_1 = area—small piston
A_2 = area—large piston

For the distance-to-diameter relationship,

$$s_1 A_1 = s_2 A_2$$
$$s_1 \frac{\pi d_1^2}{4} = s_2 \frac{\pi d_2^2}{4}$$

s_1 = distance moved—small piston
s_2 = distance moved—large piston
d_1 = diameter—small piston
d_2 = diameter—large piston

Therefore,

$$\frac{s_1}{s_2} = \frac{d_2^2}{d_1^2}$$

Because we have an input work and an output work delivered by the pump system just described, it may be considered to be a machine. As a machine, this device will also have a displacement ratio, a mechanical advantage, and an efficiency ratio. Friction and the weight of the liquid affect the mechanical advantage. Compressibility of the liquid affects the displacement ratio. (It is assumed that there will be no leakage.)

Example 6

In Fig. 13.2(c), the hydraulic press has the following specifications: diameter of the small piston, 2 in.; diameter of the large piston, 10 in.; force exerted by the small piston, 500 lb while moving 5 in. Calculate (1) the force capable of being exerted by the large piston; (2) the distance moved by the large piston; (3) the displacement ratio; (4) the mechanical advantage if the efficiency is 80%.

Solution

1. Small-piston pressure P_1:

$$P_1 = \frac{f_1}{A_1} = \frac{500}{\pi 2^2/4} = \frac{500 \times 4}{\pi 2^2}$$
$$= 160 \text{ psi}$$

$$\left\{ \begin{array}{l} d_1 = 2 \text{ in.} \\ d_2 = 10 \text{ in.} \\ f_1 = 500 \text{ lb} \\ s_1 = 5 \text{ in.} \\ \text{eff} = 80\% \end{array} \right\}$$

The force exerted by the large piston is

$$f_2 = P_2 A_2 = 160 \times \frac{\pi 10^2}{4}$$
$$= 12{,}560 \text{ lb}$$

2. The distance moved by the large piston is

$$\frac{s_1}{s_2} = \frac{d_2^2}{d_1^2}$$

Solving for s_2,

$$s_2 = \frac{s_1 d_1^2}{d_2^2} = \frac{5 \times 2^2}{10^2}$$
$$= 0.200 \text{ in.}$$

3. The displacement ratio is

$$\text{DR} = \frac{s_1}{s_2} = \frac{5}{0.2} = 25$$

4. The mechanical advantage is

$$\text{MA} = \frac{\text{force out}}{\text{force in}} = \frac{f_2}{f_1} \times \text{eff} = \frac{12{,}560}{500} \times 0.8 = 20.1$$

13.4. Manometers

In Fig. 13.3(a), the pressure at the point A is a function of the height h' and the specific weight of the liquid—but only if we assume that no additional pressure, other than atmospheric pressure, is acting on the surface of the liquid. For the purposes of this chapter, pressures are to be considered *gage* unless otherwise stated. The height times the specific weight of the liquid at point A is a constant and has nothing to do with the shape of the cylinder holding the liquid. If this reasoning is carried one step further, it can be seen that

Sec. 13.4   *Manometers*

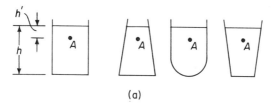

(a)

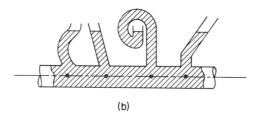

(b)

Fig. 13.3

the pressure at the bottom of the cylinders must all be equal. The forces at the bottom of the containers may not all be equal.

For the same reasons that the pressures at the bottom of the containers are equal, the pressures at the center of the horizontal pipe (Fig. 13.3b) must all be equal.

Example 7

The right leg of the U-tube in Fig. 13.4 is filled with mercury to a depth of 2 in. This mercury applies a pressure against a thin separator, as shown. How much water must be poured into the left leg of the U-tube in order to balance the pressures on both sides of the U-tube?

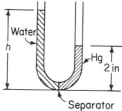

Fig. 13.4

Solution

Pressure created by the mercury:

$$P_m = 2 \times 0.491$$
$$= 0.982 \text{ psi}$$

Height of the water needed to balance:

$$h \times 0.0361 = 0.982$$
$$h = \frac{0.982}{0.0361}$$
$$= 27.2 \text{ in. of water}$$

172   *Liquids*   Chap. 13

The simplest manometer is shown in Fig. 13.5(a). In this manometer, the level of the liquid at A will be balanced by the level of the liquid at A'. This is so because the pressure (14.7 psi) against the surface of the liquid in the right-hand tube is the same as the pressure in the left-hand tube. Also, the specific weight is the same in both tubes (same liquid). Since the specific weights in both sides of the tube are equal and the tube is open at both ends (thus making the atmospheric pressure the same in both legs of the tube), it follows that $A'-A$ will stand at the same height above $x'-x$.

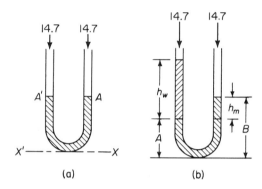

Fig. 13.5

If a quantity of water is added to the left side of the tube (see Fig. 13.5b), the mercury in the left side will be depressed and the mercury in the right side will rise until the distance $B - A = h_m$ balances the height of the water, h_w. The atmospheric pressures (14.7 psi) still balance each other, as does the height A in both tubes. Thus h_m will balance the pressure h_w.

Example 8

If Fig. 13.5(b), $h_w = 18$ in. of water. Calculate the height of the mercury column, h_m.

Solution

The pressure due to h_w must be equal to h_m:

$$18 \text{ in.} \times 0.0361 = h_m \times 0.491$$

$$h_m = \frac{18 \times 0.0361}{0.491} \qquad \left\{ \begin{array}{l} h_w = 18 \text{ in.} \\ W_w = 0.0361 \text{ lb/ft}^3 \\ W_m = 0.491 \text{ lb/ft}^3 \end{array} \right\}$$

$$= 1.32 \text{ in. of Hg}$$

Alternate Solution

The column of water is 13.6 times longer than the column of mercury. Thus

$$h_m = \frac{h_w}{13.6} = \frac{18}{13.6}$$
$$= 1.32 \text{ in. of Hg}$$

Another variation of the manometer just discussed is illustrated in Fig. 13.6(a). In this manometer, the mercury portion x–y is in equilibrium because the mercury column from y to the bottom of the tube balances the mercury column from x to the bottom of the tube. The water column from y to the top of the tube balances the water column from z to the top of the tube. Therefore, the tube z-y-x is in equilibrium. This leaves the pressure at x equal to the height of the mercury column, h_m. The pressure at P_w is equal to the height of the water column, h_w, plus the height of the mercury column, h_m. Notice that distances h and h_w are measured from the center of the mainline of the feed pipe. The sum $h_m + h_w$ is equal to the upward pressure created by the flow of water at point P.

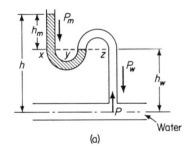 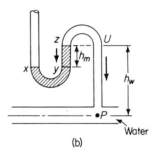

Fig. 13.6

Example 9

If Fig. 13.6(a), if $h_m = 12$ in. and the height h from the top of the mercury to the center of the pipe is 18 in., what is the pressure in the the pipe? Express the pressure in (1) psig; (2) in. of water; (3) in. of Hg.

Solution

1. The pressure due to the mercury is

$$P_m = 0.491 \times 12 = 5.892 \text{ psig}$$

The pressure due to the water is

$$P_w = (18 - 12) \times 0.0361 = 0.2166 \text{ psig}$$

The total pressure at the center of the pipe is

$$P = P_m + P_w = 5.8920 + 0.2166 = 6.1086 \text{ psig}$$

2. Inches of water:

$$h = \frac{P}{W} = \frac{6.1086}{0.0361} = 169.2 \text{ in. of water}$$

3. Inches of Hg:

$$h = \frac{P}{W} = \frac{6.1086}{0.491} = 12.44 \text{ in. of Hg}$$

Figure 13.6(b) shows the manometer of Fig. 13.6(a). The difference is that Fig. 13.6(b) indicates less than atmospheric pressure in the mainline. That is, the pressure in the tube $P-U$ yields a column of water and consequently a pressure based on h_w. Acting in the opposite direction from h_w is the pressure due to the column h_m $(z-y)$. Therefore, $h_w - h_m = P$ at the center of the pipe.

The same principles which we have been discussing hold true for siphons, Fig. 13.7(a). Once started, the column of water h_w will tend to create a vacuum at the top of the tube at A. However the atmospheric pressure on the liquid in the container pushes a column of liquid up into the left-hand side of the tube to fill the partial vacuum which h_w is attempting to create at A. Once the column of liquid is broken at A, the siphon will cease to work.

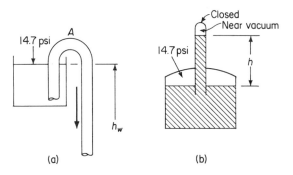

Fig. 13.7

An extension of the principle of the siphon will show that, by inverting a tube of mercury into a jar of mercury, the standard atmospheric pressure is capable of supporting a column of mercury 29.92 in. high. Thus as the tube is inverted and lowered into the container of mercury, the column of mercury in the tube drops until the pressure created by the column is equal to the atmospheric pressure. This is shown in Fig. 13.7(b).

The end of the tube not in the mercury pool must be closed. Thus as the mercury in the tube settles, a vacuum is created at the closed end. This is the principle of the *barometer*. At the base of the barometer is a well which is

Sec. 13.4 Manometers

exposed to the atmosphere. The closed end of the tube is calibrated to read inches or centimeters of mercury.

This vacuum places certain restrictions on the length of the column of liquid when a pump is used to maintain the vacuum at the top of a hose for pumping water out of a well. As this chamber is being evacuated, the atmospheric pressure is keeping the chamber filled. The maximum height of a water column, when the vacuum being created by the pump is almost perfect, is 34 ft at sea level.

$$h = 14.7 \text{ psi}$$
$$= 29.92 \text{ in. of Hg}$$
$$= 34 \text{ ft of water}$$
$$= 76 \text{ cm of Hg}$$

Figure 13.8(a) shows a barometer which can be transported from place to place without the fear of losing mercury. The principle of the *aneroid barometer* is the expansion and contraction of a bellows (Fig. 13.8b). The bellows is a vacuum chamber made of metal thin enough so that it will partially collapse

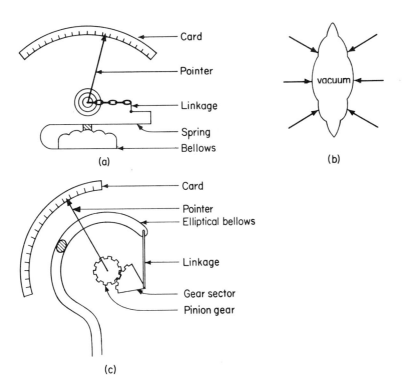

Fig. 13.8

as the pressure on the outer walls increases and expand as the pressure on the outer walls decreases. This thin-wall bellows must be supported by a flat spring, so that the walls will not collapse but rather work in opposition to the spring. The bellows is connected to a pointer through a series of levers. A card underneath the pointer is calibrated against a standard to read barometric pressure.

Figure 13.8(c) shows a *Bourdon gage*, named after its inventor. The essential feature of the gage is a hollow elliptical tube bent into an are or coil. As the pressure fills this tube, it tends to change the elliptical section into a circular section. In so doing, the coil must straighten out, causing the gear sector to rotate the pinion gear which, in turn, has a pointer attached. The dial is then calibrated against known pressures or vacuums, or both.

13.5. Surface Tension: Adhesion and Cohesion

Adhesion is the attraction which exists between the molecules of two different substances. *Cohesion* is the attraction which exists between the molecules of two like substances. Cohesion is largely responsible for the phenomenon of surface tension. Adhesion, together with surface tension, will be used in our discussion of capillarity.

Cohesive forces play a large part in holding the molecules of a piece of steel together. Breaking the steel and then trying to bring these forces back into play is impossible under ordinary conditions. Grinding and polishing the two contact surfaces will give better conditions for cohesion. Gage blocks which are lapped to a high degree of flatness and smoothness bring the molecules close enough together so that some of the cohesive forces are restored.

The forces of adhesion are sometimes relatively stronger than those of cohesion. This is so because in some cases the distance between unlike molecules need not be as close to create substantial attraction. Examples of holding power between unlike substances are glued, painted, or soldered surfaces.

If we were to select and isolate a molecule A in the center of a liquid, as shown in Fig. 13.9, we would find two things: (1) this molecule is surrounded by similar molecules, thus making the forces cohesive; (2) all the forces acting on our molecule are equal in magnitude. Since all the forces are equal, there is no resultant force acting on molecule A.

The discussion of molecule B in our liquid (Fig. 13.9) is quite different. Molecule B is at the air–liquid interface and is only partially surrounded by molecules which attract by cohesion. The upper surface is subjected to adhesive forces which are less in magnitude, partially because there are fewer molecules per unit area

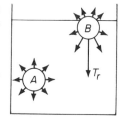

Fig. 13.9

Sec. 13.5   *Surface Tension: Adhesion and Cohesion*   177

in the air than in the liquid. In any event, this creates an unbalance and therefore a resultant force T_r into the liquid. This resultant sets up an internal force within the liquid which establishes a stronger bond at the surface of the liquid and therefore creates the "skin" effect known as *surface tension*.

This surface tension is the force per unit length and is independent of the shape or size of the surface. The purity of the liquid and the temperature do affect the surface tension of the liquid, which is constant for the entire surface of the liquid. Surface tension is measured by the force required to break through surface film. Surface-tension values are shown in Appendix Table A.9.

Figure 13.10(a) shows the standard cohesive forces which create a "film of forces" on the surface of a liquid. In Fig. 13.10(b), a wire, with a sensitive coil spring attached, has been immersed in the liquid. An external force is applied to one end of the spring in an effort to pull the wire out of the liquid surface. The attraction between the liquid molecules and the wire molecules

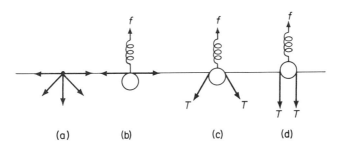

Fig. 13.10

is great enough so that the liquid molecules cling to the wire. In Fig. 13.10(c), the clinging effect creates tangents T to the surface of the liquid at the wire so that just before breakthrough (Fig. 13.10d), the two tangents may be considered as acting in opposition to the upward force f along the entire length of the wire. Thus the force required to pull the wire through the surface of the liquid is

$$f = 2T \times l \quad \left\{ \begin{array}{l} T = \text{surface tension: lb/ft} \\ l = \text{length: ft} \\ f = \text{force: lb} \end{array} \right\}$$

If the wire is in the form of a loop, the equation for surface tension would be

$$f = 2T \times \pi d$$

or, solving for T,

$$T = \frac{f}{2\pi d} \quad \left\{ \begin{array}{l} d = \text{diameter of loop: ft} \\ f = \text{force: lb} \\ T = \text{surface tension: lb/ft} \end{array} \right\}$$

Example 10

A wire frame measuring 2 in. in diameter is immersed in a bath of water. Find the reading which would appear on a sensitive scale which is pulling the loop out of the water (1) at 20 °C; (2) at 100 °C; (3) if the bath is benzene.

Solution

1. For water at 20 °C,

$$f = 2T\pi d = 2 \times 0.00499 \times 3.14 \times \tfrac{2}{12}$$
$$= 0.00524 \text{ lb}$$

2. For water at 100 °C,

$$f = 2 \times 0.00403 \times 3.14 \times \tfrac{2}{12}$$
$$= 0.00423 \text{ lb}$$

$$\left\{\begin{array}{l} d = \tfrac{2}{12} \text{ ft} \\ T_{20} = 0.00499 \text{ lb/ft} \\ T_{100} = 0.00403 \text{ lb/} \\ T_b = 0.00198 \text{ lb/ft} \end{array}\right\}$$

3. When the liquid is benzene,

$$f = 2 \times 0.00198 \times 3.14 \times \tfrac{2}{12}$$
$$= 0.00208 \text{ lb}$$

In a large container, the upper surface of the liquid is level. At the edges, where the liquid touches the container, the surface may be concave up (Fig. 13.11a), or concave down (Fig. 13.11b). This curvature is called the *meniscus*. If the adhesive forces between the liquid and the container are greater than the cohesive forces within the liquid, the liquid will be pulled up the side of the container until the height of the water column is in equilibrium with the adhesive forces. If the cohesive forces within the liquid are greater than the forces between the liquid and the walls of the container, the column will be depressed.

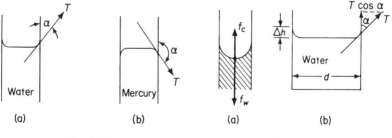

Fig. 13.11   **Fig. 13.12**

In Fig. 13.12(a), the force f_c is the *force due to capillarity* and is acting

Sec. 13.5 Surface Tension: Adhesion and Cohesion

upward. The force f_w is the force due to the column of liquid Δh. These two forces are in equilibrium:

$$f_c = f_w \quad \begin{cases} f_c = \text{capillarity force} \\ f_w = \text{liquid-column force} \\ \Delta h = \text{change in column height} \\ d = \text{diameter of tube or container} \end{cases}$$

In Fig. 13.12(b) the diameter of the tube is d and the contact ring between the liquid and the container is πd; T is the surface tension tangent to the meniscus and therefore $T \cos \alpha$ is the force acting up the wall. Thus the force of capillarity is

$$f_c = \pi d T \cos \alpha$$

This force f_c will cause the liquid column to increase, creating a force f_w due to the weight of the height of this column. Thus

$$f_w = VW = \frac{\pi d^2}{4} \Delta h W$$

Since $f_c = f_w$,

$$\pi d T \cos \alpha = \frac{\pi d^2}{4} \Delta h W \quad \begin{cases} V = \text{volume: ft}^3 \\ W = \text{sp wt: lb/ft}^3 \\ \Delta h = \text{capillarity: ft} \end{cases}$$

Solving for Δh,

$$\Delta h = \frac{4T \cos \alpha}{dW}$$

Notice that as the diameter of the tube decreases, Δh increases. The values for α may be greater or less than 90°. Some values are given in Appendix Table A.10.

Example 11

A tube 0.072 in. in diameter is connected to a water line at temperature 20°C. The gage pressure in the water line is given as 3 in. of mercury. The end of the tube is exposed. (1) What is the error due to capillarity? (2) What is the uncorrected height of the water column? (Assume the water to be slightly impure.)

Solution

1. The change in height due to capillarity (Fig. 13.13a) is

$$\Delta h_w = \frac{4T \cos \alpha}{dW_w} = \frac{4 \times 0.00499 \times 0.906}{0.006 \times 62.4} \quad \begin{cases} d = \frac{0.072}{12} \\ \quad = 0.006 \text{ ft} \\ T_{20} \text{ (table)} = 0.00499 \text{ lb/ft} \\ W_3 = 62.4 \text{ lb/ft}^3 \\ \alpha \text{ (table)} = 25° \\ \cos \alpha = 0.906 \end{cases}$$

$$= 0.0483 \text{ ft} = 0.58 \text{ in.}$$

2. The uncorrected pressure height is

$$3 \text{ in. of Hg} \times 13.6 = 40.80 \text{ in. of water}$$

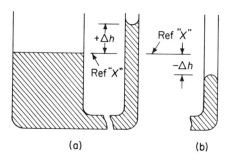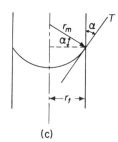

Fig. 13.13

the corrected height is

40.80 + 0.58 = 41.38 in. of water

Example 12

Assume the condition to be the same as in Example 11. (1) Find the error due to capillarity if the manometer contains mercury at 20 °C. (2) What is the equivalent error in terms of water pressure?

Solution

1. Error due to capillarity (Fig. 13.13b):

$$\Delta h_m = \frac{4T \cos \alpha}{dW_m} = \frac{4 \times 0.0319 \, (-0.766)}{0.006 \times 62.4 \times 13.6}$$

$$= -0.0192 \text{ ft} = -0.23 \text{ in. of water}$$

$$\left\{ \begin{array}{l} d = 0.006 \text{ ft} \\ T_{20} = 0.0319 \text{ lb/ft} \\ W_m = 62.4 \times 13.6 \\ \alpha = 140° \\ \cos \alpha = 0.766 \end{array} \right\}$$

(Negative sign indicates depression.)

2. The equivalent water error is

$$\Delta h_w = 0.23 \times 13.6$$
$$= 3.13 \text{ in. of water}$$

For small tubes (Fig. 13.13c), the equation for capillarity becomes

$$\Delta h = \frac{4T}{Wdm} \quad \{ d = \text{diameter of meniscus} \}$$

Problems

1. A tank 40 ft × 50 ft is used to store water. The tank is 30 ft high and is filled with water. Calculate the pressure at the bottom of the tank in (a) psf; (b) psi; (c) in. of water; (d) in. of Hg.

Problems

2. Convert each part in Prob. 1 to absolute pressure if the local atmospheric pressure is 14.5 psia.
3. Calculate the force exerted on the bottom of the tank in Prob. 1.
4. Express 22.6 psig as a pressure in (a) psia; (b) in. of water abs; (c) in. of Hg abs; (d) cm of Hg abs.
5. A large tank is filled with oil. If the pressure at the bottom of the tank is 640 psf, calculate the depth of the oil.
6. A tank has a 3-in. layer of mercury, on top of which is 5 in. of water covered with 4 in. of oil. Calculate (a) absolute pressure on the base of the tank; (b) the absolute pressure on the mercury; (c) the absolute pressure on the water. All pressures should be in psia.
7. Calculate the pressure at a point 20 ft below the surface of a lake if the atmospheric pressure is 30.4 in. of Hg. Express your answer as (a) psig; (b) psia.
8. A conical tank is 6 ft in diameter at the top and 4 ft in diameter at the bottom and 8 ft deep. Calculate (a) the pressure at the base of the tank; (b) the force at the base of the tank (the tank is filled with oil); (c) the weight of the oil in the tank.
9. Calculate the difference of pressures at a depth of 40 ft below the surface of a fresh and salt-water lake. (Salt water's sp gr = 64 lb/ft^3).
10. A pipeline is used to transmit oil (sp gr = 0.85) at a pressure of 18 psig. Express the pressure in (a) ft of oil; (b) ft of water; (c) in. of Hg.
11. A hydraulic piston 1.5 in. in diameter has a 6-in. stroke. The pressure exerted by this small piston is 90 psi. (a) What diameter piston is needed to transmit 5400 lb? (b) How far will the large piston move?
12. A hydraulic jack, arranged for lifting cars, has a system which is 70% efficient and which has a mechanical advantage of 100. The small piston, which can exert a force of 75 lb, has a diameter of 0.5 in. and moves a distance of 4 in. Find (a) the diameter of the large piston; (b) the displacement ratio; (c) the distance moved by the large piston; (d) the pressure in psi; (e) the force exerted by the large piston.
13. The barometric pressure is 29 in. of Hg at a time when a tank of water is siphoned into another water tank. The difference in water levels between the two tanks is 50 ft. The upper tank is 10 ft deep, and the upper end of the siphon is 12 ft above

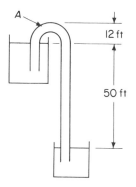

Fig. 13.14

the upper water level. (a) What is the pressure at the upper end of the siphon when the inlet is choked off? (b) What is the pressure at the upper end of the siphon when the outlet end of the siphon is choked off? (*Note*: By "upper end" is meant the highest point, *A*, Fig. 13.14.)

14. A U-tube is fastened to a pipe to measure pressure, as shown in Fig. 13.15(a). Calculate the pressure (psig) at the center of the pipe if the pipe transports water.
15. The pressure in a pipe is measured with a mercury manometer as shown in Fig. 13.15(b). The pressure is known to be 5.9 psia. Calculate the position of the mercury in the open end of the manometer with reference to the centerline of the pipe.

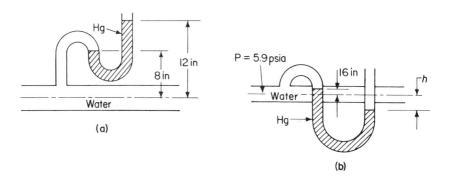

Fig. 13.15

16. A U-tube containing water is connected to a gasline. When the gas is turned on, the water level difference in both legs of the manometer is 16 in. Calculate the gas pressure.
17. A U-tube has mercury poured into it and then water. How much water is needed in the left-hand tube to make the mercury in the right-hand tube rise to a 3-in. difference in mercury levels?
18. Mercury is poured into a U-tube. A liquid (sp gr = 0.79) is poured into the tube until the difference in mercury levels is 1.5 in. What is the height of the liquid column?
19. A water manometer is connected to the bottoms of two containers. The dif-

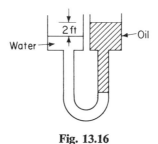

Fig. 13.16

Problems

ference in level between the liquids in the two tubes is 2 ft. If one tanks contains oil (sp gr = 0.88), calculate the height difference in tanks. (See Fig. 13.16.)

20. Three small-bore glass tubes, open at the top, are attached to a waterline. The diameters of the tubes are 0.06 in., 0.2 in., and 0.3 in. Assuming the pressure in the water line is 3.2 in. of Hg, how high will the water rise in each tube?

21. A pressure of 35 psi is measured with a water manometer (bore = 0.08 in.) and with a mercury manometer having the same bore. (a) What is the error due to capillarity in each case? (b) What is the difference in water-equivalent error between both liquids? (The temperature is 20°C and in both cases the liquids are slightly impure.)

22. Turpentine (sp gr = 0.878) rises 2.7 in. in a tube 0.012 in. in diameter. The angle α between the liquid and the glass is 18°. What is the surface tension of the liquid?

23. A wire is bent into an equilateral triangle, each leg measuring 2 in. What force would be required to pull the wire through the liquid in Prob. 22, all conditions remaining the same?

24. A piece of wire made of platinum is bent into a rectangular shape so that the sides are 3 in. and 4 in. The rectangle is pulled steadily through the surface of a mercury bath with a force of 1.187 oz. What is the surface tension of the liquid?

25. A U-tube mercury manometer has one of its legs at 60° with the horizontal. Assuming the pressure at the vertical leg is 30 psi, calculate the length of the mercury column in the bent leg.

14

Buoyancy

14.1. Archimedes' Principle

Figure 14.1 shows a block, dimensions $a \times b \times c$, submerged in a liquid to a depth such that the top of the block is a distance h below the surface. *Archimedes' principle* states that, if a body is submerged in a liquid, a force, equal to the liquid displaced, will act in an upward direction tending to support the object. If this upward force, which we shall henceforth call the *buoyant force*, is *greater* than the weight of the object, the object will float; if the buoyant force is *less* than the weight of the object, the object will sink; if the buoyant force is *equal* to the weight of the object, the object will remain suspended in the liquid.

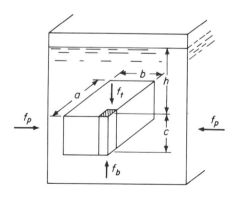

Fig. 14.1

Sec. 14.1   Archimedes' Principle   185

Note carefully that the shape of the object is irrelevant. In Fig. 14.1, the shaded rectangle is subject to a vertical buoyant force. If the entire object is cut into small rectangles and each small rectangle has a buoyant force operating on it, the sum of the volumes will equal the volume of the object; and the sum of the individual buoyant forces will equal the total buoyant force acting on the object.

This reasoning also applies to irregular objects. The buoyant force is dependent upon the force acting at the center of gravity of the displaced fluid and the specific weight of the displaced fluid, not upon the volume or specific weight of the object. The theory also holds whether the body is homogeneous or not, and whether the object floats or sinks.

We also know that pressure is equal to force per unit area:

$$P = \frac{f}{A}$$

In Fig. 14.1, all sides of the block are subjected to pressure due to the height of the column of liquid multiplied by its specific weight. The force exerted on the sides of the block is equal to the pressure multiplied by the area perpendicular (projected area) to this pressure. Notice that forces f_p, acting on both sides of the block in opposite directions, are in equilibrium and cancel each other out. Also, f_t, acting at the top of the block on the area $a \times b$, is due to the pressure and the specific weight of the liquid pressing on this area. This force f_t is opposed by a force f_b created by the weight of column of liquid from the surface of the liquid to the area at the bottom of the block. Forces f_b and f_t need not balance nor equal each other. From Fig. 14.1:

The downward pressure on the top of the block is

$$P_t = hW$$

The upward pressure on the bottom of the block is

$$P_b = (h + c)W$$

The total downward force on the area ab (top) is

$$f_t = hW \times ab$$

The total upward force on the area ab (bottom) is

$$f_b = (h + c) \times W \times ab$$

The force f_p acting on the side areas are in equilibrium. Therefore, the resultant force acting on the entire block will be the difference between f_b and f_t:

$$B = f_b - f_t$$
$$= (h + c)Wab - (h)Wab$$
$$= Wcab$$

$\left\{ \begin{array}{l} B = \text{buoyant force acting up} \\ V = \text{volume of displaced liquid} \\ W = \text{sp wt of liquid} \end{array} \right\}$

But *cab* is the volume V of the block. Therefore,

$$B = VW$$

Thus the buoyant force acting on an object in a liquid is equal to the volume of the displaced liquid multiplied by the specific weight of the liquid.

Another useful equation may be derived in terms of the specific gravity of the liquid. Specific gravity of the unknown liquid:

$$\frac{W_s}{W_w} = \text{sp gr}$$

Solving for W_s,

$\begin{Bmatrix} W_s = \text{sp wt of liquid} \\ W_w = \text{sp wt of water} \\ \text{sp gr} = \text{sp gr of liquid} \\ B = \text{buoyant force} \end{Bmatrix}$

$$W_s = W_w \times \text{sp gr}$$

Therefore,

$$B = VW_w \times \text{sp gr}$$

Since specific weight is weight per unit volume, and density is mass per unit volume, we can say that if

$$W = Dg$$

then, substituting,

$$B = VW = VDg \qquad \begin{Bmatrix} D = \text{density} \\ g = \text{gravitational acceleration} \end{Bmatrix}$$

14.2. Specific-Weight and Specific-Gravity Determinations: Solids

The specific weight of a *regular* object may be found by computing the volume from the regular dimensions and the weight by weighing the object. Since specific weight is a unit concept, it can be found by dividing the weight by the volume. The specific weight thus found, divided by the specific weight of water, yields the specific gravity of the solid.

If the object is *irregular* in shape, the volume may be found by submerging it in a fluid of known specific weight and then finding the buoyant force. This buoyant force, when divided by the specific weight of the *fluid*, yields the volume of the irregular object. Once the volume is found, and the object is weighed, dividing the weight by the volume gives the specific weight. Then, dividing the specific weight of the object by the specific weight of water yields the specific gravity.

Next, we examine three typical problems dealing with buoyancy.

Example 1

An irregular object when weighed in air is 6 lb. When submerged in water, the reading on the scale is 5.26 lb. Calculate the specific gravity of the irregular object.

Solution

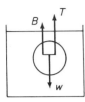

Fig. 14.2

This example deals with an object which will sink when submerged in a liquid. It can be seen from Fig. 14.2 that the scale is acting in the opposite direction from the weight of the object. Also, the buoyant force B must be helping the force T, because the scale registers a greater reading when the object is weighed in air than when it is weighed in the fluid. Since the object neither sinks any further into the liquid nor rises in the liquid, it must be in equilibrium. Therefore, the forces acting upward must equal the forces acting down, as shown in Fig. 14.2. The forces to the right and to the left cancel each other. From Fig. 14.2:

The buoyant force is

$$w = T + B$$
$$6 = 5.26 + B$$
$$B = 0.74 \text{ lb}$$

The volume of the irregular object is

$$V = \frac{B}{W_w} = \frac{0.74}{0.0361}$$
$$= 20.5 \text{ in.}^3$$

$$\left\{ \begin{array}{l} w = 6 \text{ lb (in air)} \\ T = 5.26 \text{ lb (submerged)} \\ B = \text{buoyant force} \\ W_w = \text{sp wt of water} \end{array} \right\}$$

The specific weight of the irregular object is

$$\text{sp wt} = \frac{w}{V} = \frac{6}{20.5}$$
$$= 0.293 \text{ lb/in.}^3$$

The specific gravity of the irregular object is

$$\text{sp gr} = \frac{W_s}{W_w} = \frac{0.293}{0.0361}$$
$$= 8.12$$

Example 2

It is desired to find the specific gravity of an irregular object that floats. Under these circumstances, it is necessary to add a weight to the light object so that it can be submerged. When the object is weighed in air (Fig. 14.3a), it is found to weigh 6 lb. The sinker weighs 4 lb in air. When only the sinker is submerged in water (Fig. 14.3b), the scale reads 8 lb. When both the irregular object and the sinker are submerged in water, the scale reads 1.5 lb. What is the specific gravity of the irregular object?

Buoyancy Chap. 14

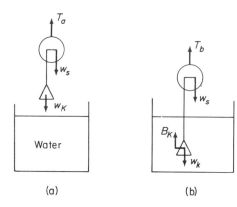

Fig. 14.3

Solution

From Fig. 14.3(a):

The weight in air is

$$T_a = 6 + 4 = 10 \text{ lb}$$

From Fig. 14.3(b):

The buoyant force B_k (loss in weight) on the sinker is

$$T_b + B_k = w_s + w_k$$

$$B_k = w_s + w_k - T_b = 6 + 4 - 8 = 2 \text{ lb}$$

From Fig. 14.3(c):

The buoyant force B_{s+k} is

$$T_c + B_{s+k} = w_s + w_k$$

$$B_{s+k} = w_s + w_k - T_c$$

$$= 6 + 4 - 1.5$$

$$= 8.5 \text{ lb}$$

$\left\{ \begin{array}{l} w_s = 6 \text{ lb} \\ w_k = 4 \text{ lb} \\ T_b = 8 \text{ lb} \\ T_c = 1.5 \text{ lb} \\ B_s = \text{buoyant force on irregular object} \\ B_k = \text{buoyant force on sinker} \\ T_a = \text{weight of both in air} \\ B_{s+k} = \text{buoyant force on both object and sinker} \end{array} \right.$

The buoyant force B_s is

$$B_s = B_{s+k} - B_k = 8.5 - 2$$

$$= 6.5 \text{ lb}$$

The weight of the water displaced by the irregular object is

$$w_w = B_s$$

$$= 6.5 \text{ lb}$$

Sec. 14.2   *Determining Solid Specific-Weight and -Gravity*   **189**

The volume of the water displaced is

$$V_s = \frac{6.5}{0.0361}$$
$$= 180 \text{ in.}^3$$

The specific weight of the irregular object is

$$W_s = \frac{w_s}{V_s} = \frac{6}{180}$$
$$= 0.0333 \text{ lb/in.}^3$$

The specific gravity of irregular object is

$$\text{sp gr} = \frac{W_s}{W_w} = \frac{0.0333}{0.0361}$$
$$= 0.922$$

Example 3

A 2-lb block of wood floats in water as shown in Fig. 14.4. If the block has the dimensions 4 in. × 6 in. × 3 in., how much of the block will float *above* water?

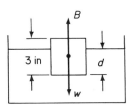

Fig. 14.4

Solution

From Fig. 14.4:

The buoyancy is

$$B = w = 2 \text{ lb}$$

The weight of the displaced water is

$$B = d \times 4 \times 6 \times 0.0361$$

The portion of the block under water is

$$d = \frac{B}{4 \times 6 \times 0.0361} = \frac{2}{0.866}$$
$$= 2.3 \text{ in.}$$

The portion of the block above water is

$$3 - 2.3 = 0.7 \text{ in.}$$

14.3. Specific-Weight and Specific-Gravity Determinations: Liquids

Case 1. SP WT AND SP GR OF LIQUIDS USING OBJECTS WHICH SINK

Using the equation for objects which sink into a liquid, the specific weight or specific gravity of an unknown liquid may be found, given the buoyant force and the volume of the displaced liquid.

Example 4

It is required to find the specific weight and the specific gravity of a liquid. An object which measures 30 in.^{3} weighs 12.330 lb in air and 11.445 lb when submerged in the liquid. Find the specific weight and specific gravity of the liquid.

Solution

Loss in weight of object when submerged:

$$B = w_a - w_s$$
$$= 12.330 - 11.445 \qquad \left\{ \begin{array}{l} V = 30 \text{ in.}^3 \\ w_a = 12.330 \text{ lb} \\ w_s = 11.445 \text{ lb} \end{array} \right\}$$
$$= 0.885 \text{ lb}$$

Specific weight of the liquid:

$$W_L = \frac{B}{V} = \frac{0.885}{30} = 0.0295 \text{ lb/in.}^3$$

Specific gravity of the liquid:

$$\text{sp gr} = \frac{0.0295}{0.0361} = 0.817$$

Case 2. SP WT AND SP GR OF LIQUIDS USING OBJECTS WHICH FLOAT

It is probably easier to calculate the specific weight and specific gravity of a liquid by using a solid object which floats in the liquid. In this method, the buoyant force can be found by weighing the object in air. The object will sink into the unknown liquid until it displaces its own weight of liquid. Thus the buoyant force is a constant, and in the equation

$$B = VW$$

the specific weight varies inversely as the volume. If the buoyant force is to remain constant, then as the specific weight of the liquid increases, the (immersed) volume of the object decreases; or as the specific weight of the liquid decreases, the (immersed) volume of the object must increase. A corollary of this is that if both B and V (of the object) are held constant, the object will sink deeper into the liquid as the specific weight of the liquid decreases. This

Determining Liquid Specific-Weight and -Gravity

is so because, in order to displace the weight of the liquid equal to the weight of the object, more liquid must be displaced. Thus, in order to displace the weight of the object, in a lighter liquid, the object must sink further into the liquid.

Example 5

A block 4 in. × 6 in. × 3 in. (see Fig. 14.4) is submerged into an unknown liquid until 0.8 in. is above the surface of the liquid. If the block weighs 2 lb, calculate the specific weight and the specific gravity of the liquid.

Solution

Area in contact with the liquid:
$$A = 4 \times 6 = 24 \text{ in.}^2$$
Depth below surface to which the block sinks:
$$d = 3 - 0.8 = 2.2 \text{ in.}$$
Volume of the liquid displaced:
$$V = 24 \times 2.2 = 52.8 \text{ in.}^3$$
Buoyant force:
$$B = \text{weight of body} = 2 \text{ lb}$$
Specific weight of the liquid:
$$W_e = \frac{B}{V} = \frac{2}{52.8} = 0.0379 \text{ lb/in.}^3$$
Specific gravity:
$$\text{sp gr} = \frac{0.0379}{0.0361} = 1.05$$

Case 3. Sp Wt and Sp Gr of Liquids Using Specific-Gravity Bottles

Still another method to determine the specific gravity of an unknown liquid is by use of the *specific-gravity bottle* shown in Fig. 14.5.

Fig. 14.5

First, the empty bottle is weighed. Then it is filled with the unknown liquid. The tapered glass stopper, which has a hole through its center, is pushed into the bottle. The excess liquid is forced through the hole until the stopper is secured. Bottle and contents are weighed and the known weight and volume of the liquid is thus found. If the bottle is emptied and the process repeated with water, the weight for equal volumes of the two liquids is known. The weight of the liquids is computed by subtracting the weight of the empty bottle from the weight of the full bottle. The specific gravity of the liquid is

$$\text{sp gr} = \frac{w_L}{w_w} \qquad \left\{ \begin{array}{l} w_L = \text{weight of liquid} \\ w_w = \text{weight of water} \end{array} \right\}$$

Example 6

Liquid is poured into a specific-gravity bottle. The weight of the empty bottle is 52.24 g. When the bottle is filled with an unknown liquid, the weight of bottle and liquid is 177.24 g. When filled with water, the weight of bottle and water is 151.24 g. What is the specific gravity and specific weight (in metric and British units) of the unknown liquid?

Solution

Weight of the unknown liquid:

$$w_L = 177.24 - 52.24 = 125 \text{ g}$$

Weight of water:

$$w_w = 151.24 - 52.24 = 99 \text{ g}$$

$$\text{sp gr} = \frac{w_L}{w_w} = \frac{125}{99} = 1.262$$

$$\left\{ \begin{array}{l} w_B = 52.24 \text{ g: weight of bottle} \\ w_{B+L} = 177.24 \text{ g: bottle and liquid} \\ w_{B+w} = 151.24 \text{ g: bottle and water} \end{array} \right\}$$

In the metric system, specific weight equals specific gravity:

$$\text{sp wt} = 1.262 \text{ g/cm}^3$$

Specific weight in the British system:

$$\text{sp wt} = 1.262 \times 0.0361 = 0.0456 \text{ lb/in.}^3$$

Case 4. Sp Wt and Sp Gr of Solids Using Specific-Gravity Bottles

The specific-gravity bottle may also be used to determine the specific weight and the specific gravity of solids.

Example 7

It is desired to find the specific gravity of some copper shot. The specific-gravity bottle used, when empty, weighs 48.8 g. The copper shot is dropped into the bottom of the bottle. Both are then weighed and found to weigh 117.6 g. The rest of the bottle is then filled with water and weighed. Bottle, water, and copper are found to weigh 209.6 g. Find the specific gravity and the specific weight in metric and British units of the copper shot.

Solution

Weight of the copper:

$$w_c = 117.6 - 48.8 = 68.8 \text{ g}$$

Weight of the water:

$$w_w = 209.6 - 117.6 = 92 \text{ g}$$

$$\left\{ \begin{array}{l} w_B = 48.8 \text{ g: weight of bottle} \\ w_{B+c} = 117.6 \text{ g: weight of bottle and copper} \\ w_{B+c+w} = 209.6 \text{ g: weight of bottle, copper, and water} \\ w = 100 \text{ g: capacity of bottle} \end{array} \right\}$$

Sec. 14.4   Hydrometry   193

Specific gravity of the copper:

$$\text{sp gr} = \frac{w_c}{w - w_w} = \frac{68.8}{100 - 92} = 8.6$$

Specific weight in the metric system:

$$\text{sp wt} = 8.6 \text{ g/cm}^3$$

Specific weight in the British system:

$$\text{sp wt} = 8.6 \times 0.0361 = 0.310 \text{ lb/in.}^3$$

14.4. Hydrometry

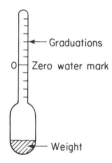

Fig. 14.6

Hydrometry is based on the theory discussed in the first part of this chapter, and more specifically, in Case 2, Example 5. It has been stated that an object which floats will sink into a liquid until it displaces the volume of liquid equal in weight to the weight of the object.

The stem of a hydrometer is usually made so that its cross-sectional area is smaller than the lower portion of the bulb (see Fig. 14.6). The slender stem, graduated, allows the hydrometer to sink into the liquid with relatively large movement for small displacement of liquid. Thus the graduations on the stem are far enough apart so that they can be read.

Example 8

You are given a hydrometer which weighs 1.024 oz. The graduated portion of the stem is 0.375 in. in diameter. If the hydrometer is placed in a liquid whose specific gravity is 1.6, how much higher will it rise above the zero water mark?

Solution

The zero water mark is the graduated reference number which appears at the liquid level when the hydrometer is put into water. Since the hydrometer floats in both the liquid and the water, the weight of the hydrometer will equal the weight of the displaced liquid.

Buoyant force:

$$B = w_h = 1.024 \text{ oz} = 0.064 \text{ lb}$$

Specific weight of liquid:

$$W_L = 1.6 \times 0.0361 = 0.05776 \text{ lb/in.}^3$$

$\left\{\begin{array}{l} w_h = \text{hydrometer weight: 1.024 oz} \\ \text{sp gr} = \text{liquid: 1.6} \\ d = \text{diameter of stem: 0.375 in.} \\ W_L = \text{sp wt of liquid} \\ W_w = \text{sp wt of water} \\ V_L = \text{volume of liquid} \end{array}\right\}$

Volume of liquid displaced:

$$V_L = \frac{B}{W_L} = \frac{0.064}{0.05776} = 1.108 \text{ in.}^3$$

Volume of displaced water:

$$V_w = \frac{B}{W_w} = \frac{0.064}{0.0361} = 1.773 \text{ in.}^3$$

$$\begin{cases} V_w = \text{volume of water} \\ B = \text{buoyant force} \\ A = \text{area} \\ \Delta l = \text{change in length} \\ \text{of stem} \end{cases}$$

Change in volume from one liquid to another:

$$\Delta V = 1.773 - 1.108 = 0.665 \text{ in.}^3$$

But $\Delta V = A\, \Delta l$.

Change in length of stem:

$$\Delta l = \frac{\Delta V}{A} = \frac{0.665}{\pi(0.1875)^2} = 6.024 \text{ in.}$$

The zero mark will be 6.024 in. higher than the second liquid (sp gr = 1.6).

Example 9

A hydrometer is 20 cm long and has a stem whose cross-sectional area is 0.3 cm². The hydrometer weighs 30 g and floats in water so that 10 cm is above the water mark. (1) Find the specific gravity of a liquid which floats the hydrometer 15 cm above the liquid. (2) Find the specific gravity of the liquid which floats the hydrometer 5 cm above the liquid.

Solution

In the metric system of units, 1 g will displace 1 cm³ of water because, by definition, the specific weight of water is 1 g/cm³. The hydrometer weighs 30 g and will therefore displace 30 cm³ of water. If the hydrometer floats higher in the new liquid than it did in water, this means that the buoyant force is still 30 g but the volume displaced is less. If the hydrometer floats deeper in the new liquid than it did in the water, the buoyant force is still 30 g, but the volume of the new liquid displaced is greater.

1. If the hydrometer floats 5 cm above the water mark of 10 cm, the hydrometer has decreased its volume displacement below the water displacement:

   New volume:

   $$V_L = V_w - (A \times \Delta l)$$
   $$= 30 \text{ cm}^3 - (0.3 \times 5) = 30 - 1.5$$
   $$= 28.5 \text{ cm}^3$$

   Specific weight of the unknown liquid:

   $$W_L = \frac{B}{V_L} = \frac{30}{28.5} = 1.053 \text{ g/cm}^3$$

Specific gravity of the unknown liquid:

$$\text{sp gr} = \frac{W_L}{W_w} = \frac{1.053}{1.00} = 1.053$$

2. If the hydrometer floats 5 cm below the 10-cm water mark, the hydrometer has increased its volume displacement above the water displacement:

New volume:

$$V_L = 30 + (0.3 \times 5) = 31.5 \text{ cm}^3$$

Specific weight of unknown liquid:

$$W_L = \frac{B}{V_L} = \frac{30}{31.5} = 0.95 \text{ g/cm}^3$$

Specific gravity of unknown liquid:

$$\text{sp gr} = \frac{0.952}{1.00} = 0.952$$

14.5. Viscosity

Viscosity is a resistance to shearing action which takes place when an attempt is made to slide one plate over another with an intervening layer of a fluid (e.g., oil). The resistance to shear is attributed to the cohesive forces which act between the molecules of the fluid. Thus viscosity is the resistance to shear within a fluid (liquid or gas). For liquids, viscosity decreases with increased temperature. For gases, viscosity increases with temperature increase.

Problems

1. A piece of brass which measures 2 in. × 3 in. × 4 in. is attached to a spring scale. When lowered into water, the scale registers 6.58 lb. Calculate the weight of the brass in air.
2. Calculate the scale reading if the steel block in Prob. 1 is lowered into an oil bath (sp gr = 0.75).
3. An object is suspended from a spring scale. When suspended in air, the object weighs 4 lb. When the object is submerged in water, the scale reads 3.5 lb. Calculate its specific gravity.
4. One ft^3 of glass weighs 160 lb. A piece of this glass loses 4.8 lb when submerged in water. What is the weight of the glass (a) in air? (b) when submerged?
5. A raft is built from pine logs (sp wt = 40 lb/ft^3). The raft is 8 ft × 5 ft × 2 ft. The 2-ft dimension sinks into the water (sp gr = 1.08) for a distance of 8 in. Calculate (a) the weight of the raft; (b) the depth to which the raft will sink in fresh

water. (c) What must the specific weight of the wood be in order for the raft to remain in equilibrium below the surface?

6. A 300-ton iceberg (sp gr = 0.94) floats in water (sp gr = 1.04). Calculate (a) the percentage of the iceberg below the water surface; (b) the weight of the ice below the water surface; (c) the volume above the water surface.
7. A box has a weight of 1.75 lb in air and a volume of 8 in.^{3}. If it is submerged in a liquid, the apparent weight is 1.05 lb. Calculate (a) the specific weight of the liquid; (b) the specific gravity of the liquid.
8. A 30-lb piece of metal is supported by a cord below water. Its apparent weight is 28.5 lb. Calculate the weight of 1 ft^3 of this metal.
9. A piece of pine has a 4-in. diameter and is 6 ft long. Assume its specific gravity is 0.65 and that it floats on end. (a) Calculate the length of the stick under water. (b) How much weight must be fastened to the end of the stick so that 8 in. will protrude above the water?
10. A 500-lb rectangular box, 2 ft × 3 ft × 4 ft is thrown into a fresh water pond such that the 4-ft side is the depth. (a) How much of the box will sink into the pond? (b) What is the least weight that must be added to the box to cause it to sink to the bottom of the pond 4 ft deep? (c) How much more will the box rise or sink into the water if 50 gal of gasoline are put into the box before it is thrown into the pond (sp gr of gasoline = 0.72)?
11. A balloon has a volume of 80,000 ft^3 and is filled with helium (sp wt = 0.0111 lb/ft^3). Calculate the force trying to raise the balloon (sp wt of air = 0.0806 lb/ft^3). Assume standard temperature and pressure (STP).
12. A balloon is filled with 4 oz of helium. Calculate the volume of helium and the buoyant force of the air. Assume STP.
13. A wooden block weighs 10 lb in air. An 8-lb sinker is attached to the block. The weight of the sinker and the block is 15.6 lb when only the sinker is submerged in water. When both the sinker and block are submerged, the apparent weight is 4 lb. Calculate the specific weight and the specific gravity of the wood.
14. A wooden block weighs 5 lb in air. A sinker attached to the block weighs 4 lb in air. The weight of sinker and block is 8.5 lb when only the sinker is immersed in water. When both sinker and block are immersed in water, the apparent weight is 2 lb. Find the specific weight and the specific gravity of the wood.
15. A specific-gravity bottle weighs 146.4 g when filled with water and 204.6 g when filled with carbon tetrachloride. The empty specific-gravity bottle weighs 30 g. Calculate, for carbon tetrachloride in the metric system, (a) the specific gravity; (b) the specific weight. (c) Calculate both in the British system.
16. A specific-gravity bottle weighs 130.4 g when filled with water. When filled with carbon tetrachloride, it weighs 191.24 g. The specific-gravity bottle weighs 29 g when empty. What are the specific gravity and specific weight of the carbon tetrachloride in the metric and British systems of units?
17. A 52-g specific-gravity bottle weighs 198 g when lead shot is put into it. The specific-gravity bottle is then filled with water. The total weight is now 335 g. If the bottle holds 150 cm^3 of water, calculate the specific gravity of the lead.
18. A hydrometer has a cylindrical stem 0.2 in. in diameter. The hydrometer weighs 0.065 lb and is to be used in a range of specific gravity 0.7–0.86. How long must the stem be in order to cover this range?

19. A hydrometer weighs 0.08 lb and has a uniform diameter of 0.44 in. (a) Where is the water mark on the hydrometer? (b) How much will the hydrometer rise above the water mark in a liquid of specific gravity 1.26?
20. Referring to Prob. 19, calculate how much the hydrometer will sink below the previous water mark in a liquid of specific gravity 0.8.
21. The stem of a hydrometer is 4 mm in diameter and 30 cm long. The hydrometer weighs 40 g and sinks into water so that its stem protrudes 12 cm above the surface of the water. (a) When the hydrometer is submerged in an unknown liquid, the stem emerges only 5 cm. (b) When the hydrometer is submerged in another unknown liquid, the stem emerges 17 cm. What is the specific gravity of each liquid?

15

Liquids: Motion

15.1. Bernoulli's Theorem

Liquids may possess potential or kinetic energy. They also possess energy as a result of static pressure. If the liquid flow is nonturbulent, these three forms of energy may all contribute to the overall pressure exerted by the system. *Bernoulli's theorem* is a statement of the sum of these energies.

Potential head (pot hd) is the potential energy divided by the weight:

$$\text{pot hd} = \frac{wh}{w} = h$$

Velocity head (vel hd) is the kinetic energy of the system divided by the weight of the liquid:

$$\text{vel hd} = \frac{\frac{1}{2}(w/g)v^2}{w} = \frac{v^2}{2g}$$

w = weight: lb
h = height of column above reference: ft
pot hd = ft
pres hd = ft
vel hd = ft
W = sp wt: lb/ft³
v = velocity: ft/sec
g = grav accel: ft/sec²

Pressure head (pres hd) is the energy exerted by the liquid as a consequence of its pressure. Since

$$P = hW$$

and since h can support a pressure P per unit weight W of liquid, then

$$\text{pres hd} = h = \frac{P}{W}$$

Sec. 15.2 Quantity Flow: The Continuity Principle

Since the principle of conservation of energy applies, the sum of these three energies must equal a constant K. Thus

$$h + \frac{v^2}{2g} + \frac{P}{W} = K$$

If the equation is applied to two points, Fig. 15.1, then

$$h_1 + \frac{v_1^2}{2g} + \frac{P_1}{W} = h_2 + \frac{v_2^2}{2g} + \frac{P_2}{W}$$

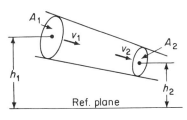

Fig. 15.1

If the initial flow is aided by a mechanism, such as a pump, then this additional energy must be added to the appropriate side of the equation (the left-hand side above). If the mechanism uses energy from the system, then this energy must be subtracted from the left-hand side of the equation:

$$h_1 + \frac{v_1^2}{2g} + \frac{h_1}{W} \pm \text{energy} = K$$

15.2. Quantity Flow: The Continuity Principle

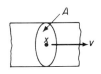

Fig. 15.2

Figure 15.2 shows a section of a pipe which has a given area A. If the liquid flowing in the pipe is pressing forward at a given velocity (no turbulence), then for every unit of time, a given volume of liquid will flow past a point x in the pipe. This volume of liquid is represented as a quantity of flow by the letter Q:

$$Q = \frac{\text{volume}}{\text{unit time}}$$

$$= \frac{As}{t} = \frac{Avt}{t}$$

$$= Av$$

$\left\{\begin{array}{l} Q = \text{quantity flow: ft}^3/\text{sec} \\ A = \text{area: ft}^2 \\ v = \text{velocity: ft/sec} \\ s = \text{distance: ft} \\ t = \text{time: sec} \end{array}\right\}$

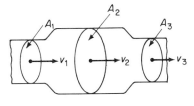

Fig. 15.3

Figure 15.3 shows a pipe having several diameters. Since liquids are incompressible and since we are dealing with nonturbulent flow, the *continuity principle* states that the quantity of liquid entering the pipe at one end per unit time must leave the other end in the same time. If less liquid leaves the pipe than enters it, the volume will build up and the pipe will break. If more liquid leaves than enters the pipe, the pipe will eventually be empty. Therefore,

$$Q_1 = Q_2 = Q_3$$

Substituting for Q,

$$Q_1 = A_1 v_1, \qquad Q_2 = A_2 v_2, \qquad Q_3 = A_3 v_3$$

the *equation of continuity* becomes

$$A_1 v_1 = A_2 v_2 = A_3 v_3 = \ldots$$

Example 1

Figure 15.4 shows a two-diameter pipe. The input diameter is 6 in. and the output diameter is 9 in. The velocity of the liquid at the input end is 10 ft/sec. Calculate (1) the quantity flow in both diameters; (2) the velocity of the liquid at the liquid at the output end.

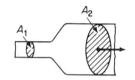

Fig. 15.4

Solution

The areas are

$$A_1 = \frac{\pi d_1^2}{4} = \frac{\pi (6/12)^2}{4} = 0.196 \text{ ft}^2$$

$$A_2 = \frac{\pi d_2^2}{4} = \frac{\pi (9/12)^2}{4} = 0.442 \text{ ft}^2$$

The quantity flow is $\left\{\begin{array}{l} d_1 = 6 \text{ in.} \\ d_2 = 9 \text{ in.} \\ v_1 = 10 \text{ ft/sec} \end{array}\right\}$

$$Q_1 = Q_2 = A_1 v_1 = \frac{\pi 6^2}{4 \times 144} \times 10 = 1.96 \text{ ft}^3/\text{sec}$$

The velocity at the large end is

$$v_2 = \frac{Q}{A_2} = \frac{1.96}{0.442} = 4.43 \text{ ft/sec}$$

Example 2

The two-diameter pipe shown in Fig. 15.5 transports a liquid (sp gr = 0.7). The two distances to the centers of cross-sections A_1 and A_2 from the reference plane are respectively 4 in. and 8 in., and their respective areas are 5 in. and 9 in. The pressure in the small section is 40 psig (lb/in.² gage), and this pressure produces a flow of liquid of 600 ft³/min. Calculate (1) the potential head; (2) the velocities; (3) the velocity head; (4) the pressure head in A_1; (5) the pressure in A_2.

Sec. 15.2 Quantity Flow: The Continuity Principle

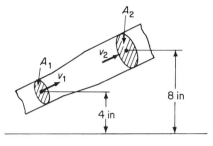

Fig. 15.5

Solution

1. Potential head:

$$(\text{pot hd})_1 = h_1 = \frac{4}{12} = \frac{1}{3} = 0.33 \text{ ft}$$

$$(\text{pot hd})_2 = h_2 = \frac{8}{12} = \frac{2}{3} = 0.67 \text{ ft}$$

2. Velocities:

$$v_1 = \frac{Q}{A_1} = \frac{(600/60)4}{\pi(5/12)^2} = 73.34 \text{ ft/sec}$$

$$v_2 = \frac{Q}{A_2} = \frac{(600/60)4}{\pi(9/12)^2} = 22.65 \text{ ft/sec}$$

3. Velocity heads:

$$(\text{vel hd})_1 = \frac{v_1^2}{2g} = \frac{(73.34)^2}{2(32)} = 84.04 \text{ ft}$$

$$(\text{vel hd})_2 = \frac{v_2^2}{2g} = \frac{(22.65)^2}{2(32)} = 8.02 \text{ ft}$$

$$\left\{\begin{array}{l} \text{sp gr} = 0.7 \\ h_1 = 4 \text{ in.} \\ h_2 = 8 \text{ in.} \\ d_1 = 5 \text{ in.} \\ d_2 = 9 \text{ in.} \\ P_1 = 40 \text{ psig} \\ Q = 600 \text{ ft}^3/\text{min} \\ g = 32 \text{ ft/sec}^2 \\ W = 62.4 \times 0.7 \end{array}\right\}$$

4. Pressure head in A_1:

$$(\text{pres hd})_1 = \frac{P_1}{W} = \frac{40 \times 144}{62.4 \times 0.7} = 131.87 \text{ ft}$$

5. Pressure in A_2:

Solving for P_2 in Bernoulli's equation,

$$P_2 = W\left(h_1 + \frac{v_1^2}{2g} + \frac{P_1}{W} - h_2 - \frac{v_2^2}{2g}\right)$$
$$= 62.4 \times 0.7(0.33 + 84.04 + 131.87 - 0.67 - 8.02) = 9065.78 \text{ psf}$$
$$= 9065.78 \text{ psf} = 63.0 \text{ psi}$$

15.3. Horsepower

The total head calculated in Sec. 15.2 may be used to find the horsepower generated by the flow of a stream of fluid. Since total head is energy per weight of liquid and since quantity of flow is energy per unit time, the power is

$$\text{power} = QW \times \text{total head}$$

If this power is divided by 500 ft-lb/sec, the equation for horsepower delivered becomes

$$\text{hp} = \frac{QW \times \text{total head}}{550}$$

Example 3

Calculate the horsepower delivered by a pump which generates energy of 25 ft-lb/lb, when the quantity flow is 5 ft^3/sec of water.

Solution

$$\text{hp} = \frac{5 \text{ ft}^3/\text{sec} \times 62.4 \text{ lb/ft}^3 \times 25 \text{ ft/lb/lb}}{550 \text{ ft-lb/sec}} = 14.2 \text{ hp}$$

15.4. Impact of Fluid Flow: Stationary Obstruction

Figure 15.6(a) shows a stream of liquid striking a stationary surface. Assuming that the liquid before impact is flowing with a velocity whose vector is perpendicular to the stationary surface, and that the liquid after impact flows parallel to the surface, it can be said that the velocity in the x-direction

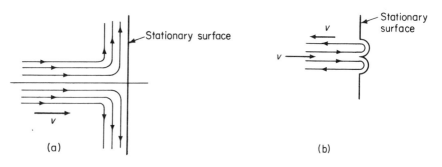

Fig. 15.6

is zero *after* impact (idealized condition). The momentum in the x-direction is completely destroyed during impact.

Figure 15.6(b) shows the case where the velocity in the x-direction before impact is equal to the velocity after impact (assuming no losses). None of the momentum present before impact is destroyed during impact; only the direction is changed.

Impacts may also occur against stationary surfaces, as shown in Fig. 15.7(a) and (b).

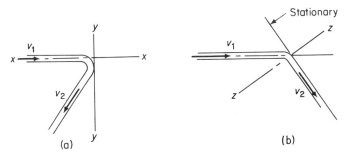

Fig. 15.7

15.5. Vena Contracta and the Coefficient of Discharge

In this chapter, we investigate the various methods for measuring fluid flow—for both liquids and gases. First, we consider the case of the orifice, (Fig. 15.8), essentially a small opening in a tank. We assume that the hole is perfectly round, that the stream of fluid is perfectly round, and that the fluid flow is nonturbulent. The surface of the liquid is subjected to atmospheric pressure (unless stated otherwise) and the reference line is taken to the center of the orifice.

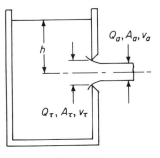

Fig. 15.8

The total discharge through the orifice is less than the theoretical value of discharge. As the stream leaves the orifice, it contracts, so that it achieves its minimum area some distance from the orifice itself. At this distance, the velocity of the liquid is at its maximum, although less than the calculated value at the orifice. This contracted area A_a is called the *vena contracta;* it is the true area of the *stream.* Thus the true flow calculations of the stream are according to the *vena contracta.*

204 *Liquids: Motion* Chap. 15

The reduction in area of water, from theoretical to actual, is approximately 40 per cent. The reduction in velocity, from theoretical to actual, is about 2 per cent. This is owning to friction between the liquid and the walls of the container and to friction within the liquid itself. Note that we are referring to the actual area A_a as the area of the stream. Therefore, the orifice becomes the theoretical area A_t of the stream.

The *theoretical* area, velocity, and quantity flow may all be converted to *actual* area, velocity, and quantity flow by multiplying the theoretical values by a coefficient. We will restrict our discussion to this *coefficient of discharge* C_d.

15.6. Flow Through an Orifice

Bernoulli's theorem may be applied to Fig. 15.9. The following reasoning is applied: (1) The velocity v_1 is taken back in the container so that it equals zero. (2) If the surface of the liquid is taken as the reference plane, then $h_1 - h_2 = 0$. (3) The area A_a is at atmospheric pressure; and since the equations are related to atmospheric pressure, A_a is equal to zero, and therefore $P_2 = 0$. Bernoulli's equation reduces to

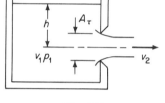

Fig. 15.9

$$\frac{P_1}{W} = \frac{v_2^2}{2g}$$

Therefore, if $P/W = h$,

$$v_2 = \sqrt{2gh}$$

From $Q = vA$ and the coefficient of discharge C_d,

$$Q_a = C_d A_t \sqrt{2gh}$$

The time required to empty a tank under *constant pressure* is

$$t = \frac{Ah}{C_d A_c \sqrt{2gh}} \quad \left\{ \begin{array}{l} A = \text{area of tank: ft}^2 \\ t = \text{time: sec} \\ C_d = \text{coef of discharge} \\ A_t = \text{theoretical area of orifice: ft}^2 \\ h = \text{height: ft} \\ g = 32 \text{ ft/sec}^2 \end{array} \right.$$

Example 4

A tank 20 ft in diameter has an orifice 45 ft below the surface of a liquid. The diameter of the orifice is 4 in. and the coefficient of discharge is 0.77. Calculate (1) the velocity of discharge; (2) the quantity of water discharged per second; (3) the time required to empty the tank under constant pressure.

Sec. 15.6   Flow Through an Orfice   205

Solution

1. The velocity at discharge is

$$v_2 = \sqrt{2gh} = \sqrt{2(32)(45)} = 53.7 \text{ ft/sec}$$

2. The quantity of water discharged is

$$Q_a = C_d A_t \sqrt{2gh} = 0.77 \left(\frac{\pi 4^2}{4 \times 144}\right)(53.7)$$

$$= 3.6 \text{ ft}^3/\text{sec}$$

$$\begin{cases} h = 45 \text{ ft} \\ g = 32 \text{ ft/sec}^2 \\ C_d = 0.77 \\ d_t = 4 \text{ in.} \\ d = 20 \text{ ft} \\ g = 32 \text{ ft/sec}^2 \end{cases}$$

3. The time required to empty the tank under constant pressure is

$$t = \frac{Ah}{C_d A_t \sqrt{2gh}} = \frac{(\pi\, 20^2/4)45}{3.6} = 3925 \text{ sec}$$

$$= 65.4 \text{ min}$$

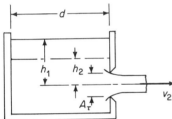

Fig. 15.10

When the *head is not constant* but varying from h_1 to h_2 as shown in Fig. 15.10, the time required to empty the tank from h_1 to h_2 is

$$t = \frac{2A(\sqrt{h_1} - \sqrt{h_2})}{C_d A_t \sqrt{2g}}$$

The time required to *empty* the tank is

$$t = \frac{2Ah_1}{C_d A_t \sqrt{2gh}}$$

Example 5

Calculate, in Example 4, (1) the time required to reduce the height of the water above the orifice from 45 ft to 10 ft; (2) the time required to empty the tank.

Solution

The area of the tank is

$$A = \frac{\pi 20^2}{4} = 314 \text{ ft}^2$$

The area of the orifice is

$$A_t = \frac{\pi 4^2}{4 \times 144} = 0.087 \text{ ft}^2$$

1. The time required to empty the water from h_1 to h_2 is

$$t = \frac{2A(\sqrt{h_1} - \sqrt{h_2})}{C_d A_t \sqrt{2g}} = \frac{2(314)(\sqrt{45} - \sqrt{10})}{0.77(0.087)(\sqrt{2 \times 32})} = 4160 \text{ sec}$$

$$= 69.3 \text{ min}$$

$$\begin{cases} h_1 = 45 \text{ ft} \\ h_2 = 10 \text{ ft} \\ v_2 = 53.7 \text{ ft/sec} \\ C_d = 0.77 \\ d_t = 4 \text{ in.} \\ d = 20 \text{ ft} \\ g = 32 \text{ ft/sec}^2 \end{cases}$$

2. The time required to empty the tank is

$$t = \frac{2Ah_1}{C_dA_t\sqrt{2gh}} = \frac{2(314)45}{0.77\,(0.087)\sqrt{2 \times 32 \times 45}} = 7850 \text{ sec}$$
$$= 130.8 \text{ min}$$

15.7. The Nozzle

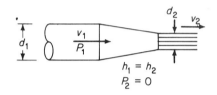

Fig. 15.11

The discharge through a nozzle may also be calculated using Bernoulli's equation and the continuity equation. In Fig. 15.11, a horizontal nozzle gives $h_1 = h_2$ and $P_2 = 0$.

The equation for the pressure head in the hose is

$$\frac{P_1}{W} = \frac{v_2^2}{2g} - \frac{v_1^2}{2g}$$

The pressure P_1 is therefore

$$P_1 = \frac{W}{2g}(v_2^2 - v_1^2)$$

The equation for actual flow is

$$Q_a = \frac{C_dA_2A_1}{\sqrt{A_1^2 - A_2^2}}\sqrt{2g\frac{P_1}{W}}$$

Example 6

The nozzle in Fig. 15.11 is subjected to a pressure of 6 psig and a velocity of 210 ft/min at $d_1 = 10$ in. If the diameter of the issuing stream is 6 in. and the coefficient of discharge is 0.7, calculate the quantity of water per minute flowing from the nozzle.

Solution

The areas are

$$A_1 = \frac{\pi d_1^2}{4} = \frac{3.14(10/12)^2}{4} = 0.545 \text{ ft}^2$$

$$A_2 = \frac{\pi d_2^2}{4} = \frac{3.14(6/12)^2}{4} = 0.196 \text{ ft}^2$$

$$\left\{ \begin{array}{l} P_1 = 6 \text{ psig} \\ v = 210 \text{ ft/min} \\ d_1 = 10 \text{ in.} \\ d_2 = 6 \text{ in.} \\ C_d = 0.7 \\ g = 32 \text{ ft/sec}^2 \\ W = 62.4 \text{ lb/ft}^3 \end{array} \right.$$

$$v = \frac{210 \text{ ft/min}}{60 \text{ sec/min}} = 3.5 \text{ ft/sec}$$

$$P_1 = 6 \text{ psig} \times 144 = 864 \text{ psfg}$$

The actual flow is

$$Q_a = \frac{C_d A_1 A_2}{\sqrt{A_1^2 - A_2^2}} \sqrt{2g \frac{P_1}{W}}$$

$$= \frac{0.7(0.545)(0.196)}{\sqrt{(0.545)^2 - (0.196)^2}} \sqrt{2(32)\frac{864}{62.4}}$$

$$= 4.38 \text{ ft}^3/\text{sec} \times 60 = 262.8 \text{ ft}^3/\text{min}$$

15.8. The Venturi Meter

The *Venturi meter*, Fig. 15.12(a), is an instrument used for the very accurate measurement of full fluid flow under pressure. The liquid enters area A_1 and is very rapidly constricted by a tapered portion, included angle approximately 20°, into an area A_2, which is smaller than area A_1. Thus, in order that the same quantity of liquid per unit time may pass area A_2 as entered A_1, the velocity must *increase* from v_1 to v_2. But as the velocity increases, the pressure drops. Therefore, P_2 will be less than P_1 because the velocity increase requires an increase in kinetic energy. The kinetic energy comes from the pressure. The more liquid that flows, the greater the kinetic energy absorbed, and therefore the greater the difference in pressures. The included angle on the downstream end of the Venturi meter is about 5°, and it affects the pressure very slightly.

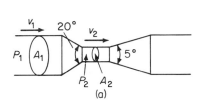
(a)

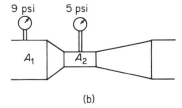
(b)

Fig. 15.12

The equation for velocity in A_1, Fig. 15.12(a), may be obtained from the equation

$$v_1 = \frac{A_2}{\sqrt{A_1^2 - A_2^2}} \sqrt{\left(\frac{2g}{W}\right)(P_1 - P_2)}$$

The actual liquid flow is

$$Q_a = C_d A_1 v_1$$

208 Liquids: Motion Chap. 15

The velocity in A_2 is

$$v_2 = \frac{A_1}{A_2} v_1$$

Example 7

A Venturi meter has an input diameter of 6 in. and a small diameter of 3 in., as shown in Fig. 15.12(b). The pressures are $P_1 = 9$ psi and $P_2 = 5$ psi. Calculate (1) the velocity of the liquid in the mainline; (2) the liquid flow; (3) the velocity in the small diameter. The coefficient of discharge is 0.95.

Solution

The areas are

$$A_1 = \frac{\pi (6/12)^2}{4} = 0.196 \text{ ft}^2$$

$$A_2 = \frac{\pi (3/12)^2}{4} = 0.049 \text{ ft}^2$$

$$P_1 = 9 \times 144 = 1296 \text{ psf}$$

$$P_2 = 5 \times 144 = 720 \text{ psf}$$

1. The velocity in the mainline is

$$v_1 = \frac{A_2}{\sqrt{A_1^2 - A_2^2}} \sqrt{\frac{2g}{W}(P_1 - P_2)}$$

$$= \frac{0.049}{\sqrt{(0.196)^2 - (0.049)^2}} \sqrt{\frac{2(32)}{62.4}} (1296 - 720)$$

$$= 6.25 \text{ ft/sec}$$

$\left\{\begin{array}{l} d_1 = 6 \text{ in.} \\ d_2 = 3 \text{ in.} \\ P_1 = 9 \text{ psi} \\ P_2 = 5 \text{ psi} \\ C_d = 0.95 \\ W = 62.4 \text{ lb/ft}^3 \\ g = 32 \text{ ft/sec}^2 \end{array}\right\}$

2. The liquid flow is

$$Q_a = C_d A_1 v_1 = 0.95(0.196)(6.25) = 1.163 \text{ ft}^3/\text{sec}$$

3. The velocity in A_2 is

$$v_2 = \frac{A_1}{A_2} v_1 = \frac{0.196}{0.049} (6.25) = 25 \text{ ft/sec}$$

15.9. The Pitot Tube for Measuring Pressures

A tube inserted into a fluid-carrying pipe may measure various conditions. The tube in Fig. 15.13(a) measures static pressure only. In Fig. 15.13(b), the end of the tube is open in the direction of the flow of the liquid and registers static and velocity pressure. This is called a *Pitot tube*. The tube in Fig. 15.13(c)

Sec. 15.9 The Pilot Tube for Measuring Pressures

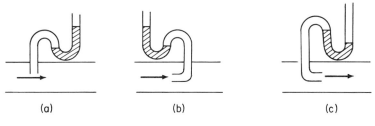

Fig. 15.13

faces in the direction of flow (downstream). Since a vacuum is created at the tube opening as a result of the velocity of the liquid, this tube measures vacuum and static pressure.

A combination of the Pitot tube and the static-pressure tube, as shown in Fig. 15.14(a), measures velocity and static pressure. Thus from Bernoulli's equation,

$$\frac{v_1^2}{2g} + \frac{P_1}{W} = \frac{v^2}{2g} + \frac{P_2}{W}$$

If $P_1 = 13.6h_1 W$, $P_2 = 13.6h_2 W$, and $v_1 = 0$, the equation reduces to

$$v = \sqrt{2g(h_1 - h_2) \times \text{sp gr of liquid in manometer}}$$

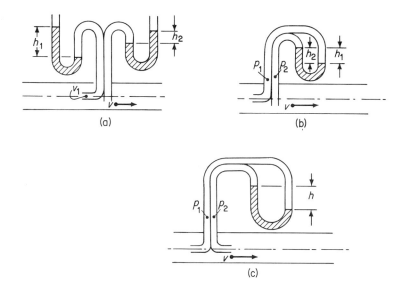

Fig. 15.14

210 Liquids: Motion Chap. 15

Example 8

Calculate the velocity of a liquid equipped with two mercury manometers as shown in Fig. 15.14(a) if $h_1 = 12$ in. and $h_2 = 3$ in.

Solution

$$v = \sqrt{2g\,(h_1 - h_2)13.6} = \sqrt{2(32)\left(\frac{12}{12} - \frac{3}{12}\right)13.6}$$

$$= 25.5 \text{ ft/sec}$$

If the table arrangement is as shown in Fig. 15.14(b), the equation for a mercury manometer measuring the velocity of a liquid in the pipeline is

$$v = \sqrt{2g\,(\Delta h)\left(\frac{13.6}{\text{sp gr}} - 1\right)}$$

Example 9

Assume the difference in mercury levels of a manometer, Fig. 15.14(b), is 9 in. and that the liquid in the pipe is gasoline (sp gr = 0.7). Calculate the velocity of the gasoline.

Solution

$$v = \sqrt{2(32)\left(\frac{9}{12}\right)\left(\frac{13.6}{0.7} - 1\right)}$$

$$= 29.7 \text{ ft/sec}$$

The upstream end of the *pitometer*, Fig. 15.14(c), measures pressure and velocity, whereas the downstream end of the tube measures vacuum and pressure. Thus

$$\frac{P_1}{W} + \frac{v^2}{2g} = \frac{P_2}{W} - \frac{v^2}{2g}$$

If $P_1 = h_1 W$ and $P_2 = 13.6 W h_2/(\text{sp gr})_L$, the equation reduces to*

$$v = \sqrt{gh\left[\frac{13.6}{(\text{sp gr})_L} - 1\right]}$$

Example 10

The pitometer, Fig. 15.14(c), registers a difference in mercury level of 18 in. when subjected to a flow of a liquid (sp gr = 1.2). Calculate the theoretical velocity.

---

* The actual velocity v in the following equation must be multiplied by a coefficient, usually 0.8.

Solution

$$v = \sqrt{gh\left[\frac{13.6}{(\text{sp gr})_L} - 1\right]} = \sqrt{32\left(\frac{18}{12}\right)\left(\frac{13.6}{1.2} - 1\right)}$$

$$= 22.34 \text{ ft/sec}$$

Problems

1. A horizontal section of pipe has two diameters. The first diameter is 3 in. and the second diameter is 6 in. If the liquid flow through the first diameter is 300 ft³/min, calculate (a) the velocity in the small diameter; (b) the velocity in the second diameter.
2. Assume three horizontal sections of pipe with the following diameters: $d_1 = 4$ in.; $d_2 = 6$ in.; $d_3 = 5$ in. If the velocity of the fluid in d_1 is 30 ft/sec, calculate (a) the velocity through d_2; (b) the velocity through d_3; (c) the quantity flow.
3. A constant-diameter vertical pipe has a water flow of 60 psi. A plug is removed from the top of the pipe, thus creating a jet. (a) How high will the jet go? (b) Calculate the velocity of the escaping water at the opening.
4. A tank of water is 100 ft above the ground. If no water is flowing from the tank, calculate (a) the pressure head; (b) the velocity head; (c) the potential head. All calculations are to take place at the surface of the water.
5. In Prob. 4, calculate the following at the *discharge point*: (a) the pressure head; (b) the velocity head; (c) the potential head if the water is discharged from the bottom of the tank to a street 20 ft below the bottom of the tank.
6. In Fig. 15.15, the pressure in the 5-in.-diameter section is 20 psig and the flow is 180 ft.³/min. If the specific gravity of the liquid is 0.8, calculate the pressure at the small end.

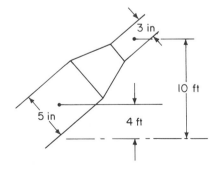

Fig. 15.15

7. Calculate the pressure at the small diameter in Fig. 15.16 if the pressure at the large diameter is 20 psi.

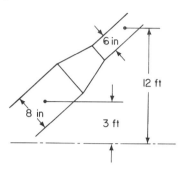

Fig. 15.16

8. In Fig. 15.17, the flow of a liquid is 120 ft³/min and its specific gravity is 0.9. If the pressure in the large-diameter pipe is 50 psig, calculate the velocity at the two areas and the pressure at the small area.

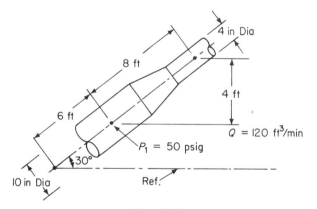

Fig. 15.17

9. Figure 15.18 shows a tank containing water to a depth of 4.6 ft to the connecting-pipe centerline. The water in the first tube stands at 3.45 ft; in the second pipe,

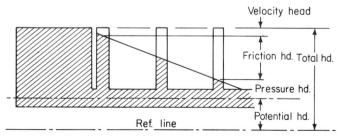

Fig. 15.18

at 2.3 ft; in the third pipe, at 1.15 ft. The centerline is 2 ft above the reference line. Find the total head by two methods, neglecting capillarity.
10. A pipe, diameter 10 in., enters the input end of a pump. When the water flow is 240 ft^3/min, a vacuum gage reads 14 in. of mercury. The output pipe is 8 ft vertically underneath the input pipe. The output pipe is 6 in. in diameter and records a pressure reading of 40 psig. Find the horsepower output delivered by the pump.
11. The input opening of a turbine is 8 in.; the output opening is 10 in. The input pressure is 50 psi; the output pressure is 8 psi. If the input end is 4 ft above the output end and the quantity flow is 6 ft^3/sec, find the horsepower that the turbine takes from the water.
12. A 50-ft-diameter tank has an orifice 20 ft below the surface of the liquid. The diameter of the orifice is 6 in. and the coefficient of discharge is 0.8. Calculate (a) the velocity of discharge; (b) the actual quantity flow; (c) the time required to empty the tank under constant pressure.
13. Assume the pressure in Prob. 12 is not constant. How long will it take to empty the tank?
14. Assume the pressure in Prob. 12 is not constant and the tank is emptied from 20 ft to 8 ft above the orifice. Calculate the time required.
15. A tank whose diameter is 20 ft holds water so that the depth of the liquid to the centerline of a 2-in.-diameter orifice is 16 ft. The coefficient of discharge is 0.9. (a) How much water is left in the tank after 2 hr, 5 min? (b) How much water is discharged?
16. An orifice whose diameter is 1 in. has its centerline 30 ft below the surface of a liquid. The orifice is capped and air is pumped into the tank to a pressure of 50 psi. If the coefficient of discharge is 0.8, find the rate of liquid discharged when the plug is pulled from the orifice.
17. A nozzle is subjected to a pressure of 8 psig and a velocity of 300 ft/min at $d_1 = 6$ in. in Fig. 15.11. If the diameter of the issuing stream is 4 in. and the coefficient of discharge is 0.8, calculate the quantity of water per minute flowing from the nozzle.
18. A water hose 1 in. in diameter has a nozzle $\frac{3}{8}$ in. in diameter fastened to it. When the water is turned on and the hose is held horizontally so the nozzle is 4 ft off the ground, the stream strikes the ground 40 ft away. (a) What is the velocity of the water? (b) What is the pressure in the hose?
19. Referring to Fig. 15.12(b), the gage pressures are $P_1 = 12$ psi and $P_2 = 9$ psi. The diameters of the Venturi meter are 5 in. and 2 in., respectively. If the coefficient of discharge is 0.9, calculate (a) the velocity in the mainline; (b) the liquid flow; (c) the velocity in the small diameter.
20. A Venturi meter is inserted into the horizontal section of a pipeline whose entrance diameter is 18 in. Find the flow of water if the throat diameter is 12 in., the difference in pressures is 30 psi, and the coefficient of discharge is 0.97.
21. Given the two mercury manometers as shown in Fig. 15.14(a), if $h_1 = 20$ in. and $h_2 = 8$ in., calculate the velocity of the liquid in the pipe.
22. The velocity of water in a pipe is known to be 40 ft/sec. What should be the difference in height in a mercury Pitot-tube static-pressure manometer of the type shown in Fig. 15.14(a)?

23. Assume the difference in mercury levels in a manometer Fig. 15.14(b) to be 6 in. and the liquid flowing in the pipe has a specific gravity of 0.87. Calculate the velocity of the fluid.
24. The pitometer, Fig. 15.14(c), measures the flow of a liquid (sp gr = 0.85). If the difference in mercury levels is 12 in., calculate the velocity in the pipe.
25. A pitometer, Fig. 15.14(c), registers a difference in mercury levels of 8 in. when inserted into a pipe, coefficient of velocity 0.9. The velocity in the pipeline is known to be 8 ft/sec. Calculate the specific gravity of the liquid.

16

The Expansion of Materials

16.1. Temperature

Temperature is a concept which is indirectly associated with kinetic energy through the concept of *heat*. The measurement or detection of temperature is a subjective concept insofar as the human body is concerned. Things seem "hotter" if the heat flows from the object to one's hand, and "colder" if the heat flows from one's hand to the object. Therefore, the human body is very unreliable for determining the temperature of a substance. Even when it is necessary to check comparative hotness or coldness, the human body is inefficient and many times will give the wrong answer, especially when the temperature differential is small. It was because of the unreliability of the human senses that the fundamental principle that "heat always flows from the hot to the cold area" was established.

Many devices utilize the physical property of a substance that responds to temperature change with some accuracy. These devices may be used as temperature indicators. The *thermometer* is one of these devices. It employs a column of mercury in a glass tube and relies on the principle that the addition of heat will expand the mercury in the tube. The range through which such a thermometer may be used depends upon the linear relationship between the degree rise in temperature and the unit increase in length of the mercury column.

When the thermometer is placed in contact with a mixture of ice and water at standard pressure, the mercury will be at one level. When the thermometer is placed in contact with a mixture of steam and water at constant pressure, the mercury will stand at another level. These two levels, called *fixed points*, are taken as reference points, and the space between them may be divided into any convenient number of divisions.

The *Fahrenheit scale* fixes the ice point at 32 °F and the steam point at 212 °F, thus giving 180 °F between these two points. The *centigrade*, or *Celsius*, *scale* fixes a value of 0 °C as the ice point and 100 °C as the steam point, and gives a range of 100 °C between the two fixed points. (The foregoing conditions prevail at standard pressure.)

The use of the notations "degree Fahrenheit" and "Fahrenheit degree" should be noted very carefully. *Degree Fahrenheit* simply refers to a temperature, whereas *Fahrenheit degree* refers to a *difference* in two temperatures. The same reasoning applies to degree centigrade and centigrade degree.

16.2. Fahrenheit and Centigrade Conversions

The ratio of the centigrade interval to the Fahrenheit interval yields the conversion equation from the Fahrenheit to the centigrade scale, or from the centigrade to the Fahrenheit scale. Thus

$$\frac{°C}{100°C} = \frac{°F - 32}{212° - 32}$$

Solving for centigrade (°C),

$$C = \tfrac{5}{9}(°F - 32) \qquad \left\{ \begin{array}{l} °C = \text{centigrade temp} \\ °F = \text{Fahrenheit temp} \end{array} \right\}$$

Solving for Fahrenheit (°F),

$$F = \tfrac{9}{5}°C + 32$$

When dealing with a temperature difference, or interval, the equations are

$$\frac{C°}{100} = \frac{F°}{180}$$

Solving for C°,

$$C = \tfrac{5}{9} F°$$

Solving for F°,

$$F = \tfrac{9}{5} C°$$

Example 1

A heat-treating furnace increases its temperature from 70 °F to 1500 °F. Calculate (1) both temperatures as centigrade temperatures; (2) the temperature difference. (3) Verify using the temperature-degree-conversion equation.

Solution

1. Conversion to centigrade-temperature scale:

$$C = \tfrac{5}{9}(°F - 32) = \tfrac{5}{9}(70° - 32°)$$
$$= 21.1°C$$
$$C = \tfrac{5}{9}(°F - 32) = \tfrac{5}{9}(1500° - 32)$$
$$= 815.5°C$$

2. Temperature difference:

$$\Delta T = 815.5°C - 21.1°C = 794.4C°$$

3. Temperature difference using conversion equation:

$$C = \tfrac{5}{9}F° = \tfrac{5}{9}(1500° - 70°) = \tfrac{5}{9} \times 1430F°$$
$$= 794.4C°$$

16.3. Absolute-Temperature Scales

If the lowest temperature possible is used as one of the fixed points, a scale may be established which uses this point as a reference. This type of scale is called an *absolute-zero scale* and the zero is referred to as *absolute zero*.

On the Fahrenheit scale, the absolute-zero temperature is at $-460°F$, or at $0°R$, where "R" stands for the *Rankin* scale, which is the absolute Fahreneeit scale. The centigrade-scale absolute zero is at $-273°C$, or $0°K$, where "K" stands for *Kelvin*.

Thus $0°C$ would read $273°K$ on a Kelvin thermometer, and $0°F$ would read $460°R$ on a Rankin thermometer, as shown in Fig. 16.1. Note that

$$32°F = 492°R = 0°C = 273°K$$

Thus to change from the Fahrenheit to Rankin scale,

$$R = °F + 460°$$

To change from the centigrade to the Kelvin scale,

$$K = °C + 273$$

To change from the Kelvin to the Rankin scale,

$$R = \tfrac{9}{5}°K$$

To change from the Rankin to the Kelvin scale,

$$K = \tfrac{5}{9}°R$$

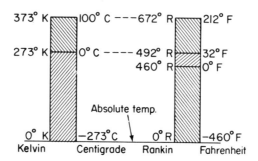

Fig. 16.1

Example 2

Convert 86°F to (1) centigrade degrees; (2) Kelvin degrees; (3) Rankin degrees.

Solution

1. Conversion to the centigrade scale:

$$C = \tfrac{5}{9}(°F - 32) = (86°F - 32)$$
$$= 30°C$$

2. Conversion to the Kelvin scale:

$$K = 273° + 30° = 303°K$$

3. Conversion to the Rankin scale:

$$R = 460° + 86° = 546°R$$

Example 3

Convert −45°C to (1) Fahrenheit degrees; (2) Kelvin degrees; (c) Rankin degrees.

Solution

1. Conversion to the Fahrenheit scale:

$$F = \tfrac{9}{5}°C + 32 = \tfrac{9}{5}(-45°C) + 32$$
$$= -49°F$$

2. Conversion to the Kelvin scale:

$$K = 273 + (-45) = 228°K$$

Sec. 16.4 Linear Expansion of Materials 219

3. Conversion to the Rankin scale:

$$R = 460 + (-49) = 411°R$$

16.4. Linear Expansion of Materials

When a rod is heated, a change in the length of the rod takes place. This change in length is called *linear expansion*. Linear expansion is proportional to the change in temperature, Δt, and to the length of the rod at the initial temperature. Thus

$$\Delta l = l_f - l_i \quad \text{and} \quad \Delta T = T_f - T_i$$

and

$$\Delta l \propto l_i \, \Delta T$$

The equation is

$$\Delta l = \alpha l_i \, \Delta T$$

Solving for the *coefficient of linear expansion*,

$\begin{pmatrix} T_f = \text{final temp} \\ T_i = \text{initial temp} \\ \Delta T = \text{change in temp} \\ l_i = \text{initial length} \\ l_f = \text{final length} \\ \Delta l = \text{change in length} \\ \alpha = \text{coef of expansion} \end{pmatrix}$

$$\alpha = \frac{\Delta l}{l_i \Delta T} = \frac{l_f - l_i}{l_i(T_f - T_i)} = \frac{\cancel{ft}}{\cancel{ft} \times \text{deg.}} = \text{per deg.}$$

The coefficient of linear expansion for various solids and liquids is given in Appendix A.11.

The units indicate that the change in length per unit length is based entirely on the change in temperature. Thus the units of length, if they are the same, will cancel, and the coefficient of linear expansion is a function of temperature only. Thus steel expands $0.063 \times 10^{-4}/F°$, or $0.0000063/F°$, for every degree increase in Fahrenheit temperature. Since we are dealing with a change in temperature, the foregoing coefficient may be changed to the equivalent centigrade units by writing

$$\alpha = \tfrac{9}{5}(6.3 \times 10^{-6}/F°) = 11.36 \times 10^{-6}/C°$$

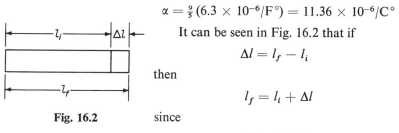

Fig. 16.2

It can be seen in Fig. 16.2 that if

$$\Delta l = l_f - l_i$$

then

$$l_f = l_i + \Delta l$$

since

$$\Delta l = \alpha l_i \, \Delta T$$

The equation for length at the final temperature may be written

$$l_f = l_i + \Delta l = l_i + \alpha l_i \, \Delta T = l_i(1 + \alpha \, \Delta T)$$

Example 4

An aluminum rod is 3 ft long at a temperature of 40°F. Calculate the change in length of the rod when the temperature goes to 400°F. The temperature coefficient of expansion is 13.33 × 10⁻⁶/°F.

Solution

$$\Delta l = \alpha l_i \Delta T = \alpha l_i(T_f - T_i)$$
$$= (13.33 \times 10^{-6}/°F) \times 3(400 - 40)$$
$$= 0.0144 \text{ ft}$$

$\left(\begin{array}{l} \Delta T = \text{change in temp} \\ T_f = 400°F \\ T_i = 40°F \\ l_i = 3 \text{ ft} \\ \alpha = 13.33 \times 10^{-6}/°F \end{array} \right)$

Example 5

In Example 4, calculate the new temperature if the rod expanded by 0.005 ft from 40°F.

Solution

$$\Delta l = \alpha l_i (T_f - T_i)$$
$$0.005 = (13.33 \times 10^{-6} \times 3)(T_f - 40°F)$$
$$T_f = \frac{0.005}{(13.33 \times 10^{-6}) \times 3} + 40°$$
$$= 165°F$$

16.5. Area Expansion of Materials

Area expansion deals with the change in area as a result of expansion of two dimensions (Fig. 16.3). The *coefficient of area expansion* may be found

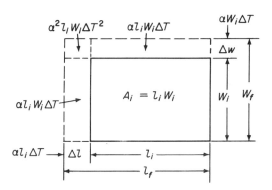

Fig. 16.3

Sec. 16.6   Volume Expansion of Materials   221

by simply multiplying the coefficient of linear expansion by 2. If we let $\beta = 2\alpha$, the equation for area expansion becomes

$$A_f = A_i(1 + \beta \Delta T) \quad \begin{cases} A_i = \text{initial area} \\ A_f = \text{final area} \\ \Delta T = \text{temp change} \\ \beta = \text{coef of area expansion} \end{cases}$$

Example 6

A circular hole cut into aluminum is 3.500 in. in diameter at 60°F. Calculate the area of the hole at 300°F.

Solution

The initial area is

$$A_i = \frac{\pi(3.5)^2}{4} = 9.62 \text{ in.}^2$$

The coefficient of area expansion is

$$\beta = 2\alpha = 2(13.33 \times 10^{-6}) = 26.66 \times 10^{-6}/°F \quad \begin{Bmatrix} d_i = 3.500 \text{ in.} \\ \alpha = 13.33 \times 10^{-6}/°F \end{Bmatrix}$$

The new area is

$$A_f = A_i(1 + \beta \Delta T) = 9.62[1 + (26.66 \times 10^{-6})(300 - 60)]$$
$$= 9.682 \text{ in.}^2$$

16.6. Volume Expansion of Materials

Volume expansion results from an increase in volume as a result of expansion through three dimensions. The *coefficient of volume expansion* may be found by multiplying the coefficient of linear expansion by 3. If we let (small *delta*) $\delta = 3\alpha$, then the equation for volume expansion is

$$V_f = V_i(1 + \delta \Delta T)$$

Example 7

Calculate (1) the new volume of an aluminum ball at 200°F which has a diameter of 24 in. at −20°F; (2) the new diameter.

Solution

1. The initial volume is

$$V_i = \tfrac{4}{3} \pi r^3 = \tfrac{4}{3} \pi (1)^3 = 4.18 \text{ ft}^3$$

The coefficient of volume expansion is

$$\delta = 3\alpha = 3(13.33 \times 10^{-6}) = 39.99 \times 10^{-6}$$

The change in temperature is

$$\Delta T = T_f - T_i = [200 - (-20)] = 220 F°$$

The new volume is

$$V_f = V_i(1 + \delta \Delta T) = 4.18[1 + (39.99 \times 10^{-6})(220)]$$
$$= 4.2168 \text{ ft}^3$$

$\begin{Bmatrix} T_f = 200°F \\ T_i = -20°F \\ d_i = 24 \text{ in.} = 2 \text{ ft} \\ \alpha = 13.33 \times 10^{-6}/°F \end{Bmatrix}$

2. The new diameter is

$$V_f = \frac{\pi d^3}{6} = \frac{\pi 2^3}{6} = 4.2168$$

Solving for d,

$$d = \sqrt[3]{\frac{V_f \times 6}{\pi}} = \sqrt[3]{\frac{4.2168 \times 6}{\pi}}$$
$$= 2.0049 \text{ ft} = 24.059 \text{ in.}$$

Note that when dealing with a substance which is in a container, when both substance and container are subjected to a temperature increase, both will expand. Thus if one desires to find out how much liquid overflows the sides of the container on raising the temperature, it is not enough to calculate the expansion of the liquid alone. The expansion of the container—volume expansion—is also necessary. The procedure is as follows:

Obtain the new volume for the liquid:

$$V_{fL} = V_i(1 + \delta_L \Delta T)$$

Obtain the new volume for the container:

$$V_{fc} = V_1(1 + \delta_c \Delta T)$$

Subtract, $V_{fL} - V_{fc}$, to get ΔV:

$$V_{fL} - V_{fc} = V_i(1 + \delta_L \Delta T) - V_i(1 + \delta_c \Delta T)$$
$$\Delta V = V_i \Delta T(\delta_L - \delta_c)$$

$\begin{Bmatrix} d = \text{diameter of container} \\ \delta_L = \text{cubic coef of expansion of the liquid} \\ \delta_c = \text{cubic coef of expansion of container} \\ \Delta T = \text{change in temp, } T_f - T_i \end{Bmatrix}$

But

$$\Delta V = \frac{\pi d^2}{4}(h)$$

Substitute and solve for h:

$$h = \frac{4}{\pi d^2}(V_i \Delta T)(\delta_L - \delta_c)$$

$\begin{Bmatrix} V_i = \text{volume at initial temp} \\ V_{fL} = \text{final volume of liquid} \\ V_{fc} = \text{final volume of container} \\ \Delta V = \text{change in volume} \end{Bmatrix}$

Sec. 16.6 Volume Expansion of Materials

Example 8

A thermometer bulb is filled with 10 cm³ of mercury at a temperature of 5°C. The bore of the tube attached to the bulb is 0.04 cm in diameter. How high will the mercury rise in the tube if the temperature increases to 90°C? The coefficient of cubical expansion of mercury is 0.1×10 M³/F° and the coefficient of linear expansion for glass is $4.9 \times 10^{-6}/°F$.

Solution

1. Convert the coefficients to the centigrade scale:

$$\delta_m = \tfrac{9}{5}(0.1 \times 10^{-3}/°F) = 0.18 \times 10^{-3}/°C \text{ (mercury)}$$
$$\alpha_g = \tfrac{9}{5}(4.9 \times 10^{-6}/°F) = 8.82 \times 10^{-6}/°C \text{ (glass)}$$
$$\delta_g = 3\alpha = 3(8.82 \times 10^{-6}) = 26.46 \times 10^{-6}/°C \text{ (glass)}$$
$$\Delta T = T_f - T_i = 90 - 40 = 50°C$$

$\left\{ \begin{array}{l} T_f = 90°C \\ T_i = 40°C \\ d_1 = 0.04 \text{ cm} \\ \delta_m = 0.1 \times 10^{-3}/°F \\ \alpha_g = 4.9 \times 10^{-6}/°F \\ V_i = 10 \text{ cm}^3 \end{array} \right\}$

2. The final volume of the mercury is

$$V_{fm} = V_i(1 + \delta_m \Delta T) = 10[1 + (0.18 \times 10^{-3} \times 50)]$$
$$= 10.090 \text{ cm}$$

3. The final volume of the glass is

$$V_{fg} = V_i(1 + \delta_g \Delta T) = 10[1 + (26.46 \times 10^{-6} \times 50)]$$
$$= 10.013 \text{ cm}$$

4. The change in volume is

$$\Delta V = 10.090 - 10.013 = 0.077 \text{ cm}^3$$

5. The height to which the mercury will rise is

$$h = \frac{\Delta V}{A} = \frac{\Delta V}{\pi d^2/4} = \frac{0.077 \times 4}{\pi 0.04^2}$$
$$= 61.3 \text{ cm}$$

We now show a most useful equation for making barometric correction for error due to the expansion of the brass scale and to the decrease in specific weight of mercury as the temperature increases. A decrease in the specific weight of mercury will yield a higher column because the pressure will be able to support a longer column of mercury. This will give an error for a given atmospheric pressure.

Since brass expands linearly and mercury expands according to its volume, the true height may be found by equating the expansion of mercury to the expansion of the brass. Thus

224 The Expansion of Materials Chap. 16

Hence
$$h_c[1 + \delta(T_f - T_m)] = h_f[1 + \alpha(T_f - T_B)]$$

$$h_c = \frac{h_f[1 + \alpha(T_f - T_B)]}{1 + \delta(T_f - T_m)}$$

$\begin{cases} h_f = \text{barometric height at } t_f \\ h_c = \text{corrected height of mercury} \\ \qquad \text{column: in. of Hg} \\ T_f = \text{room temp} \\ T_m = \text{reference temp: } 32°\text{F}; 0°\text{C} \\ \qquad \text{standard temp at which} \\ \qquad \text{mercury has standard density} \\ T_B = \text{ref temp } (62°\text{F } or \text{ } 17°\text{C}) \text{ at} \\ \qquad \text{which the brass was calibrated} \\ \delta = \text{cubical coef of expansion for Hg} \\ \alpha = \text{linear coef for brass} \end{cases}$

16.7. Expansion of Gases

Since most gases follow the same pattern of volume expansion as liquids, we can use the equations which utilize the coefficient of cubic expansion to find change in volume. It can be shown that, when the pressure is not too high, most gases expand at the rate of $\frac{1}{273}$ of the volume per degree rise in centigrade temperature, or $\frac{1}{492}$ per degree rise in Fahrenheit temperature, starting in each case at the freezing point (0°C or 32°F). It is necessary first to convert the expansion at a temperature to the freezing point and then to relate the freezing volume to the second desired temperature. The expansion is calculated at constant pressure.

We know that

$$\delta = \frac{\Delta V}{V_i \Delta T}$$

We also know that gases, under constant pressure, change their volume for every degree rise in temperature by $\frac{1}{273}$ the volume at 0°C and $\frac{1}{492}$ the volume at 32°F. Therefore, for Fahrenheit temperatures,

$$\frac{1}{492} = \frac{\Delta V}{V_i \Delta T} \quad \text{or} \quad V_f - V_i = \frac{1}{492}(V_i \Delta T)$$

and

$$V_f = V_i + \frac{1}{492}(V_i \Delta T)$$
$$= V_i\left[1 + \frac{1}{492}(T_f - T_i)\right]$$

$\begin{cases} V_f = \text{final volume} \\ V_i = \text{volume at 0°C } or \\ \qquad 32°\text{F} \\ T_f = \text{final temp} \\ T_i = \text{temp at 0°C } and \\ \qquad 32°\text{F} \end{cases}$

For centigrade temperatures,

$$V_f = V_i\left[1 + \frac{1}{273}(T_f - T_i)\right]$$

Example 9

A certain amount of gas occupies 30 ft^3 at 50°F. Calculate the volume at 180°F, assuming the pressure remains constant.

Solution

For the volume at T_1,

$$V_1 = V_i \left(1 + \frac{T_1 - T_i}{492}\right)$$

$$30 = V_i \left(1 + \frac{50 - 32}{492}\right)$$

$T_1 = 50°F$
$T_2 = 180°F$
$V_1 =$ volume at $t_1 = 30$ ft^3
$V_2 =$ volume at t_2
$V_i =$ volume at 0°C *or* 32°F

$$V_i = 28.94 \text{ ft}^3$$

For the volume at T_2,

$$V_2 = V_i \left(1 + \frac{T_2 - T_i}{492}\right) = 28.94 \left[1 + \left(\frac{180 - 32}{492}\right)\right]$$

$$= 37.65 \text{ ft}^3$$

16.8. Instruments Used as Temperature Indicators

There are various methods for checking temperature. The property of expansion with increase of temperature is utilized in the mercury and alcohol thermometers. The mercury or alcohol is put into a glass capillary, which is scaled and graduated. The expansion of the material must be relatively uniform within the range of operation of the instrument.

In some cases, materials which melt at a given temperature may be used as temperature indicators. Wax cones, which melt at their own characteristic temperature, show the effects upon their solid state. Pure metals or salts which maintain a fixed temperature as they go through a solid-to-liquid-phase change are used to calibrate other temperature-indicating instruments.

Bimetallic strips, consisting of two strips riveted together, bend toward the metal strip having the smaller coefficient of expansion when heated. This occurs because one strip is expanding more per unit rise (or decrease) in temperature than the other, causing both strips to bend in the direction of the strip that bends the least. If the combination of materials bends uniformly per degree change in temperature, the amount of bending can be used to indicate temperature, once the unit has been calibrated.

The *constant-volume gas thermometer* utilizes the change in pressure which takes place when the confined gas expands to support a column of

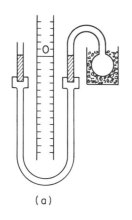 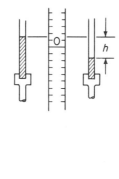 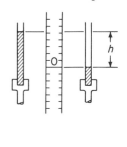

(a)   (b)   (c)

Fig. 16.4

mercury (see Fig. 16.4). The bulk is packed in ice. The left-hand tube is raised or lowered until the mercury levels in both tubes are equal. The bulb is removed from the ice and heated (in hot water). The gas expands, causing the mercury to show a difference in mercury levels, h (see Fig. 16.4b). The left-hand tube is now raised to bring the mercury in the right-hand tube to the zero point of constant volume as shown in Fig. 16.4(c). This maintains the gas volume as it was in Fig. 16.4(a). The volume of gas is now supporting a column of mercury which, when calibrated at the ice and steam points, may be used as a thermometer.

Optical pyrometers use a loop of wire attached to a battery. The voltage fed into the wire is rheostat controlled and connected to a millivoltmeter whose readings may be translated into temperature readings. The loop of wire is heated by the electric current. The amount of current sent through the wire is directly related to the glow emitted by the wire. As the temperature of the wire increases, the color of the wire goes from a dull red, through cherry red. yellow, and white. The glow of the loop is compared to the object being checked by sighting through a telescope to the loop at the hot object. If the temperature of the wire is hotter than the object being checked, the wire will be seen clearly as a glow, having brightness. If the temperature of the loop is less than the temperature of the object, the loop will appear as a dark line. If the temperature of the loop and the temperature of the object are the same, the loop will disappear when viewing the object through the loop. At this point, the temperature–millivolt chart is referred to. Sometimes the rheostat dial is calibrated to read temperature directly.

A *thermoelectric pyrometer* has two dissimilar wires fused at one end, called the *hot end*. Each wire enters a millivoltmeter calibrated to read temperature. The fused wires as a unit are called a *thermocouple*. A difference in thermoelectric potential is set up when the hot end is exposed to the tem-

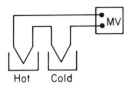

Fig. 16.5

perature to be checked. Since the current flow is different for different metals for the same temperature interval, the cold end of the couple will register a potential difference on the millivoltmeter. This millivoltmeter may be calibrated to read temperature directly, or it may be attached to some other instrument which will register temperature, e.g., a strip chart (see Fig. 16.5).

Materials used as commercial combinations are chromel–alumel, used for temperatures not to exceed 2000°F; platinum–platinum-rhodium, used for temperatures not to exceed 3000°F; iron–iron-constantan for temperatures about 1600°F; and copper–constantan, which is limited to 600°F. Several thermocouples connected in series constitute a *thermopile*, which is extremely sensitive to measuring small temperature changes at very large distances, e.g., the temperature of the stars.

Resistance thermometers are resistance wires wound on a long wire mandrel. The resistance of the wire varies as the temperature. Therefore, if the resistance of the wire at various temperatures is plotted as a calibration curve, the resistance coil may be used as a thermometer.

Problems

1. Change the following readings as indicated: (a) 15°C to °F; (b) 47°F to °C; (c) −60°C to °F; (d) −40°F to °C; (e) 25°C to °F; (f) 72°F to °C.
2. Change the following as indicated: (a) 50°F to Rankin; (b) 50°C to Kelvin; 50°R to °F; (d) 50°K to centigrade.
3. Convert the following as indicated: (a) 20°F to °K; (b) 20°C to °R; (c) 20°K to °F; (d) 20°R to °C; (e) 20°R to °K; (f) 20°F to °R.
4. A steel bar is 20 ft long at 60°F. Calculate its change in length at 90°F.
5. A brass bar is 12 ft long at a temperature of −10°F. Calculate its length at 100°F.
6. If steel rails are 40 ft long and if they are laid when the temperature is 60°F, how much space should be left between the rails so that they will not buckle at 120°F.
7. A brass bar 8 ft long and a steel bar 8.06 ft long are at 80°F. At what temperature would both bars be the same length?
8. A brass bushing at 70°F has a diameter of 1 in. Calculate its diameter when heated to 800°F.
9. The inside diameter of a large steel pipe is 24 in. at −10°F. The pipe is 10 ft long. Calculate the following when the temperature is 100°F: (a) the length of the pipe; (b) the new cross-sectional area; (c) the new diameter; (d) the new internal volume.
10. An aluminum forging is at a temperature of 120°F as a result of the temperature of the cutting oil used. The blueprint calls for a finished diameter of a machined

hole of 4.1255 in. at 70 °F. To what size must the hole be machined so that it will meet the print specifications?

11. A steel-plug gage calibrated at 70 °F is used to check the accuracy of the hole turned in Prob. 10. What should be the size of the gage? (Assume the forging and the gage are at 90 °F when the check is made.)

12. A copper bushing is bored to a diameter of 1.995 in. in a machine when the temperature is 70 °F. This bushing must be a press fit on a shaft whose diameter is 2.000 in. at 70 °F. To what temperature must the bushing be heated so that it will just pass over the shaft?

13. A steel bushing is to have a shrink fit, so that it may be heated and just pushed onto a shaft. The shaft is 1.752 in. in diameter and the bushing is 1.748 in. in diameter. Both are at 72 °F. What must be the temperature of the bushing, if heated, so that it will just slip over the shaft?

14. Alcohol has a coefficient of cubical expansion of $1.13 = 10^{-3}/°F$. If 500 cm³ of alcohol is put into a glass container so that the container is holding its maximum at 10 °C, how much will overflow the sides when the temperature increases to 70 °C?

15. A long glass tube, sealed at one end, is filled with mercury to a depth of 100 cm at 10 °C. The glass tube is graduated and has a diameter of 0.5 cm. What will be the reading on the graduations at 100 °C?

16. The bulb of a mercury thermometer is attached to a tube with a bore diameter of 0.004 in. The bulb holds 0.003 in.³ of mercury at 0 °F when full. (a) Neglecting the expansion of the glass, how far from the 0 °F mark will the top of the mercury be when the temperature is 100 °F? (b) How much error was involved by not considering the expansion of the glass? (c) Where does the mercury stand now?

17. A thermometer bulb is filled with 10 cm³ of mercury at a temperature of 10 °C. The bore of the tube attached to the bulb is 0.05 cm in diameter. How high will the mercury rise in the tube when the temperature increases to 90 °C? The coefficient of cubical expansion is $0.1 \times 10^{-3}/F°$ and the linear coefficient of expansion for glass is $0.045 \times 10^{-4}/F°$.

18. A 20-ft aluminum rod is measured with a steel tape at a temperature of 20 °F. The rod and tape are brought into the plant where the temperature is 80 °F. If the rod is measured in the plant, what should the tape read?

19. A quantity of 800 cm³ of mercury is poured into a brass container 6 in. in diameter so that the container is just filled when the temperature is 20 °C. The temperature of the contents and beaker is then raised to 150 °C so that some of the mercury spills off. Then the temperature is lowered to 80 °C. (a) How much mercury overflows the container? (b) How far below the surface of the container is the surface of the mercury at the last temperature? (Linear coefficient of the brass is $0.155 \times 10^{-4}/°F$; volume coefficient of mercury is $0.1 \times 10^{-4}/°F$.)

20. The volume of a gas is 100 ft³ at 80 °F. Calculate the volume of the gas at 400 °F at standard pressure.

21. A sample of 6 ft³ of gas is expanded to 30 ft³ by raising the temperature from 32 °F. Calculate the new temperature.

22. A gas is expanded from 100 cm³ at 20 °C to 150 cm³ at a constant pressure. Calculate the final temperature.

17

Specific Heat, Heat Transfer

17.1. Specific Heat

It has been shown that heat "flows" from an object at high temperature to an object at low temperature. If there is a transfer of heat, it is said that there is a transfer of kinetic energy.

The unit of heat in the British system of units is the Btu, which stands for *British thermal unit*. One Btu is defined as the heat required to raise the temperature of 1 pound of water 1 Fahrenheit degree. One *calorie* is the heat required to raise the temperature of 1 gram of water 1 centigrade degree.

Specific heat (sp ht) is a proportionality constant and represents the heat required to raise 1 pound of a substance through a temperature difference of 1 Fahrenheit degree; or the heat, in calories, required to raise 1 gram of a substance 1 centigrade degree.

Whereas the Btu and the calorie are related to water and represent a fixed quantity of heat, specific heat varies from substance to substance. Experiments show that 0.033 calorie will change the temperature of 1 gram of mercury 1 centigrade degree. Also, 0.117 Btu will change the temperature of 1 pound of steel 1 Fahrenheit degree. The student should take note of the fact that the magnitude of specific heat is the same in the British and the metric systems. This is shown in Appendix A.12.

If the heat rise Q is proportional to the mass m and the temperature change ΔT, then the proportion may be written as

$$Q \propto m \, \Delta T$$

The equation for change in heat—sometimes called *sensible heat* because it results from a change in temperature which is detectable by the sense of touch—is

$$Q = cm \, \Delta T \quad \left\{ \begin{array}{l} m = \text{mass: g } or \text{ lb} \\ Q = \text{change in heat: cal } or \text{ Btu} \\ \Delta T = \text{change in temp: C}° \text{ } or \text{ F}° \\ c = \text{sp ht: cal/g–C}° \text{ or Btu/lb–F}° \end{array} \right\}$$

Or

$$c = \frac{Q}{m \, \Delta T} = \frac{\text{Btu}}{\text{lb-F}°}$$

Equating specific heat in the British system with the specific heat in the metric system,

$$\frac{1 \text{ Btu}}{\text{lb-F}°} \times \frac{\text{lb}}{453.6 \text{ g}} \times \frac{180 \text{F}°}{100 \text{C}°} = \frac{1 \text{ cal}}{\text{g-C}°}$$

$$1 \text{ Btu} = \frac{453.6 \times 100}{180} \text{ cal} = 252 \text{ cal}$$

Specific heats are related to the definition of the Btu and the calorie and consequently to the mass, or weight, of water. If a substance requires a certain amount of heat to raise its temperature 1 degree, this amount of heat represents the heat capacity of that substance. The quantity of water which has the same heat capacity as the substance is called the *water equivalent.* Thus

$$\text{water equivalent} = cm = \text{heat capacity} \quad \left\{ \begin{array}{l} c = \text{sp ht} \\ m = \text{mass: g } or \text{ lb} \end{array} \right\}$$

The specific heat of water is not a constant for all temperatures. It varies only slightly from its average value of 1.000, in the 32°F–212°F range. It is 1.000 at 59°F and at 149°F. At 32°F, the specific heat of water is 1.008; at 75°F, it is 0.998; at 212°F, it is 1.007.

Example 1

How many Btu should be added to a 20 lb piece of steel to raise its temperature from 70°F to 1450°F?

Solution

$$Q = cm \, \Delta t = cm \, (T_f - T_i)$$
$$= 0.117 \times 20(1450° - 70°)$$
$$= 3229 \text{ Btu}$$

$$\left\{ \begin{array}{l} m = 20 \text{ lb} \\ T_i = 70°\text{F} \\ T_f = 1450°\text{F} \\ c = 0.117 \end{array} \right\}$$

Sec. 17.2   *Equilibrium through Mixing*   231

Example 2

What is the final temperature if 220 Btu are added to 10 lb of aluminum? The aluminum is at an initial temperature of 70°.

Solution

$$Q = cm(T_f - T_i)$$
$$220 = 0.22 \times 10 (T_f - 70)$$
$$T_f = 170°F$$

$$\left\{ \begin{array}{l} c = 0.22 \\ T_i = 70°F \\ Q = 220 \text{ Btu} \\ m = 10 \text{ lb} \end{array} \right\}$$

Example 3

Calculate the water equivalent of 50 lb of 60:40 brass.

Solution

$$\text{water equivalent} = cm = 0.092 \times 50$$
$$= 4.6 \text{ Btu/F°}$$

$$\left\{ \begin{array}{l} m = 50 \text{ lb} \\ c = 0.092 \end{array} \right\}$$

17.2. Equilibrium through Mixing

The law of conservation of energy applies when a system of objects is so insulated that no energy can escape or be absorbed by the system. A system of objects at different temperatures, when brought into contact with each other, will eventually reach an equilibrium temperature. The sum of the energies gained by the "colder" objects will equal the sum of the energies lost by the "hotter" objects. Thus

$$\text{heat gained} = \text{heat lost}$$

The statement holds for solids as well as liquids. Since pressure affects the volume of the solid or liquid only slightly, the method of mixtures may be applied to solids as well as liquids, but not to gases.

Example 4

A 5-lb aluminum pail and 20 lb of water are at 72°F. A 12-lb copper block is heated to 200°F and then dropped into the water. Calculate the resulting temperature of the system, assuming no losses.

Solution

The copper block will lose heat to the water and the aluminum. The final equilibrium temperature will be between 72°F and 200°F.

the water: $T_f > 72\,°F$

the aluminum: $T_f > 72\,°F$

the copper: $T_f < 200\,°F$

Heat gained by the water + heat gained by the aluminum = heat lost by the copper:

$$c_w m_w \,\Delta T_w + c_a m_a \,\Delta T_a = c_c m_c \,\Delta T_c$$

$\left\{\begin{array}{l} c_w = \text{Btu/lb–F}° \\ m_w = 20\text{ lb} \\ \Delta T_w = T_f - 72 \\ c_a = 0.22\text{ Btu/lb–F}° \\ m_a = 5\text{ lb} \\ \Delta T_a = T_f - 72 \\ c_c = 0.093\text{ Btu/lb–F}° \\ m_c = 12\text{ lb} \\ \Delta T_c = 200 - T_f \end{array}\right\}$

Substituting,

$$[1 \times 20(T_f - 72)] + [0.22 \times 5(T_f - 72)] = 0.093 \times 12(200 - T_f)$$

$$20 T_f - 1440 + 1.1 T_f - 79.2 = 224.0 - 1.1 T_f$$

$$22.2\, T_f = 1743.2$$

$$T_f = 78.5\,°F$$

17.3. Calorimetry

Early in this chapter, we referred to a transfer of heat which is accompanied by a change in temperature as *sensible heat*. Under other circumstances, a body may gain or lose heat without any temperature change. This happens when a material is going through a change of its physical homogeneity, which is referred to as a *change of phase*. When ice, on being heated, melts, the mixture of ice and water will remain at 32 °F (0 °C) until all the ice has melted.

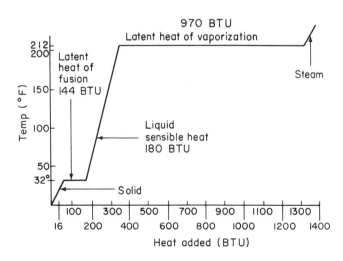

Fig. 17.1

Sec. 17.3 *Calorimetry* 233

The resulting water will then pick up sensible heat at the rate of 1 Btu per pound for every increase of 1 Fahrenheit degree until the water reaches 212°F. Once again, the heat input will be present but the temperature will remain at 212°F until all the water has turned to steam. Again, the phase change—water to steam—must be complete before the steam will pick up sensible heat so that the temperature of the steam will increase according to its specific heat (see Fig. 17.1) *Note:* The pressure at 212°F is 14.7 lb/in.^{2}.

If heat is added to ice, it takes 144 Btu to melt each pound of ice, or 80 cal to melt each gram. Of course, the reverse is true if it becomes necessary to freeze water to ice. This amount of heat (144 Btu/lb or 80 cal/g) is called the *latent heat of fusion* of water. If heat is added to water to turn it to steam, it takes 970.4 Btu to convert 1 lb of water to steam, or 540 cal to convert 1 g of water to steam. Here again, the process can be reversed. This amount of heat (970.4 Btu/lb or 540 cal/g) is called the *latent heat of vaporization* of water. Other substances have characteristic values for their latent heats of fusion and vaporization (see Appendix A.12).

For our calculations, we shall use the following *approximations* for water at standard pressure:

Specific heat of ice (c_i) = 0.5 Btu/lb-F° = 0.5 cal/g-C°
Latent heat of fusion (L_f) = 144 Btu/lb = 80 cal/g
Specific heat of water (c_w) = 1 Btu/lb-F° = 1 cal/g-C°
Latent heat of vaporization (L_v) = 970 Btu/lb = 540 cal/g
Specific heat of steam (c_s) = 0.5 Btu/lb-F° = 0.5 cal/g-C°

The latent heat required to complete the fusion-phase change is

$$Q = L_f m$$

The latent heat required to complete the vaporization-phase change is

$$Q = L_v m \quad \begin{cases} Q = \text{Btu } or \text{ cal} \\ L_f = \text{latent heat of fusion} \\ L_v = \text{latent heat of vaporization} \\ m = \text{mass: lb} \end{cases}$$

Example 5

One hundred pounds of ice at 10°F are to be changed to steam at 225°F, standard pressure. How much heat must be supplied to consummate the change? Assume no heat losses.

Solution

To change 100 lb of ice at 10°F to 100 lb of ice at 32°F,

$$Q_i = c_i m_i \,\Delta T = 0.5 \times 100(32 - 10)$$
$$= 1100 \text{ Btu}$$

To change 100 lb of ice at 32°F to 100 lb of water at 32°F,

$$Q_f = L_f m = 144 \times 100 = 14{,}400 \text{ Btu}$$

To change 100 lb of water at 32°F to 100 lb of water at 212°F,

$$Q_w = c_w m_w \Delta T = 1 \times 100(212 - 32)$$
$$= 18{,}000 \text{ Btu}$$

To change 100 lb of water at 212°F to 100 lb of steam at 212°F,

$$Q_v = L_v m = 970 \times 100 = 97{,}000 \text{ Btu}$$

To change 100 lb of steam at 212°F to 100 lb of steam at 225°F,

$$\left\{\begin{array}{l} m = 100 \text{ lb} \\ T_i = 10°\text{F} \\ T_s = 225°\text{F} \\ L_f = 144 \text{ Btu/lb} \\ L_v = 970 \text{ Btu/lb} \\ c_i = 0.5 \\ c_w = 1 \\ c_s = 0.5 \end{array}\right\}$$

$$Q_s = c_s m_s \Delta T = 0.5 \times 100(225 - 212) = 650 \text{ Btu}$$

The total heat needed to raise 100 lb of ice at 10°F to steam at 225°F, standard pressure, is

$$Q = Q_i + Q_f + Q_w + Q_v + Q_s$$
$$= 1100 + 14{,}400 + 18{,}000 + 97{,}000 + 650$$
$$= 131{,}150 \text{ Btu}$$

Now, assume it is desired that a quantity of ice be melted, so that the final condition of the mixture at equilibrium temperature is ice and water. Let us first make a wrong assumption and see what happens. The following example will illustrate. Assume first that the final equilibrium temperature is T (other than at the fusion temperature).

Example 6

A 10-lb copper container holding 15 lb of water is at 70°F. If 5 lb of ice at 10°F are added to the mixture, will all the ice melt?

Solution (First Assumption)

Assume the final condition to be liquid at temperature T.
Heat lost by the container:

$$Q_c = 0.093 \times 10(70 - T)$$

Heat lost by water:

$$Q_w = 1 \times 15(70 - T)$$

Heat gained by ice:

$$Q_i = 0.5 \times 5(32 - 10) + (144 \times 5) + 1 \times 5(T - 32)$$

$$\left\{\begin{array}{l} c_c = 0.093 \\ m_c = 10 \text{ lb} \\ T_c = 70°\text{F} \\ c_w = 1 \\ m_w = 15 \text{ lb} \\ T_w = 70°\text{F} \\ c_i = 0.5 \\ m_i = 5 \text{ lb} \\ T_i = 10°\text{F} \end{array}\right\}$$

Sec. 17.3   *Calorimetry*   235

Heat lost = heat gained:

$$0.093 \times 10(70 - T) + 1 \times 15(70 - T)$$
$$= 0.5 \times 5(32 - 10) + (144 \times 5) + 1 \times 5(T - 32)$$
$$15T + 0.93T + 5T = 65.1 + 1050 - 55 - 720 + 160$$
$$T = 23.9°F$$

The first assumption was not good because, at standard pressure, liquid water cannot exist below 32°F.

Solution (Second Assumption)

Assume the final condition to be a mixture of ice and water at 32°F, and m_i the unknown mass of ice.

Heat lost by container:

$$Q_c = 0.093 \times 10(70 - 32)$$

Heat lost by water:

$$Q_w = 1 \times 15(70 - 32)$$

Heat gained by ice:

$$Q_i = 0.5 \times 5(32 - 10) + 144 m_i$$

Heat lost = heat gained:

$$0.093 \times 10(70 - 32) + 1 \times 15(70 - 32) = 0.5 \times 5(32 - 10) + 144 m_i$$

Ice melted:

$$m_i = \frac{550.34}{144} = 3.82 \text{ lb}$$

Ice left in mixture:

$$5 - 3.82 = 1.18 \text{ lb}$$

Total water in mixture:

$$3.82 + 15 = 18.82 \text{ lb}$$

Notice very carefully that the container and the water lose 605.34 Btu. This is all the Btu available to melt the ice. But at 10°F, the ice needs 55 + (144 × 5) = 775 Btu to melt completely. Therefore, there is not enough heat energy available to melt all the ice.

If it becomes desirable to find the percentage Z of each phase present, the equation may be rewritten as follows:

$$0.093 \times 10(70 - 32) + 1 \times 15(70 - 32) = 0.5 \times 5(32 - 10) + 144 \times 5 \times Z$$
$$Z = 0.764 = 76.4\%$$

Check:

76.4% × 5 lb ice = 3.82 lb melted
5 − 3.82 = 1.18 lb of ice left

17.4. Heat of Combustion

Heat of combustion is the heat energy generated when burning a unit mass or unit volume of a fuel. The heat of combustion of a material is determined by burning a known mass of the fuel in a *calorimeter* and recording the temperature rise of a predetermined quantity of water. Appendix Table A.13 shows some values for heat of combustion of some common materials.

Example 7

An 1800-g sample of water in a copper calorimeter goes through a temperature increase of 4.9C° when 1 g of fuel oil is burned. The water equivalent of the calorimeter is 400 g. What is the heat of combustion of the fuel oil in cal/g and Btu/lb? Assume no change in pressure.

Solution

$Q = c_w m_e \Delta T = 1(1800 + 400)(4.9)$

= 10,780 cal/g

$\left\{ \begin{array}{l} c_w = 1 \\ m_e = \text{total equivalent} \\ Q = \text{heat of combustion} \end{array} \right\}$

or

$Q = 10,780 \dfrac{\text{cal}}{\text{g}} \times \dfrac{1 \text{ Btu}}{252 \text{ cal}} \times \dfrac{454 \text{ g}}{\text{lb}}$

= 19,421 Btu/lb

17.5. Heat Energy

We have been talking about heat as energy. In this respect, 1 Btu is equivalent to 778 ft-lb of energy; or 1 calorie is equivalent to 4.18 joules of energy. These are the bridges that fill the gap between heat and mechanical energy. These constants have been determined experimentally and the description, below, of Joule's experiment shows one method for establishing and verifying this constant.

Once this relationship has been established, other forms of energy may be related to heat energy, because these other forms have been correlated to mechanical energy.

Sec. 17.5 Heat Energy

Thus, by definition, 1 joule of work per second will be delivered when 1 *ampere* (amp) of current flows through a resistance which will create 1 *volt* of potential difference across its terminals. This definition ties together mechanical and electrical energy. This amount of work is defined as the *watt:*

$$1 \text{ amp} \times 1 \text{ volt} = 1 \text{ watt}$$

and 1 watt-sec = 1 joule = 10^7 ergs = 1 newton-meter (nt-m).
Also, since

$$1 \text{ watt-sec} = 0.737 \text{ ft-lb}$$

then

$$1 \text{ Btu} = 778 \, \cancel{\text{ft-lb}} \times \frac{1 \text{ watt-sec}}{0.737 \, \cancel{\text{ft-lb}}} = 1055 \text{ watt-sec}$$

Also,

$$1 \text{ hp} = 550 \frac{\cancel{\text{ft-lb}}}{\cancel{\text{sec}}} \times \frac{1 \text{ watt-sec}}{0.737 \, \cancel{\text{ft-lb}}} = 746 \text{ watts}$$

Since

$$1 \text{ watt} = 252 \frac{\text{cal}}{\cancel{\text{Btu}}} \times \frac{\cancel{\text{Btu}}}{778 \, \cancel{\text{ft-lb}}} \times 0.737 \frac{\cancel{\text{ft-lb}}}{\text{sec}} = 0.238 \frac{\text{cal}}{\text{sec}}$$

then

$$\frac{1 \text{ watt}}{0.238} = \frac{1 \text{ cal}}{\text{sec}}$$

$$1 \text{ cal} = 4.18 \text{ watt-sec}$$

Also, it can be shown that

$$1 \text{ hp-hr} = 2545 \text{ Btu}$$
$$1 \text{ kilowatt hour (kw-hr)} = 3412 \text{ Btu}$$

In a series of papers James Joule presented to science, the results of many experiments covering many conversions from mechanical to heat energy, electrical to heat energy, etc., all very closely correlated. His experiments lead to the conversion methods discussed above. The conversion factors which establish the equivalence of mechanical, electrical, and heat energies are sometimes referred to as the first law of thermodynamics.

For the present, we shall describe Joule's experiment leading to the determination of the mechanical equivalent of heat (778 ft-lb/Btu). The apparatus shown in Fig. 17.2 consists of an inner cup holding a thermometer immersed in a known quantity of water. The outer cup is revolved by a motor, through a pair of bevel gears, and the number of revolutions per unit time is observed on the counter and recorded. If the outer cup is revolved, the inner cup may be held stationary by putting weights on the weight hanger, its supporting string being wrapped around a pulley of known radius. Thus a torque

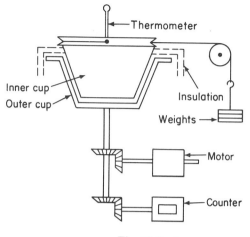

Fig. 17.2

is exerted which puts the friction force between the two cups into equilibrium. This friction will register any rise in temperature on the thermometer. The water equivalent of the cup and the water and the temperature rise will yield the mechanical equivalent of heat.

Example 8

The apparatus (Fig. 17.2) is being used to verify the mechanical equivalent of heat. The inner cup is filled with 1 lb of water. The water equivalent of the two cups is 0.06 lb. The radius of the pulley is 6 in., and the force required to hold the pulley steady is $\frac{1}{2}$ lb; 2000 revolutions of the outer cup cause an increase in temperature from 70°F to 73.8°F.

Solution

Mechanical energy:

$$\text{wk} = 2\pi r N f$$
$$= 2 \times 3.14 \times \tfrac{1}{2} \times 2000 \times \tfrac{1}{2}$$
$$= 3140 \text{ ft-lb}$$

Heat energy in Btu:

$$Q = cm\,\Delta T = 1 \times 1.06 \times 3.8\text{F}° = 4.0 \text{ Btu}$$

Mechanical equivalent of heat = ft-lb/Btu:

$$= \frac{\text{mechanical energy (ft-lb)}}{\text{heat energy (Btu)}}$$

$$= \frac{3140}{4.03} = 779.1 \frac{\text{ft-lb}}{\text{Btu}}$$

$\left\{ \begin{array}{l} r = 6 \text{ in.} = \tfrac{1}{2} \text{ ft} \\ N = 2000 \text{ rev} \\ f = \tfrac{1}{2} \text{ lb} \\ m = m_w + m_c = 1 + 0.06 = 1.06 \text{ lb} \\ T = T_f - T_i = 73.8 - 70 = 3.8°\text{F} \end{array} \right\}$

Sec. 17.5   *Heat Energy*   239

Example 9

A bulb is immersed into 30 lb of water in a copper container weighing 8 lb. The bulb operates on 1 amp and 120 volts and takes 1.5 hr to heat the water and container from 70°F to 90°F. Find the electrical equivalent of heat in the metric system.

Solution

Heat needed for water:

$$Q_w = 1 \times 30(90 - 70)$$
$$= 600 \text{ Btu}$$

Heat needed for calorimeter:

$$Q_c = 0.093 \times 8(90 - 70)$$
$$= 14.88 \text{ Btu}$$

Total heat needed:

$$Q = Q_w + Q_c$$
$$= 600 + 14.88 \text{ Btu}$$
$$= 614.88 \text{ Btu}$$

Electrical energy put into the system:

$$P = VI = 120 \times 1 = 120 \text{ watts}$$

Total watts:

$$120 \times 5400 \text{ sec} = 648{,}000 \text{ elec watt-sec}$$

$\left\{\begin{array}{l} m_w = 30 \text{ lb} \\ m_c = 8 \text{ lb} \\ c_w = 1 \\ c_c = 0.093 \\ \Delta T = 90-70 = 20\text{F}° \\ V = 120 \text{ volts} \\ I = 1 \text{ amp} \\ W = 120 \text{ watts} \\ t = 1.5 \text{ hr} \times 3600 = 5400 \text{ sec} \\ E_e = \text{elec equivalent: British units} \\ E_m = \text{elec equivalent: metric units} \end{array}\right\}$

Value of E_e in British units:

$$648{,}000 E_e = 614.88 \text{ Btu}$$
$$E_e = \frac{614.88 \text{ Btu}}{648{,}000 \text{ watt-sec}} = \frac{0.000949 \text{ Btu}}{\text{watt-sec}}$$

Value of E_m in metric units:

$$E_m = 0.000949 \frac{\text{Btu}}{\text{watt-sec}} \times 252 \frac{\text{cal}}{\text{Btu}}$$
$$= 0.2392 \frac{\text{cal}}{\text{watt-sec}}$$

17.6. Heat Transfer: Convection

Convection is the process by which heat energy is transported from one place to another in a material medium, which we may characterize as a *carrier*. The carrier will transport the heat energy to a new location, at which point the medium gives up its energy. The motion of the carrier is in the direction of the "colder" region, where it gives up its heat energy and then continues to circulate. Remember that heat is energy; in terms of convection, it needs a carrier.

If we assume the carrier to be a ball of molecules, agitated by the energy the molecules pick up—say, from the top of a hot stove—the additional agitation reduces the density (volume increase for same number of molecules) and the ball floats toward the ceiling. The molecules inside are still in an agitated condition. The unit, on striking the ceiling, gives up its energy, the motion, is reduced, the density increased, and the unit starts toward the floor again. In its circulation near the floor, it will eventually fill a vacancy left by some other unit which has started for the ceiling. The cycle starts over again as long as the stove has heat energy to give up, the molecules have energy to gain, and there is a differential in temperature.

Note that a lake freezes according to the foregoing principle. It freezes on top first because water is at maximum density at 4°C. Thus, as the water at the top of the lake reaches 4°C, it sinks to the bottom, forcing the less dense water at the bottom to rise and in turn be cooled to 4°C. Now assume the entire lake to be at 4°C; as the top cools below 4°C, its density becomes less. Thus the water with a temperature less than 4°C will remain on top and freeze first.

17.7. Heat Transfer: Conduction

Conduction resembles convection in that it also needs a transporting material. In this case, the materials must be in contact so that the molecules when agitated will collide with one another. The collisions per unit area are greater at the "hot" end than at the "cold" end. Thus there is no transporting of material as in convection. The agitated molecule —the more agitated it gets, the more heat energy it has "absorbed"—is indicative of heat energy. By colliding with other molecules, it gives up some of its energy.

The equation for conduction depends upon

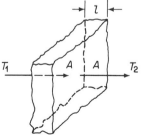

Fig. 17.3

Sec. 17.7 Heat Transfer: Conduction 241

the material, the area involved, and the temperature gradient change (temperature per unit length). Figure 17.3 shows a block of material where temperature T_1 must be greater than T_2; otherwise, the direction of heat flow, indicated by the arrow, would be incorrect. The distance through which the heat travels is taken as l; the area A involved is also indicated on the sketch.

The rate of flow is proportional to the temperature change ΔT and the area A, and inversely proportional to the length l. Thus

$$\frac{Q}{t} \propto \frac{A(T_1 - T_2)}{l}$$

The heat flow is also proportional to the time t of flow:

$$Q \propto \frac{tA(T_1 - T_2)}{l}$$

Multiply the right-hand side by a proportionality constant, called the *coefficient of thermal conductivity*.* The equation becomes

$$Q = \frac{KAt(T_1 - T_2)}{l} = \frac{KAt\,\Delta T}{l}$$

Solving for K_f, the units for the coefficient in the British system are

$$K_f = \frac{Ql}{At\,\Delta T} = \frac{\text{Btu-in.}}{\text{ft}^2\text{-hr-F}°}$$

In metric units,

$$K_c = \frac{Ql}{At\,\Delta T} = \frac{\text{cal-cm}}{\text{cm}^2\text{-sec-C}°}$$

The temperature gradient $\Delta T/l$ is

$$\frac{\Delta T}{l} = \frac{Q}{KAt} = \frac{\text{F}°}{\text{in.}} \quad \text{or} \quad \frac{\text{C}°}{\text{cm}}$$

$\left\{\begin{array}{l} Q = \text{heat energy conducted for the distance } l \text{ Btu } or \text{ cal} \\ A = \text{cross-sectional area of material:} \\ \quad \text{ft}^2 \text{ or cm}^2 \\ T_1 = \text{higher temp: }°\text{F } or \text{ }°\text{C} \\ T_2 = \text{lower temp: }°\text{F } or \text{ }°\text{C} \\ L = \text{length of flow path in direction of flow: in. } or \text{ cm} \\ t = \text{time of flow: hr } or \text{ sec} \\ K_f = \text{coef of thermal conductivity: of thermal conductivity: British units} \\ K_c = \text{coef of thermal conductivity: metric units} \end{array}\right\}$

Example 10

An aluminum plate 1 in. thick measuring 26 in. × 24 in. is subjected to a 200°F temperature on one side and a 70°F temperature on the other. Calculate the time required to pass 5.6×10^6 Btu through the plate.

Solution

The heat equation is

$$Q = \frac{KAt\,\Delta T}{l}$$

Solve for t:

$\left\{\begin{array}{l} Q = 5.6 \times 10^6 \text{ Btu} \\ K_f = 1450 \frac{\text{But-in.}}{\text{ft}^2\text{-hr-F}°} \\ A = \frac{36}{12} \times \frac{24}{12} = 6 \text{ ft}^2 \\ T_1 = 200°\text{F} \\ T_2 = 70°\text{F} \\ l = 1 \text{ in.} \end{array}\right\}$

*Values in Appendix Table A.14.

$$t = \frac{Ql}{KA\,\Delta T} = \frac{5.6 \times 10^6 \times 1}{1450 \times 6(200 - 70)}$$
$$= 4.95 \text{ hr}$$

17.8. Heat Transfer: Radiation

Radiation, or *radiant energy*, differs from conduction or convection in that it does not need a carrier. It traverses space, and when absorbed it agitates the molecules of the absorbing material so that it is converted into heat energy.

Radiant energy lies just beyond the red end of the electromagnetic-wave spectrum and is no different from any other electromagnetic energy. It travels at the speed of light, 186,000 mi/sec, or 3×10^{10} cm/sec. It has a wavelength from 4×10^{-2} cm to 8×10^{-5} cm. Since 1 micron $(\mu) = 10^{-4}$ cm, it can be stated as $4 \times 10^{-2} (1\,\mu/10^{-4}) = 4 \times 10^2\,\mu$ to $8 \times 10^{-1}\,\mu$. In *angstrom* units, it would be 4×10^{-2} cm $\times (1\,A/10^{-8}$ cm$) = 4 \times 10^6$ A to 8×10^3 A.

All bodies radiate various forms of energy at different peak frequencies. The plot of various absolute temperatures—wavelength against energy per second released—will yield typical sets of curves, as shown in Fig. 17.4.

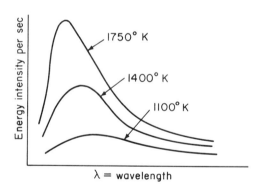

Fig. 17.4

Kirchoff's law indicates that a good radiator is a good absorber; for the law states that at any wavelength there exists a proportionality between emission and absorption at any temperature. Thus cast iron is a good absorber of heat energy when heated, and by the same token it will also radiate well. White cloth reflects well, while dark cloth absorbs well. The dark cloth should radiate well also. In the case of clear glass, radiant energy passes right through and thus there is no energy in the glass to be radiated. Do not confuse radiation with *reflection*. If the radiant energy is reflected or passes through the material, it can be neither absorbed nor radiated.

Sec. 17.8 Heat Transfer: Radiation

The Stefan–Boltzman law states that the total area under each curve (Fig. 17.4) is proportional to the fourth power of the absolute temperature. It applies to a "black-body"* radiator because it emits radiation according to its temperature. It is a perfect absorber. Stefan showed experimentally and Boltzman showed theoretically that energy radiated is proportional to the fourth power of the absolute temperature:

$$E = \sigma T^4$$

And when radiated from a body having an area A,

$$E = \sigma A T^4$$

Since it also receives energy from the surrounding temperature level,

$$E = \sigma A(T^4 - T_0^4)$$

$\left\{ \begin{array}{l} E = \text{energy: watts} \\ \sigma = \text{Stefan constant} \\ = 5.67 \times 10^{12} \text{ watts/cm}^2\text{-}(°K^4) \\ A = \text{area: cm}^2 \\ T = \text{temp of body: °K} \\ T_o = \text{temp of surroundings} \\ c = \text{an experimental constant: Newton's law of cooling} \\ E_m = \text{nt: energy if area is in meters,} \\ \sigma = 5.67 \times 10^{-8} \frac{\text{watt}}{\text{m}^2\text{-(K°)}^4} \end{array} \right\}$

Over a small change in temperature, the preceding law, even though true, becomes difficult to handle and so Newton's approximation—called *Newton's law of cooling*— is applied. It states that the rate of cooling is proportional to the temperature difference; or

$$E_m = cA(T - T)$$

The foregoing discussion applies to "black-body radiation." But not all materials absorb (or radiate) as black bodies. As the temperature increases, the wavelength becomes shorter for the most agitated waves and the peak of the curve displaces toward the left. *Wein's law* states that this most agitated wavelength is inversely proportional to the absolute temperature and that the product of wavelength and temperature is equal to a constant value. This constant is called *Wein's constant* and has a value of 0.2898 when the shortest wavelength is in centimeters and T is the absolute temperature; λ is the highest point on the curve. The Wein equation becomes

$$\lambda T = 0.2898 \quad \left\{ \begin{array}{l} \lambda = \text{wavelength: cm} \\ T = \text{absolute temp: °K} \\ \text{constant} = 0.2898 \text{ cm-°K} \end{array} \right\}$$

Thus if the wavelength of the radiant energy considered can be found, the absolute temperature may be calculated.

Example 11

A tungsten filament 25.4 cm long with a diameter of 0.0005 in. operates inside a lamp at 2000°K. If the efficiency of the lamp is 25%, what is the power radiated?

*A black body is one which absorbs all the radiant heat which falls upon it.

Solution

Surface area of the filament:

$$A = \pi Dl = 3.14 \times 0.00127 \times 25.4$$
$$= 0.1013 \text{ cm}^2$$

When

$$\sigma = 5.67 \times 10^{-12} \text{ watt/cm}^2\text{-(K}°)^4$$

$$\begin{Bmatrix} d = 0.0005 \times 2.54 \\ = 0.00127 \text{ cm} \\ l = 25.4 \text{ cm} \\ T = 2 \times 10^3 \text{ °K} \\ \textit{eff} = 25\% \end{Bmatrix}$$

Energy radiated:

$$E = \sigma A T^4 (25\%) = 5.67 \times 10^{-12} \times 0.1013 \times (2 \times 10^3)^4 \times 0.25$$
$$= 2.3 \text{ watts}$$

Problems

1. Discuss "convection," "conduction," and "radiation" as you understand them.
2. Calculate the water equivalent of a 6-lb steel bushing.
3. An 8-lb copper liner is pressed into a 3-lb aluminum bushing. Calculate the water equivalent of the combination.
4. How many Btu must be added to 40 lb of cast iron to raise its temperature from 70°F to 220°F?
5. In Prob. 3, how many Btu would be required to raise the temperature of the combination from 70°F to 150°F?
6. Calculate the final temperature if 1200 Btu are added to 20 lb of steel at 80°F.
7. A steel ball is dropped into 30 lb of water at 34°F. The steel ball has been heated to 300°F and weighs 24 lb. What is the final equilibrium temperature? Neglect the container, etc.
8. Solve Prob. 7 if the water is in an aluminum container which weighs 10 lb. A copper rod used as a stirrer weighs 3 oz, but only one-third of the rod is in contact with the water. A thermometer which weighs 12 oz, one-fourth of which is immersed in the water, is used to check the final temperature. What is the equilibrium reading on the thermometer?
9. An aluminum container has a mass of 200 g and has 450 g of water poured into it at a temperature of 24°C. A 400-g block of copper is at a temperature of 110°C when dropped into the calorimeter. If the temperature increases to 300°, what is the specific heat of the copper block?
10. A 30-lb. steel casting at 1450°F is quenched in a water bath at 80°F. If the final temperature of the system is 150°F, how much water was used? Assume a 12-lb tin container.
11. It is desired to find the temperature of a heat-treating furnace. A brass ball weighing 12 lb is put into the furnace and heated to the furnace temperature. A steel container weighs 10 lb and contains 30 lb of water at a temperature of 70°F. The ball is removed rapidly and plunged into the water. The rise in temperature of the water and the container is 50°F. What was the furnace temperature?

Problems 245

12. Calculate the equilibrium temperature when 50 g of lead shot at 100°C are dropped into 150 g of water at 25°C. The water is in a 10-g copper vessel.
13. Calculate the heat required to change 30 lb of ice at 5°F to steam at 230°F, standard pressure.
14. A 20 lb steel crucible contains 40 lb of molten lead at 1000°F. If 20 lb of molten zinc at a temperature of 800°F is poured into the lead, calculate the final equilibrium temperature.
15. Eight lb of mercury at −50°F are heated to 900°F at standard pressure. How many Btu are required? The melting point of mercury is −38°F; the boiling point is 675°F; the latent heat of fusion is 5 Btu/lb; and the latent heat of vaporization is 127 Btu/lb. Specific heat of solid mercury is 0.034; of the liquid, 0.033; and of the vapor, 0.025.
16. One gram of fuel is burned in a copper calorimeter, water equivalent 0.9 lb, which contains 2000 g of water. If the temperature of the water increases by 45°C, what is the material being burned?
17. A large drill bit is used to drill a hole in a large cast-iron plate weighing 150 lb. It takes 12 min and 2 hp at the drill point to drill the hole. (a) What is the heat production? (b) If the room temperature is 70°F, what is the highest temperature reached by the block? Assume 90% of the heat is effective in raising the temperature of the block.
18. The apparatus shown in Fig. 17.2 is used in the Joule experiment. The inner cup is filled with 1.5 lb of water. The water equivalent of the two cups is 0.4 lb. The diameter of the pulley, opposing friction, is 14 in. and the force required to hold the pulley is $\frac{3}{4}$ lb. If room temperature is 85°F and 3000 revolutions of the cup cause an increase in temperature, what is the final temperature registered on the thermometer?
19. A $\frac{1}{2}$-in. diameter aluminum rod is 1 ft long. If one end of the rod is in boiling water and the other end in ice, how much ice will be melted in 1 hr as a result of the thermal conductivity alone?
20. The dimensions of an ice box are 30 in. × 30 in. × 48 in. high. If the insulation consists of 2.5-in. glass-wool batts, coefficient of thermal conductivity of 0.29 Btu-in./ft²-hr-F°, how much ice water leaves the ice box in 8 hr? The temperature inside the ice box is 40°F and the average room temperature is 75°F. Assume there is no heat loss through the bottom or top of the box but only through the walls.
21. A hot body, area 100 cm² is at a temperature of 5000°K. If the temperature of the body's surroundings is 800°F, what is its radiant-energy output in watts?

18

Gases

18.1. The Gas Laws: Boyle's Law and Charles' Law

In Chapter 17, pressure was taken as standard and as having little effect upon the expansion of solids or liquids. Gases, however, are greatly affected by pressure. A change in the volume of a gas occurs whenever the absolute temperature or the absolute pressure increases or decreases. There can be no linear or area change since the gas takes the shape of its container and expands or contracts in all directions.

Boyle's law states that if the mass and absolute temperature of a gas are held constant, the product of the absolute pressure and the volume is equal to a constant; or

$$VP = K \quad \left\{ \begin{array}{l} P = \text{abs pressure} \\ V = \text{volume} \end{array} \right\}$$

A given mass of a gas at a fixed temperature will be a function of that constant. Hence, the product of the absolute pressure and volume at any instant will equal the product of the absolute pressure and volume at any other instant; or

$$P_1 V_1 = P_2 V_2$$

Thus, since $P_1 \times V_1$ equals a constant K, if P_2 increases, V_2 must decrease to maintain the value of the constant.

Example 1

A mass of gas, at constant temperature, occupies 15 ft³ at a pressure of 30 psig.

Sec. 18.1 *The Gas Laws: Boyle's Law and Charles Law* 247

If the pressure is increased to 45 psig, calculate the new volume. Atmospheric pressure is 14.7 psia.

Solution

$$P_1V_1 = P_2V_2 \quad \left\{ \begin{array}{l} P_1 = 30+14.7 = 44.7 \text{ psia} \\ P_2 = 45+14.7 = 59.7 \text{ psia} \\ V_1 = 15 \text{ ft}^3 \end{array} \right\}$$

Solve for V_2 and substitute:

$$V_2 = \frac{P_1V_1}{P_2} = \frac{44.7 \times 15}{59.7} = 11.2 \text{ ft}^3$$

Jacques Charles treated volume and temperature; *Charles' law* states that the volume of a gas varies as the absolute temperature. Hence, the ratio of the volume to the absolute temperature is equal to a constant if the mass and absolute pressure are constant:

$$\frac{V}{T} = K \quad \left\{ \begin{array}{l} V = \text{volume} \\ T = \text{abs temp} \end{array} \right\}$$

Thus the ratio of volume to absolute temperature at any instant equals the ratio of the volume to the absolute temperature at any other time, if the mass and absolute pressure are held constant; or

$$\frac{V_1}{T_1} = \frac{V_2}{T_2}$$

Example 2

The volume of a gas is 15 ft³ when the temperature is 40°F. If the pressure is held constant, calculate the volume of the gas when the temperature is 100°F.

Solution

$$\frac{V_1}{T_1} = \frac{V_2}{T_2} \quad \left\{ \begin{array}{l} V_1 = 15 \text{ ft}^3 \\ T_1 = 40 + 460 = 500°\text{R} \\ T_2 = 100 + 460 = 560°\text{R} \end{array} \right\}$$

Solve for V_2 and substitute:

$$V_2 = \frac{T_2V_1}{T_1} = \frac{560 \times 15}{500} = 16.8 \text{ ft}^3$$

The foregoing holds true for the ratio of absolute pressure to absolute temperature, if the volume and the mass are held constant; or

$$\frac{P}{T} = K \quad \left\{ \begin{array}{l} P = \text{abs pressure} \\ T = \text{abs temp} \end{array} \right\}$$

At two different instances,

$$\frac{P_1}{T_1} = \frac{P_2}{T_2}$$

Example 3

The volume of a gas is held constant. If the pressure is 10 psig at 40°F, calculate the pressure at 100°F. The atmospheric pressure is 14.7 psia.

Solution

$$\frac{P_1}{T_1} = \frac{P_2}{T_2} \qquad \left\{ \begin{array}{l} P_1 = 10 + 14.7 = 24.7 \text{ psia} \\ T_1 = 40 + 460 = 500°R \\ T_2 = 100 + 460 = 560°R \end{array} \right\}$$

Solve for P_2 and substitute:

$$P_2 = \frac{T_2 P_1}{T_1} = \frac{560 \times 24.7}{500} = 27.7 \text{ psia}$$

The gage pressure is

$$27.7 - 14.7 = 13 \text{ psig}$$

The previous statements may be combined and stated as

$$\frac{PV}{T} = K \qquad \left\{ \begin{array}{l} P = \text{abs pressure} \\ T = \text{abs temp} \\ V = \text{volume} \end{array} \right\}$$

At two different instances,

$$\frac{P_1 V_1}{T_1} = \frac{P_2 V_2}{T_2}$$

Example 4

A balloon is partially filled with 30,000 ft³ of gas at a temperature of 70°F and a pressure of 14.7 psia. If the temperature changes to 20°F and the pressure to 5.8 psia, calculate the new volume.

Solution

$$\frac{P_1 V_1}{T_1} = \frac{P_2 V_2}{T_2}$$

Solve for V_2:

$$V_2 = \frac{P_1 V_1 T_2}{P_2 T_1} = \frac{14.7 \times 30,000 \times 480}{5.8 \times 530}$$

$$= 68,900 \text{ ft}^3$$

$$\left\{ \begin{array}{l} V_1 = 30,000 \text{ ft}^3 \\ T_1 = 70 + 460 = 530°R \\ P_1 = 14.7 \text{ psia} \\ T_2 = 20 + 460 = 480°R \\ P_2 = 5.8 \text{ psia} \end{array} \right\}$$

18.2. The Universal Gas Law and Its Constant R

Avogadro's law deals with, and leads to, the proposition that the volume of a gas is proportional to its mass at constant pressure and temperature. Thus

Sec. 18.2 The Universal Gas Law and Its Constant R

and
$$V \propto m$$
$$\frac{V}{m} = K$$
$\{m = \text{mass or weight}, V = \text{volume}\}$

Under varying conditions,
$$\frac{V_1}{m_1} = \frac{V_2}{m_2}$$

Also, for a given mass of a gas (at constant pressure and temperature), the volume of a gas is inversely proportional to its molecular weight. Thus

$$V \propto \frac{1}{M} \quad \{M = \text{mol wt}\}$$

and, therefore,
$$VM = K$$

All the preceding statements may be combined into the *universal gas law:*

$$\frac{PV}{T} = R\frac{m}{M}$$

Solving this equation for R, the *universal gas constant*,

$$R = \frac{PVM}{Tm}$$

$\left\{\begin{array}{l} P = \text{abs pressure} \\ T = \text{abs temp} \\ V = \text{volume} \\ m = \text{mass or weight} \\ M = \text{mol wt} \\ R = \text{universal gas const} \end{array}\right\}$

Experiments have shown that 1 *gram-molecule* (g-mol) of any gas occupies 22,400 cm³ when the gas is at 0 °C and standard pressure of 76 cm of mercury. By definition, *molecular weight* is equal to the mass of 1 g-mol. Thus the molecular weight M of helium is 4, and 1 g-mol of helium has a mass m of 4 g from

$$\frac{m}{M} = \frac{4}{4} = 1 \text{ g-mol}$$

The pressure may be stated as a product of 1 atmosphere of pressure resting on a 1-cm² area or

$$P = 76 \times 13.6 \times 1 \times 980 = 1.013 \times 10^6 \text{ dynes/cm}^2$$

The absolute temperature is 273 °K.

Combining the above values, the metric value for the universal gas constant is

$$R = \frac{PVM}{Tm} = 1.013 \times 10^6 \text{ dynes/cm}^2 \times 22,400 \text{ cm}^3 \times \frac{1}{1 \text{ g-mol}} \times \frac{1}{273 °K}$$

$$= 8.32 \times 10^7 \frac{\text{dyne-cm}}{\text{g-mol-°K}}$$

$$= 8.32 \times 10^7 \frac{\text{erg}}{\text{g-mol-°K}}$$

$\left\{\begin{array}{l} P = \text{abs pressure: dynes/cm}^2 \\ V = \text{volume: cm}^3 \\ T = \text{abs temp: °K} \\ m = \text{mass: g} \\ M = \text{mol wt: g/g-mol} \\ \frac{m}{M} = \text{g-mol} \end{array}\right\}$

Since there are 10^7 ergs per joule,

$$R = 8.32 \frac{\text{joule}}{\text{g-mol-}°K}$$

In the British system of units,

$$R = 8.32 \frac{\cancel{\text{joule}}}{\cancel{\text{g-mol-}°K}} \times 0.737 \frac{\text{ft-lb}}{\cancel{\text{joule}}} \times 453.6 \frac{\cancel{\text{g-mol}}}{\text{lb-mol}} \times \frac{5°\cancel{K}}{9°R}$$

$$= 1542 \frac{\text{ft-lb}}{\text{lb-mol-}°R} \quad \left\{ \begin{array}{l} P = \text{abs pressure: lb/ft}^2 \\ V = \text{volume: ft}^3 \\ T = \text{abs temp:}°R \\ m = \text{mass: lb} \\ M = \text{mol wt: lb/lb-mol} \\ \frac{m}{M} = \text{lb-mol} \\ R* = \text{universal gas const} \end{array} \right\}$$

Example 5

A tank holds 2 lb of air at 150 psia when the temperature is 90°F. Calculate the volume of the gas if the molecular weight of air is 28.8 lb/lb mol.

Solution

The pressure in pounds per square foot is

$$P = 150 \text{ psia} \times 144 = 21{,}600 \text{ psfa}$$

The absolute temperature is

$$T = 90 + 460 = 550°R$$

Thus

$$V = \frac{mRT}{MP} = \frac{2 \times 1542 \times 550}{28.8 \times 21{,}600} = 2.73 \text{ ft}^3 \quad \left\{ \begin{array}{l} m = 2 \text{ lb} \\ M = 28.8 \text{ lb/lb/mol} \\ R = 1542 \text{ ft-lb/lb-mol-}°R \end{array} \right\}$$

The last equation in the above example, $V = m\,RT/MP$, is an *equation of state*. It represents the conditions which prevail at a particular instant for an ideal gas; however, we shall apply these ideal-gas equations to "real gases."

When several gases are included in a system, if volume and temperature are held constant, the total pressure is the sum of the partial pressures exerted by each of the gases involved. This is *Dalton's law of partial pressures*.

Thus the total pressure is

$$P_t = P_1 + P_2 = \frac{m_1 RT}{M_1 V} + \frac{m_2 RT}{M_2 V}$$

$$= \frac{RT}{V}\left(\frac{m_1}{M_1} + \frac{m_2}{M_2}\right)$$

---

*Engineers sometimes use 55 ft-lb/lb for the value of the constant R. When this is used temperature must be in degrees Rankin.

18.3. Gases and Specific Heat

It is now necessary to define *specific heat* of gases more thoroughly because, when heat is applied to a gas, it responds by changing its volume. If heat is applied to a gas, there will be some internal work done in opposition to any external pressure that may exist in order that the gas may expand. There may also be an increase in temperature due to the increase in internal energy.

Thus we redefine specific heat when it applies to gas: c_v is defined as the *specific heat at constant volume* (the subscript denoting "constant volume"). If heat is added to a unit mass of gas and that heat increases the temperature of the gas through 1 K° without affecting the volume, that heat represents the specific heat of that gas while holding the volume constant.

We define c_p: as the *specific heat* of a gas *at constant pressure* (the subscript p denotes "constant *pressure*." If heat is added to a unit mass of a gas and the heat increases the temperature of the gas through 1° without changing the pressure, that heat represents the specific heat of the gas while holding the pressure constant (see Appendix A.15).

When discussing specific heats of gases, the work done against any external pressure becomes important. When discussing c_v (specific heat at constant volume), we see that all the heat added goes to increasing the internal energy as pressure. None of the heat need be used for increasing the volume, because there is no increase of volume for c_v. Therefore, as the temperature increases as a result of the addition of heat, all the energy goes toward increasing the pressure, and $P/T = K$.

When a process takes place at constant pressure (c_p), some of the energy goes toward increasing the internal energy as well as aiding in the change of volume. Again, none of this heat goes toward increasing the internal pressure because the process takes place at constant pressure. The expansion of the gas (increase in volume) takes place as a result of work:

$$\text{wk} = P \, \Delta V \quad \left\{ \begin{array}{l} P = \text{abs pressure} \\ \Delta V = \text{change in volume} \end{array} \right\}$$

If the gas in the chamber (Fig. 18.1) expands and moves a distance h against the pressure, the force on the area of the piston will be

$$f = PA$$

Since the work is equal to force × distance,

$$\text{wk} = f \, \Delta h = PA \, \Delta h$$

But $A \, \Delta h = \Delta V$. Therefore, when the piston moves a distance Δh,

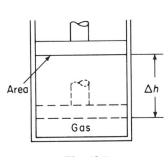

Fig. 18.1

252   *Gases*   Chap. 18

$$\text{wk} = P \Delta V$$

The conditions used in setting up c_p and c_v were for a 1° rise in temperature and for constant mass. $c_p - c_v$ represents external work per unit mass and unit temperature where c_p represents heat energy which affects volume change of a system and where c_v is the heat needed to change the *internal* temperature of the system by 1°. Thus the difference in specific heats represents the external work for unit mass and temperature. The net energy required to do work against external pressure is $c_p - c_v$. Therefore, $\text{wk} = (c_p - c_v)m \, \Delta T$.

The volume V_1 and the temperature T_1 may also be changed by the addition of heat energy to V_2 and T_2 without changing the pressure P. Therefore,

$$PV_1 = \frac{m}{M} RT_1 \quad \text{and} \quad PV_2 = \frac{m}{M} RT_2$$

Subtracting,

$$PV_1 - PV_2 = \frac{m}{M} RT_1 - \frac{m}{M} RT_2$$

Factoring,

$$P(V_1 - V_2) = \frac{m}{M} R(T_1 - T_2)$$

If $V_1 - V_2 = \Delta V$ and $T_1 - T_2 = \Delta T$, then

$$P \Delta V = \frac{m}{M} R \, \Delta T$$

But $P \Delta V =$ work. Thus, equating both work equations,

$$\frac{m}{M} R \, \Delta T = (c_p - c_v) m \, \Delta T$$

Solving for R,

$$R = M(c_p - c_v)$$

Since M is the mass per molecule, the universal gas constant R represents the work done against a constant external pressure when heating 1 mole of gas for a temperature increase of 1°. (*Important:* remember that all units must be in the same system.) Rewriting the equation and dividing by the mechanical equivalent of heat, J, the equation becomes

$$\frac{R}{MJ} = c_p - c_v \qquad \left\{ \begin{array}{l} R = \text{universal gas const} \\ \quad = 8.32 \times 10^7 \text{dyne-cm/g-mol-°K} \\ \quad = 1542 \text{ ft-lb/lb-mole-°R} \\ J = 4.18 \times 10^7 \text{ dyne-cm/cal} \\ \quad = 778 \text{ ft-lb/Btu} \\ M = \text{mol wt} \\ c_v = \text{cal/g-°C} = \text{Btu/lb-°F} \\ c_p = \text{cal/g-°C} = \text{Btu/lb-°F} \end{array} \right.$$

Example 6

From the values of specific heat at constant pressure and constant volume, derive the molecular weight of helium.

Solution

$$M = \frac{R}{J(c_p - c_v)} = \frac{1542}{778(1.250 - 0.755)} = 4 \qquad \left\{\begin{array}{l} c_p = 1.250 \\ c_v = 0.755 \\ J = 778 \\ R = 1542 \end{array}\right\}$$

18.4. The Kinetic Theory of Heat

The *kinetic theory* demands that gas molecules be governed by the laws of elastic impact in their collision with one another and with the walls of their container. It is further assumed that the path of a molecule between collisions is a straight line and that, therefore, its direction of motion can be changed only by collision with another molecule. During this collision, the molecule gives up some of its energy of motion. It may get back more or less energy during the next collision with another molecule. In any event, over a period of time the energy is divided equally throughout the entire gas without any overall loss.

These impacts may be direct hits or glancing blows. Therefore, not only is energy exchanged between two colliding molecules, but their direction may also be changed, as just indicated. This causes a zigzag path reminiscent of the Brownian movement of fine particles suspended in water or of particles of smoke in motionless air. There is, however, a *mean free path* which is actually an average distance between impacts.

With reference to Fig. 18.2, according to probability theory, a molecule m_1 will strike two opposite sides of the container one-third of the total times that it strikes all the sides. In this case, sides Z_1 and Z_2 will together receive one-third of all the collisions of the molecule with the sides of the box.

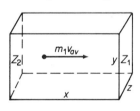

Fig. 18.2

If the molecule m_1 strikes Z_1 with a momentum of $m_1 v_{av}$ and rebounds with perfect elasticity, it will rebound with a momentum $-m_1 v_{av}$ away from Z_1. The change in momentum demands that

$$m_1 v_{av} - (-m_1 v_{av}) = 2m_1 v_{av}$$

The distance traveled by our molecule will be x before impact and x after impact; or a total distance of $2x$. We have said that v_{av} is the average velocity in distance units per second. Therefore, dividing the average velocity by $2x$, we get the number of impacts per second:

$$\text{impacts per second} = \frac{v_{av}}{2x}$$

The reciprocal of the preceding equation will give the time for every impact, or

$$\text{average time for each impact} = \frac{2x}{v_{av}}$$

The average force exerted by each molecule is the ratio of the changes in momentum to the time of each impact, or

$$\text{impact force exerted by each molecule} = \frac{2m_1 v_{av}}{2x/v_{av}}$$

$$= \frac{m_1 v_{av}^2}{x}$$

If there are a total of N molecules, then the force exerted on one face of the container will be $N/3$ and

$$\text{force in one direction} = \frac{N m_1 \bar{v}^2}{3x}$$

This is the force against the area of either side yz, the area. By definition,

$$P = \frac{f}{A} = \frac{N m_1 \bar{v}^2}{3x(yz)}$$

If xyz equals the volume V and if Nm_1 equals the total mass m of all the molecules, the equation becomes

$$P = \frac{m\bar{v}^2}{3V} \quad \text{or} \quad PV = \frac{1}{3} m\bar{v}^2$$

It has been shown that

$$PV = \frac{m}{M} RT$$

$\begin{cases} R = \text{universal gas const} \\ T = \text{abs temp} \\ M = \text{mol wt} \\ v = \text{av velocity} \\ V = \text{volume} \\ m = \text{mass (Note: if energy is in ft-lb then } m \text{ is in slugs)} \end{cases}$

Equating both PV equations,

$$\frac{m}{M} RT = \frac{1}{3} m\bar{v}^2$$

Solving for RT,

$$RT = \tfrac{1}{3} M\bar{v}^2$$

Multiplying by 3 and dividing by 2,

$$\frac{3}{2} RT = \frac{1}{2} M\bar{v}^2$$

This indicates the proportionality between the absolute temperature and the kinetic energy of the molecules.

Sec. 18.4   The Kinetic Theory of Heat   255

Example 7

Calculate (1) the average velocity, and (2) the kinetic energy of the nitrogen molecule if the temperature is 80°F. (3) Calculate the average velocity of the molecule in the metric system and check it against the answer in part (1).

Solution

1. Average velocity:

$$\frac{3}{2} RT = \frac{1}{2} \frac{M}{g} \bar{v}^2$$

Solve for v:

$$\bar{v} = \sqrt{\frac{3RTg}{M}} = \sqrt{\frac{3 \times 1542 \times 540 \times 32}{28}}$$
$$= 16.9 \times 10^2 = 1690 \text{ ft/sec}$$

$\left\{ \begin{array}{l} M = \text{mol wt of nitrogen} = 28 \\ T = \text{temp} = 80 + 460 = 540°R \\ R = 1542 \text{ ft-lb/lb-mol-°R} \\ g = 32 \text{ ft/sec}^2 \end{array} \right\}$

2. The kinetic energy is

$$\frac{3}{2} RT = \frac{3}{2}(1542)(540) = 1.25 \times 10^6 \text{ ft-lb}$$

3. The average velocity in the metric system is

$$v = \sqrt{\frac{3RT}{M}} = \sqrt{\frac{3 \times 8.3 \times 10^7 \times 300}{28}}$$
$$= 5.16 \times 10^4 \text{ cm/sec}$$

$\left\{ \begin{array}{l} M = \text{mol wt} = 28 \\ T = 540 \times \frac{5}{9} = 300°K \\ R = 8.3 \times 10^7 \text{ erg/g-mol-°K} \end{array} \right\}$

Check:

$$5.16 \times 10^4 \frac{\text{cm}}{\text{sec}} \times \frac{1 \text{ in.}}{2.54 \text{ cm}} \times \frac{1 \text{ ft}}{12 \text{ in.}} = 1690 \text{ ft/sec} \quad (check)$$

The major assumption to be drawn from this most important equation is that the kinetic energy of a gas is directly related to its absolute temperature. Therefore when the velocity of the molecule is at zero, the *absolute* temperature is zero.

For a monatomic gas, if the volume is constant, the heat goes toward an increase of the internal energy. Thus the heat added per unit mass to change the temperature 1° is its specific heat at constant volume.

$$c_v = \frac{3R}{2MJ}$$

Since $c_p - c_v = R/MJ$, then, solving for c_p,

$$c_p = \frac{R}{MJ} + c_v = \frac{R}{MJ} + \frac{3R}{2MJ} = \frac{5}{2} \frac{R}{MJ}$$

Taking the ratio c_p/c_v,

$$\frac{c_p}{c_v} = \frac{\frac{5}{2}(R/MJ)}{\frac{3}{2}(R/MJ)} = \frac{5}{3} = 1.67 \quad (monatomic\ gas)$$

It can be shown that a monatomic gas has three translational degrees of freedom but no rotational degrees of freedom. Note that the concept of degrees of freedom is one of the fundamental assumptions of kinetic theory. If, as stated, $\frac{3}{2}(R/MJ)$ represents three degrees of freedom, it can be said that

$$\frac{R}{2MJ}$$

represents *one* degree of freedom and the following relationships result, which further support the kinetic theory.

For a *diatomic gas*, which has five degrees of freedom—one rotational degree of freedom is missing,

$$c_v = \frac{5}{2}\frac{R}{MJ}$$

$$c_p = \frac{7}{2}\frac{R}{MJ}$$

$$c_p = \frac{7}{5} = 1.4 \qquad (\textit{diatomic gas})$$

For a *polyatomic gas*, which has six degrees of freedom—three translational and three rotational,

$$c_v = \frac{6}{2}\frac{R}{MJ}$$

$$c_p = \frac{8}{2}\frac{R}{MJ}$$

$$c_p = \frac{8}{6} = 1.33 \qquad (\textit{polyatomic gas})$$

The error in the calculations of c_v and c_p may be high for the higher molecular weights owing to the heat generated from the vibration of the atom in the molecule. This addition of heat for the higher weights cannot be overlooked.

Example 8

Calculate the values of c_v, c_p, and c_p/c_v for (1) helium; (2) hydrogen; (3) ammonia.

Solution

1. Helium is monatomic ($M = 4$):

$$c_v = \frac{3}{2}\frac{R}{MJ} = \frac{3}{2} \times \frac{8.3}{4 \times 4.18} = 0.7446$$

$$c_p = \frac{5}{2}\frac{R}{MJ} = \frac{5}{2} \times \frac{8.3}{4 \times 4.18} = 1.241$$

$$\frac{c_p}{c_v} = \frac{1.241}{0.7446} = 1.667$$

2. Hydrogen is diatomic ($M = 2$): $\begin{cases} R = 8.3 \text{ joule/g-mol-}°K \\ J = 4.18 \text{ joule/cal} \end{cases}$

$$c_v = \frac{5}{2}\frac{R}{MJ} = \frac{5}{2} \times \frac{8.3}{2 \times 4.18} = 2.482$$

$$c_p = \frac{7}{2}\frac{R}{MJ} = \frac{7}{2} \times \frac{8.3}{2 \times 4.18} = 3.475$$

$$\frac{c_v}{c_p} = \frac{3.475}{2.482} = 1.4$$

3. Ammonia is polyatomic ($M = 17$):

$$c_v = \frac{6}{2}\frac{R}{MJ} = \frac{6}{2} \times \frac{8.3}{17 \times 4.18} = 0.35$$

$$c_p = \frac{8}{2}\frac{R}{MJ} = \frac{8}{2} \times \frac{8.3}{17 \times 4.18} = 0.467$$

$$\frac{c_p}{c_v} = \frac{0.467}{0.35} = 1.33$$

Problems

1. A gas is at a constant temperature; 60 ft³ fills a tank at a pressure of 50 psia. Calculate the pressure if the gas is released into a 90-ft³ tank.
2. Ten ft³ of a gas at 70°F is heated to 180°F. Calculate the new volume if the pressure is constant.
3. Six ft³ of a gas is compressed by a piston to 2.5 ft³ at a constant temperature. If the initial pressure was 12 psig, calculate the new gage pressure.
4. Ten ft³ of air is at 1 atmosphere of pressure. Calculate the final volume if the air is compressed by 4 atm at a constant temperature.
5. Thirty ft³ of a gas at 32°F is expanded to a volume of 50 ft³ at a constant pressure. Calculate the new temperature.
6. A piston is used to compress a gas. If the pressure is 150 psia when the volume of the chamber is 3 ft³, what is the pressure when the piston closes the chamber to 1.5 ft³? The temperature is held constant.
7. A gas occupies 10 ft³ at 90°F and at a pressure of 60 psig. Assume the volume is increased to 12 ft³ and the temperature is decreased to 70°F. Calculate the new gage pressure.
8. The volume of a tire is 1200 in.³ at 18 psig and at a temperature of 70°F. Assuming negligible change in volume, calculate the pressure if the temperature of the tire increases to 110°F.
9. A gas occupies 6 ft³ at 80°F at a pressure of 40 psig. If the volume is decreased to 4 ft³ as the temperature is dropped to 40°F, calculate the new gage pressure.
10. In Prob. 9, assume a volume increase to 10 ft³. Calculate the new gage pressure.
11. Three hundred ft³ of a gas (mol wt = 17) are at a pressure of 10 cm of mercury gage and a temperature of 20°C. Calculate the weight of the gas.

12. In Prob. 11, calculate the temperature if the pressure drops to 8 cm of Hg gage and the volume increases to 400 ft.
13. Oxygen (mol wt = 32) is heated, yielding an increase in pressure from 100 psia to 120 psia. (a) What is the weight of the oxygen, if the initial temperature is 50°F and the initial volume is 48 ft^3? (b) What is the new volume at constant temperature?
14. Forty ft^3 of carbon dioxide at a pressure of 100 psia and a temperature of 30°F are heated to a temperature of 90°F at constant pressure. Determine the final volume and the weight of the carbon dioxide.
15. Eighteen lb of helium at a pressure of 400 psia and 40°F are heated at constant pressure from a volume of 60 ft^3 to a volume of 65 ft^3. Find the final temperature and the molecular weight of the gas.
16. A gas has values of c_p = 3.4 and c_v = 2.4. What is the gas?
17. A gas has values of c_p = 0.248 and c_v = 0.174. What is the gas?
18. Calculate the average velocity and the kinetic energy for the carbon dioxide molecule at a temperature of 70°F.
19. (a) How many Btu must be added to 6 lb of nitrogen when the pressure is increased from 50 psia to 70 psia, if the temperature is 100°F at the initial pressure? (b) What is the volume throughout the reaction?
20. Calculate the values of c_p and c_v for argon, oxygen, and ammonia, using the equations and method of the kinetic theory; check with the reference table (Appendix A.15).
21. Repeat Prob. 20 for helium, nitrogen, and carbon dioxide.
22. Calculate the c_p/c_v values for Probs. 20 and 21.
23. What are the average velocity and the kinetic energy of the molecules of 10 ft of helium gas at a temperature of 90°F and a pressure of 130 psia?
24. Eight lb of a gas occupy 300 ft^3 when the pressure is 5 psig. If the gas is carbon monoxide, what is the average velocity of the molecules?

19

P-V Diagrams

19.1. P-V Diagrams

Up to this point, we have been working with the mathematics of pressure–volume–temperature relationships. The state of a given mass of gas at a number of different time intervals can be plotted and a curve obtained which will yield valuable information. If the pressures are plotted on the y-axis (ordinate) and the volumes on the x-axis (abscissa), a curve is obtained such that a point on the curve will represent a temperature, which will show its corresponding pressure and volume (see Fig. 191a). These curves are called *P-V diagrams*. Plotting the temperature on the curve usually eliminates the need for three-dimensional plotting.

Notice that the shaded area in Fig. 19.1(a) represents a small rectangle having a width of $V_2' V_2'' = \Delta V$ and an altitude P_2 at the center of the rectangle. The area of this rectangle (Fig. 19.1b) is $P \Delta V$. But we have seen in Chapter 18 that $P \Delta V$ represents *work*. Therefore, if the area under the entire curve $(V_1 V_3)$ is divided into a large number of areas, the summation of all these areas will represent *external work*.

In Fig. 19.1(b), it can be seen that the small area A and the small area B are not quite equal. If $V_2' V_2''$ is taken small enough—that is, the larger the number of small areas taken—then the summation of all these areas will approach the actual area under the entire curve or the area between the limits desired. This summation may be accomplished with the calculus, or close approximations may be arrived at with algebra. Also, the area under a plotted curve may be obtained with a *planimeter*, an instrument that measures an enclosed area by tracing the area outline.

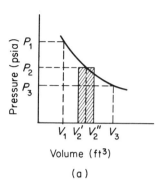 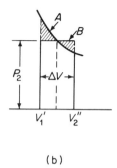

(a)   (b)

Fig. 19.1

The important fact is that the area bounded by the curve and the abscissa represents *external work*.

19.2. Constant Volume: Isometric Change

We have seen, in Chapter 18, that if heat is added to a system, the heat goes toward increasing the internal kinetic energy and the pressure changes according to the absolute temperature. Thus, if the change in internal energy equals ΔW_i, when the volume is constant,

$$Q = \Delta W_i$$

But for constant volume,

$$Q = c_v m \, \Delta T$$

Therefore, the internal heat (in Btu) is

$$\Delta W_i = c_v m \, \Delta T$$

But the external work is

$$\Delta W_e = P \, \Delta V = 0$$

$\left\{\begin{array}{l} Q = \text{heat added} \\ \Delta W_i = \text{change in internal energy} \\ \Delta W_e = \text{change in external energy} \\ c_v = \text{sp ht at constant volume} \\ m = \text{mass} \\ \Delta T = \text{change in temp} \\ P = \text{pressure} \\ \Delta V = \text{change in volume} \end{array}\right\}$

The total heat added is therefore

$$Q = c_v m \, \Delta T + 0 = c_v m \, \Delta T$$

Thus where the volume is constant and the pressure and temperature vary as shown in Fig. 19.2—this is called an *isometric change*. The external work must be zero. This can be seen in Fig. 19.2, because the area under the curve, which is a straight line parallel to the y-axis, is zero. The area bounded by the y-axis and the curve represents internal work. That is, $P_1/T_1 = P_2/T_2$ with the volume constant.

Sec. 19.3 Constant Pressure: Isobaric Change

Example 1

A fixed quantity of nitrogen gas weighing 4 lb is at a temperature of 140°F and a pressure of 40 psia. The final volume is the same as the initial volume but the gas, after heat is added, is under a new pressure of 60 psia. Find (1) the volume; (2) the final temperature; (3) the increase in internal energy; (4) the total energy added.

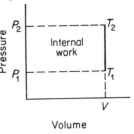

Fig. 19.2

Solution

1. The volume of 4 lb of nitrogen under these conditions is

$$P_1 V = \frac{m}{M} R T_1$$

Solving for V,

$$V = \frac{m}{M} \frac{R T_1}{P_1}$$

Substituting,

$$V = \frac{4 \times 1542 \times 600}{28 \times 40 \times 144} = 22.95 \text{ ft}^3$$

2. The final temperature is

$$\frac{P_1}{T_1} = \frac{P_2}{T_2}$$

Solving for T_2 and substituting,

$$T_2 = \frac{60 \times 600}{40} = 900°R = 440°F$$

$\left\{ \begin{array}{l} m = 4 \text{ lb} \\ M = 28 \\ R = 1542 \\ P_1 = (40 \text{ psia} \times 144) \text{ psfa} \\ P_2 = (60 \text{ psia} \times 144) \text{ psfa} \\ T_1 = 140°F + 460 = 600°R \\ c_v = 0.176 \\ W_i = \text{internal work} \\ W_e = \text{external work} \\ Q = \text{total heat added} \end{array} \right\}$

3. The increase in internal energy is

$$W_i = c_v m \, \Delta T = 0.176 \times 4(900 - 600) = 211.2 \text{ Btu}$$

4. The total energy added is

$$W_e = 0$$
$$Q = W_e + W_i = 0 + 211.2 = 211.2 \text{ Btu}$$

19.3. Constant Pressure: Isobaric Change

When the pressure is constant, and the temperature and volume change—this is called *isobaric change*—the gas law becomes $V_1/T_1 = V_2/T_2$ and the area under the curve, as shown in Fig. 19.3, becomes $P \, \Delta V$—this is called *isobaric change*—which again represents external work.

Since a volume change is involved, there must be work against external pressure, which remains constant. If the pressure remains constant, the volume change must be a result of the heat added, which contributes to the internal energy as well as to the external work. When P is constant, the external work (in Btu) is

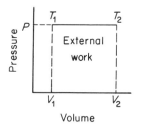

Fig. 19.3

$$W_e = \frac{P \, \Delta V}{J}$$

The internal work W_i is

$$W_i = c_v m \, \Delta T$$

The total heat added is

$$Q = W_e + W_i$$

$$= \frac{P \, \Delta V}{J} + c_v m \, \Delta T$$

J = 778 ft-lb/Btu
P = pressure
V = volume
c_v = sp ht at const volume
c_p = sp ht at const pressure
m = mass
T = change in temp
W_e = external work
W_i = internal energy
Q = total heat shield

Since the process takes place at constant pressure, the total heat added may also be found from

$$Q = c_p m \, \Delta T$$

Example 2

A fixed quantity of nitrogen gas weighing 4 lb is at a temperature of 140 °F and a constant pressure of 40 psia. The initial volume is 22.95 ft³. If the temperature of the system increases to 260 °F, calculate: (1) the final volume; (2) the external work; (3) the internal energy; (4) the total heat added. (5) Check the answer in (4) with $Q = c_v m T$.

Solution

1. Final volume:

$$\frac{V_1}{T_1} = \frac{V_2}{T_2}$$

Solve for V_2:

$$V_2 = \frac{V_1 T_2}{T_1} = \frac{22.95 \times 720}{600}$$

$$= 27.54 \text{ ft}^3$$

Sec. 19.4   Constant Temperature: Isothermal Change   263

2. External work:

$$W = \frac{P \Delta V}{J} = \frac{P(V_2 - V_1)}{J}$$

$$= \frac{40 \times 144(27.54 - 22.95)}{778}$$

$$= 34 \text{ Btu}$$

$\left\{ \begin{array}{l} m = 4 \text{ lb} \\ T_1 = 140 + 460 = 600°R \\ V_1 = 22.95 \text{ ft}^3 \\ T_2 = 260 + 460 = 720°R \\ P_1 = (40 \text{ psia} \times 144) \text{ psfa} \\ R = 1542 \\ M = 28 \\ J = 778 \\ c_v = 0.176 \\ c_p = 0.248 \end{array} \right.$

3. Internal energy:

$$W_i = c_v m (T_2 - T_1)$$
$$= 0.176 \times 4(720 - 600)$$
$$= 85 \text{ Btu}$$

4. Total energy:

$$Q = W_i + W_e = 85 + 34 = 119 \text{ Btu}$$

5. Total heat added (method using c_p):

$$Q = c_p m \Delta T = 0.248 \times 4(720 - 600)$$
$$= 119 \text{ Btu} \quad (check)$$

19.4. Constant Temperature: Isothermal Change

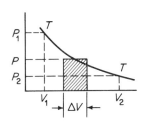

Fig. 19.4

If an expansion takes place at constant temperature and the pressure and volume change as shown in Fig. 19.4—it is called an *isothermal change*. The gas law is $P_1 V_1 = P_2 V_2$ and the curve is shown in Fig. 19.4. Here the internal energy remains constant—T is constant—and the internal work is zero:

$$\Delta W_i = 0$$

The heat added (at constant temperature) will go toward the work of expanding the volume against the external pressure. Thus the external work is the area under the curve. To find this area—external work—we add all the areas $P \Delta V$ between V_1 and V_2. The equation for external work is*

$$W_e = P_1 V_1 \ln \left(\frac{V_2}{V_1}\right) = P_1 V_1 \times 2.31 \log_{10} \left(\frac{V_2}{V_1}\right)$$

* This derivation is omitted and left to a formal course in thermodynamics.

Since $W_i = 0$, the total heat added, Q, is (in Btu units)

$$Q = W_i + W_e = 0 + \frac{P_1 V_1}{J} \ln\left(\frac{V_2}{V_1}\right)$$

Notice that, because the reaction takes place at constant temperature, the equation $c_p m \, \Delta T$ is useless; i.e., $\Delta T = 0$. Also, neither the pressure nor the volume is constant; therefore, c_p and c_v cannot be used.

Example 3

At a temperature of 140°F, 4 lb of nitrogen gas have a volume of 22.95 ft³ at a pressure of 40 psia. The temperature remains constant as the volume increases to 45.9 ft³. Find (1) the new pressure; (2) the external work of expansion; (3) the internal energy of expansion; (4) the total heat added.

Solution

1. The new pressure:

$$P_1 V_1 = P_2 V_2$$

Solve for P_2:

$$P_2 = \frac{P_1 V_1}{V_2} = \frac{40 \times 144 \times 22.95}{45.9}$$

$$= 2880 \text{ psfa} = 20 \text{ psia}$$

2. The external work of expansion:

$$W_e = \frac{P_1 V_1}{J} \ln\left(\frac{V_2}{V_1}\right)$$

$$= \frac{40 \times 144 \times 22.95}{778} \ln\left(\frac{45.9}{22.95}\right)$$

$$= 169.9 \ln 2 = 117.6 \text{ Btu}$$

$\left\{\begin{array}{l} P_1 = (40 \text{ psia} \times 144) \text{ psfa} \\ V_1 = 22.95 \text{ ft}^3 \\ V_2 = 45.9 \text{ ft}^3 \\ T = 140 + 460 = 600\,°\text{F} \\ R = 1542 \\ m = 4 \text{ lb} \\ M = 28 \end{array}\right\}$

3. The internal energy of expansion:

$$W_i = 0$$

4. The total heat added:

$$Q = W_e + W_i = 117.6 + 0 = 117.6 \text{ Btu}$$

19.5. Constant Heat: Isentropic Change—Adiabatic Expansion

In *adiabatic expansion*, there are several versions of the gas law $PV^\gamma = K$, where γ (*gamma*) $= c_p/c_v$. Three of the useful forms are

Sec. 19.5 Constant Heat: Isentropic Change-Adiabatic Expansion

$$P_1 V_1^\gamma = P_2 V_2^\gamma$$
$$T_1 V^{\gamma-1} = T_2 V_2^{\gamma-1}$$
$$\frac{T_1}{T_2} = \frac{P_1}{P_2}^{(\gamma-1)/\gamma}$$

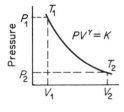

Fig. 19.5

This expansion takes place in an insulated chamber so that no heat can get in or out—thus it is an *isentropic change*. Figure 19.5 shows the curve for adiabatic expansion. Figure 19.6 shows all four curves plotted on one side of coordinate axes, so that the student can see how they differ from one another. Thus if the gas is compressed, the temperature must increase. The external work is at the expense of the internal energy during the expansion and the equation reduces to*

$$W_e = \frac{P_1 V_1 - P_2 V_2}{J(\gamma - 1)}$$

where $\gamma = c_p/c_v$.

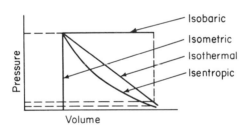

Fig. 19.6

The internal energy is

$$W_i = c_v m \, \Delta T$$

But the total heat added is

$$Q = 0$$

In the equation $PV^\gamma = K$, when $\gamma = 1$, the expansion becomes isothermal; or $PV = K$, where the temperature is constant. The difference in P–V diagrams is one of slope.

Another useful equation is

$$c_p - c_v = \frac{R}{MJ}$$

*Derivation left to a formal course in thermodynamics.

Example 4

Four lb of nitrogen gas, occupying 22.95 ft^3 at a temperature of 140°F and a pressure of 40 psia, expand adiabatically. If the new volume is 45.9 ft^3, find (1) the new pressure; (2) the new temperature; (3) the external work; (4) the increase in internal energy; (5) the total heat added to the system.

Solution

1. The new pressure is

$$P_1 V_1^\gamma = P_2 V_2^\gamma$$

$$(40 \times 144)(22.95)^{1.409} = P_2(45.9)^{1.409}$$

Solve for P_2:

$$P_2 = \frac{40 \times 144(22.95)^{1.409}}{(45.9)^{1.409}}$$

$$= \frac{2169.648 \text{ psfa}}{144}$$

$$= 15.067 \text{ psia}$$

2. The new temperature may be found from

$$\frac{P_1 V_1}{T_1} = \frac{P_2 V_2}{T_2}$$

Solve for T_2:

$$T_2 = \frac{P_2 V_2 T_1}{P_1 V_1} = \frac{15.067 \times 144 \times 45.9 \times 600}{40 \times 144 \times 22.95}$$

$$= 452°R = -8°F$$

$\left\{\begin{array}{l} m = 4 \text{ lb} \\ V_1 = 22.95 \text{ ft}^3 \\ V_2 = 45.9 \text{ ft}^3 \\ P_1 = (40 \text{ psia} \times 144) \text{ psfa} \\ T_1 = 140 + 460 = 600°R \\ c_p = 0.248 \\ c_v = 0.176 \\ \gamma = \frac{c_p}{c_v} = 1.409 \end{array}\right\}$

3. The external work is

$$W_e = \frac{P_1 V_1 - P_2 V_2}{J(\gamma - 1)} = \frac{(40 \times 144 \times 22.95) - (15.067 \times 144 \times 45.9)}{778(1.409 - 1)}$$

$$= 104 \text{ Btu}$$

4. The internal energy is

$$W_i = c_v m(T_2 - T_1) = 0.176 \times 4(452 - 600)$$

$$= -104 \text{ Btu}$$

5. Total heat added by definition is

$$Q = W_e + W_i = 104 - 104 = 0 \quad (\textit{no heat added})$$

$$W_i = -W_e = -104 \text{ Btu}$$

19.6. The Carnot Cycle and the P–V Diagram

The ideal cycle is a closed system which has no losses due to friction and in which the potential and kinetic energies, as a whole, of the mass of gas remain constant. This latter is referred to as a *nonflow process*. If these conditions are present (and of course this is not possible) we have a reversible process. That is, the various steps which go to make up a cycle may be retraced.

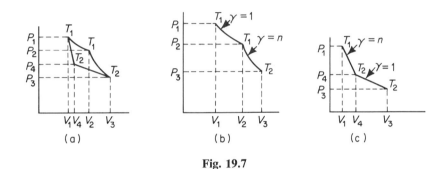

Fig. 19.7

Figure 19.7(a) shows an ideal cycle called the *Carnot cycle*. The cycle consists of two expansions (Fig. 19.7b), and of two contractions (Fig. 19.7c), where the process returns to its original state of $P_1 V_1 T_1$.

Referring to Fig. 19.7(a), during the first expansion ($\overline{T_1 T_1}$), heat is added without any change in temperature T_1, so that the volume increases from V_1 to V_2 as the pressure decreases from P_1 to P_2 (the temperature remaining constant). This is possible because heat is added to maintain the temperature at T_1 as V_1 expands to V_2 and as P_1 contracts to P_2. This part of the cycle is isothermal. The internal energy remains constant, since all the heat added is used to increase the volume and decrease the pressure (external work).

The second expansion ($\overline{T_1 T_2}$) takes place without any addition of heat and is adiabatic. Since no heat is added, any work accomplished by the system must be at the expense of the temperature, or internal energy. In this case, as in the previous expansion, the volume increases from V_2 to V_3 as the pressure drops from P_2 to P_3. Since no heat is added to cause these changes, the energy must come from the internal energy of the system; therefore, the temperature drops from T_1 to T_2 along curve $\overline{T_1 T_2}$.

If we use the same reasoning, the compression which takes place at constant temperature T_2, as V_3 contracts to V_4, and P_3 increases to P_4, is an isothermal contraction. It takes place with a loss of heat without any change of temperature along curve $\overline{T_2 T_2}$.

The contraction $\overline{T_2T_1}$ is adiabatic and takes place without any loss of heat, but returns it to T_2, which increases to T_1 (the original temperature). At the same time, V_4 is contracting to V_1 and P_4 is increasing to P_1. We are back at the starting point with the original conditions $P_1V_1T_1$, and the ideal cycle has been completed.

The net work done against the external pressure is represented by the area enclosed by these four processes (Fig. 19.7a).

The efficiency of a Carnot cycle (see Section 19.7) is

$$\text{eff} = \frac{\text{work output}}{\text{heat input}} = \frac{\text{wk}}{Q_a} = \frac{Q_a - Q_s}{Q_a}$$

But for $\overline{T_1T_1}$, $\quad\quad\quad\quad\quad\quad\quad\quad \left\{ \begin{array}{l} Q_a = \text{heat added} \\ Q_s = \text{heat subtracted (lost)} \end{array} \right\}$

$$Q_a = \frac{P_1V_1}{J} \ln\left(\frac{V_2}{V_1}\right)$$

$$Q_s = \frac{P_3V_3}{J} \ln\left(\frac{V_4}{V_3}\right) = -\frac{P_3V_3}{J} \ln\left(\frac{V_3}{V_4}\right)$$

Note: $\ln\dfrac{V_4}{V_3} = -\ln\left(\dfrac{V_3}{V_4}\right)$

The efficiency becomes

$$\text{eff} = \frac{Q_a - Q_s}{Q_a} = \frac{(P_1V_1/J) \ln(V_2/V_1) - (P_3V_3/J) \ln(V_3/V_4)}{(P_1V_1/J) \ln(V_2/V_1)}$$

But it can be shown that the ratio of the volumes is $V_2/V_1 = V_3/V_4$.* There-

Proof:

$$P_2V_2^\gamma = P_3V_3^\gamma = K$$
$$P_1V_1^\gamma = P_4V_4^\gamma = K$$

Also,

$$\frac{P_2V_2}{T_1} = \frac{P_3V_3}{T_2} \quad \text{and} \quad \frac{P_1V_1}{T_1} = \frac{P_4V_4}{T_2}$$

Divide the adiabatic equation by the universal gas law equation:

$$\frac{P_2V_2^\gamma}{P_2V_2/T_1} = \frac{P_3V_3^\gamma}{P_3V_3/T_2} \quad \text{and} \quad \frac{P_1V_1^\gamma}{P_1V_1/T_1} = \frac{P_4V_4^\gamma}{P_4V_4/T_2}$$

$$\frac{T_1V_2^\gamma}{V_2} = \frac{T_2V_3^\gamma}{V_3} \quad \text{and} \quad \frac{T_1V_1^\gamma}{V_1} = \frac{T_2V_4^\gamma}{V_4}$$

$$T_1V_2^{\gamma-1} = T_2V_3^{\gamma-1} \quad \text{and} \quad T_1V_1^{\gamma-1} = T_2V_4^{\gamma-1}$$

Therefore,

$$\frac{T_2}{T_1} = \frac{V_2^{\gamma-1}}{V_3} \quad \text{and} \quad \frac{T_2}{T_1} = \frac{V_1^{\gamma-1}}{V_4}$$

Equating,

$$\frac{V_2}{V_3} = \frac{V_1}{V_4} \quad \text{or} \quad \frac{V_2}{V_1} = \frac{V_3}{V_4}$$

fore, replacing V_3/V_4 with V_2/V_1 in the preceding equation, the efficiency equation reduces to

$$\text{eff} = \frac{\cancel{(1/J)\ln(V_2/V_1)}(P_1V_1 - P_3V_3)}{\cancel{(1/J)\ln(V_2/V_1)}P_1V_1}$$

If for P_1V_1 and P_3V_3 the values $(m/M)RT_1$ and $(m/M)RT_2$ are substituted, the efficiency of a heat engine becomes

$$\text{eff} = \frac{\cancel{[(m/M)R]}T_1 - \cancel{[(m/M)R]}T_2}{\cancel{[(m/M)R]}T_1} = \frac{T_1 - T_2}{T_1} \quad \left\{ \begin{array}{l} T_1 = \text{high temp: °R or °K} \\ T_2 = \text{low temp: °R or °K} \end{array} \right\}$$

No engine can be more efficient than a Carnot cycle working between the limits of temperature T_1 and T_2. This leads to the statement of the *second law of thermodynamics*, which says that under no conditions can heat flow from a low temperature to a higher temperature unless outside heat is added; heat can flow only from hot to cold. To utilize this heat flow, a source of heat must have a colder reservoir toward which it can flow.

Example 5

A Carnot engine operates from heat delivered at 350°F and coming out of the engine at 160°F. What is the upper efficiency of the engine?

Solution

$$\text{eff} = \frac{T_1 - T_2}{T_1} = \frac{810 - 620}{810} \quad \left\{ \begin{array}{l} T_1 = 350°F + 460 = 810°R \\ T_2 = 160°F + 460 = 620°R \end{array} \right\}$$
$$= 0.2346 = 23.46\%$$

19.7. The Thermodynamic Temperature Scale

Figure 19.8 shows two adiabatic curves A_1 and A_2, crossed by three

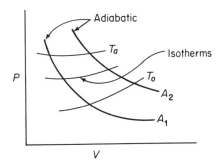

Fig. 19.8

isothermal curves. The area under each isothermal curve represents the work output corresponding to heat input—if heat is added. Therefore,

$$\text{eff} = \frac{\text{work output}}{\text{heat input}}$$

$$= \frac{Q_a - Q_s}{Q_a} = \frac{T_a - T_s}{T_a} \qquad \begin{cases} Q_a = \text{heat added} \\ Q_s = \text{heat lost} \\ T_a = \text{temp when heat is added} \\ T_s = \text{temp at completion} \end{cases}$$

or

$$\frac{Q_s}{Q_a} = \frac{T_s}{T_a}$$

Since the absolute temperatures are *proportional* to the heat added, it is possible, as long as the two isothermal curves cross the same two adiabatic curves, to set up a temperature scale. The two isothermal curves may be chosen arbitrarily. Assume two isothermal curves, one at the steam point and one at the ice point. Let $T_0 = 32°F$ and $T_a = 212°F$ be taken as the two isothermal curves. Then

$$\frac{T_0}{T_a} = \frac{Q_0}{Q_a} \quad \text{and} \quad \frac{T_0}{T_a - T_0} = \frac{Q_0}{Q_a - Q_0} \qquad \begin{cases} T_0 = \text{lower isotherm} \\ T_a = \text{upper isotherm} \\ Q_0 = \text{output heat} \\ Q_a = \text{input heat} \end{cases}$$

Solving for T_0,

$$T_0 = (T_a - T_0)\frac{Q_0}{Q_a - Q_0}$$

If $T_a - T_0$ is defined as equal to 100°K, then, in terms of the ratio of heat input to heat output,

$$T_0 = 100 \left(\frac{Q_0}{Q_a - Q_0}\right)$$

But in general, since $T_a - T_0$ may be any temperature; and since $Q_a - Q_0$ is the net area enclosed by the two adiabatic and the two isothermal curves, the general equation becomes

$$T = Q \left(\frac{T_0}{Q_0}\right) \qquad \begin{cases} T = \text{any upper-level temperature} \\ T_0 = \text{ice-point temperature} \\ Q = \text{heat added at upper level} \\ Q_0 = \text{heat at ice-point temperature} \end{cases}$$

Problems

1. Interpret the *P–V* diagram for (a) isobaric expansion; (b) isometric expansion; (c) isothermal expansion; (d) adiabatic expansion.
2. Three lb of ammonia gas is at a constant pressure of 50 psia and 100°F. Calculate, fr when the temperature increases to 175°F, (a) the internal work; (b) the external work; (c) the total heat added.

Problems

3. Two lb of helium gas at constant volume is at a temperature of 70°F and a pressure of 30 psia. The pressure changes to 50 psia. Calculate (a) the volume; (b) the final temperature; (c) the increase in internal energy; (d) the total energy added to the system.
4. Five lb of oxygen is at a pressure of 70 psia. The volume is the same at the initial temperature as at 300°F. If the total heat required to consumate the process is 125 Btu, calculate (a) the initial temperature; (b) the volume; (c) the pressure at the final temperature.
5. Four lb of air expand at constant pressure from a temperature of 70°F. (a) Determine final temperature if the increase in internal energy is 1000 Btu and the volume becomes 20 ft^3. Calculate (b) the initial volume; (c) the pressure; (d) the work done in ft-lb.
6. The temperature of a quantity of hydrogen remains constant while the volume increases from 100 ft^3 to 150 ft^3. The initial pressure is 30 psia when the temperature is 70°F. Find (a) the mass of the gas; (b) the new pressure; (c) the external energy in ft-lb/lb; (d) the total heat added.
7. A gas is at a pressure of 300 psia when the volume is 3.6 ft^3. If the gas expands isothermally to a volume of 10 ft^3, find the total heat added.
8. Ten lb of air at 350 psia are compressed isothermally until the pressure is 450 psia. If the original volume was 15 ft^3, find (a) the temperature; (b) the final volume; (c) the external work; (d) the total heat added to compress the gas.
9. Eight lb of helium gas occupy a volume of 30 ft^3 at a pressure of 480 psia. If the gas expands adiabatically to a new volume of 35 ft^3, find (a) the new temperature; (b) the new pressure; (c) the external work; (d) the internal energy.
10. A certain gas has values of $c_p = 0.238$ and $c_v - 0.170$ and weighs 4 lb at a constant temperature of 110°F. If 600 Btu are added to the gas, the final pressure becomes 15 psia. Find (a) the molecular weight; (b) the final and initial volumes; (c) the initial pressure; (d) the total work.
11. A Carnot cycle ($\gamma = 1.6$) starts at 140 psia and 3 ft^3 and expands isothermally to 6 ft^3 and then adiabatically to 8 ft^3. The cycle then starts on its compressional phase. It compresses to 4 ft^3 isothermally and then adiabatically to 3 ft^3, the starting point. Find all the pressures.
12. With the data given in Prob. 11, find (a) the total heat added; (b) the net heat; (c) the efficiency.
13. Heat at 250°F is delivered to a Carnot engine and ejected into a cold receptacle at a temperature of 145°F. Find the maximum efficiency of the engine.

20

Electrostatics

20.1. The Electric Charge

Neutral objects have equal amounts of positive and negative charges. A neutral rubber or ebonite rod rubbed with a piece of wool picks up electrons lost by the wool. The rod becomes *negatively charged* because it now possesses an excess number of electrons. The wool, which gave up these electrons, is said to have a deficiency of electrons and therefore is said to be *positively charged*.

A neutral glass rod rubbed with a piece of silk loses electrons to the silk. Since the glass rod possesses a deficiency of electrons, it is said to be positively charged. The silk now has an excess of electrons and is said to be negatively charged.

Objects which carry the same charge will repel each other. Objects which carry unlike charges will attract each other. Thus in Fig. 20.1(a), two pith balls possessing equal numbers of positive and negative charges distributed at random will neither attract nor repel each other.

If a glass rod is rubbed with silk, the glass rod will become positively charged. If the rod is brought near one of the pith balls, Fig. 20.1(b), the charges in the pith ball will redistribute so that the right side of the pith ball becomes negative and the left side becomes positive. Note that in Fig. 20.1(c), the reverse happens when a negatively charged rubber rod is brought near the pith ball. If the rods are removed, the charges redistribute inside the pith balls.

Now assume the glass rod touches the pith ball in Fig. 20.1(b). The negative charges will move to the glass rod and leave a deficiency of electrons

Sec. 20.1   *The Electric Charge*   273

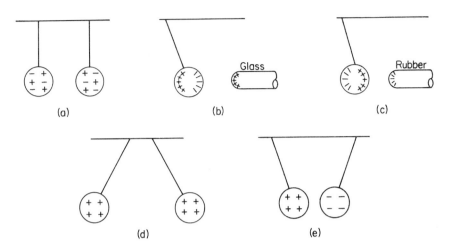

Fig. 20.1

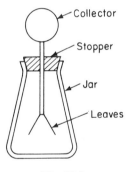

Fig. 20.2

in the pith ball. That is, the pith ball will be said to be charged positively.

If the second pith ball is also charged positively with a charged glass rod, the two pith balls will repel as shown in Fig. 20.1(d). If the second pith ball is charged negatively with a rubber rod, the two pith balls will attract each other as shown in Fig. 20.1(e).

The gold-leaf *electroscope*, Fig. 20.2, may be charged by *contact* or by *induction*. Charging by *contact* has been described in the preceding paragraphs. Charging by induction will be described in the succeeding discussion.

If a negatively charged rubber rod is brought near the collector, the electroscope charges separate so that the negative charges are driven into the leaves, thus leaving the collector positive, Fig. 20.3(a), and the leaves separate.

If the rod touches the collector, some of the negative charges from the rod transfer to the collector. When the rod is removed, the leaves remain separated, because the electroscope now possesses an excess of negative charges.

In Fig. 20.3(b), a positively charged glass rod causes the negative charges to accumulate in the collector, which makes the leaves positive. If the rod touches the collector, it drains off negative charges. Thus when the rod is removed, there remains a deficiency of negative charges in the electroscope. The leaves will remain separated.

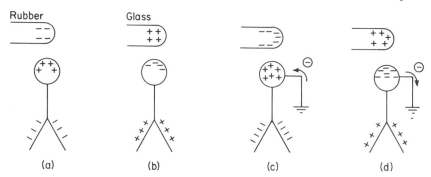

Fig. 20.3

In Fig. 20.3(c), the electroscope may be charged negatively by grounding the collector. Electrons will flow from ground to make up the *collector's* deficiency of negative charges. When the ground is broken and the rubber rod removed, the electroscope remains negative. Notice that the rod need not touch the collector. This is called charging by induction.

In Fig. 20.3(d), the electrons will leak off to ground. When the ground is broken, the electroscope remains positively charged.

Figure 20.4 shows Faraday's ice-pail experiment. The charges on the pail may be separated as shown in Fig. 20.4(a). If the inside of the pail is touched with the rod, the negative charges will neutralize the rod (Fig. 20.4b) and leave the inside of the pail neutral as shown in Fig. 20.4(c). When the rod is removed, the net charge on the pail is positive.

If the rod in Fig. 20.4(a) is *not* permitted to touch the side of the pail, and the outside of the pail is grounded (Fig. 20.4d), negative charges will neutralize the outside of the pail. If the ground is broken and the rod removed, in that order, the net charge on the pail will be negative.

The above procedure can be repeated with a negatively charged rod. The student should make this analysis.

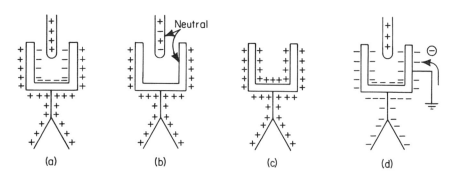

Fig. 20.4

20.2. Units

The next problem which must be faced in the study of electricity is an understanding of units. The system of units which will be used in electricity is the *meter–kilogram–second–ampere* (mksa) *system*.

In the preceding section, it was established that like charges repel each other and that unlike charges attract each other. Figure 20.5 shows two single positive charges. It becomes necessary to define the magnitude of a unit charge to some reference level. This can be done by saying that, if a force of 1 *newton* (nt) is exerted by one charged body when repelling another like-charged body, when these bodies are 1 meter apart in a vacuum, then the magnitude of each charged body is said to be 1 *coulomb* (coul).

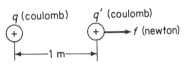

Fig. 20.5

Coulomb's law states that the force of attraction or repulsion between two charges is directly proportional to the products of the magnitude of the two charges and inversely proportional to the square of the distance between them. Thus

$$f \text{ (newtons)} \propto \frac{qq' \text{ (coulombs)}^2}{s^2 \text{ (meters)}^2}$$

$\begin{cases} q = \text{elec charge: coul} \\ q' = \text{elec charge: coul} \\ s = \text{distance: m} \\ f = \text{force: nt} \end{cases}$

The equation is written

$$f = K \frac{qq'}{s^2}$$

Through experimentation, it has been established that K is very close to 9×10^9. Thus, solving for K,

$$K = \frac{fs^2}{qq'} = 9 \times 10^9 \text{ nt-m}^2/\text{coul}^2$$

To eliminate the quantity 4π (which appears in the formula for the area of a sphere, see Section 20.5), a new constant ε_0 (*epsilon*) is defined as

$$\varepsilon_0 = 8.85 \times 10^{-12} \text{ coul}^2/\text{n-m}^2 = \frac{1}{4\pi(9 \times 10^9)}$$

Since $K = 9 \times 10^9$,

$$\varepsilon_0 = \frac{1}{4\pi K}$$

Solving for K,

$$K = \frac{1}{4\pi\varepsilon_0} \times 9 \times 10^9 \text{ nt-m}^2/\text{coul}^2$$

Electrostatics

Therefore, Coulomb's law in the mksa system of units becomes

$$f = \frac{Kqq'}{s^2} = \frac{1}{4\pi\varepsilon_0} \times \frac{qq'}{s^2} = 9 \times 10^9 \frac{qq'}{s^2}$$

Example 1

Two positive charges are 50 cm apart. One charge is 1×10^{-6} coul and the other is 1.6×10^{-6} coul. Calculate the force of repulsion between them.

Solution

The force is

$$f = 9 \times 10^9 \times \frac{(1 \times 10^{-9})(1.6 \times 10^{-9})}{(0.5)^2}$$

$$= 57.6 \times 10^{-9} \text{ nt}$$

$$\left\{ \begin{array}{l} K = 9 \times 10 \text{ nt-m}^2/\text{coul}^2 \\ q = +1 \times 10^{-9} \text{ coul} \\ q' = +1.6 \times 10^{-9} \text{ coul} \\ s = 0.50 \text{ m} \end{array} \right\}$$

Coulomb showed the above relationship to be true with his torsion balance, Fig. 20.6. One of the pith balls on the torsion rod was charged and a like charge was place on the fixed rod. The torsion head was then adjusted to create an angle between the rotating and the stationary pith ball. The force between the two pith balls and the distance between them yields Coulomb's law.

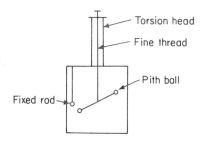

Fig. 20.6

Another system of units (cgs) may still be used in some quarters. If this should be necessary, the conversions may be made using the following factors:

1 coulomb = 3×10^9 electrostatic units of charge

1 newton = 10^5 dynes

1 meter = 10^2 centimeters

Example 2

Three bodies are charged respectively $+25 \times 10^{-6}$ coul, -15×10^{-6} coul, and $+8 \times 10^{-6}$ coul. If the three charges are placed in a straight line, as shown in Fig. 20.7, calculate the resultant force and the direction of this force on the $+25 \times 10^{-6}$ coul charge.

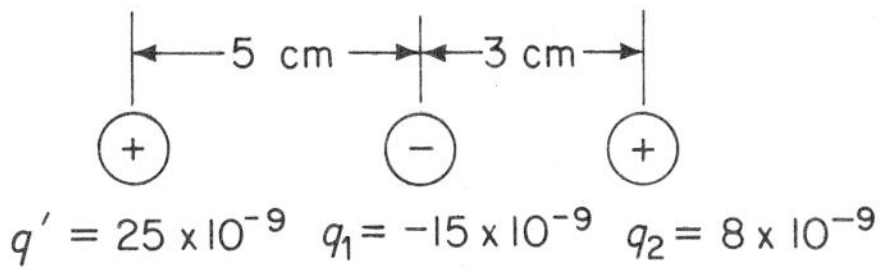

Sec. 20.2   *Units*   277

```
|←—5 cm—→|←—3 cm—→|
   (+)        (−)        (+)
q' = 25 x 10⁻⁹   q₁ = −15 x 10⁻⁹   q₂ = 8 x 10⁻⁹
```

Fig. 20.7

Solution

The force of attraction between $+q'$ and $-q_1$ is

$$f_1 = K\frac{q_1 q'}{s^2} = 9 \times 10^9 \times \frac{(25 \times 10^{-9})(-15 \times 10^{-9})}{(0.05)^2}$$

$= 135 \times 10^{-5}$ nt   (q_1 *attracts* q' *to right*)

The force of repulsion between $+q'$ and $+q_2$ is

$$f_2 = K\frac{q_2 q'}{s^2} = 9 \times 10^{-9} \times \frac{(25 \times 10^{-9})(8 \times 10^{-9})}{(0.08)^2}$$

$= 28.1 \times 10^{-5}$ nt   (q^2 *repels* q' *to left*)

$$\left\{\begin{array}{l} q' = +25 \times 10^{-9} \text{ coul} \\ q_1 = -15 \times 15^{-9} \text{ coul} \\ q_2 = +8 \times 10^{-9} \text{ coul} \\ s_1 = 5 \text{ cm} = 0.05 \text{ m} \\ s_2 = 3 + 5 = 8 \text{ cm} = 0.08 \text{ m} \\ K = 9 \times 10^{-9} \text{nt-m}^2/\text{coul}^2 \end{array}\right\}$$

The net force is

$$f = f_1 - f_2 = (135 \times 10^{-5}) - (28.1 \times 10^{-5})$$

$= 106.9 \times 10^{-5}$ nt   (*net force to right*)

Example 3

The same three charges as used in Example 2 are placed so that they form a triangle as shown in Fig. 20.8(a). Calculate the magnitude and direction of the resultant force on the $+25 \times 10^{-6}$ coul charge.

Solution

1. The force of repulsion on the $+25 \times 10^{-6}$ coul charge by the $+8 \times 10^{-6}$ coul charge is

$$f_1 = 9 \times 10^9 \times \frac{(8 \times 10^{-9})(25 \times 10^{-9})}{(0.10)^2} = 18 \times 10^{-5} \text{ nt} \quad (\textit{repel})$$

2. The force of attraction on the $+25 \times 10^{-6}$ coul charge by the -15×10^{-6} coul charge is

$$f_2 = 9 \times 10^{-9} \times \frac{(15 \times 10^{-9})(25 \times 10^{-9})}{(0.05)^2} = 135 \times 10^{-5} \text{ nt} \quad (\textit{attract})$$

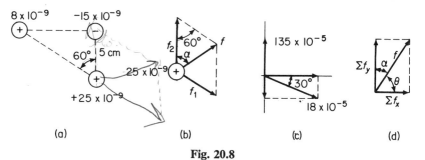

Fig. 20.8

3. The net force on the $+25 \times 10^{-6}$ coul charge is the *vector sum* of the f_1 and f_2 forces (see Fig. 20.8b). This resultant may be found from the law of cosines. Thus

$$f^2 = f_1^2 + f_2^2 - 2f_1 f_2 \cos 60°$$
$$= (18 \times 10^{-5})^2 + (135 \times 10^{-5})^2 - 2(18 \times 10^{-5})(135 \times 10^{-5}) \cos 60°$$
$$f = 127 \times 10^{-5} \text{ nt}$$

The angle that f makes with the y-axis is, using the law of sines,

$$\frac{18 \times 10^{-5}}{\sin \alpha} = \frac{127 \times 10^{-5}}{\sin 60°}$$
$$\sin \alpha = 0.123$$
$$\alpha = 7$$

Alternate Solution

Calculate the resultant and its direction by the resolution of the forces. Thus in Fig. 20.8(c), the vector sums in the x- and y-directions are

$$\sum f_x = 18 \times 10^{-5} \times \cos 30° = 15.6 \times 10^{-5} \text{ nt}$$
$$\sum f_y = (135 \times 10^{-5}) - (18 \times 10^{-5}) \sin 30° = 126 \times 10^{-5} \text{ nt}$$

Figure 20.8(d) shows the resultant:

$$f = \sqrt{(126 \times 10^{-5})^2 + (15.6 \times 10^{-5})^2}$$
$$= 127 \times 10^{-5} \text{ nt} \quad (check)$$

The angle θ is

$$\sin \theta = \frac{126 \times 10^{-5}}{127 \times 10^{-5}} = 0.992$$
$$\theta = 83°$$
$$\alpha = 90° - 83° = 7 \quad (check)$$

Sec. 20.2 Units

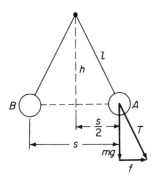

Fig. 20.9

The following method is used to measure the charge on a pith ball when one pith ball repels another:

In Fig. 20.9, ball A and ball B repel each other mutually. The mass of A is determined very carefully. From similar triangles in Fig. 20.9,

$$\frac{mg}{f} = \frac{h}{s/2}$$

Solving for f,

$$f = \frac{mgs}{2h}$$

Coulomb's law states

$$f = K\frac{qq'}{s^2}$$

Equating,

$$\frac{mgs}{2h} = K\frac{qq'}{s^2}$$

$$q = \sqrt{mgs^3/2hK}$$

If q equals q', then the charge on one pith ball is

Example 4

Two equally charged pith balls are separated by a distance of 24 cm. If the length of the string is 30 cm and the mass of the pith balls is 0.04 g, calculate the charge on each pith ball.

Solution

The altitude h is

$$h = \sqrt{(0.24)^2 - \left(\frac{0.24}{2}\right)^2}$$
$$= 0.208 \text{ m}$$

$$\begin{cases} L = 24 \text{ cm} = 0.24 \text{ m} \\ s = 20 \text{ cm} = 0.20 \text{ m} \\ m = 4 \times 10^{-2} \text{ g} = 4 \times 10^{-5} \text{ kg} \\ g = 9.8 \text{ m/sec}^2 \\ K = 9 \times 10^{-9} \text{ nt-m}^2/\text{coul}^2 \end{cases}$$

The charge is

$$q = \sqrt{\frac{mgs^3}{2hK}} = \sqrt{\frac{4 \times 10^{-5} \times 9.8 \times (0.2)^3}{2 \times 0.208 \times 9 \times 10^9}}$$
$$= 28.94 \times 10^{-9} \text{ coul}$$

20.3. The Electric Field

If the *electric-field strength* at a point is defined as a force exerted on a *unit positive charge* at that point, then electric-field intensity is a vector quantity and Coulomb's equation becomes

$$E = K\frac{q \times 1}{s^2} \frac{\text{newtons}}{\text{coulombs}}$$

where q' is the unit positive charge. Therefore, if it is desired to find the field strength at a point in space, the method employed is to place a unit charge at that point and test. The charge should be a *unit plus charge*. By definition, then,

$$E = \frac{f}{q'} = K\frac{q}{s^2} = \frac{1}{4\pi\varepsilon_0} \times \frac{q}{s^2}$$

Rewritten,

$$E = \frac{q}{\varepsilon_0 4\pi s^2}$$

Since $4\pi r^2$ equals area A,

$$E = \frac{q}{\varepsilon_0 A}$$

Letting σ (*sigma*) equal charge per unit area, the equation becomes

$$E = \frac{\sigma}{\varepsilon_0}$$

Before we work illustrative problems, some conventions must be adopted. Figure 20.10(a) and (b) shows imaginary *electric flux*, or *electric lines of force*,

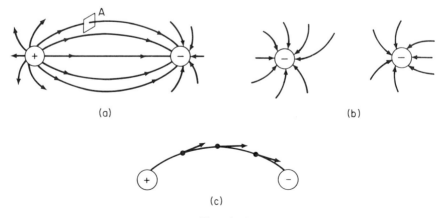

Fig. 20.10

Sec. 20.4 *Potential Energy, Difference, and Equipotential Surfaces*

which emanate from a plus (+) charge and end in a negative (−) charge. Also, they are continuous and are conceived as not crossing each other. Also, the direction of the electric field at any point is the direction of the tangent vector to the curve at that point. This is shown in Fig. 20.10(c), where several tangents are drawn.

In the mksa system, the *unit* electric-field strength is represented as 1 *line per square meter*. The area must be perpendicular to the direction of the lines of flux and is shown in Fig. 20.10(a).

20.4. Potential Energy, Potential, Potential Difference, and Equipotential Surfaces

If 1 *joule* of energy is required to move a positive unit charge from a point x to a point y, against a field effective from y to x, then the charge at y is said to be at a higher *potential* of 1 *volt* than when it was at x. If the definition of work is the displacement in the direction of motion multiplied by the force component acting in the same direction as the displacement, then

$$V_y - V_x = fs \cos \theta$$

It is probably more practical to compare a potential at a point with the zero potential at infinity and then calculate the work required to move a unit positive charge from the reference point at infinity, against the field, to the point being considered. This is called *electric potential energy*.

It should be stressed that the path through which this charge is moved is of no consequence, since it is the *displaced* distance from infinity to the point in question which is important. The work done is dependent only upon the starting point and the final point. Therefore, the work required to move a + charge from x to y in Fig. 20.11 against a field created by q is

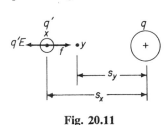

Fig. 20.11

$$W^*_{x \to y} = Kqq' \left(\frac{1}{s_y} - \frac{1}{s_x} \right)$$

If the point x is at infinity, then $1/s_x$ equals zero, and drops out. The equation for work required to move a charge, against a field, from infinity to y becomes

$$W_{\infty \to y} = \frac{Kqq'}{s_y} \quad \left\{ \begin{array}{l} q = \text{charge creating field: coul} \\ q' = \text{charge moving in field: coul} \\ s = \text{distance moved in the field:} \\ W = \text{work: joules} \end{array} \right\}$$

* $W_{x \to y}$ may best be derived using the calculus.

It is also true that if the total field is created by several fields, the work required to move a charge against the net field is the *algebraic sum* of the several fields. Thus

$$W_{\infty \to y} = Kq' \sum \frac{q}{s} \quad (algebraic\ sum)$$

The *potential* of the charge q' place into the field is equal to the work per unit charge, or the joule per coulomb. Thus

$$V_x = \frac{W}{q'} = K \sum \frac{q}{s} \quad (algebraic\ sum) \quad \{V_x = \text{potential at point } x\}$$

If a charge q' is brought from infinity to a point, and it requires 1 joule of work for every coulomb to move this charge against an electrostatic field, the potential at the point is said to be 1 volt; or

$$1\ \text{joule/coul} = 1\ \text{volt}$$

The equation for work can therefore be stated as

$$W = Vq'$$

Carrying the reasoning one step further, it can be said that the potential difference between two points x and y, is at a higher potential than x and possesses more potential energy at y than at x, is 1 volt when the positive charge, moved from the point x to point y requires 1 joule of work for every coulomb moved; so that

$$W_{x \to y} = V_{yx} = V_y - V_x$$

Example 5

Two charges are placed 20 cm apart as shown in Fig. 20.12(a). The charges have magnitudes of $+50 \times 10^{-6}$ coul and -50×10^{-6} coul. Calculate (1) the field strength at points x, y, and z; (2) the potential at points x, y, and z; (3) the potential difference between x and y, y and z, x and z. (4) Calculate the

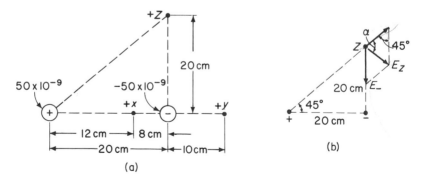

Fig. 20.12

Sec. 20.4 Potential Energy, Difference, and Equipotential Surfaces   283

potential energy, assuming a charge with a magnitude of $+2 \times 10^{-6}$ coul is placed at x, y, and z. (5) Calculate the work required to move the charges from x to y, y to x, y to z, x to z.

Solution

1. The field intensity at x, at y, and at z.

   (a) At x:

   $$E_+ = K\frac{q}{s^2} = 9 \times 10^9 \times \frac{50 \times 10^{-9}}{(0.12)^2}$$
   $$= 3.125 \times 10^4 \text{ nt/coul} \quad (\textit{repels } x\rightarrow)$$
   $$E_- = 9 \times 10^9 \times \frac{50 \times 10^{-9}}{(0.08)^2}$$
   $$= 7.031 \times 10^4 \text{ nt/coul} \quad (\textit{attracts } x\rightarrow)$$
   $$E_x = E_+ + E_- = (3.125 \times 10^4) + (7.031 \times 10^4)$$
   $$= 10.156 \times 10^4 \text{ nt/coul} \quad (\textit{to the right}\rightarrow)$$

   (b) At y:

   $$E_+ = 9 \times 10^9 \times \frac{50 \times 10^{-9}}{(0.30)^2}$$
   $$= 0.5 \times 10^4 \text{ nt/coul} \quad (\textit{repels } y\rightarrow)$$
   $$E_- = 9 \times 10^9 \times \frac{50 \times 10^{-9}}{(0.10)^2}$$
   $$= 4.5 \times 10^4 \text{ nt/coul} \quad (\textit{attracts } y\leftarrow)$$
   $$E_y = (4.5 \times 10^4) - (0.5 \times 10^4) = 4 \times 10^4 \text{ nt/coul} \quad (\textit{to left}\leftarrow)$$

   (c) The field intensity at z is the vector sum of the two vectors E_+ and E_- as shown in Fig. 20.12(b).

   $$E_+ = 9 \times 10^9 \times \frac{50 \times 10^{-9}}{(0.283)^2}$$
   $$= 0.562 \times 10^4 \text{ nt/coul}$$
   $$E_- = 9 \times 10^9 \times \frac{50 \times 10^{-9}}{(0.20)^2}$$
   $$= 1.125 \times 10^4 \text{ nt/coul}$$
   $$E_z^2 = (0.562 \times 10^4)^2 + (1.125 \times 10^4)^2$$
   $$\quad - 2(0.562 \times 10^4)(1.125 \times 10^4) \cos 45°$$
   $$E_z = 0.828 \times 10^4 \text{ nt/coul}$$

From the law of sines,

$$\frac{0.828 \times 10^4}{\sin 45°} = \frac{1.125 \times 10^4}{\sin a}$$

$$a = 73°50'$$

2. The potentials at points x, y, and z

   (a) At x (the signs $+$ and $-$ must be adhered to):

   Due to the $+$ charge:

   $$V_+ = K\frac{q}{s} = 9 \times 10^9 \times \frac{+50 \times 10^{-9}}{0.12}$$
   $$= +0.375 \times 10^4 \text{ volts}$$

   Due to the $-$ charge:

   $$V_- = 9 \times 10^9 \times \frac{-50 \times 10^{-9}}{0.08}$$
   $$= -0.562 \times 10^4 \text{ volts}$$

   The potential at x (algebraic sum):

   $$V_x = (+0.375 \times 10^4) - (0.562 \times 10^4)$$
   $$= -0.187 \times 10^4 \text{ volts}$$

   (b) At y:

   Due to the $+$ charge:

   $$V_+ = 9 \times 10^9 \times \frac{+50 \times 10^{-9}}{0.3}$$
   $$= 0.15 \times 10^4 \text{ volts}$$

   Due to the $-$ charge:

   $$V_- = 9 \times 10^9 \times \frac{-50 \times 10^{-9}}{0.1}$$
   $$= -0.45 \times 10^4 \text{ volts}$$

   The potential at y is

   $$V_y = (0.15 \times 10^4) - (0.45 \times 10^4)$$
   $$= -0.3 \times 10^4 \text{ volts}$$

   (c) At z:

   Due to the $+$ charge:

   $$V_+ = 9 \times 10^9 \times \frac{+50 \times 10^{-9}}{0.283}$$
   $$= +0.159 \times 10^4 \text{ volts}$$

Sec. 20.4 Potential Energy, Difference, and Equipotential Surfaces

Due to the − charge:

$$V_- = 9 \times 10^9 \times \frac{-50 \times 10^{-9}}{0.2}$$
$$= -0.225 \times 10^4 \text{ volts}$$

The potential at z is

$$V_z = (0.159 \times 10^4) - (0.225 \times 10^4)$$
$$= -0.066 \times 10^4 \text{ volts}$$

3. The potential difference between x and y, y and x, y and z, x and z.

   (a) The potential difference between x and y is

   $$V_{xy} = V_x - V_y = (-0.187 \times 10^4) - (-0.3 \times 10^4)$$
   $$= +0.113 \times 10^4 \text{ volts}$$

   (b) The potential difference between y and x is

   $$V_{yx} = V_y - V_x = (-0.3 \times 10^4) - (-0.187 \times 10^4)$$
   $$= -0.113 \times 10^4 \text{ volts}$$

   (c) The potential difference between y and z is

   $$V_{yz} = V_y - V_z = (-0.3 \times 10^4) - (-0.066 \times 10^4)$$
   $$= -0.234 \times 10^4 \text{ volts}$$

   (d) The potential difference between x and z is

   $$V_{xz} = V_x - V_z = (-0.187 \times 10^4) - (-0.066 \times 10^4)$$
   $$= -0.121 \times 10^4 \text{ volts}$$

4. The potential energy at points x, y, and z, if a point charge of 2×10^{-6} coul is placed at each of the points.

   (a) At x:

   $$\text{PE}_x = q'V_x = 2 \times 10^{-9} \times (-0.187 \times 10^4)$$
   $$= -0.375 \times 10^{-5} \text{ joule}$$

   (b) At y:

   $$(\text{PE})_y = q'V_y = 2 \times 10^{-9} \times (-0.3 \times 10^4)$$
   $$= -0.6 \times 10^{-5} \text{ joule}$$

286   *Electrostatics*   Chap. 20

(c) At z:

$$(PE)_z = q'V_z = 2 \times 10^{-9} \times (-0.066 \times 10^4)$$
$$= -0.132 \times 10^{-5} \text{ joule}$$

5. The work required to move a charge of 2×10^{-6} coul from x to y, from y to x, from y to z, from x to z.

(a) From x to y:

$$W_{xy} = q'V_{xy} = 2 \times 10^{-9} (-0.113 \times 10^4)$$
$$= -0.226 \times 10^{-5} \text{ joule}$$

(b) From y to x:

$$W_{yx} = q'V_{yx} = 2 \times 10^{-9} \times (+0.113 \times 10^4)$$
$$= +0.226 \times 10^{-5} \text{ joule}$$

(c) From y to z:

$$W_{yz} = q'V_{zy} = 2 \times 10^{-9} \times (+0.234 \times 10^4)$$
$$= +0.468 \times 10^{-5} \text{ joule}$$

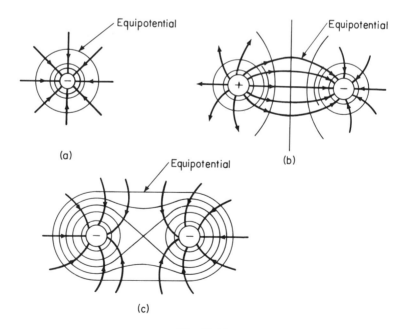

Fig. 20.13

(d) From x to z:

$$W_{xz} = q'V_{zx} = 2 \times 10^{-9} \times (+0.121 \times 10^4)$$
$$= +0.242 \times 10^{-5} \text{ joule}$$

Equipotential surfaces are surfaces where the potentials at a distance from a point are equal (see Fig. 20.13). The vector component is at right angles to the lines of force emanating from the charged surface. Notice that to move a charge from one point on an equipotential surface to another point on the same equipotential surface requires no work, because the potential is the same at both points.

20.5. The Electric Field for a Sphere and a Charged Wire

If a point charge is placed at the center of a sphere, the electric flux associated with the point charge would be the same as the flux which extends outward from the surface of the sphere carrying the same charge. The potential is the same throughout the inside of the sphere. As a result, all the charges may be considered as concentrated at the center of the sphere. If the field strength is zero inside the sphere, the potential of the sphere at the surface or inside the surface may be obtained with the equation for potential.

$$V = K\frac{q}{s} \quad \left\{ \begin{array}{l} s = \text{distance from the center of the sphere: m} \\ q = \text{charge on the sphere: coul} \\ K = \frac{1}{4\pi\varepsilon_0} = 9 \times 10^9 \end{array} \right\}$$

No value of s may be less than the radius of the sphere.

The *work* required to move a charge from a point x to a point y against a field E, along a given distance s, has the same magnitude as the potential difference between the two points. Thus

$$V = Es$$

Substituting this equation into the potential equation gives the *field strength* as

$$E = K\frac{q}{s^2}$$

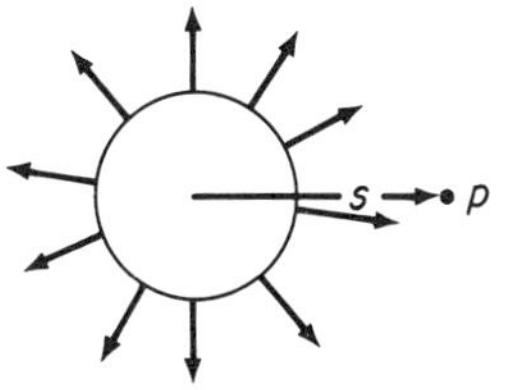

This can be shown by considering the number of lines of flux which radiate out from the surface of a sphere (Fig. 20.14).

Fig. 20.14

$$E = \frac{1}{4\pi\varepsilon_0} \times \frac{q}{s^2}$$

288   *Electrostatics*   Chap. 20

Multiplying through by ε_0,

$$\varepsilon_0 E = \frac{q}{4\pi s^2}$$

But $4\pi r^2$ is the area of a sphere, so that

$$\varepsilon_0 E = \frac{q}{A}$$

If the units are chosen properly, q may be replaced by N, the number of lines of flux per unit area. Thus

$$N = q = \varepsilon_0 EA$$

The electric field at a point P outside a long straight wire is

$$E = \frac{q}{2\pi s l \varepsilon_0}$$

Since $E = V/s$,

$$V = \frac{q}{2\pi l \varepsilon_0}$$

$\left\{\begin{array}{l} N = \text{number of lines of flux} \\ q = \text{charge} \\ E = \text{elec field strength} \\ \varepsilon_0 = \text{constant} = 8.85 \times 10^{-12} \\ A = \text{area of sphere} \\ l = \text{length of wire} \\ s = \text{distance from a wire to point } P \end{array}\right\}$

Example 6

A 4-cm-diameter sphere has a charge of 15×10^{-6} coul. Calculate the field strength and potential (1) along an extended radius anywhere inside the sphere; (2) at a point 3 cm outside the sphere.

Solution

1. Field strength and potential inside the sphere:

$$E = 0$$

$$V = 9 \times 10^9 \times \frac{15 \times 10^{-9}}{2 \times 10^{-2}} \quad \left\{\begin{array}{l} s = 2 \text{ cm} = 0.02 \text{ m} \\ q = 15 \times 10^{-9} \text{ coul} \end{array}\right\}$$

$$= 6750 \text{ volts}$$

2. The field strength and potential 3 cm outside the sphere:

$$E = 9 \times 10^9 \times \frac{15 \times 10^{-9}}{(5 \times 10^{-2})^2}$$

$$= 5.4 \times 10^4 \text{ nt/coul}$$

$$V = 9 \times 10^9 \times \frac{15 \times 10^{-9}}{5 \times 10^{-2}}$$

$\{ s = 3 + 2 = 5 \text{ cm} = 0.05 \text{ m} \}$

$$= 2700 \text{ volts}$$

20.6. The Electric Field Associated with Two Parallel Plates

In Fig. 20.15(a), a charged negative particle between two plates in an electric field is subjected to an upward force which balances the gravitational force acting down. Equating,

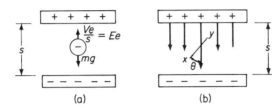

Fig. 20.15

$$\frac{Ve}{s} = mg$$

To calculate the equilibrium voltage, solve for V:

$$V = \frac{mgs}{e}$$

$\left\{ \begin{array}{l} m = \text{mass of electron:} \\ \quad 9.11 \times 10^{-31} \text{ kg} \\ e = \text{charge on electron:} \\ \quad 1.6 \times 10^{-19} \text{ coul} \\ g = 9.8 \text{ m/sec}^2 \end{array} \right\}$

Example 7

Calculate the equilibrium voltage necessary to suspend an electron in an electric field between two plates 3 cm apart.

Solution

The voltage is

$$V = \frac{9.11 \times 10^{-31} \times 9.8 \times 3 \times 10^{-2}}{1.6 \times 10^{-19}}$$

$= 1.67 \times 10^{-12}$ volts $= 1.67$ $\mu\mu v$ (*micromicrovolts*)

On the assumption that the electric field between two large oppositely charged parallel plates is uniform, it can be shown (see Fig. 20.15b) with the calculus that

$$E = \frac{V_{xy}}{s}$$

when carrying a charge from x to y against the field.

Millikan's oil-drop experiment employs the force on a negatively charged particle of oil sprayed between two charged plates. If one of the drops is permitted to fall between the plates when no field exists, it reaches a terminal velocity which is a function of the buoyant force of the air and the density of the drop. Thus the force acting up on the drop will be equal to the force acting down. For a falling drop, unaffected by any field, the net force acting down is

$$\frac{4}{3}\pi r^3 g(\rho_d - \rho_a)$$

The net force acting up, due to the viscosity of the air, is

$$6\pi\eta r v$$

The net force of the field is

$$f = qE$$

Therefore,

$$qE = \frac{4}{3}\pi r^3 g(\rho_d - \rho_a) = 6\pi\eta r v$$

$\left\{\begin{array}{l} q = \text{charge on drop} \\ E = \text{elec field intensity} \\ g = \text{gravitational acc: } 9.8 \text{ m/sec}^2 \\ r = \text{radius of drop} \\ \rho_d = \text{density of oil} \\ \rho_a = \text{density of air} \\ \eta = \text{viscosity of air} \\ v = \text{terminal velocity} \end{array}\right\}$

Example 8

Given an oil drop, radius 3×10^{-6} m, the number of electrons is to be found using Millikan's experiment. The density of the oil is 800 kg/m³, the density of the air is 1.3 kg/m³ and the viscosity of the air is 180×10^{-7} nt/m²-sec. (1) Find the nearest whole number of electrons. The velocity of the drop is 6.29×10^{-3} m/sec when an electric field of 4×10^4 nt/coul is present. (2) What is the terminal velocity of the oil drop when no field is present?

Solution

1. The number of electrons in the oil drop is

$$v = \frac{NEe}{6\pi\eta r}, \quad \text{where } N = q$$

Solve for N:

$$N = \frac{6\pi\eta r}{Ee}$$

$\left\{\begin{array}{l} \eta_{\text{air}} = 18 \times 10^{-7} \text{nt/m}^2\text{-sec} \\ \rho_d = 800 \text{ kg/m}^3 \\ \rho_a = 1.3 \text{ kg/m}^3 \\ g = 9.8 \text{ m/sec}^2 \\ r = 3 \times 10^{-6} \text{ m} \\ v = 6.3 \times 10^{-3} \text{ m/sec} \\ E = 4 \times 10 \text{ nt/coul} \end{array}\right\}$

$$= \frac{6 \times 3.14 \times (180 \times 10^{-7}) \times (3 \times 10^{-6}) \times (6.3 \times 10^{-3})}{(4 \times 10^4) \times (1.6 \times 10^{-19})}$$

$$= 1000 \text{ electrons}$$

2. The terminal velocity of the drop when no electric field is present is

$$6\pi\eta r v = \frac{4}{3}\pi m^3 g(\rho_d - \rho_a)$$

Solve for v:

$$v = \frac{(\frac{4}{3})\pi r^3 g(\rho_d - \rho_a)}{6r}$$

$$= \frac{2r^2 g(\rho_d - \rho_a)}{9\eta}$$

$$= \frac{2(3 \times 10^{-6})^2 \times 9.8(800 - 1.3)}{9 \times 180 \times 10^{-7}}$$

$$= 8.7 \times 10^{-4} \text{ m/sec}$$

$\left\{ \begin{array}{l} r = 3 \times 10^{-6} \text{ m} \\ E = 4 \times 10 \text{ nt/coul} \\ v = 6.3 \times 10^{-3} \text{ m/sec} \\ \quad \text{(velocity in field)} \\ \rho = 1.6 \times 10^{-19} \text{ coul} \\ \quad \text{(charge on electron)} \end{array} \right\}$

Problems

1. Calculate the force on the 20×10^{-9} coul charge in Fig. 20.16.
2. Calculate the force on the 30×10^{-9} coul charge in Fig. 20.16.
3. Calculate the force on the 15×10^{-9} coul charge in Fig. 20.16.

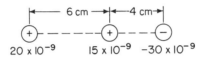

Fig. 20.16

4. Calculate the force on the 30×10^{-9} coul charge in Fig. 20.17.
5. Calculate the force on the 15×10^{-9} coul charge in Fig. 20.17.

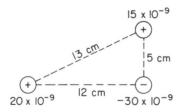

Fig. 20.17

6. Calculate the force on the -12×10^{-9} coul charge in Fig. 20.18.

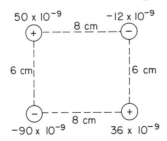

Fig. 20.18

7. Assume charge A to be $+20 \times 10^{-9}$ coul. Calculate the force on charge A in Fig. 20.19.

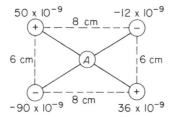

Fig. 20.19

8. Two 0.05-g pith balls (Fig. 20.9) with charges of 6×10^{-8} coul repel each other. If the distance $h = 10$ cm, calculate the distance between them.
9. Two 0.06-g pith balls are separated by a distance of 16 cm. The strings holding the pith balls are each 45 cm long. Calculate the charge on each pith ball.
10. Calculate the electric-field strength and direction midway between two charges $+25 \times 10^{-9}$ coul and $+15 \times 10^{-9}$ coul if the two charges are 12 cm apart.
11. Calculate the electric field strength 4 cm above the 15×10^{-9} coul charge in Prob. 10.
12. In Fig. 20.19, assume A is a point charge. Calculate the field strength at A.
13. Two charges are placed 12 cm apart, as shown in Fig. 20.20. The magnitudes of the charges are $+60 \times 10^{-9}$ coul and -40×10^{-9} coul. Find the potential (a) midway between the charges; (b) at a point 4 cm to the right of the -40×10^{-9} coul charge; (c) at a point directly above the -40×10^{-9} coul charge and 24 cm from the $+60 \times 10^{-9}$ coul charge.
14. Calculate the potential at A in Fig. 20.19.
15. Two equal but oppositely charged particles of 50×10^{-9} coul are separated by 25 cm. A positive point charge is placed so that it forms an equilateral triangle with the two charges. Calculate (a) the electric intensity at this point; (b) the potential.
16. Two charges, magnitude -18×10^{-9} coul and $+24 \times 10^{-9}$ coul, are placed 24 cm apart, as shown in Fig. 20.20. Calculate the potential difference V_{LM} between the two points.

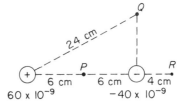

Fig. 20.20

17. If a charge of $+12 \times 10^{-9}$ coul is placed at L and at M in Fig. 20.21, calculate (a) the potential energy at each point; (b) the work required to move the charge from point L to point M.

Fig. 20.21

18. A 9-cm-diameter spherical conductor has a charge of 40×10^{-9} coul. Find the field strength and potential (a) at a point 3 cm from the center of the sphere; (b) at a point 3 cm outside the sphere.
19. A long charged wire creates an electric field. If the charge on the wire is 15×10^{-9} coul per unit length, how far from the wire is the electric field 3×10^3 m/coul?
20. An oil drop, radius 4×10^{-6} m, contains 150 electrons. The density of the oil sprayed is 1000 kg/m³. (a) What field strength is needed to just balance the electron? (b) What is the terminal velocity of the electron when the field is removed?

21

Capacitance

21.1. Capacitance: Parallel-Plate

We have seen that placing a sphere into the field of another *charged* object will cause the charges on the sphere to be separated according to the charge on the object. Thus, in Fig. 21.1(a), if sphere 1 has a negative charge on it, the charges on sphere 2 will be separated as shown.

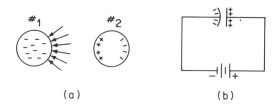

Fig. 21.1

If, instead of two spheres, two plates are substituted and connected to the terminals of a battery as shown in Fig. 21.1(b), the plate connected to the negative terminal of the battery will assume a negative potential with respect to the positive terminal of the battery. As the electrons flow to the negative plate from the battery, they are replaced in the battery by electrons from the other plate, which then has a deficiency of electrons; we say this plate is positive. There is actually a build-up of potential difference between the two

Sec. 21.1 Capacitance: Parallel-Plate

plates. If the *air* between the two plates does not break down, the potential difference between the plates will eventually be the same as the potential difference between the battery terminals. In effect, charges are being stored on these two plates. This is a *capacitor*.

The charge per unit area on one of the plates is designated by σ (the Greek letter *sigma*). In Sec. 20.3, it was shown that the charge per unit area is equal to the electric-field intensity times a constant:

$$\sigma \text{ (vacuum)} = \frac{q}{A} = \varepsilon_0 E$$

Solving for E,

$$E = \frac{q}{\varepsilon_0 A}$$

If s is the distance between the plates, the potential difference between the plates is

A = area of one plate
q = charge on one plate
E = elec field intensity
σ = charge per unit area
ε_0 = 8.85×10^{-12} coul/nt-m^2
s = distance between plates
V = potential difference between plates

$$V = Es \quad \text{or} \quad V = \frac{q}{\varepsilon_0 A} s$$

Since s, A, and ε_0 are constants for a given capacitor, the charge on the capacitor in a vacuum will be proportional to the potential difference applied to the plates:

$$q \propto V$$

Multiply V by a constant C, called *capacitance*, and the proportion is transformed into an equation such that

$$q = CV$$

or

$$C = \frac{q}{V} = \frac{\varepsilon_0 A}{s} \quad (vacuum)$$

In the mksa system, where the unit of charge q is the coulomb and the potential difference is the volt, the unit of capacitance C is the *farad*. Thus if 1 coulomb of charge is put on a capacitor by applying 1 volt of potential, the capacitance is said to be 1 farad. In the mksa system of units, A is in square meters and s is in meters.

Assuming that the air between the plates will break down and become conducting if a stronger charge is placed on the capacitor plates, then it may become necessary to put some *dielectric* (nonconducting material) between the two plates. Thus, depending upon the material inserted between the plates, the capacitance may be increased. The ratio of ε_0 for a vacuum to ε for the material inserted may be represented by the small letter k. Therefore,

$$\frac{\varepsilon}{\varepsilon_0} = k \quad (\textit{no units}), \qquad \varepsilon \text{ and } \varepsilon_0 = \frac{\text{coul}^2}{\text{nt-m}^2}$$

or

$$\varepsilon = k\varepsilon_0$$

The symbol ε (*epsilon*) stands for the *permittivity* for a material, ε_0 is the permittivity of a vacuum, and k is the *dielectric coefficient* (or *dielectric constant*). Some representative values for k are given in Appendix A.16. The permittivity ε may be obtained by multiplying k from the table by 8.85×10^{-12}.

If it takes two plates to construct one capacitor, then N plates will give $N - 1$ capacitors. For multiple-plate capacitors, the equation becomes

$$C = \frac{k\varepsilon_0 A(N - 1)}{s}$$

Since $\varepsilon = k\varepsilon_0$,

$$C = \frac{\varepsilon A(N - 1)}{s}$$

$\left\{ \begin{array}{l} C = \text{capacitance: farads} \\ A = \text{area} \\ s = \text{distance between plates} \\ \varepsilon_0 = \text{permittivity of air} \\ \varepsilon = \text{permittivity of material} \\ k = \text{dielectric const} \end{array} \right\}$

Example 1

A 200-plate capacitor is separated by thin layers of shellac. The dielectric constant for shellac is 3 and each layer is 0.0005 cm thick. If the plates are 2 cm × 5 cm, find the capacitance.

Solution

The capacitance is

$$C = \frac{K\varepsilon_0 A(N - 1)}{s}$$

$$= \frac{3 \times (8.85 \times 10^{-12}) \times (10 \times 10^{-4}) \times (200 - 1)}{5 \times 10^{-6}}$$

$\left\{ \begin{array}{l} N = 2000 \\ s = 5 \times 10^{-6} \text{ m} \\ A = 10 \times 10^{-4} \text{ cm}^2 \\ k = 3 \\ \varepsilon_0 = 8.85 \times 10^{-12} \end{array} \right\}$

$$= 1.057 \times 10^{-6} \text{ farad}$$

21.2. Capacitance: Spherical and Tubular

It was shown in Sec. 20.5 that the potential on a sphere is

$$V = \frac{Kq}{s}$$

Thus in Fig. 21.2, where s_1 is the radius of the inner sphere and s_2 is the radius of the outer sphere,

Fig. 21.2

Capacitors in Series and Parallel

the potential of the system is

$$V_{1-2} = \frac{q}{4\pi\varepsilon_0 s_1} - \frac{q}{4\pi\varepsilon_0 s_2}$$

Dividing through by q,

$$\frac{V}{q} = \frac{1}{4\pi\varepsilon}\left(\frac{1}{s_1} - \frac{1}{s_2}\right)$$

But $q/V = C$ and, therefore,

$$\frac{1}{C} = \frac{1}{4\pi\varepsilon_0}\left(\frac{1}{s_1} - \frac{1}{s_2}\right) = \frac{1}{4\pi\varepsilon_0}\left(\frac{s_2 - s_1}{s_1 s_2}\right) \qquad \begin{cases} s_1 = \text{radius of small sphere} \\ s_2 = \text{radius of large sphere} \\ C = \text{capacitance} \\ k = \text{dielectric const} \end{cases}$$

Solving for C and multiplying by k,

$$C = \frac{4\pi\varepsilon_0 s_1 s_2 k}{s_2 - s_1} = \frac{k s_1 s_2}{(9 \times 10^9)(s_2 - s_1)}$$

For a cylinder capacitor, Fig. 21.3, it can be shown with the calculus that

$$C = \frac{4\pi\varepsilon_0 lk}{2\ln(s_1/s_2)} = \frac{lK}{18 \times 10^9 \ln(s_1/s_2)}$$

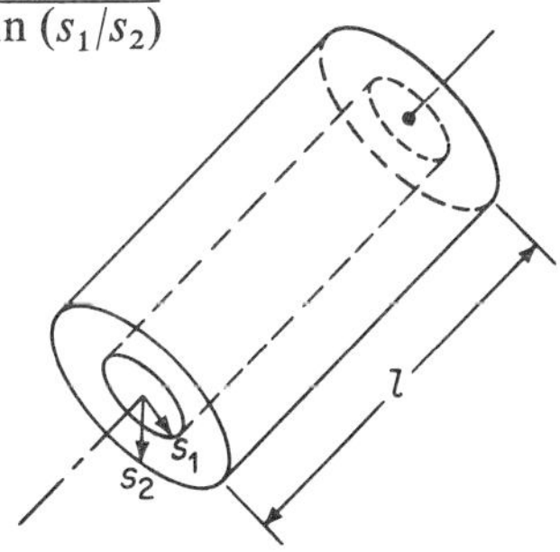

Example 2

Calculate the capacitance of a spherical capacitor if the inner radius is 4 cm and the outer radius is 6 cm. The spheres are separated by glycerin (dielectric const = 55).

Fig. 21.3

Solution

$$C = \frac{k s_1 s_2}{(9 \times 10^9)(s_2 - s_1)} = \frac{55 \times 0.04 \times 0.06}{(9 \times 10^9)(0.06 - 0.04)} \qquad \begin{cases} s_2 = 0.06 \text{ m} \\ s_1 = 0.04 \text{ m} \\ k = 55 \end{cases}$$

$$= 0.733 \times 10^{-9} = 7.33 \times 10^{-4} \,\mu\text{f} \qquad (microfarad)$$

21.3. Capacitors in Series and Parallel

Capacitors may be connected in series, parallel, or in combination series–parallel circuits. Figure 21.4(a) shows three capacitors connected in series. If these capacitors are connected to the terminals of a battery, they will all have the same charge, because the plates of each will attract an equal charge on its adjacent plates. Thus

$$q = q_1 = q_2 = q_3 = \ldots q_n$$

The potential across each capacitor must be

$$V_1 = \frac{q}{C_1}, \qquad V_2 = \frac{q}{C_2}, \qquad V_3 = \frac{q}{C_3} \ldots V_n = \frac{q}{C_n}$$

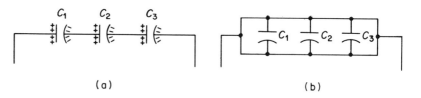

(a)   (b)

Fig. 21.4

The total potential drop is equal to the sum of the potential drops across all the capacitors:

$$V = V_1 + V_2 + V_3 + \ldots + V_n$$

Substituting the values for V_1, V_2, etc.,

$$V = \frac{q}{C_1} + \frac{q}{C_2} + \frac{q}{C_3} + \ldots + \frac{q}{C_n}$$

Factoring q,

$$V = q\left(\frac{1}{C_1} + \frac{1}{C_2} + \frac{1}{C_3} + \ldots + \frac{1}{C_n}\right)$$

Dividing both sides of the equation by q,

$$\frac{V}{q} = \frac{1}{C_1} + \frac{1}{C_2} + \frac{1}{C_3} + \ldots + \frac{1}{C_n}$$

But $V/q = 1/C$; therefore,

$$\frac{1}{C} = \frac{1}{C_1} + \frac{1}{C_2} + \frac{1}{C_3} + \ldots + \frac{1}{C_n}$$

If three capacitors are connected in parallel as shown in Fig. 21.4(b), it can be seen that each of the capacitors will have the same potential difference across its plates. Therefore,

$$V = V_1 = V_2 = V_3 = \ldots V_n$$

The total charge will be the sum of the charges on each capacitor, or

$$q = q_1 + q_2 + q_3 + \ldots + q_n$$

Since

$$q_1 = C_1 V, \quad q_2 = C_2 V, \quad q_3 = C_3 V, \ldots q_n = C_n V$$

we may substitute these values in the sum-of-the-charges equation:

$$q = C_1 V + C_2 V + C_3 V + \ldots + C_n V$$

Divide both sides of the equation by V_1:

$$q/V = C_1 + C_2 + C_3 + \ldots + C_n$$

Sec. 21.3 Capacitors in Series and Parallel

But $q/V = C$; therefore,
$$C = C_1 + C_2 + C_3 + \ldots + C_n$$

Example 3

Calculate (1) the equivalent capacitance for the circuit in Fig. 21.5(a); (2) the charge on the circuit.

Solution

1. The equivalent capacitance for the lower loop is calculated by first considering the 12-μf and the 6-μf series combination. Thus

$$\frac{1}{C} = \frac{1}{C_1} + \frac{1}{C_2} = \frac{1}{12} + \frac{1}{6}$$

$$C = 4 \ \mu f \quad (Fig.\ 21.5b)$$

The two 4-μf capacitors are in parallel; therefore,

$$C = 4 + 4 = 8 \ \mu f \quad (Fig.\ 21.5c)$$

This equivalent capacitance (8 μf) is in series with the rest of the circuit (Fig. 21.5d). Thus

$$\frac{1}{C} = \frac{1}{8} + \frac{1}{16} + \frac{1}{24} + \frac{1}{32} = \frac{25}{96}$$

$$C = 3.84 \ \mu f \quad (Fig.\ 21.5e)$$

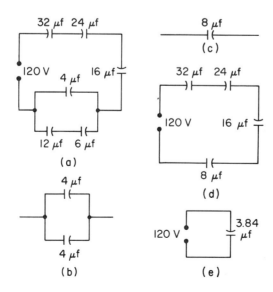

Fig. 21.5

2. The charge is

$$q = CV = (3.84 \times 10^{-6}) \times 120$$
$$= 460.8 \; \mu\text{coul}$$

21.4. Dielectric Strength

If a high enough potential difference is placed on the plates of a capacitor, the dielectric between the plates will break down and become conducting. If the dielectric material breaks down, it is permanently damaged. Although some materials, such as oil, are able to reestablish their dielectric strength, generally, if the dielectric material breaks down, it becomes nonconducting. The breakdown of the dielectric characteristics constitutes a separation of the electrons from the atoms in the material and therefore permanent damage results.

Dielectric strength and the dielectric constant should not be confused with each other. They are two separate values and have no relation to each other. The dielectric strength depends upon the thickness of the materials and is a measure of the potential difference applied to the plates before breakdown. The maximum field strength which causes breakdown is the dielectric strength of the material and the units are kv/m (kilovolts per meter in the mks system) and volt/mm (cgs).

Example 4

Two plates are separated by a dielectric mica 2 mm thick. If a voltage of 0.6 kv causes the dielectric material to break down, what is the maximum field strength (dielectric strength) which may be applied to the plates?

Solution

The dielectric strength, or breakdown point, is

$$V_{max} = E_{max} \, s$$

$$E_{max} = \frac{V_{max}}{s} = \frac{0.6}{0.2} \quad \left\{ \begin{array}{l} s = \text{thickness of dielectric between} \\ \quad \text{plates} = 2 \text{ mm} = 0.2 \text{ cm} \\ V_{max} = 0.6 \text{ kv} \end{array} \right\}$$

$$= 3 \text{ kv/cm} \times 10^2 \text{ cm/m} = 300 \text{ kv/m}$$

21.5. Energy of a Charged Capacitor

Work is involved in moving charges from one plate to another through a circuit. Starting from zero potential to a maximum potential, the potential used to calculate the work must be the average potential. Thus

Sec. 21.5 Energy of a Charged Capacitor 301

$$W = q\left(\frac{1}{2}V\right) = \frac{1}{2}qV \quad \begin{Bmatrix} W = \text{joules} \\ q = \text{coul} \\ V = \text{volts} \\ C = \text{farads} \end{Bmatrix}$$

For the equation in terms of capacitance and voltage, substitute $CV = q$:

$$W = \frac{1}{2}(CV)V = \frac{1}{2}CV^2$$

For the equation in terms of charge and capacitance, substitute $q/C = V$:

$$W = \frac{1}{2}(q)\left(\frac{q}{C}\right) = \frac{1}{2}\frac{q^2}{C}$$

Example 5

Calculate (1) the equivalent capacitance in Fig. 21.6; (2) the charge on the equivalent capacitance; (3) the charge across each capacitor; (4) the voltage across each capacitor; (5) the energy required to charge each capacitor; (6) the total energy.

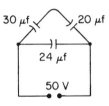

Fig. 21.6

Solution

1. The equivalent capacitance for the series combination is

$$\frac{1}{C} = \frac{1}{30} + \frac{1}{20}$$

$$C = 12 \ \mu\text{f}$$

The equivalent capacitance is

$$C = 12 + 24 = 36 \ \mu\text{f}$$

2. The charge on the equivalent capacitance is

$$q = CV = 36 \times 50 = 1800 \ \mu\text{coul} = 1800 \times 10^{-6} \ \text{coul}$$

3. The charges on capacitors in series are equal and the charge on parallel capacitors equals the sum of the separate charges. The voltages across the 12-μf equivalent 24-μf capacitor is

$$q_{24} = CV = 24 \times 50 = 1200 \ \mu\text{coul} = 1200 \times 10^6 \ \text{coul}$$

and the charge on the equivalent 12-μf capacitance is

$$q_{12} = CV = 12 \times 50 = 600 \ \mu\text{coul} = 600 \times 10^{-6} \ \text{coul}$$

In parallel,

$$q = q_{24} + q_{12} = 1200 + 600 = 1800 \ \mu\text{coul} = 1800 \times 10^{-6} \ \text{coul}$$

4. The total voltage across the series capacitors is the sum of the separate voltages. The charge on each capacitor is the same. Thus

$$V_{20} = \frac{q}{C} = \frac{600 \times 10^{-6}}{20 \times 10^{-6}} = 30 \text{ volts}$$

$$V_{30} = \frac{q}{C} = \frac{600 \times 10^{-6}}{30 \times 10^{-6}} = 20 \text{ volts}$$

$$V_{24} = 50 \text{ volts}$$

5. The energy required to charge each capacitor is

$$W_{20} = \frac{1}{2} CV^2 = \frac{1}{2}(20 \times 10^{-6})(30)^2 = 0.009 \text{ joule}$$

$$W_{30} = \frac{1}{2} CV^2 = \frac{1}{2}(30 \times 10^{-6})(20)^2 = 0.006 \text{ joule}$$

$$W_{24} = \frac{1}{2} CV^2 = \frac{1}{2}(24 \times 10^{-6})(50)^2 = 0.03 \text{ joule}$$

6. Total work is

$$W_t = \frac{1}{2}(36 \times 10^{-6})(50)^2 = 0.045 \text{ joule}$$

Add the separate energy values as a check:

$$W_t = W_{20} + W_{30} + W_{24} = 0.009 + 0.006 + 0.030 = 0.045 \text{ joule}$$

Problems

1. Two parallel plates, 12 cm × 16 cm, carry a charge of 8×10^{-9} coul. What is the electric-field intensity created by this charge?
2. Calculate the potential difference for the data in Prob. 1 if the plates are 0.0006 in. apart.
3. A parallel-plate capacitor has alternate layers of foil and paper, dielectric constant 2. If the capacitor has 150 plates, the paper thickness is 0.006 cm, and the plates are 6 cm wide × 400 cm long, calculate the capacitance.
4. A 12-μf mica capacitor has plate dimensions of 3 cm × 600 cm. The mica is 0.002 cm thick. Calculate the number of plates in the capacitor.
5. A spherical capacitor has a capacitance of 0.19 $\mu\mu f$ (micromicrofarads). If the inner radius is 2 cm, the outer radius 2.6 cm, and the dielectric constant 2, calculate the capacitance.
6. A spherical capacitor is to have a capacitance of 28.4 $\mu\mu f$. If the inner radius is 4 cm, find the thickness of the maximum dielectric (dielectric const = 2.3) necessary to yield this capacitance.
7. Two capacitors are connected in series to a 60-volt source. The capacitance, of the capacitors are 20 μf and 30 μf. Calculate (a) the equivalent capacitance; (b) the charge on each capacitor; (c) the voltage across each capacitor.
8. Three capacitors—12 μf, 18 μf, 30 μf—are connected in series to a 220-volt source. Calculate (a) the equivalent capacitance; (b) the charge on each capacitor; (c) the voltage across each capacitor.

Problems 303

9. If the two capacitors in Prob. 7 are connected in parallel, calculate (a) the equivalent capacitance; (b) the voltage across each capacitor; (c) the charge on each capacitor.
10. If the three capacitors in Prob. 8 are connected in parallel, calculate (a) the equivalent capacitance; (b) the voltage across each capacitor; (c) the charge on each capacitor.
11. Two capacitors, 12 μf and 18 μf, are connected in parallel. The parallel combination is connected in series to a 30-μf capacitor and then to a 220-volt battery. Calculate (a) the equivalent capacitance; (b) the voltage across each capacitor; (c) the charge on each capacitor.
12. Calculate the charge on the equivalent capacitor in Fig. 21.7.

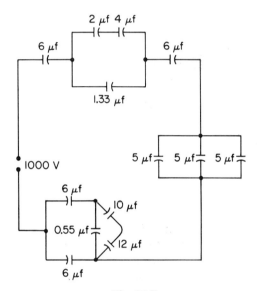

Fig. 21.7

13. Two capacitors are connected in series. If the voltage across the pair is 80 $\times$ 10^{-2} volt and the dielectric strength is 500 kv/m, what must be the thickness of the dielectric?
14. Three capacitors are connected in series. Their capacitances are 8 μf, 12 μf, and 20 μf. The charge on the capacitors is 30 μcoul. Calculate the energy for the systems.
15. Calculate the energy in each of the capacitors in Prob. 14.
16. A two-plate parallel capacitor, having plates of 6 cm $\times$ 10 cm, is connected to a 120-volt source. A layer of fine glass 0.2 cm thick (dielectric const = 7) is inserted between the plates. Find (a) the capacitance of the plates; (b) the energy stored in the plates; (c) the field strength between the plates; (d) the charge on one of the plates.

22

Electric Current: Simple Circuits

22.1. Conductors, Insulators, and Dielectrics

In order to understand an electric current, we need some understanding of the structure of the atom. If we look at the hydrogen atom, we see a *nucleus* which has a comparatively large mass, 1.67×10^{-27} kg; and in orbit about the nucleus we find an *electron* whose mass is 9.1×10^{-31} kg, about 1840 times lighter than the nucleus. Since almost all the mass is concentrated in the nucleus, and since a hydrogen atom must be very close to 1.67×10^{-27} kg.

The electron is said to be *in orbit* about the nucleus. This circular orbit can exist only if there is a centripetal force acting on the electron, balanced by the tendency of the moving electron to leave its circular orbit and fly off into space. The *centripetal force* is the attraction which exists between the electron and the proton. A *neutral* atom will have a nucleus whose magnitude of plus charge just equals the magnitude of minus charge in the electron. Thus the *centrifugal* force trying to separate the charges is just balanced by the centripetal force of attraction between the two charges. (This and elliptical orbits are discussed more fully in Chapter 36.) A neutral atom will always possess balanced charges. The charge on a proton and on an electron is 1.6×10^{-19} coul.

The next element in the periodic table is helium, Fig. 22.1(b). Here the atom is composed of two protons in the nucleus and two electrons in orbit

Sec. 22.1   *Conductors, Insulators, and Dielectrics*   305

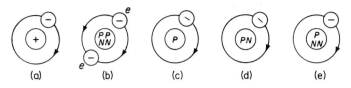

Fig. 22.1

about the nucleus. In the nucleus, we also find two additional particles (*neutrons*) which are charged neutrally but have about the same mass as the proton. Thus the mass of the helium atom is essentially the mass of all the protons and all the neutrons in the nucleus.

The charge on the nucleus is the total charge on the protons in the nucleus; the number of protons in the nucleus is designated as the *atomic number*. The number of protons and neutrons in the nucleus is called the *mass number*. These inhabitants of the nucleus are called *nucleons*. The foregoing is symbolized as

$$_{\text{(atomic number)}} {}_2\text{He}^4 \, {}^{\text{(mass number)}}$$

Thus for neutral helium, subscript 2 indicates the number of protons in the nucleus; superscript 4 indicates the total number of nucleons. The number of protons subtracted from the total number of nucleons must yield the number of neutrons in the nucleus:

$$4 \text{ nucleons} - 2 \text{ protons} \rightarrow 2 \text{ neutrons}$$

Example 1

How many neutrons, protons, and electrons are present in the atom whose symbol is

$$_{11}\text{Na}^{23}$$

Solution

$$11 \text{ protons}$$
$$11 \text{ electrons}$$
$$23 - 11 = 12 \text{ neutrons}$$

Note very carefully that this discussion applies to the neutral atom. Atoms may vary in several ways from the neutrally charged atom.

Assume the $+$ and $-$ charges to be electrically balanced in an atom. It is possible to have atoms which are electrically neutral (the $+$ and $-$ charges balance) and furthermore possess the same number of protons and electrons, but their neutron population may differ. For instance, the hydrogen atom has one proton in its nucleus and one orbital electron. But it was found that hydrogen could also have one or two neutrons inside its nucleus. This additional mass would have no effect upon the electron charge. Elements with additional neutrons are called *isotopes*. We will hear more about isotopes later. Figure 22.1(c), (d), and (e) shows the three isotopes of hydrogen: Fig. 22.1(c) is hydrogen with its one electron and one proton; Fig. 22.1(d) is deuterium, with its one electron, one proton, and one neutron; Fig. 22.1(e) is tritium, with its one electron, one proton, and two neutrons.

Of even greater importance to the study of electricity is the ability of the atom to release electrons from its outer structure. We shall see that these families of electrons circling the nucleus create *energy shells*—both circular and ellipitcal. Once the inner orbit is fully populated, another orbit is started to create a second shell, etc. Thus, some elements may have one or two electrons in a newly created shell which may be considered "extra" electrons; others may have partially completed shells which may be considered as having a "deficiency" of electrons. The elements with extra electrons will release them, if the proper conditions prevail. When these proper conditions exist, elements with a deficiency of electrons will attempt to complete their shells by picking up electrons. The element which has "extra" electrons has a *plus valence number*. The element with a "deficiency" of electrons has a *negative valence number*. If two elements—one having an excess of one electron and one having a deficiency of one electron—come together, both shells may be completed and a *molecule* will be formed.

At this point, it is even more important to our discussion that electrons may become detached from their orbits and drift through the structure by being readily released and readily picked up. (In some cases, orbits overlap.) In any event, in a metal, large numbers of electrons may move about freely in any direction, although there need be no net drift in any one direction.

Assume now that, to the right end of the wire in Fig. 22.2, we attach a generator which puts electrons into the wire at the right while electrons are being removed at the left end of the wire. The field will be *toward the right*, which is at a negative potential (electrons are being added), and away from the left, which is at a positive potential (electrons are being removed). Thus the electron will be directed in the opposite direction from the field, as shown in Fig. 22.2. We shall soon distinguish between negative-current (electron) and positive-current (field) flow.

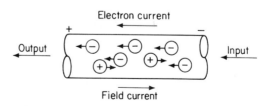

Fig. 22.2

It is important to realize that the electrons are in motion in all directions, but that only the *net* motion is from one end of the wire to the other. Thus the average drift of electrons from right to left is very little. But the *energy wave* created by collision between electrons may move through the wire at almost the speed of light. In other words, the electron entering at the right is not the same electron which leaves the wire at the left. It is rather the energy which has moved through the wire. The electron has picked up kinetic energy and released some or all of it during collision. Since the net velocity is in one direction (right to left), the energy flow is in one direction.

When a force is applied (voltage) to counteract the attracting force of the nucleus on the electron and the atom easily releases electrons to take part in this drift motion, the material is said to be a *good conductor*. If the material is of such a nature that the nucleus holds its electrons very strongly and there are only a few electrons which can take part in this drift, the material is said to be an *insulator*. Between these two extremes are many materials having varying quantities of "free electrons" available in the structure. Thus there are good conductors, fair conductors, fair insulators, and good insulators. Notice that *resistors* may be included in the "fair-conductor–fair-insulator" category. This *resistance* to almost free flow or almost complete stoppage of electron drift is the exact quality needed in a resistor.

22.2. Ohm's Law

The ampere has been defined, by the International Committee on Weights and Measures in 1935, as a current set up in two parallel conductors in a vacuum 1 meter apart and infinitely long which causes a wire to undergo a force of 2×10^{-7} newton for every meter of length of wire. The coulomb per second is equal to 1 ampere and is based on the number of electrons transported past an area in unit time.

Thus

$$q = It \quad \left\{ \begin{array}{l} q = \text{charge: coul} \\ I = \text{current: amp} \\ t = \text{time: sec} \end{array} \right\}$$

and therefore the current is

$$\text{amperes} = \frac{\text{coulombs}}{\text{seconds}}$$

The voltage applied to a wire creates the force necessary to cause electron drift and therefore may be defined in terms of the charges moved and the resistance encountered in attempting to move these charges. To overcome this resistance, *work* must be done. Figure 22.2 showed how a difference in the number of electrons was created when electrons were supplied at one end of the wire and energy was impressed on these electrons. The difference in the quantity of electrons created a negative and a positive end to the wire, or a potential difference was created which we refer to as a *voltage*. It was seen in Sec. 20.4 that a potential difference (PD) of 1 volt is said to exist if it takes 1 joule of energy to move 1 coulomb of charge from one point to another. And, therefore,

$$V = \frac{W}{q} = \frac{\text{joules}}{\text{coulombs}} \qquad \left\{ \begin{array}{l} W = \text{energy: joules} \\ q = \text{charge: coul} \\ V = \text{PD: volts} \end{array} \right\}$$

It is possible to express resistance in terms of voltage and amperage. If the voltage is proportional to the amperage, then

$$V \propto I$$

This proportion is converted into an equation by multiplying I by the proportionality constant R (resistance). *Ohm's law* becomes

$$V = RI$$

Solving for R, $\qquad \left\{ \begin{array}{l} R = \text{resistance: ohms} \\ V = \text{PD: volts} \\ I = \text{current: amp} \end{array} \right\}$

$$R = \frac{V}{I} = \text{volt-amp}$$

The unit of resistance is the *ohm* (the unit is named for Georg S. Ohm, a German physicist). The resistance is 1 ohm when the potential difference is 1 volt and the current is 1 amp. The ohm is the volt per ampere. The symbol for resistance is the Greek upper-case letter *omega* (Ω); when drawing circuit diagrams, -⋀⋀⋀- is generally used.

Example 2

A potential difference of 15 volts is applied to a wire which has a resistance of 5 ohms. Calculate the current in the wire:

Solution

$$I = \frac{V}{R} = \frac{15}{5} = 3 \text{ amp} \qquad \left\{ \begin{array}{l} V = 15 \text{ volts} \\ R = 5 \text{ ohms} \end{array} \right\}$$

Ohm's law may be used to take care of several resistances. If they are arranged as shown in Fig. 22.3(a), they are said to be in *series*. If resistors are in series, any energy flow which occurs must flow from one resistor to the next, in the order in which they are connected. That is, if n number of electrons enter at x, then n number of electrons must leave y. Otherwise, if more enter x than leave y, a build-up must take place somewhere in the wire, etc. Thus the current I is dependent upon the electrons (charge) per second fed into the

Sec. 22.2   Ohm's Law   309

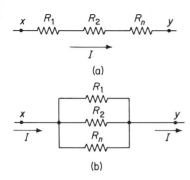

Fig. 22.3

left side; once the number of electrons entering x and leaving y is stabilized, the current is said to be *constant* throughout the line. Therefore, the current passing through each resistor is equal and

$$I_T = I_1 = I_2 = I_3 = \cdots = I_n$$

Each resistor creates a blocking effect, or voltage drop. The sum of all these voltage drops should be equal to the total voltage input. The total voltage drop for resistors in series is

$$V_T = V_1 + V_2 + V_3 + \cdots + V_n$$

However,

$$V_1 = IR_1, \quad V_2 = IR_2, \quad V_3 = IR_3 \cdots V_n = IR_n$$

Substituting in the former equation,

$$V_T = IR_1 + IR_2 + IR_3 + \cdots + IR_n$$

Factor out an I:

$$V_T = I(R_1 + R_2 + R_3 + \cdots + R_n)$$

Then, dividing both sides of the equation by I,

$$\frac{V_T}{I} = R_1 + R_2 + R_3 + \cdots + R_n$$

Since $R_T = V_T/I$,

$$R_T = R_1 + R_2 + R_3 + \cdots + R_n$$

and R_T is referred to as the *equivalent resistance*.

In a *parallel* arrangement, Fig. 22.3(b), it is still necessary that the same number of electrons leave y as entered at x per unit time. In a parallel arrangement, however, there are n possible paths which the electrons can take when moving from x to y. They, of course, will take the path which offers the least resistance. The greater the resistance, the less the flow. The sum total of all the electrons per unit time that get through all the resistors and arrive at y will equal the total number of electrons that enter at x. Therefore,

$$I_T = I_1 + I_2 + I_3 + \cdots + I_n$$

The voltage which impresses itself on the x-side of *all* resistors is the same. Thus

$$V_T = V_1 = V_2 = V_3 = \cdots = V_n$$

The voltage for each resistor is

$$V_T = I_1 R_1, \quad V_T = I_2 R_2, \quad V_T = I_3 R_3 \cdots V_T = I_n R_n$$

Solving each of these voltage equations for I,

$$I_1 = \frac{V_T}{R_1}, \quad I_2 = \frac{V_T}{R_2}, \quad I_3 = \frac{V_T}{R_3} = \cdots I_n = \frac{V_T}{R_n}$$

Substituting these values into the sum of the current equation,

$$I_T = \frac{V_T}{R_1} + \frac{V_T}{R_2} + \frac{V_T}{R_3} + \cdots + \frac{V_T}{R_n}$$

Factor out a V_T and divide both sides of the equation by V_T:

$$\frac{I_T}{V_T} = \frac{1}{R_1} + \frac{1}{R_2} + \frac{1}{R_3} + \cdots + \frac{1}{R_n}$$

Since $I_T/V_T = 1/R_T$,

$$\frac{1}{R_T} = \frac{1}{R_1} + \frac{1}{R_2} + \frac{1}{R_3} + \cdots + \frac{1}{R_n}$$

Example 3

Three resistors—R_1, R_2, and R_3—have values of 10, 18, 30 ohms, respectively, and are connected as shown in Fig. 22.3(a) and then in parallel as shown in Fig. 22.3(b). Calculate the equivalent resistance (1) in series; (2) in parallel.

Solution

1. The equivalent resistance in series is

$$R = R_1 + R_2 + R_3 = 10 + 18 + 30 = 58 \text{ ohms} \qquad \begin{cases} R_1 = 10 \text{ ohms} \\ R_2 = 18 \text{ ohms} \\ R_3 = 30 \text{ ohms} \end{cases}$$

2. The equivalent resistance in parallel is

$$\frac{1}{R} = \frac{1}{10} + \frac{1}{18} + \frac{1}{30} = \frac{9 + 5 + 3}{90} = \frac{17}{90}$$

$$R = 5.3 \text{ ohms}$$

22.3. Simple Circuits

A battery, or any other source of voltage, supplies voltage and current to the line. The symbol for a battery is ⊣|⊢

At this point, a distinction must be made between "electron" and "conventional" current flow. The first is associated with negative flow; the second, with positive flow. We shall use "conventional" flow in our circuits and indicate whenever electron flow is used.

In Fig. 22.4(a), the battery shows voltage and current effective *internally* from x to y.

In Fig. 22.4(a), the long line will put the battery at a plus potential with

Sec. 22.3   *Simple Circuits*   311

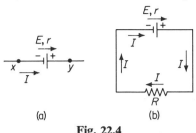

Fig. 22.4

the short line. Thus, the *conventional* flow internally will be minus to plus ($- \rightarrow +$).

Also it should be recognized that batteries will have *internal resistance*. This internal resistance is shown as a small r. The symbol E is the *ideal voltage* which the battery could put out if there were no internal resistance r in the battery. It represents the work per unit charge and is called the *electromotive force*(emf). The units are joules per coulomb.

The terminal voltage V, not the emf E, will be

$$V = E - Ir$$

Figure 22.4(b) shows a battery in a circuit. Remember there can be no current flow unless the circuit is complete. That is, there must be a source of current and a way back to the source. The *conventional* flow is from $-$ to $+$ internally (inside the battery) and from $+$ to $-$ externally (in the line). Thus in Fig. 22.4(b), the current flows from the plus side of the battery through the resistor R to the negative side of the battery. The current flow enters the negative side of the battery and flows back to the plus side.

Example 4

Three resistors are connected in series, as shown in Fig. 22.5(a), to an emf of 11 volts. Calculate (1) the equivalent resistance; (2) the current in the line; (3) the voltage drop across each resistor.

Solution

1. The total resistance is

$$R_T = R_1 + R_2 + R_3$$
$$= 3 + 6 + 2$$
$$= 11 \text{ ohms}$$

$\left\{ \begin{array}{l} R_1 = 3 \text{ ohms} \\ R_2 = 6 \text{ ohms} \\ R_3 = 2 \text{ ohms} \\ E = 11 \text{ volts} \end{array} \right\}$

2. The amperage in the circuit is

$$E = I \sum R$$

Solve for I and substitute:

$$I = \frac{E}{\sum R} = \frac{11}{11} = 1 \text{ amp}$$

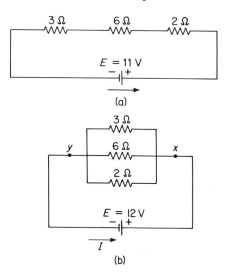

Fig. 22.5

3. The voltage drop across each resistor is

$$V_3 = IR = 1 \times 3 = 3 \text{ volts}$$
$$V_6 = 1 \times 6 = 6 \text{ volts}$$
$$V_2 = 1 \times 2 = 2 \text{ volts}$$

The sum of the voltage drops equals 11 volts (*check*).

Example 5

Figure 22.5(b) shows 3 ohms, 6 ohms, and 2 ohms connected in parallel to a source of emf of 12 volts. Find (1) the equivalent resistance; (2) the amperage in the circuit; (3) the amperage in each resistor; (4) the voltage drop across each resistor.

Solution

1. The equivalent resistance is

$$\frac{1}{R} = \frac{1}{R_1} + \frac{1}{R_2} + \frac{1}{R_3}$$
$$= \frac{1}{3} + \frac{1}{6} + \frac{1}{2}$$
$$= \frac{2 + 1 + 3}{6}$$
$$R = 1 \text{ ohm}$$

Sec. 22.3   Simple Circuits   313

2. The amperage in the circuit is

$$I = \frac{E}{R} = \frac{12}{1} = 12 \text{ amp}$$

3. The voltage impressed across the parallel arrangement is 12 volts. Since the amperage through each resistor is a function of the magnitude of each of the resistors:

$$I_3 = \frac{V}{R_3} = \frac{12}{3} = 4 \text{ amp}$$

$$I_6 = \frac{V}{R_6} = \frac{12}{6} = 2 \text{ amp}$$

$$I_2 = \frac{V}{R_2} = \frac{12}{2} = 6 \text{ amp}$$

the total amperage from x to y is

$$I = I_3 + I_6 + I_2 = 4 + 2 + 6$$
$$= 12 \text{ volts} \quad (check)$$

4. The voltage drop across each resistor is 12 volts. (Why?)

Combinations of parallel and series arrangements are possible. In Fig. 22.6(a), resistors R_1 and R_2 are in series with each other. Resistors R_3, R_4, and R_5 are in parallel with each other. The parallel combination is also in series with the series combination.

In Fig. 22.6(b), resistors R_1 and R_2 are in series with each other and the series combination is in parallel with the resistor R_3.

In Fig. 22.6(c), R_1 and R_2 are in parallel with each other. The combination is in series with R_3, which is in parallel with R_4.

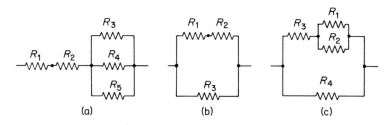

Fig. 22.6

Example 6

Refer to Fig. 22.7(a) and calculate (1) the equivalent resistance; (2) the total resistance in the circuit; (3) the amperage in the circuit; (4) the voltage drop across R_5.

314 Electric Current: Simple Circuits Chap. 22

Solution

1. The equivalent external resistance is found as follows:
   (a) Combine R_5 and R_6 in parallel:

   $$\frac{1}{R_{5,6}} = \frac{1}{R_5} + \frac{1}{R_6} = \frac{1}{2} + \frac{1}{4}$$

   $$R_{5,6} = 1\tfrac{1}{3} \text{ ohms}$$

   (b) Resistor $R_{5,6}$ is in series with R_4; R_3 is in series with R_2:

   $$R_{4,5,6} = R_4 + R_{5,6} = 2\tfrac{2}{3} + 1\tfrac{1}{3}$$
   $$= 4 \text{ ohms}$$

   $\left\{\begin{array}{l} E = 8 \text{ volts} \\ r = 0.4 \text{ ohm} \\ R_1 = 2 \text{ ohms} \\ R_2 = 3 \text{ ohms} \\ R_3 = 1 \text{ ohm} \\ R_4 = 2\tfrac{2}{3} \text{ ohms} \\ R_5 = 2 \text{ ohms} \\ R_6 = 4 \text{ ohms} \\ R_7 = 8 \text{ ohms} \end{array}\right\}$

   and

   $$R_{2,3} = R_2 + R_3 = 3 + 1 = 4 \text{ ohms}$$

   (c) Resistors $R_{2,3}$, $R_{4,5,6}$, and R_7 are in parallel [Fig. 22.7(c)]:

   $$\frac{1}{R_{2,3,4,5,6,7}} = \frac{1}{R_{2,3}} + \frac{1}{R_{4,5,6}} + \frac{1}{R_7}$$
   $$= \frac{1}{4} + \frac{1}{4} + \frac{1}{8}$$

   $$R_{2,3,4,5,6,7} = 1.6 \text{ ohms}$$

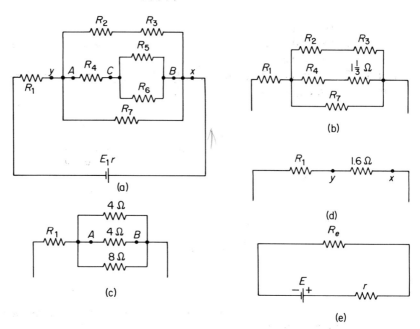

Fig. 22.7

(d) And finally, $R_{2,3,4,5,6,7}$ is in series with R_1 [Fig. 22.7(d)]:

$$R_e = R_{2,3,4,5,6,7} + R_1 = 1.6 + 2.0 = 3.6 \text{ ohms}$$

2. The total resistance in the circuit must include the resistance in the battery. This is seen to be in series in Fig. 22.7(e):

$$R = R_e + r = 3.6 + 0.4 = 4 \text{ ohms}$$

3. The amperage in the circuit is

$$I = \frac{E}{R} = \frac{8}{4} = 2 \text{ amp}$$

4. The voltage drop across R_5 may be found as follows: (It is important to remember that, in a series arrangement of resistors, all have equal amperages; in a parallel arrangement, all have equal voltages.)

   (a) The resistance across xy [Fig. 22.7(d)] is 1.6 ohms.

   (b) The voltage drop across xy is

   $$V_{xy} = IR = 2 \times 1.6 = 3.2 \text{ volts}$$

   Note: the amperage across xy is the same as the amperage in the entire circuit.

   (c) The resistance across AB [Fig. 22.7(c)] is 4 ohms. Therefore, the amperage across AB is

   $$I_{AB} = \frac{V}{R} = \frac{3.2}{4} = 0.8 \text{ amp}$$

   Note: The voltage across AB is the same as the voltage across $R_{2,3}$ and R_7.

   (d) The equivalent resistance of C and B [Fig. 22.7(b)] is

   $$R_{CB} = 1\tfrac{1}{3} \text{ ohms}$$

   (e) The voltage drop for R_{CB} is

   $$V_{CB} = IR = 0.8 \times 1\tfrac{1}{3} = 1.07 \text{ ohms}$$

   Note: The amperage is the same for AC as for CB.

   (f) Therefore, the amperage across R_5 is

   $$I = \frac{V}{R_5} = \frac{1.07}{2} = 0.535 \text{ amp}$$

   Note: The voltage is the same for R_4 and R_5.

22.4. Batteries in Series and Parallel

Several cells may be connected in series, as in Fig. 22.8(a), and in parallel, as shown in Fig. 22.8(b).

Cells arranged so that the positive terminal of one is connected to the negative terminal of another, etc., are said to be *in series*. The *equivalent emf* is the sum of the separate emf's. The internal resistances are in series and they are added to find their equivalent resistances. The current in each cell is the same as the current in the entire circuit.

For cells in parallel [Fig. 22.8(b)] the negative terminal of one cell is connected to the negative terminal of the other, and the positive terminal of one is connected to the positive terminal of the other. The emf of each of the cells is the same as the overall voltage; the total resistance is the sum of the reciprocal resistances. The total current is the sum of the individual currents.

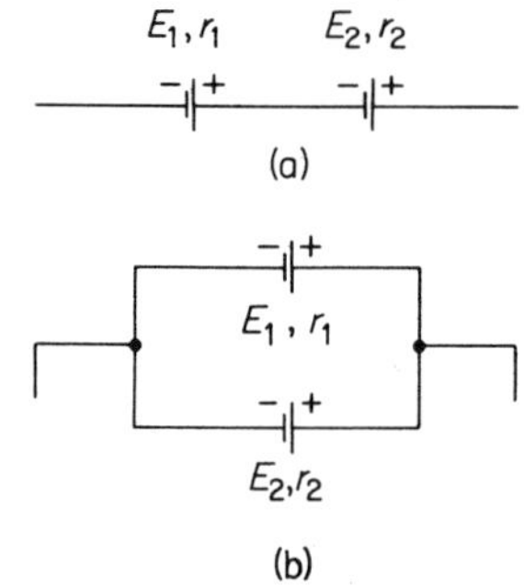

Usually when cells are connected in parallel, their values should be exactly alike; otherwise there is local current flow which does not contribute to the overall current in the circuit.

Fig. 22.8

Example 7

Find (1) the equivalent voltage for four $4\frac{1}{2}$-volt cells in series and connected through a 5.1-ohm resistor (the resistances of each of the cells are as follows: 0.28, 0.3, 0.22, 0.1 ohm); (2) the equivalent resistance in the cells; (3) the equivalent resistance for the circuit; (4) the amperage in the system; (5) the terminal voltage in the 0.3-ohm-resistance cell.

Solution

1. Voltage of the cells in series:

$$E = 4 \times 4\frac{1}{2} = 18 \text{ volts}$$

2. The equivalent resistance for the batteries is

$$r_T = 0.28 + 0.3 + 0.22 + 0.1 = 0.9 \text{ ohm}$$

3. The equivalent resistance of the circuit is

$$R_T = R + r_T = 5.1 + 0.9 = 6 \text{ ohms}$$

4. The amperage in the circuit is

$$I_T = \frac{E}{R_T} = \frac{18}{6} = 3 \text{ amp}$$

5. The terminal voltage for the 0.3-ohm-resistance cell is

$$V = E - Ir = 4.5 - (3 \times 0.3) = 3.6 \text{ volts}$$

Example 8

Two 4-volt batteries (internal resistance = 0.2) are connected in parallel. The combination is connected in series through a 6-ohm resistor. Calculate (1) the

Sec. 22.5　　　　　　　　　*Resistivity*　　　　　　　　　317

equivalent resistance for the circuit; (2) the equivalent voltage for the batteries; (3) the current in the circuit; (4) the current supplied by each cell; (5) the terminal voltage for each cell; (6) the voltage drop in each cell.

Solution

1. The equivalent resistance is

$$\frac{1}{r} = \frac{1}{0.2} + \frac{1}{0.2} = \frac{2}{0.2}$$

$$r = 0.1 \text{ ohm}$$

$$R = 0.1 + 6 = 6.1 \text{ ohms}$$

2. The equivalent voltage for the parallel circuit is 4 volts.
3. The current in the circuit is

$$I = \frac{V}{R} = \frac{4}{6.1} = 0.66 \text{ amp}$$

4. The current supplied by each cell is 0.33 amp.
5. The terminal voltage for each cell.

$$V = E - Ir = 4 - (0.33 \times 0.2) = 3.934 \text{ volts}$$

6. Notice that the *Ir* drops in all the cells must be equal in a parallel arrangement. Therefore,

$$V - Ir - 0.33 \text{ amp} \times 0.2 = 0.066 \text{ volt}$$

and the currents in each cell are

$$I_{0.2} = \frac{0.066}{0.2} = 0.33 \text{ amp} \quad (check)$$

22.5. Resistivity

The longer the conductor, the greater the resistance to current flow. The larger the diameter, the less the resistance to current flow. Thus

$$R \propto \frac{l}{A}$$

This proportion may be converted to an equation by multiplying the right side by a constant ρ (*rho*). This constant is called *resistivity*. Values for resistivity are shown in Appendix A.17.

$$R = \rho \frac{l}{A} \qquad \left\{ \begin{array}{l} R = \text{resistance: ohms} \\ A = \text{area: m}^2 \\ l = \text{length: m} \\ \rho = \text{resistivity: ohm-m} \end{array} \right\}$$

318   *Electric Current: Simple Circuits*   Chap. 22

The unit for ρ in the mksa system is

$$\rho = \frac{RA}{l} = \text{ohm-m}$$

Example 9

A platinum wire has a resistivity of 11.0×10^{-8} ohm-m and a diameter of 0.030 cm. If the wire is 800 cm long, calculate the resistance of the wire.

Solution

The resistance is

$$R = \frac{\rho l}{A} = \frac{(11 \times 10^{-8}) \times 8 \times 4}{\pi (3 \times 10^{-4})^2}$$
$$= 12.45 \text{ ohms}$$

$$\left\{ \begin{array}{l} \rho = 11 \times 10^{-8} \text{ ohm-m} \\ d = 0.03 \text{ cm} = 3 \times 10^{-4} \text{ m} \\ l = 800 \text{ cm} = 8 \text{ m} \end{array} \right\}$$

The reciprocal of resistivity is *conductivity*. Therefore, the conductance of a wire may be stated as

$$\rho = \frac{1}{\sigma} \quad \text{or} \quad R = \frac{l}{\sigma A}$$

Therefore,

$$G = \frac{1}{R} = \frac{\sigma A}{l}$$

$$\left\{ \begin{array}{l} \rho = \text{resistivity} \\ \sigma = \text{conductivity} \\ R = \text{resistance} \\ G = \text{conductance} \end{array} \right\}$$

In the British system of units, in order to avoid the use of π which appears when calculating the area of a round wire, the *circular mil* was defined. A mil is equal to one-thousandth of an inch. Thus a wire whose diameter is 0.001 in. is 1 mil in diameter. Since the diameter–area ratio is the *square* of the diameter, the area, in circular mils, is equal to the diameter squared in mils.

Thus, assuming a wire of diameter 0.050 in.,

$$d = 0.050 \text{ in.} = 50 \text{ mils}$$

The area, represented in circular mils (CM) is

$$d^2 = 50^2 \text{ CM}$$

Therefore, the resistance equation in the British system is

$$R = \frac{\rho l}{A}$$

Solving for ρ,

$$\rho = \frac{RA}{l} = \frac{\text{ohm-CM}}{\text{ft}}$$

$$\left\{ \begin{array}{l} R = \text{resistance: ohms} \\ d = \text{diameter: in. } or \text{ mils} \\ A = \text{area: (mils)}^2 \\ l = \text{length: ft} \\ \rho = \text{resistivity: ohm-CM/ft} \end{array} \right\}$$

Sec. 22.6   Temperature Coefficient of Resistance   319

Example 10

The resistivity of a copper wire is 10.4 ohm-CM/ft. The wire is $\frac{1}{32}$ in. in diameter and 72 in. long. Calculate the resistance of the wire in British units.

Solution

The resistance is

$$R = \frac{\rho l}{A} = \frac{10.4 \times 6}{(31.25)^2}$$

$$\left\{ \begin{array}{l} d = \frac{1}{32} \text{ in.} = 0.03125 \text{ in.} \\ A = (31.25)^2 \text{ CM} \\ l = 72 \text{ in.} = 6 \text{ ft} \\ \rho = 10.4 \text{ ohm-CM/ft} \end{array} \right\}$$

$$= 0.064 \text{ ohm}$$

If it should be desired to change square mils to circular mils, use

$$\text{CM} = \text{square mils} \times \frac{4}{\pi}$$

Example 11

Convert a rectangular wire, $\frac{1}{4}$ in. $\times \frac{1}{8}$ in. to circular mils.

Solution

$$A = \tfrac{1}{4} \text{ in.} \times \tfrac{1}{8} \text{ in.} = 0.250 \text{ in.} \times 0.125 \text{ in.} = 25 \text{ mil} \times 125 \text{ mil}$$

$$= 31{,}250 \text{ sq mils} \times \frac{4}{\pi} = 39{,}812 \text{ CM}$$

22.6. Temperature Coefficient of Resistance

The electrical resistance of all materials varies with every degree change in temperature and is related to the resistance at 0 °C. For most metals, the resistance increases for every degree change in temperature (except manganin, which remains rather constant). For nonmetals, generally, the resistance decreases for every degree change in temperature—that is, the *temperature coefficient of resistance* α is negative.

Experimentally, if resistance is plotted as the ordinate and temperature as the abscissa (see Fig. 22.9), the resulting curve is close to a straight line between T_0 and T_1. Since it is very important that the resistance be related to 0 °C, it is possible to extrapolate the curve back to the ordinate (under labo-

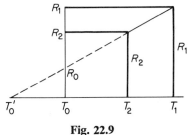

Fig. 22.9

ratory conditions) and find the resistance at 0 °C.

Ideally, the curve should be extrapolated back to the abscissa, so that the temperature T_0, where the resistance is *zero*, is found as in Fig. 22.9. From similar triangles,

$$\frac{R_1}{T'_0 + T_1} = \frac{R_0}{T'_0 + T_0}$$

But if $T_0 = 0$, the equation becomes

$$R_1 = \frac{R_0(T'_0 + T_1)}{T'_0} = R_0\left(1 + \frac{1}{T'_0}T_1\right)$$

If $\alpha = 1/T'_0$, then

$$R_1 = R_0(1 + \alpha T_1)$$

$\left\{\begin{array}{l} T_0 = 0\,°C \\ R_0 = \text{resistance at } 0\,°C \\ T_1 = \text{new temp in }°C \\ R_1 = \text{resistance at } T_1 \\ T'_0 = \text{temp where } R = 0 \\ \alpha = \text{temp coef of} \\ \quad \text{resistance at } 0\,°C \end{array}\right\}$

The student is cautioned that T is a change in temperature related to 0 °C. Note Example 12 very carefully.

Example 12

A certain nickel wire has a temperature coefficient of resistance of 0.006/ °C. If the resistance is 70 ohms at 120 °C, calculate its resistance at 250 °C.

Solution

Since a is related to 0 °C, the equation must be applied twice.

The resistance at 0 °C is

$$R_2 = R_0(1 + \alpha T_2)$$

Solving for R_0,

$$R_0 = \frac{R_2}{1 + \alpha T_2} = \frac{70}{1 + (0.006 \times 120)}$$
$$= 40.7 \text{ ohms}$$

$\left\{\begin{array}{l} T_2 = 120\,°C \\ T_1 = 250\,°C \\ R_0 = \text{resistance at }°C \\ \alpha = 0.006/°C \text{ at } 0\,°C \\ R_2 = 70 \text{ ohms} \\ R_1 = \text{resistance at } 250\,°C \end{array}\right\}$

The resistance at 250 °C is

$$R_1 = 40.7\,[1 + (0.006 \times 250)]$$
$$= 101.7 \text{ ohms}$$

Example 13

Using the relationship $T'_0 = 1/\alpha$, solve Example 12.

Solution

1. The reciprocal of the temperature coefficient of resistance is

$$T'_0 = \frac{1}{\alpha} = \frac{1}{0.0006} = 0.167 \times 10^3 = 167\,°C$$

Sec. 22.7   *Power*   321

2. The similar-triangle relationship from Fig. 22.9 is

$$\frac{R_1}{T'_0 + T_1} = \frac{R_2}{T'_0 + T_2} \qquad \left\{ \begin{array}{l} R_1 = \text{resistance at 250°C} \\ R_2 = 70 \text{ ohms at 120°C} \\ T_1 = 250\text{°C} \\ T_2 = 120\text{°C} \\ T'_0 = 167\text{°C} \end{array} \right\}$$

Solve for R_1:

$$R_1 = \frac{R_2(T'_0 + T_1)}{T'_0 + T_2} = \frac{70(167 + 250)}{167 + 120}$$

$$= 101.7 \text{ ohms} \qquad (check)$$

22.7. Power

We have defined potential difference as the work necessary to move 1 coulomb of charge from one point to another against a field:

$$W = qV \qquad \left\{ \begin{array}{l} W = \text{work: joules} \\ V = \text{PD: volts} \\ q = \text{charge: coul} \end{array} \right\}$$

But work per unit time is defined as *power;* or 1 joule per second of energy expended is equal to 1 *watt:*

$$P = \frac{W}{t} \qquad \left\{ \begin{array}{l} P = \text{energy: watts} \\ t = \text{time: sec} \end{array} \right\}$$

Substitute qV for W:

$$P = \frac{qV}{t} = \frac{q}{t} V$$

But $q/t = I$. Therefore, power in terms of amperage and voltage is

$$P = IV = \frac{\text{coul}}{\text{sec}} \times \frac{\text{joules}}{\text{coul}} = \frac{\text{joules}}{\text{sec}} = \text{watts}$$

Substituting IR for V, power in terms of amperage and resistance is

$$P = IV = I(IR) = I^2 R = (\text{amp})^2 \times \text{ohms} = (\text{amp})^2 \frac{\text{volts}}{\text{amp}} = \text{watts}$$

Substituting V/R for I, power in terms of voltage and resistance is

$$P = IV = \frac{V}{R}(V) = \frac{V^2}{R} = \frac{(\text{volts})^2}{\text{ohms}} = \frac{(\text{volts})^2}{\text{volts/amp}} = \text{volt-amp} = \text{watts}$$

Example 14

A 120-ohm resistor is rated at 10 watts. (1) What amperage flows? (2) What is the voltage?

Solution

1. The amperage is calculated from

$$P = I^2 R$$

Solve for I and substitute:

$$I = \sqrt{\frac{P}{R}} = \sqrt{\frac{10}{120}} = 0.289 \text{ amp} \qquad \left\{ \begin{array}{l} P = 10 \text{ watts} \\ R = 120 \text{ ohms} \end{array} \right\}$$

2. The voltage is

$$P = VI$$

Solve for V and substitute:

$$V = \frac{P}{I} = \frac{10}{0.289} = 34.6 \text{ volts}$$

Alternate Solution

$$P = \frac{V^2}{R}$$

Solve for V and substitute:

$$V = \sqrt{PR} = \sqrt{10 \times 120} = 34.6 \text{ volts} \qquad (check)$$

The relationship between horsepower and electrical energy is

$$1 \text{ hp} = 746 \text{ watts}$$

Example 15

An electric motor is rated to operate at 110 volts and 4 amp. What is the operating horsepower of the motor? (Neglect efficiency.)

Solution

The power involved is

$$P = VI = 110 \times 4 = 440 \text{ watts}$$

The horsepower is

$$\text{hp} = 440 \text{ watts} \times \frac{1}{746 \text{ watts/hp}} = 0.59 \text{ hp}$$

To find the energy per unit time, we have seen that

$$P = \frac{W}{t}$$

The equations for power become

$$\frac{W}{t} = VI, \qquad \frac{W}{t} = I^2 R, \qquad \frac{W}{t} = \frac{V^2}{R}$$

Sec. 22.7   Power   323

Solving for W, the work energy is

$$W = VIt = I^2Rt = \frac{V^2t}{R}$$

Example 16

(1) What is the energy supplied to an electric iron in $\frac{1}{2}$ hr if the iron is using 115 volts and drawing 6 amp? (2) What is the resistance?

Solution

1. Energy:

$$W = VIt = 115 \times 6 \times 1800$$
$$= 1{,}242{,}000 \text{ joules}$$

2. Resistance (from $W = I^2Rt$): $\left\{ \begin{array}{l} t = \frac{1}{2} \text{ hr} \times 3600 = 1800 \text{ sec} \\ V = 115 \text{ volts} \\ I = 6 \text{ amp} \end{array} \right\}$

$$R = \frac{W}{I^2 t} = \frac{1{,}242{,}000}{6^2 \times 1800}$$
$$= 19.2 \text{ ohms}$$

It has been experimentally established that 1 calorie (cal) is equal to 4.18 joules and that 1 Btu is equal to 778 ft-lb. Mechanical and electrical energy are related by these constants. Thus work energy is proportional to heat energy:

$$W \propto H \quad \text{or} \quad W = JH \qquad \left\{ \begin{array}{l} J = \text{conversion factor} \\ W = \text{energy: joules } or \text{ ft-lb} \\ H = \text{heat energy: Btu } or \text{ cal} \end{array} \right\}$$

Substituting for W and solving for H, the equation becomes

$$H = \frac{VIt}{J} \quad \text{or} \quad H = \frac{I^2Rt}{J} \quad \text{or} \quad H = \frac{V^2t}{RJ}$$

Example 17

Find the heat energy developed in Example 16 in (1) metric units; (2) British units.

Solution

1. Metric units:

$$H = \frac{W}{J} = \frac{1{,}242{,}000}{4.18} \qquad \{ W = 1{,}242{,}000 \text{ joules} \}$$
$$= 297{,}130 \text{ cal}$$

324 Electric Current: Simple Circuits Chap. 22

2. British units:

$$W = 1{,}242{,}000 \; \cancel{\text{joules}} \times 0.737 \frac{\text{ft-lb}}{\cancel{\text{joules}}} = \frac{915{,}354 \; \cancel{\text{ft-lb}}}{778 \; \cancel{\text{ft-lb}}/\text{Btu}} = 1177 \; \text{Btu}$$

Alternate Solution

$$297{,}130 \; \cancel{\text{cal}} \times \frac{1 \; \text{Btu}}{252 \; \cancel{\text{cal}}} = 1177 \; \text{Btu} \qquad (check)$$

Problems

1. A wire conducts 2000 coul every $1\frac{1}{2}$ min. Calculate the amperage in the wire.
2. Calculate the equivalent resistance in Fig. 22.10.

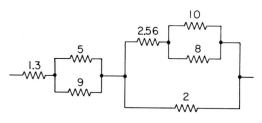

Fig. 22.10

3. Calculate the equivalent resistance in Fig. 22.11.

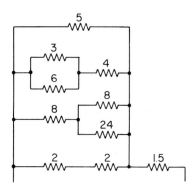

Fig. 22.11

Problems

4. Calculate the equivalent resistance for Fig. 22.12.

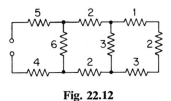

Fig. 22.12

5. Calculate the amperage in Fig. 22.13.

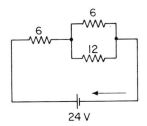

Fig. 22.13

6. In Fig. 22.14, calculate (a) the current in the circuit; (b) the amperage in each resistor; (c) the voltage drop across each resistor.

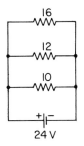

Fig. 22.14

7. In Fig. 22.15, calculate (a) the equivalent resistance; (b) the current in each resistor; (c) the voltage across each resistor.

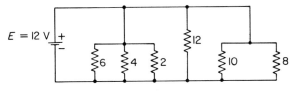

Fig. 22.15

8. In Fig. 22.16, calculate (a) the equivalent resistance; (b) the current output of the battery; (c) the current in each resistor; (d) the voltage across each resistor.

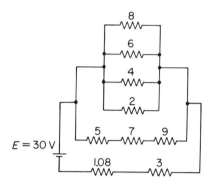

Fig. 22.16

9. In Fig. 22.17, calculate the voltage and amperage in each resistor.

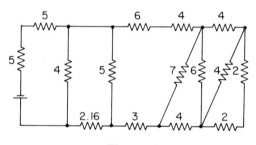

Fig. 22.17

10. In Fig. 22.18, calculate (a) the equivalent resistance; (b) the total current in the curcuit; (c) the current in each of the 4-ohm resistors; (d) the voltage drop across each resistor.

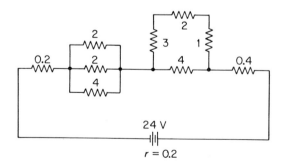

Fig. 22.18

Problems

11. Calculate, in Fig. 22.19, (a) the equivalent resistance; (b) the total current.

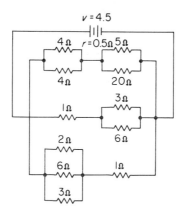

Fig. 22.19

12. Calculate (a) the equivalent voltage for 5 cells in series, each rated at 2 volts; (b) the equivalent resistance if two cells have internal resistances of 0.4 ohm each, two have internal resistances of 0.2 ohm each, one has an internal resistance of 0.5 ohm, and the batteries are connected in series to a 5-ohm external resistance. (c) Calculate the current in the circuit. (d) Calculate the terminal voltage for each cell.
13. Three batteries, emf 12 volts each, are connected in series. The batteries have internal resistances of 0.4 ohm, 0.6 ohm, and 0.5 ohm and are connected to an external resistance of 8.5 ohms. Calculate (a) the equivalent resistance; (b) the line current; (d) the terminal voltage for each cell; (d) the current supplied by each cell.
14. Assume the internal resistance for each cell in Prob. 13 is 0.5 ohm. If the cells are connected in parallel, calculate (a) the equivalent resistance; (b) the line current; (c) the amperage in each cell; (d) the voltage drop in each cell; (e) the voltage output of each cell.
15. A copper wire (resistivity = 1.7×10^{-8} ohm-m) has a diameter of 0.08 cm and is 2000 cm long. Calculate the resistance in the wire.
16. A rectangular wire has dimensions of 0.2 cm and 0.4 cm and a resistivity of 16×10^{-8} ohm-m. How long must the wire be to have a resistance of 6 ohms?
17. A motor has ¼ mi of copper wire wound around the armature. Calculate the diameter of the wire if the maximum resistance permitted is a current flow of 10 amp at 120 volts. The coefficient of resistivity is 10.4 ohm-CM/ft.
18. The resistivity of a certain wire is 600 ohm-CM/ft. A roll of this wire, diameter 0.006 in., has a resistance of 200 ohms. Calculate the length of the wire in the roll.
19. Assume the wire in Prob. 18 to be aluminum (resistivity = 17 ohm-CM/ft). How long is the wire?
20. A nichrome wire has a resistance of 200 ohms at 20°C. Calculate the resistance at 80°C. ($\alpha = 0.00045/°C$ at 20°C)

21. A copper wire has a resistance of 40 ohms at 200 °C. (a) Calculate its resistance at 20 °C. (b) If the diameter of the wire is 0.2 cm, how long is the wire? ($\alpha = 0.004/°C$ at 20 °C; $\rho = 1.54 \times 10^{-8}$ ohm-m)
22. Assume the wire in Prob. 21 is platinum. Calculate the resistance at 20 °C and the length of the wire. ($\alpha = 0.037/°C$ at 0 °C; $\rho = 10 \times 10^{-8}$ ohm-m)
23. An electric iron has a heating coil of monel wire, $\alpha = 0.002/°C$ at 0 °C. The line voltage is 110 volts when the temperature of the iron is 20° C and the amperage is 4 amp. When the iron reaches equilibrium temperature, the current is 2.5 amp. What is the final temperature of the iron?
24. A copper wire has a resistance of 100 ohms at 20 °C. What is the resistance at 120 °C? ($\alpha = 0.004$)
25. An electric toaster draws 20 amp for a period of 1.5 min. If the toaster is connected to a 110-volt line, calculate the work done (in ft-lb) by the toaster.
26. A resistor is rated at 40 watts at 110 volts. Calculate its resistance rating.
27. A heater is rated at 1320 watts at 110 volts. Find (a) the current rating; (b) the resistance. Calculate the heat developed over a period of 2 min (c) in calories; (d) in Btu.
28. Seven and one-half gal of water are heated by an immersion heater from 20 °C to 70 °C. Find the kilowatt-hours consumed by the heater.
29. A calorimeter for checking the mechanical equivalent of heat in the British system yields the following data: the weight of the calorimeter is 2.5 lb; the weight of the water is 20 lb; the sp ht of water is 1; the sp ht of the calorimeter is 0.218; the initial temperature is 50 °F. Find the final temperature if the mechanical equivalent of heat is 778 ft-lb/Btu and an immersion heater draws 2 amp at 110 volts over a period of 1 hr.

23

Circuits and Measuring Instruments

23.1. Multiple-Battery Circuits

Figure 23.1(a) shows two batteries in a circuit connected so that they support each other. The total emf is equal to the sum of the separate emf's, so that

$$E_T = E_1 + E_2$$

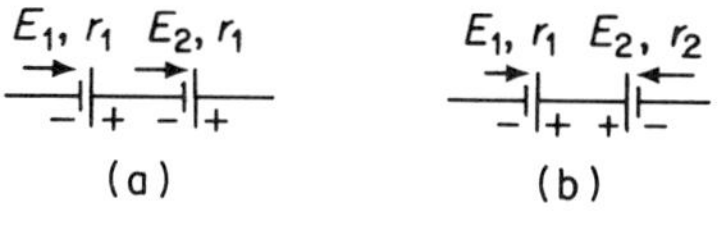

In Fig. 23.1(b), battery E_1 is putting out from left to right while battery E_2 is putting out from right to left. If E_1 is larger than E_2,

$$E_T = E_1 - E_2$$

Now let us establish several guide rules and use Fig. 23.2 as a model. The student should master the following very carefully.

1. Assume a + direction for tracing the circuit. Assume a counterclockwise direction as the tracing direction. In this case, tracing direction is the same

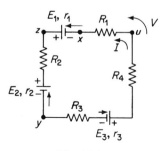

Fig. 23.1

Fig. 23.2

direction as the current I. (This need not be the case. We could have assumed a tracing direction opposite to I.)
2. The current I is either $+$ or $-$, depending upon its relation to the tracing direction.
3. The voltage E is either $+$ or $-$, depending upon its relation to the tracing direction.
4. The resistance R is always $+$.
5. The sum of the voltage inputs (batteries, etc.) must equal the sum of all the voltage drops; or

$$\sum E = \sum IR$$

Therefore, from Fig. 23.2, the separate components based on the *counterclockwise* tracing direction (V) will have the following signs:

Consider the batteries first. The E_1 battery output is in the same direction as the tracing direction, thus E_1 is positive ($+$). Battery E_2 is putting out in a *clockwise* direction, but the tracing direction is counterclockwise; therefore, E_2 is negative ($-$). Since E_3 is putting out in the same direction as V, E_3 is positive. The sum of the voltages is

$$E_T = E_1 + (-E_2) + E_3$$

Remembering that R is always $+$ and that I is in the same direction (in this case) as the tracing direction V, the Ir drops (voltage drops) are

$$\sum IR = Ir_1 + IR_1 + IR_2 + Ir_2 + IR_3 + Ir_3 + IR_3$$

Therefore, since $\Sigma E = \Sigma IR$,

$$E_1 - E_2 + E_3 = I(r_1 + R_1 + R_2 + r_2 + R_3 + r_3 + R_4)$$

It is also possible to find the potential difference between two points using this method, by applying subscripts of the two points to the tracing direction V, including this voltage quantity as one of the components of ΣE, and then applying the rules stated earlier. For example, in Fig. 23.2, V_{xzy} indicates a tracing direction from x through z to y. If V_{xzy} is positive, then x is at a higher potential than y. If V_{xzy} is negative, then x is at a lower potential than y.

Example 1

In Fig. 23.3, calculate the potential difference between points a and c.

Solution

1. To calculate I, assume a tracing counterclockwise direction, or V_{abc}. Thus E_1 is $+$; E_2 is $+$; E_3 is $-$.

$$\sum E = 10 - 12 + 6 = 4 \text{ volts}$$

2. The sum of the IR drops will be $+$ because I has been chosen in the same direction as V_{abc} and R is always $+$. Therefore,

$$\sum IR = I(4 + 0.5 + 9 + 0.7 + 7 + 0.8 + 10) = I(32.0)$$

Sec. 23.1 *Multiple-Battery Circuits* 331

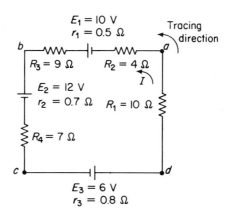

Fig. 23.3

3. Therefore,

$$I = \frac{\sum E}{\sum R} = \frac{4}{32} = \frac{1}{8} \text{ amp}$$

Notice that if the sign of I had been minus, the assumed direction of I would have been incorrect. If I is negative, the current direction should be changed before continuing and this new current direction should be maintained throughout the remainder of the solution, even though the tracing direction may be changed.

4. The potential difference V_{abc} is then calculated by equating

$$V_{abc} + \sum E = \sum IR$$

or

$$V_{abc} + E_1 - E_2 = IR_2 + Ir_1 + IR_3 + Ir_2 + IR_4$$

Substituting,

$$V_{abc} + 10 - 12 = \tfrac{1}{8}(4) + \tfrac{1}{8}(0.5) + \tfrac{1}{8}(9) + \tfrac{1}{8}(0.7) + \tfrac{1}{8}(7)$$
$$V_{abc} = +2 + 0.50 + 0.06 + 1.10 + 0.09 + 0.90 = +4.65 \text{ volts}$$

Since the answer is $+$, a is at a higher potential than c.

5. As a check, let us change the tracing direction to a counterclockwise direction: V_{abc}. Thus E_3 is negative, and I is negative, and

$$V_{adc} - E_3 = -IR_1 - Ir_3$$

Substituting,

$$V_{adc} - 6 = -\tfrac{1}{8}(10) - \tfrac{1}{8}(0.8)$$
$$V_{adc} = 6 - 1.25 - 0.1 = 4.65 \text{ volts} \qquad (check)$$

23.2. Kirchhoff's Laws

Kirchhoff's current law states that the sum of the currents approaching a junction is equal to the sum of the currents leaving that junction. *Kirchhoff's voltage law* states that the sum of the voltages and the sum of the *IR* drops should equal zero. Note that there can be neither more nor less current entering a junction than leaving it. In Fig. 23.4(a), x and y are junction points. The currents entering a junction are considered $+$ and those leaving are considered $-$, or

$$\sum I = 0$$

Kirchhoff's voltage law results in the general expression

$$\sum E = \sum IR$$

The general method for solving Fig. 23.4(a) is as follows:

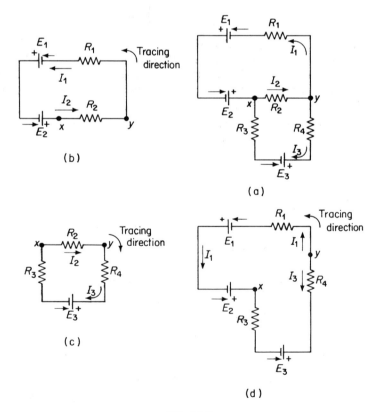

Fig. 23.4

Sec. 23.2   *Kirchhoff's Laws*   333

1. Assume a general current direction for each branch. There are three branches: I_1 and I_2 have been assumed to act counterclockwise, and I_3 clockwise. If the current directions are assumed incorrectly, the answer will be negative. However, the magnitudes of the current would be correct.
2. Write the current equations for all but one of the junction points. If there are two junction points, write one current equation; if three, write two current equations; etc. Thus at junction x, where I_1 and I_3 are entering and I_2 is leaving,

$$I_1 + I_3 - I_2 = 0$$

Since there are two junctions, x and y, only one current equation is needed. However, if junction y had been selected instead of junction x,

$$I_2 - I_1 - I_3 = 0$$

3. The number of current equations plus the number of voltage equations should equal the number of unknowns in the problem. The voltage equation is written by selecting *one* closed loop at a time. A tracing direction V is assumed to be positive. If a *battery output* is in the same direction as the tracing direction, it is assumed to be $+$; if in the opposite direction, it is assumed $-$. If the assumed *current direction* is in the same direction as the tracing direction, the current is $+$; if in the opposite direction, it is $-$. Since the resistance is taken as $+$; the IR drops assume the sign of the current.

Thus in Fig. 23.4(b), tracing from x in a counterclockwise direction, $\Sigma E = \Sigma IR$, and the voltage equation is

$$E_1 + E_2 = I_1 R_1 + I_2 R_2$$

If we choose the loop in Fig. 23.4(c) and trace in a clockwise direction from x, then E_3 is negative, and I_2 and I_3 are positive. This voltage equation becomes

$$-E_3 = I_2 R_2 + I_3 R_4 + I_3 R_3$$

If we choose the only other possibility—the outside loop, Fig. 23.4(d)—and assume a counterclockwise direction starting at y, then E_1 is $+$, E_2 is $+$, E_3 is $+$, I_1 is $+$, and I_3 is $-$. The voltage equation becomes

$$E_1 + E_2 + E_3 = I_1 R_1 + (-I_3 R_3) + (-I_3 R_4)$$

Therefore, for this circuit there are two current equations and three voltage equations from which to choose. From his mathematics, the student should know that to have a complete solution, if he has three unknowns, he needs three equations, etc.

Example 2

Some values have been inserted into Fig. 23.5. Solve for the unknown currents, using Kirchhoff's laws.

Solution

1. Assume a current direction for each loop. Note that this has already been done in Fig. 23.5.
2. There are two junctions (x and y); therefore, we need one current equation at x:

$$I_1 - I_2 + I_3 = 0$$

3. The problem has three unknowns; therefore, for complete solution we need two voltage equations.

   (a) Assume a counterclockwise tracing direction for the upper loop:

$$6 + 2 = 1I_1 + 3I_2$$
$$8 = I_1 + 3I_2$$

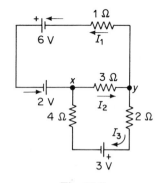

Fig. 23.5

   (b) Assume a counterclockwise tracing direction for the outside loop:

$$6 + 2 + 3 = 4(-I_3) + 2(-I_3) + I_1$$
$$11 = -6I_3 + I_1$$

4. The three equations which must be solved for the three unknowns I_1, I_2, and I_3 are

$$I_1 - I_2 + I_3 = 0 \quad \text{(a)}$$
$$I_1 + 3I_2 = 8 \quad \text{(b)}$$
$$I_1 - 6I_3 = 11 \quad \text{(c)}$$

If the current equation (a) is solved for I_2 and substituted into $I_1 + 3I_2 = 8$, the I_2 drops out, and we are left with two equations and two unknowns. These may be solved by the methods of simultaneous equations, or determinants. Therefore,

$$I_2 = I_1 + I_3$$

Substituting,

$$I_1 + 3(I_1 + I_3) = 8$$

The remaining equations in I_1 and I_3 may be solved by simultaneous equations. They are

$$4I_1 + 3I_3 = 8$$
$$I_1 - 6I_3 = 11$$

Multiplying the lower equation by 4 and subtracting,

$$I_3 = -1\tfrac{1}{3} \text{ amp}$$

Sec. 23.2   *Kirchhoff's Laws*   **335**

Substituting this value back into either of the two equations and solving for I_1,

$$4I_1 + 3(-1\tfrac{1}{3}) = 8$$

$$I_1 = 3 \text{ amp}$$

Substituting I_1 and I_3 into $I_2 = I_1 + I_3$,

$$I_2 = 3 + (-1\tfrac{1}{3}) = 1\tfrac{2}{3} \text{ amp}$$

5. A good check is to write the third voltage equation and insert the values for I. Therefore, the lower-loop voltage equation, assuming a clockwise tracing direction, is

$$-3 = 4I_3 + 3I_2 + 2I_3$$

$$-3 = 4(-1\tfrac{1}{3}) + 3(1\tfrac{2}{3}) + 2(-1\tfrac{1}{3})$$

$$-3 = -3 \quad (\textit{check})$$

Notice very carefully that $I_3 = -1\tfrac{1}{3}$ amp. The minus sign would indicate that the clockwise direction chosen for I_3 was incorrect. Be very sure that you do not confuse the current direction with the assumed tracing direction. Even though the current direction was chosen incorrectly, the magnitude of the current is correct. To avoid redoing the problem, it is recommended that the minus sign be retained and carried until the solution to the problem is complete, at which time the direction may be corrected.

Alternate Solution

An alternate method for solving the three equations is to use determinants. Thus

$$I_1 - I_2 + I_3 = 0$$

$$I_1 + 3I_2 + 0 = 8$$

$$I_1 + 0 - 6I_3 = 11$$

$$I_1 = \frac{\begin{vmatrix} 0 & -1 & 1 \\ 8 & 3 & 0 \\ 11 & 0 & -6 \end{vmatrix}}{\begin{vmatrix} 1 & -1 & 1 \\ 1 & 3 & 0 \\ 1 & 0 & -6 \end{vmatrix}} = \frac{-81}{-27} = +3 \text{ amp}$$

$$I_2 = \frac{\begin{vmatrix} 1 & 0 & 1 \\ 1 & 8 & 0 \\ 1 & 11 & -6 \end{vmatrix}}{-27} = \frac{-45}{-27} = +1\tfrac{2}{3} \text{ amp}$$

$$I_3 = \frac{\begin{vmatrix} 1 & -1 & 0 \\ 1 & 3 & 8 \\ 1 & 0 & 11 \end{vmatrix}}{-27} = \frac{36}{-27} = -1\tfrac{1}{3} \text{ amp}$$

23.3. The d'Arsonval Galvanometer

Most electrical instruments react according to the quantity of charge which passes through a sensitive coil. It will be seen later that when a current passes through a coil, the coil is polarized. A polarized permanent magnet has a fixed strength. As the variable flux, created by the current through the coil, interacts with the permanent-magnet flux, a torque is created. If the coil is suspended so that it may rotate, as the coil itself becomes north–south polarized, it will align itself with the south–north poles of the permanent magnet. This is the basic principle of the d'Arsonval galvanometer, shown in Fig. 23.6(a)

The *d'Arsonval galvanometer* has a wire-wound rectangular coil suspended with a flat wire ribbon which acts as a conductor from the impulse source. A coil spring anchors the lower end of the rectangular coil and also acts as a reaction force against the torque created by the interaction of the coil and the

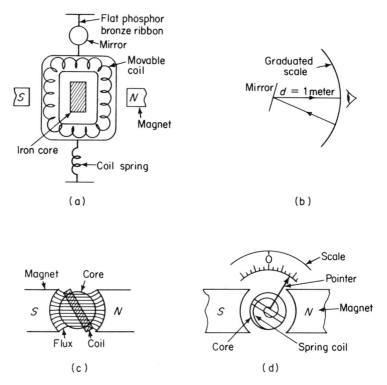

Fig. 23.6

Sec. 23.3   The d'Arsonval Galvanometer   337

permanent magnetic field. Thus, the rotation of the coils is caused by the current and restricted by the coil spring. The more current, the more torque, and the greater the deflection of the mirror.

Figure 23.6(b) shows the graduated scale (in mm) and the reflection of the light from the scale, through the mirror, to the eye. The rotation of the mirror is proportional to the amount of current that flows through the coil. Thus the rotation of the mirror may be calibrated to read current flow, because

$$I \propto L$$

To convert this proportion to an equation, multiply the current by a constant K. Hence

$$L = KI \qquad \left\{ \begin{array}{l} K = \text{current sensitivity} \\ I = \text{current} \\ L = \text{torque} \end{array} \right\}$$

In Fig. 23.6(c), an iron core is set into the coil to create a radial field which will create uniform torque on the coil. Thus, no matter what the position of the coil, the flux from the *permanent* magnet will be uniform, and the deflection of the mirror will be a function of the flux set up in the *movable* coil.

If the distance the coil deflects is proportional to the current input:

$$I \propto S \quad \text{and} \quad I = KS$$

then the current sensitivity of the galvanometer is $\left\{ \begin{array}{l} K = \text{currentsensitivity} \\ I = \text{current} \\ S = \text{deflection} \end{array} \right\}$

$$K = \frac{I}{S} = \frac{\text{current}}{\text{deflection}}$$

Example 3

The distance from a mirror to the scale on a galvanometer is 1 meter. If a current of 4.5 μa causes a deflection of 1.5 cm, calculate (1) the current sensitivity; (2) the potential difference across the coil if the resistance of the coil is 120 ohms; (3) the maximum current for full-scale deflection of 40 divisions.

Solution

1. The current sensitivity is $\{ S = 1.5 \text{ cm} = 1.5 \times 10 \text{ mm} \}$

$$K = \frac{I}{S} = \frac{4.5 \times 10^{-6}}{1.5 \times 10} = 0.3 \times 10^{-6} \text{ amp/unit mm deflection}$$

2. The potential difference across the coil is

$$V = IR = (0.3 \times 10^{-6}) \times 120 = 36 \times 10^{-6} \text{ volt}$$

3. The maximum current is

$$I = KS = (0.3 \times 10^{-6}) \times 40 = 12 \times 10^{-6} \text{ amp}$$

The Westonmeter is a variation of the d'Arsonval galvanometer. It is a portable instrument and is not as sensitive as the d'Arsonval galvanometer. Instead of the mirror, a pointer is fastened to the coil, underneath which is a graduated card as shown in Fig. 23.6(d).

23.4. The Voltmeter

A *voltmeter* is a galvanometer with an external resistance large enough to block large voltages and permit only that portion of the current to pass through for which the meter has been constructed.

Thus in Example 3, the coil of the galvanometer has a sensitivity of 0.3×10^{-6} amp per division and an internal resistance (R_v) of 120 ohms. It takes 12×10^{-6} amp for full-scale deflection of 40 divisions. Thus the maximum voltage this meter can take is 12×10^{-6} amp $\times$ 120 ohms $= 1.44 \times 10^{-3}$ volt. Both the current and voltage capacity of this coil are far below the requirements for checking circuits.

By putting a suitable blocking resistance R_e in *series* with the galvanometer, the meter becomes a voltmeter. This is shown in Fig. 23.7. Terminals are usually provided for switching more resistance into the circuit.

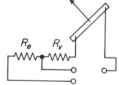

Fig. 23.7

Example 4

A galvanometer is to be used in a 120-volt circuit as a voltmeter. If the coil has a resistance of 40 ohms and requires 5 ma (milliamperes) for full-scale deflection, calculate (1) the additional series resistance required for full-scale deflection; (2) the sensitivity of the voltmeter; (3) the external resistance required to convert the galvanometer to read 20 volts.

Solution

1. The additional series resistance required for full-scale deflection is

$$R_t = \frac{V}{I} = \frac{120}{0.005} = 24{,}000 \text{ ohms}$$

Since 40 ohms is already in the meter and is to be in series with the external resistance,

$$R_t = R_e + R_v$$

Solve for R_e:

$$R_e = R_t - R_v = 24{,}000 - 40 = 23{,}960 \text{ ohms}$$

$$\left\{ \begin{array}{l} V = 120 \text{ volts} \\ I = 0.005 \text{ amp} \\ R_v = 40 \text{ ohms, internal } R \\ R_e = ? \text{ ohms, external } R \\ R_t = ? \text{ ohms, total } R \end{array} \right\}$$

2. The sensitivity of the voltmeter is

$$K_v = \frac{\text{ohms}}{\text{volts}} = \frac{24{,}000}{120} = 200 \text{ ohms/volt} \qquad \{K_v = \text{sensitivity of voltmeter}\}$$

3. The total resistance for a 20-volt voltmeter is

$$R_t = 200 \frac{\text{ohms}}{\text{volts}} \times 20 \text{ volts} = 4000 \text{ ohms}$$

$$R_e = 4000 - 40 = 3960 \text{ ohms} \qquad (\textit{external resistance})$$

23.5. The Ammeter

Remembering that a low resistance will allow more current to pass than a high resistance when both are connected in parallel with a source, a low resistance may be connected in *parallel* with a galvanometer to convert it to an *ammeter*. This is shown in Fig. 23.8. The appropriate resistance R_e will allow only currents up to the permitted maximum necessary for the operation of the instrument. In Fig. 23.8, it can be seen that

$$I = I_a + I_e$$

and I_e is controlled to the point where I_a is permissible through the meter. The IR drop in the shunt is

$$I_e R_e$$

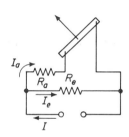

Fig. 23.8

The IR drop in the ammeter is

$$I_a R_a \qquad \left\{\begin{array}{l} R_a = \text{resistance in ammeter} \\ R_e = \text{resistance in parallel} \\ I_a = \text{current in ammeter} \\ I_e = \text{current in shunt} \\ I = \text{line current} \end{array}\right\}$$

Since the voltage drop across both branches is the same,

$$I_e R_e = I_a R_a \quad \text{and} \quad \frac{R_a}{R_e} = \frac{I_e}{I_a}$$

Since $I_e = I - I_a$,

$$\frac{R_a}{R_e} = \frac{I_e}{I_a} = \frac{I - I_a}{I_e}$$

Example 5

Assume the same galvanometer as in Example 4(a). If the current flow in the mainline is 8 amp, calculate (1) the amperage to be bypassed; (2) the permissible voltage in the meter; (3) the shunt resistance needed. (4) If 20 amp flow, compare the shunt needed with part (3).

Solution

1. The amperage bypassed, from $I = I_a + I_e$ is

$$I_e = I - I_a = 8.000 - 0.005 = 7.995 \text{ amp}$$

2. The permissible voltage in the meter is

$$V_a = I_a R_a = 0.005 \times 40 = 0.200 \text{ volt}$$

3. The voltages in both branches of the parallel circuit are equal. Thus from $I_a R_a = I_e R_e$,

$$R_e = \frac{I_a R_a}{I_e} = \frac{0.200}{7.995} = 0.025 \text{ ohm} \quad (shunt) \qquad \left\{ \begin{array}{l} V = 120 \text{ volts} \\ I_a = 0.005 \text{ amp} \\ R_a = 40 \text{ ohms} \\ I = 8 \text{ amp} \end{array} \right\}$$

4. If 20 amp flow, the shunt used should be

$$I_e = I - I_a = 20 - 0.005 = 19.995 \text{ amp}$$

and

$$R_e = \frac{I_a R_a}{I_e} = \frac{0.005 \times 40}{19.995} = 0.01 \text{ ohm} \quad (shunt)$$

Therefore, as the mainline amperage increases, the shunt resistance must decrease to permit more current to bypass the instrument.

23.6. Current-Voltage Measurement

In Fig. 23.9, the voltmeter is shown connected in parallel in such a manner that the *conventional* current flow is from + to −. The ammeter is shown connected in series in the line. The *plus* terminals of both the voltmeter and the ammeter are connected to the *plus* terminal of the battery.

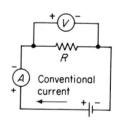

Example 6

Fig. 23.9

The circuit in Fig. 23.9 shows the following data:
battery output, 30 volts; resistor, 5 ohms; voltmeter sensitivity, 100 ohm/volt. Calculate (1) the current in the resistor; (2) the current in the voltmeter; (3) the total current in the ammeter.

Solution

1. The current in the resistor is

$$I = \frac{V}{R} = \frac{30}{5} = 6 \text{ amp}$$

Sec. 23.7   *Resistance Measurement: Ammeter–Voltmeter Method*   341

2. The voltage is impressed across the voltmeter as across the resistor. The current in the voltmeter is $\begin{cases} K = 100 \text{ ohms/volt} \\ V = 30 \text{ volts} \\ R = 5 \text{ ohms} \end{cases}$

$$I_v = \frac{V}{R_v} = \frac{30}{100 \times 30} = 0.01 \text{ amp}$$

3. The current through the ammeter is very small. The current through the ammeter and ammeter shunt is the total current in the line. Since the resistor and the voltmeter are in parallel, their currents add. Thus

$$I = I_R + I_v = 6 + 0.01 = 6.01 \text{ amp}$$

23.7. Resistance Measurement: Ammeter-Voltmeter Method

Resistance may be calculated by using Ohm's law after taking voltmeter and ammeter readings in Fig. 23.10.

When measuring large currents and small voltage drops, the connections are made as shown in Fig. 23.10(a). When measuring small currents and large voltage drops, the connections are made as shown in Fig. 23.10(b).

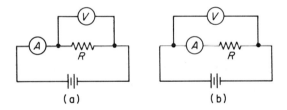

Fig. 23.10

Example 7

Show the errors in measuring resistance with an ammeter and voltmeter when connected as shown in Fig. 23.10(a) and (b). The following data applies: voltmeter internal resistance, 24,000 ohms; ammeter internal resistance, 0.03 ohm; voltmeter reading, 120 volts; ammeter reading, 0.4 amp.

Solution

1. The hookup is shown in Fig. 23.10(a).

    (a) The amperage through the ammeter is the total amperage in the circuit. But since some of this amperage is needed to operate the volt-meter, there will be an inaccuracy when the resistance in the load is calculated from $V = IR$.

(b) The amperage through the voltmeter is

$$I_v = \frac{V}{R_v} = \frac{120}{24{,}000} = 0.005 \text{ amp}$$

(c) The amperage through load is

$$I_a = I_L + I_v$$

or

$$I_L = I_a - I_v = 0.4 - 0.005$$
$$= 0.395 \text{ amp}$$

$$\left\{\begin{array}{l} V = 120 \text{ volts} \\ R_v = 24{,}000 \text{ ohms} \\ R_a = 0.03 \text{ ohm} \\ I_a = 0.4 \text{ amp} \\ I_L = \text{amperage through load} \\ I_v = \text{amperage through voltmeter} \end{array}\right\}$$

(d) Corrected resistance in the load is

$$R = \frac{V}{I_L} = \frac{120}{0.395} = 303.8 \text{ ohms}$$

(e) Uncorrected resistance is

$$R = \frac{V}{I} = \frac{120}{0.4} = 300 \text{ ohms}$$

(f) The error is

$$\frac{I_v}{I_L} = \frac{0.005(100)}{0.395} = 1.27\% \text{ error}$$

2. In Fig. 23.10(b), the ammeter reads only the current through the load, or 0.395 amp. Now the voltmeter reads too high, because the voltage required to operate the ammeter is included in the voltmeter reading.

   (a) The voltage in the ammeter is a function of the current through the ammeter and the resistance.

   $$V_a = I_a R_a = 0.395 \times 0.03 = 0.012 \text{ volt}$$

   (b) The voltage in the load is the difference of the two voltages:

   $$V = V_L + V_a$$

   or

   $$V_L = V - V_a = 120 - 0.012$$
   $$= 119.988 \text{ volts}$$

   $$\left\{\begin{array}{l} V = 120 \text{ volts} \\ V_L = \text{load voltage} \\ V_a = \text{ammeter voltage} \\ \quad = 0.012 \text{ from part 1.} \end{array}\right\}$$

   (c) The error is

   $$\frac{V_a}{V_L} = \frac{0.012(100)}{119.988} = 0.0001 = 0.01\% \text{ error}$$

This verifies the fact that Fig. 23.10(b) should be used, since the line voltage is large and the line current is small.

23.8. Resistance Measurement: The Ohmmeter

Figure 23.11 shows an ammeter with a variable resistance R_e and a resistance R protecting the voltage from the battery E. The battery, coil, and resistance R are all enclosed within a case and are in series. The resistance R_e is variable and provides the shunt resistance. The current through R_e yields the minimum resistance through the coil. Thus when R_e is a minimum, all the current flows through the parallel shunt and the needle registers zero when the external resistance is zero. As the resistance R_e increases, the current flow decreases and varies inversely as the resistance in the meter. Since the resistance in the coil is fixed, the current is dependent upon the adjustment of R_e and therefore the instrument in Fig. 23.11 is called an ohmmeter.

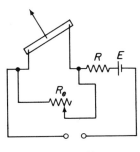

Fig. 23.11

23.9. Resistance Measurement: The Slide-Wire Wheatstone Bridge

In Fig. 23.12(b), the current splits at junction A and the current I_1 moves through R_k and R_x (if galvanometer G is balanced). At the same time, I_2 moves through l_1 and l_2. The resistance in l_1 is proportional to the resistance R_1. The resistance in l_2 is proportional to R_2. Therefore,

$$R_1 = \frac{\rho l_1}{A} \quad \text{and} \quad R_2 = \frac{\rho l_2}{A}$$

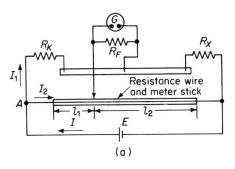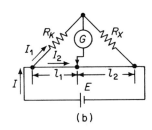

Fig. 23.12

Dividing R_1 by R_2,

$$\frac{R_1}{R_2} = \frac{l_1}{l_2}$$

The lengths l_1 and l_2 may be substituted for the resistances in the respective lengths of wire. The student should make a close comparison between Fig. 23.12(a) and Fig. 23.12(b).

When the drop in potential in l_1 equals the drop in potential R_k, the drop in potential l_2 is equal to the drop in potential in R_x and the galvanometer will show no current. Thus

$$l_1 I_2 = R_k I_1$$

Also,

$$l_2 I_2 = R_x I_1$$

Since no current flows through the galvanometer, the current in R_k equals the current in R_x and the current in l_1 equals the current in l_2. Therefore, dividing one equation by the other,

$$\frac{l_1 I_2}{l_2 I_2} = \frac{R_k I_1}{R_x I_1}$$

or

$$\frac{l_1}{l_2} = \frac{R_k}{R_x}$$

Solving for R_x,

$$R_x = \frac{l_2}{l_1} R_k$$

Example 8

In Fig. 23.12, assume $l_1 = 55$ cm and $l_2 = 45$ cm. If the resistance $R_k = 100$ ohms, what resistance has the unknown R_x?

Solution

$$R_x = \frac{R_k l_2}{l_1} = \frac{100 \times 45}{55} = 81.82 \text{ ohms} \qquad \left\{ \begin{array}{l} R_k = 100 \text{ ohms} \\ l_1 = 55 \text{ cm} \\ l_2 = 45 \text{ cm} \end{array} \right\}$$

23.10. Resistance Measurement: The Box Wheatstone Bridge

If, in Fig. 23.12(b), the two lengths l_1 and l_2 are replaced by their respective resistances R_1 and R_2, the circuit will become the standard *box Wheatstone bridge* circuit shown in Fig. 23.13.

Sec. 23.10 Resistance Measurement: The Box Wheatstone Bridge

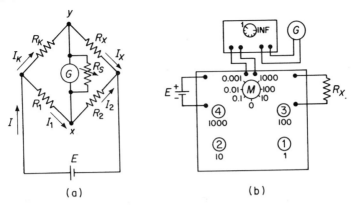

Fig. 23.13

In Fig. 23.13(a), R_k, R_1, and R_2 are adjustable resistances. Resistance R_k is adjusted until there is no deflection in the galvanometer when the unknown resistance R_x is inserted into the circuit as in Fig. 23.13(b). When the circuit is balanced (no deflection in G), junctions x and y are at the same potential. If x and y are *not* at the same potential, G will register a deflection.

Therefore, the potential difference across R_x is equal to the potential difference across R_2 and the potential difference across R_k is equal to the potential difference across R_1. If the current in G equals zero, then the current I_1 equals I_2, and the current I_k equals the current I_x. Thus

$$I_1 R_1 = I_k R_k \quad \text{and} \quad I_2 R_2 = I_x R_x$$

Since $I_1 = I_2$ and $I_k = I_x$, the division of one equation by another becomes

$$\frac{I_1 R_1}{I_1 R_2} = \frac{I_k R_k}{I_k R_x}$$

Solving for R_x,

$$R_x = \frac{R_2}{R_1} R_k$$

Note that the ratio R_2/R_1 is calibrated to the powers of 10. The resistance R_k is set on the four dials in Fig. 23.13(b) and manipulated until the galvanometer reads zero deflection. Then,

$$R_k \times \frac{R_2}{R_1} = R_x$$

The ratio R_2/R_1 is set on dial M and is called the "multiplier ratio."

Example 9

(1) What is the multiplier ratio if R_1 is 1000 ohms, R_2 is 100 ohms, and R_k is 2750 ohms? (2) What is the value of R_x?

Solution

1. The multiplier ratio is

$$\frac{R_2}{R_1} = \frac{100}{1000} = 0.1$$

2. The unknown resistance is

$$R_x = R_k \times \frac{R_2}{R_1} = 2750 \times 0.1 = 275 \text{ ohms}$$

23.11. Potentiometer Measurement: emf, Resistance, Current

In Fig. 23.14(a), the battery E (which must be greater than E_1) is putting out a voltage which is the same in all branches of a parallel circuit. Note that, if you wish to consider voltage as a pressure, be sure you understand that the pressure throughout each branch must be equal. The current I would divide into I_1 and I_2 at junction A and reunite into I at junction D.

Assume that the battery E_1 is inserted into the circuit in place of BC as shown, in opposition to the voltage in $ABCD$. When E_1 is adjusted to be

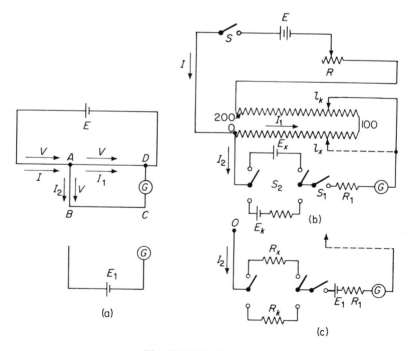

Fig. 23.14(a), (b) and (c)

Sec. 23.11   *Potentiometer Measurement: emf, Resistance, Current*   347

exactly equal to the potential difference in AD, no current will flow in $ABCD$ and there can be *no* voltage drop due to internal resistance in the battery. Thus if I_2 is zero, $I_2 r$ equals zero, and the *emf* of cell E_1 balances the voltage in AD. This is the principle of the *potentiometer*. A voltmeter requires current for its operation. A potentiometer requires no current. Thus if the emf of a battery were desired and a voltmeter is used, the internal resistance will be reflected in the equation

$$V = E - Ir \quad \text{or} \quad E = V + Ir$$

whereas, with a potentiometer, E is measured directly.

Case 1. The Measurement of emf

Figure 23.14(b) applies the principle of the potentiometer to the comparison of a cell whose voltage is unknown, E_x, to a cell whose voltage is known, E_k, usually about 1.0183 volts. The resistance R_1 is a protective resistance for the galvanometer.

With E_x in the circuit through the switch S_2, R_1 is set at its maximum resistance, 10,000 ohms, and l is moved to the midpoint of the 2-m wire. Then, R is adjusted for no deflection in G. The resistance of R_1 is then reduced for maximum sensitivity. Once balance has been achieved, the process is repeated for E_x. Therefore,

$$\frac{E_x}{E_k} = \frac{l_x}{l_k}$$

Example 10

In Fig. 23.14(b), E is 6 volts and the pointer registers 100 cm when E_x is in the circuit. When E_k is equal to 1.0183 volts and is in the circuit, and the length of the wire at balance is 92 cm, find the emf of the unknown battery.

Solution

$$E_x = E_k \frac{l_x}{l_k} = 1.0183 \times \frac{92}{100} \qquad \left\{ \begin{array}{l} E_k = 1.0183 \text{ volts} \\ l_x = 92 \text{ cm} \\ l_k = 100 \text{ cm} \end{array} \right\}$$

$$= 0.937 \text{ volts}$$

Case 2. The Measurement of Resistance

Figure 23.14(c) shows the potentiometer circuit adapted to find the unknown resistance R_x. The current I may be adjusted through R so that the current I_2 will be constant throughout the experiment and the relationship is one of resistance to length; or, from $\rho l_1 / A = R_1$, $R_1 = l_1$ because ρ and A are constant. Therefore,

$$I_2 R_k = I_1 l_k \quad \text{and} \quad I_2 R_x = I_1 l_x$$

Dividing $I_2 R_x$ by $I_2 R_k$,

$$\frac{R_x}{R_k} = \frac{l_x}{l_k}$$

This is the same equation as that used for finding unknown potentials, with the important exception that R_x and R_k are substituted for E_x and E_k and the current I_2 must be constant for the two resistors R_x and R_k.

Example 11

In Fig. 23.14(c), R_x is to be found. When R_x is in the circuit, l_x is equal to 138 cm. When R_k is in the circuit, l_k is equal to 120 cm. What is the value of R_x when R_k is 100 ohms?

Solution

$$R_x = \frac{l_x}{l_k} R_k = \frac{138}{120} \times 100 = 115 \text{ ohms}$$

Case 3. THE MEASUREMENT OF CURRENT

To find the current in a circuit, Fig. 23.14(d), using a standard cell and a known resistance, the principle that the potential difference at both ends of a resistor acts like a battery is used. This potential difference is compared to a

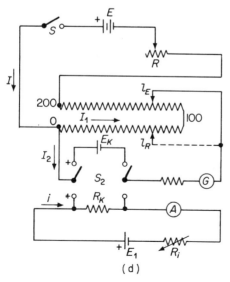

(d) Fig. 23.14(d)

known E_k. If the resistor R_k is inserted into the circuit first and l_R adjusted so that no current flows in the galvanometer, then

$$iR_k = \frac{I_1 \rho l_R}{A}$$

But when $I_2 = 0$,

$$I = I_1 \quad \text{and} \quad iR_k = \frac{I\rho l_R}{A}$$

When E_k is in the circuit and $I_2 = 0$, then

$$I_1 = I \quad \text{and} \quad E_k = \frac{I\rho l_E}{A}$$

Dividing iR_k by E_k,

$$\frac{iR_k}{E_k} = \frac{l_R}{l_E}$$

where R_k and E_k are known and l_R and l_E are measured.

If an ammeter is placed into the circuit, it may be calibrated using this method. The ammeter would be inserted into the i-circuit.

Example 12

Given are a standard resistance of 1 ohm and a standard cell of 1.0183 volts. The length of the wire which will valance the standard resistance is 115 cm. The length of the wire which will balance the standard cell is 90 cm. What is the current i in the line in Fig. 23.14(d)?

Solution

The equation for this circuit is

$$\frac{iR_k}{E_k} = \frac{l_R}{l_E}$$

Solving for i,

$$i = \frac{l_R E_k}{l_E R_k} = \frac{115 \times 1.0183}{90 \times 1}$$

$$= 1.3 \text{ amp}$$

$\left\{ \begin{array}{l} E_k = 1.0183 \text{ volts} \\ l_E = 90 \text{ cm} \\ R_k = 1 \text{ ohm} \\ l_R = 115 \text{ cm} \end{array} \right\}$

Problems

1. Calculate the voltage E_1 in the circuit Fig. 23.15. The current in the circuit is 2 amp.

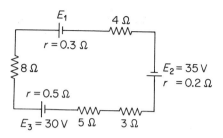

Fig. 23.15

2. Calculate the current in Fig. 23.16.

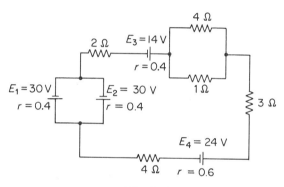

Fig. 23.16

3. In Fig. 23.17, calculate (a) the current in the circuit; (b) the voltage V_{xzy}; (c) the voltage V_{xuy}; (d) the voltage V_{zyu}.

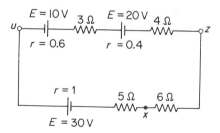

Fig. 23.17

4. In Fig. 23.18, calculate (a) the current in the circuit; (b) the voltage V_{xy}; (c) the voltage V_{zg}.

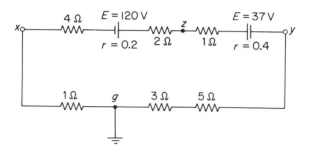

Fig. 23.18

5. In Fig. 23.19, calculate (a) the current in the line; (b) V_{ad}; (c) V_{ab}; (d) V_{bc}; (e) V_{ac}.

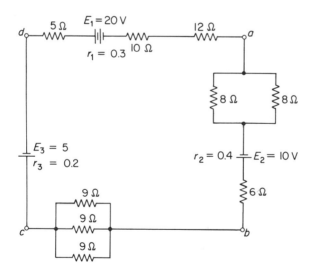

Fig. 23.19

6. Calculate the current in each branch of Fig. 23.20.

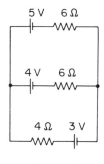

Fig. 23.20

7. Solve Fig. 23.21 for the current in each branch, using Kirchhoff's laws.

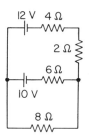

Fig. 23.21

8. Solve Fig. 23.22 for the current in each branch, using Kirchhoff's laws.

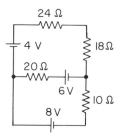

Fig. 23.22

9. Calculate the three currents in the circuit in Fig. 23.23, using Kirchhoff's laws.

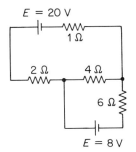

Fig. 23.23

10. Calculate I_2, I_3, and E_2 in Fig. 23.24, using Kirchhoff's laws.

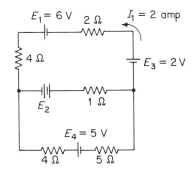

Fig. 23.24

11. Calculate all the currents in Fig. 23.25.

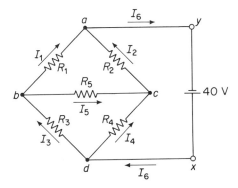

Fig. 23.25

12. When the current is 3.5 μa, a galvanometer deflects 2 cm. Calculate (a) the current sensitivity of the galvanometer; (b) the potential difference across the coil when the resistance is 40 ohms; (c) the maximum current necessary for full-scale deflection of 36 divisions.
13. A galvanometer deflects 12 divisions when the current is 4 μa. If the coil resistance is 600 ohms, calculate the potential difference when the deflection is 30 divisions.
14. A galvanometer is used in a 45-volt circuit. It requires 2 ma (milliamperes) for full-scale deflection when the coil resistance is 50 ohms. Calculate (a) the resistance needed in series for full-scale deflection; (b) the sensitivity; (c) the resistance required to read 120 volts.
15. The sensitivity of a galvanometer is 0.5×10^{-6}. The coil resistance is 200 ohms and the scale has 40 divisions. Calculate (a) the full-scale amperage; (b) the full-scale voltage; (c) the shunt needed to indicate full-scale deflection when the reading is 15 mv (millivolts). (d) Solve part (c) for 4 ma (milliamperes).
16. A 120-volt circuit has inserted into its line two voltmeters connected in series with each other. If they have scales of 150 divisions each and resistances of 20,000 ohms and 24,000 ohms, find the amount each will deflect.
17. A 300-volt voltmeter is to be converted to read 1000 volts maximum. What multiplier is needed if the resistance for 300 volts is 21,000 ohms?
18. What will be the readings on two voltmeters in series connected across a 240-volt line if their resistances are 15,000 and 25,000 ohms?
19. An ammeter has a resistance of 0.5 ohm and measures 15 amp maximum. Calculate the shunt needed to convert this meter so that it will read 50 amp maximum.
20. The shunt resistance for an ammeter is 0.005 ohm and the internal resistance is 0.8 ohm. If the meter is connected to a 20-amp line, how much current will flow through the meter and how much will be shunted?
21. A 100-ma ammeter has a resistance of 20 ohms. Calculate the size of the shunt to convert the instrument to a 500-ma instrument.
22. Calculate the error involved in measuring resistance when an ammeter is connected as in Fig. 23.10(a) and (b), using the following data: the internal resistance of the voltmeter is 100,000 ohms; the internal resistance of the ammeter is 0.02 ohm; the voltmeter reading is 220 volts; the ammeter reading is 0.4 amp.
23. Two resistances, 100 ohms and 115 ohms, are connected to a slide-wire bridge. If the 100-ohm resistance balances out at 52 cm, where will the 115-ohm resistor balance the bridge?
24. Assume a slide-wire bridge balances at 56 cm when the resistance is 220 ohms. What is the unknown resistance which balances at 46 cm?
25. Calculate (a) the multiplier ratio if R_1 is 10,000 ohms and R_2 is 10 ohms; (b) the value of R_x when R_k is 2500 ohms.
26. A box-type Wheatstone bridge is balanced so that R_1 is 54,000 ohms and R_2 is 1500 ohms. If the known resistance is 500 ohms, calculate the unknown resistance.
27. A potentiometer balances out so that a galvanometer reads zero at 40 cm when 1.0183 volts are inserted into the circuit. When an unknown cell is inserted, the potentiometer balances out at 60 cm. What is the voltage in the unknown cell?
28. A potentiometer is used to measure resistance. If a known resistance of 100

ohms balances out at 154 cm, what is the value of the unknown resistance, if it balances out at 90 cm?

29. Given are a standard resistance of 0.1 ohm and a standard cell of 1.0183 volts. The wire which will balance the standard resistance is 250 cm long. The wire which just balances the standard cell is 140 cm long. What is the amperage in the line as shown in Fig. 23.14(d)?

24

Magnetism

24.1. Magnets, Magnetic Poles, and Forces

Many centuries ago, in Asia Minor, a material known as *magnetite* was discovered. It was found to have the peculiar property of attracting small bits of iron to itself. If elongated and suspended from a string, it was discovered, it assumes a position approximately parallel to a north–south line. Its peculiar properties of attraction and repulsion can be transferred to pieces of iron so that their ends become polarized. The end of such a *bar magnet* which points toward the north is called the north-seeking pole, or simply *north pole*. The end which points toward the south is called the south-seeking pole, or *south pole*. Thus a piece of iron which is polarized is called a permanent magnet. In fact, Earth itself, because of its polarization, can be considered a large magnet.

Figure 24.1 shows a bar magnet with north (N) and south (S) poles. Note carefully that if this magnet were suspended by a string, the end marked N would point toward the magnetic north pole. Thus the north polarized end of a compass needle points toward the magnetic north pole of the Earth.

It should be pointed out that magnetic poles must exist in pairs. It is not possible to have a north pole without having a matching south pole. Both poles must have equal strength.

If iron filings are sprinkled onto a sheet of paper under which a magnet is placed, distinct patterns will form which appear to be continuous from one pole to the other. By agreement, let us say that these patterns constitute unbroken lines of force which thread through the north pole into the atmosphere, bend back through the south pole, and then through the magnet to

Sec. 24.1 *Magnets, Magnetic Poles, and Forces* 357

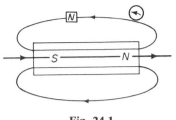

Fig. 24.1

the north pole. This is a complete magnetic circuit. The direction of these lines of force may be established and plotted using a small magnet as shown in Fig. 24.1. These lines of force are taken as unbroken and also as never crossing each other.

These lines of force are called *flux lines* and their direction is shown as acting from the north pole to the south pole *externally* and from the south pole to the north pole *internally*.

Figure 24.2(a) and (b) shows several effects of magnetic poles placed near each other. If the two magnets in Fig. 24.2(b) are brought into close contact, the north and south center poles would disappear; the two magnets would become one magnet with the same south and north poles at the ends of the bar.

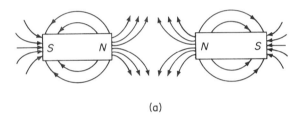

(a)

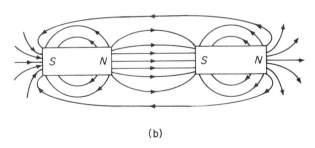

(b)

Fig. 24.2

The reverse is also true: a magnet broken into many pieces becomes little magnets, each with its own north and south poles. If these magnets are separated into smaller and smaller pieces, each becomes and remains a little magnet in its own right. The fundamental unit—the unit which retains all the characteristics of a magnet—is called a *domain*. These domains, according to modern theory, each contain about 10^{12} atoms. The original theory was that these domains are oriented in random fashion (see Fig. 24.3a) when the material is

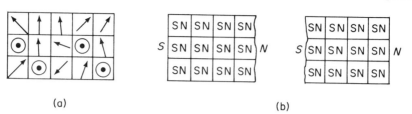

(a) (b)

Fig. 24.3

unmagnetized, but when placed into a magnetic field align themselves according to the direction of the force of the field (Fig. 24.3b). In a permanent magnet, this orientation remains even if the magnet is separated into two pieces.

This domain theory has been extended so that it now states that the electron orbit about the nucleus, as well as the electron spin about its own axis, becomes oriented in one direction when a material is magnetized (see Fig. 24.4a and b). These domains are not regular in shape, nor do all of them line up as shown. Furthermore, the domains themselves do not change position when a magnetic field is present; but the orientations of the axes of spin change their positions.

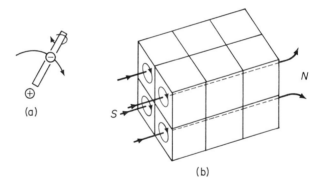

Fig. 24.4

24.2. Coulomb's Law

Coulomb's law applies to the forces between the two magnetic poles. The law states that the force f between two poles, m and m', is directly proportional to the product of the magnitude of their pole strengths and inversely proportional to the square of the distance between them. Thus

$$f \propto mm' \quad \text{and} \quad f = \frac{1}{r^2}$$

Sec. 24.2 Coulomb's Law

If the proportionalities are combined and multipled by a constant, they form an equation which will yield the force between both poles:

$$f = K\frac{mm'}{s^2}$$

where

$\begin{cases} f = \text{force: newtons} \\ m = \text{pole strength: webers} \\ s = \text{distance between } m \text{ and} \\ \quad m': \text{meters} \\ K = \text{constant} \end{cases}$

$$K = \frac{1}{4\pi\mu_0}$$

In the mks system of units, the factor μ_0 has the value $4\pi \times 10^{-7}$ in a vacuum. This constant will be developed in the next chapter. The units of m are webers (wb). The units of s are meters. The units of K are

$$K = \frac{fs^2}{m^2} = \frac{\text{newtons} \times (\text{meters})^2}{(\text{webers})^2} = \text{nt-m}^2/\text{wb}^2$$

Example 1

Two bar magnets have their north and south poles next to each other as shown in Fig. 24.5. If the pole strength of the large magnet is $800\pi \times 10^{-8}$ wb and the pole strength of the small magnet is $600\pi \times 10^{-8}$ wb, calculate the force between the two magnets.

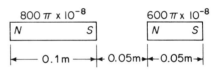

Fig. 24.5

Solution

The force equation is

$$f = K\frac{mm'}{s^2} = \frac{1}{4\pi\mu_0} \times \frac{mm'}{s^2} \qquad \begin{cases} m = 800 \times 10^{-8} \text{ wb} \\ m' = 600 \times 10^{-8} \text{ wb} \\ K = \frac{1}{4\pi\mu_0} \end{cases}$$

1. The force of attraction between both sets of north–south poles:
The south pole of the large magnet will attract the north pole of the small magnet:

$$f_{SN} = \frac{1}{4\pi \times (4\pi \times 10^{-7})} \times \frac{(800\pi \times 10^{-8})(600\pi \times 10^{-8})}{(0.05)^2}$$

$$= 1.2 \times 10^{-2} \text{ nt} \qquad (attract)$$

The north pole of the large magnet will also attract the south pole of the small magnet:

$$f_{NS} = \frac{1}{4\pi \times (4\pi \times 10^{-7})} \times \frac{(800\pi \times 10^{-8})(600\pi \times 10^{-8})}{(0.2)^2}$$

$$= 0.075 \times 10^{-2} \text{ nt} \qquad (attract)$$

The net force of attraction between the two sets of north–south poles is

$f = (1.2 \times 10^{-2}) + (0.075 \times 10^{-2}) = 1.275 \times 10^{-2}$ nt   (*attraction*)

2. The force of repulsion between both sets of like poles:
The south pole of the large magnet will repel the south pole of the small magnet:

$$f_{SS} = \frac{1}{4\pi \times (4\pi \times 10^{-7})} \times \frac{(800\pi \times 10^{-8})(600\pi \times 10^{-8})}{(0.1)^2}$$

$= 0.3 \times 10^{-2}$ nt   (*repel*)

The north pole of the large magnet will repel the north pole of the small magnet:

$$f_{NN} = \frac{1}{4\pi \times (4\pi \times 10^{-7})} \times \frac{(800\pi \times 10^{-8})(600\pi \times 10^{-8})}{(0.15)^2}$$

$= 0.133 \times 10^{-2}$ nt   (*attract*)

The net force of repulsion between both sets of like poles is

$f = (0.3 \times 10^{-2}) + (0.133 \times 10^{-2}) = 0.433 \times 10^{-2}$ nt   (*repulsion*)

3. The resultant force is

$f_r = (1.275 \times 10^{-2}) - (0.433 \times 10^{-2}) = 0.842 \times 10^{-2}$ nt   (*attraction*)

24.3. Magnetic-Field Intensity

Assume a unit "point" north pole $\boxed{N}$ is placed in the magnetic field of the magnet shown in Fig. 24.6. If the pole strength of the magnet is m and the pole strength of the point pole is m', then m will exert a force upon m' of

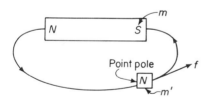

Fig. 24.6

$$f = K \frac{mm'}{s^2}$$

But if we define magnetic *field strength* H as the force on a unit test $\boxed{N}$ pole in that field, then

$$H = \frac{f}{m'} = K \frac{m}{s^2}, \quad \text{where} \quad K = \frac{1}{4\pi\mu_0}$$

The quantity H is a vector quantity and therefore the direction of the field strength is the same direction as the force on the unit test north pole $\boxed{N}$. The unit for H in the mks system is the *ampere-turn per meter* (amp-turn/m), or just ampere per meter. It is defined as the force of 1 newton exerted on a unit pole 1 meter away.

Sec. 24.3   *Magnetic-Field Intensity*   361

Example 2

A bar magnet is 12 cm long. If the pole strengths of the magnet are each $150\pi \times 10^{-7}$ wb, calculate the field strength of the point shown in Fig. 24.7(a).

Solution

In Fig. 24.7(a), a test pole $\boxed{N}$ is placed in the field as shown. The effects of the north and south poles of the magnet on the unit test pole $\boxed{N}$ are examined:

The north pole of the magnet exerts a force of repulsion on the unit north pole $\boxed{N}$ and the magnetic intensity vector H_N acts as shown. The south pole of the magnet exerts a force of attraction on the unit north pole $\boxed{N}$ and the magnetic-intensity vector H_S acts as shown. If these two vectors are added, a resultant vector H_r will act as shown. This vector H_r will give the intensity and direction of the resultant field at the point $\boxed{N}$ and will consequently be tangent to the field as shown.

Thus

$$H_N = \frac{1}{4\pi\mu_0} \times \frac{150\pi \times 10^{-7}}{(0.09)^2} = 370 \text{ amp-turns/m}$$

and $\{m = 150\pi \times 10^{-7} \text{ wb}\}$

$$H_S = \frac{1}{4\pi\mu_0} \times \frac{150\pi \times 10^{-7}}{(0.15)^2} = 133 \text{ amp-turns/m}$$

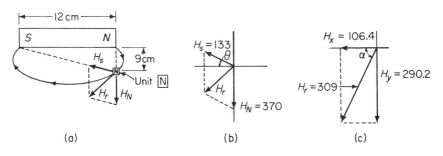

Fig. 24.7

The sum of the *y*-components in Fig. 24.7(b) is

$$\sum H_y = -370 + 133 \sin\theta$$
$$= -370 + 133 \times 0.600$$
$$= -290.2 \text{ amp-turns/m}$$
$$\sum H_x = -133 \cos\theta = -133 \times 0.800$$
$$= -106.4 \text{ amp-turns/m}$$

The resultant field strength is as shown in Fig. 24.7(c):

$$H_r = \sqrt{\sum H_y + \sum H_x} = \sqrt{(290.2)^2 + (106.4)^2}$$
$$= 3.09 \times 10^2 \text{ amp-turns/m}$$

The angle α is

$$\tan \alpha = \frac{-290.2}{-106.4} = 2.727$$
$$\alpha = 69°52'$$

24.4. Flux Density

Figure 24.8 shows an area, 1 meter square, with several flux lines penetrating it. Assume the Greek letter *phi* (ϕ) is used to represent the total flux penetrating the area. The *flux density B* is defined as the flux per unit area A. Thus

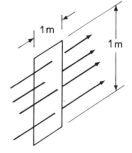

Fig. 24.8

$$B = \frac{\phi}{A} = \text{wb/m}^2$$

The number of flux lines which are capable of threading a square meter is dependent upon the material and this is directly proportional to the magnetic intensity; therefore,

$$B = \mu_0 H$$

where μ_0 is the proportionality constant of the material. Substituting for H,

$$B = \mu_0 H = \mu_0 \times \frac{1}{4\pi\mu_0} \times \frac{m}{s^2} = \frac{1}{4\pi} \times \frac{m}{s^2} = \text{wb/m}^2$$

Example 3

A magnet 8 cm long has a pole strength of $240\pi \times 10^{-7}$ wb. Calculate (1) the flux density at the point P shown in Fig. 24.9(a); (2) the intensity of the field H.

Solution

1. The point P is a point north pole. The south pole will attract the unit north pole, and the north pole will repel the unit north pole. Thus

Flux Density

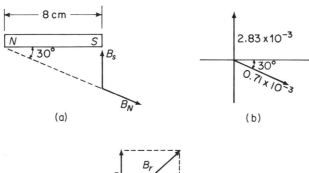

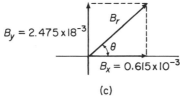

(c)

Fig. 24.9

$$B_S = \frac{1}{4\pi} \times \frac{240\pi \times 10^{-7}}{(0.046)^2} = 2.83 \times 10^{-3} \text{ wb/m}^2$$

$$B_N = \frac{1}{4\pi} \times \frac{240\pi \times 10^{-7}}{(0.092)^2} = 0.71 \times 10^{-3} \text{ wb/m}^2$$

$\{m = 240\pi \times 10^{-7} \text{ wb}\}$

From Fig. 24.9(b), the sum of the vectors in the x- and y-directions is

$$\sum B_x = 0.71 \times 10^{-3} \cos 30° = 0.615 \times 10^{-3} \text{ wb/m}^2$$

$$\sum B_y = -0.71 \times 10^{-3} \sin 30° + (2.83 \times 10^{-3}) = 2.475 \times 10^{-3} \text{ wb/m}^2$$

From Fig. 24.9(c), the resultant vector is

$$\tan \theta = \frac{2.475 \times 10^{-3}}{0.615 \times 10^{-3}}$$

$$= 4.024$$

$$\theta = 76°3'$$

$$\sin \theta = \frac{2.475 \times 10^{-3}}{B_r}$$

$$B_r = \frac{2.475 \times 10^{-3}}{\sin 76°3'} = 2.549 \times 10^{-3} \text{ wb/m}^2$$

2. The field intensity H is

$$B = \mu_0 H$$

$$H = \frac{B}{\mu_0} = \frac{2.549 \times 10^{-3}}{4\pi \times 10^{-7}} = 2.03 \times 10^{-3} \text{ amp-turns/m}$$

24.5. Torque on a Magnet

Section 24.3 showed that the force f of a field on a magnetic pole may be written as

$$f = Hm'$$

If the bar magnet in Fig. 24.10 is d meters long and the angle the magnet makes with the field is α, then the distance between the lines of action of both forces is

$$l \sin \alpha$$

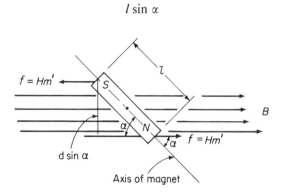

Fig. 24.10

This system is a couple, and the *torque* L is

$$L = fl \sin \alpha$$

Substituting Hm' for f,

$$L = Hm'l \sin \alpha$$

$\left\{ \begin{array}{l} L = \text{torque: nt-m} \\ H = \text{field strength: amp-turns/m} \\ m' = \text{pole strength: wb} \\ l = \text{distance: m} \\ \alpha = \text{angle between centerline of magnet and direction of field} \end{array} \right\}$

Substituting B/μ_0 for H,

$$L = \frac{B}{\mu_0} m'l \sin \alpha$$

Thus when the axis of the magnet is perpendicular to the field, $\sin \alpha$ is a maximum and the torque is a maximum. When the axis of the magnet is parallel to the field, the torque is zero because $\sin \alpha$ is zero.

The *magnetic moment* I_M is defined as magnetic moment per unit volume. If $m'l$ is the magnetic moment M and the volume of the magnet is $V = Al$, the intensity of magnetization becomes

$$I_M = \frac{M}{V} = \frac{m'l}{Al} = \frac{m'}{A} \qquad \{\ I_M = \text{wb/m}^2\ \}$$

Example 4

The length of a magnet is 12 cm. It has an area of 8 cm² and a pole strength of $4\pi \times 10^{-7}$ wb. Calculate (1) the torque on this magnet if it is placed into an external field whose density is 2×10^{-4} wb/m² at an angle of 30° with the direction of the field; (2) the intensity of magnification; (3) the magnetic-field intensity; (4) the magnetic movement; (5) the force on the magnetic poles.

Solution

1. The torque:

$$L = \frac{B}{\mu_0} m'l \sin \alpha$$

$$= \frac{2 \times 10^{-4}}{4\pi \times 10^{-7}} \times (4\pi \times 10^{-7}) \times 0.12 \sin 30°$$

$$= 0.12 \times 10^{-4} \text{ nt-m}$$

2. The intensity of magnification:

$$I_M = \frac{m'}{A} = \frac{4\pi \times 10^{-7}}{8 \times 10^{-4}} = 1.57 \times 10^{-3} \text{ wb/m}^2$$

$$\left\{ \begin{array}{l} m' = 4\pi \times 10^{-7} \text{ wb} \\ B = 2 \times 10^{-4} \text{ wb/m}^2 \\ \alpha = 30° \\ l = 0.12 \text{ m} \\ a = 8 \text{ cm}^2 = 8 \times 10^{-4} \text{ m}^2 \end{array} \right\}$$

3. The magnetic-field intensity:

$$H = \frac{B}{\mu_0} = \frac{2 \times 10^{-4}}{4\pi \times 10^{-7}} = 0.159 \times 10^3 \text{ amp-turns/m}$$

4. The magnetic moment:

$$M = m'l = 4\pi \times 10^{-7} \times 0.12 = 1.51 \times 10^{-7} \text{ wb-m}$$

5. The force on the magnetic poles:

$$f = Hm' = (0.159 \times 10^3) \times (4\pi \times 10^{-7}) = 2 \times 10^{-4} \text{ nt}$$

24.6. The Ampere

We now define the *ampere* in terms of the force between two current-carrying wires: if two infinitely long wires are parallel to each other and are 1 meter apart in a vacuum, they will exert a force on each other of 2×10^{-7} newton per meter of wire length when 1 ampere of current flows through each wire.

In the laboratory, if a small compass is placed near a current-carrying wire, it will be seen that a magnetic field exists. The direction of this field may be established with the "right-hand rule" (Fig. 24.11). If the wire is grasped with the right hand so that the thumb points in the direction of the current flow, the fingers will point in the direction of the flux. Notice

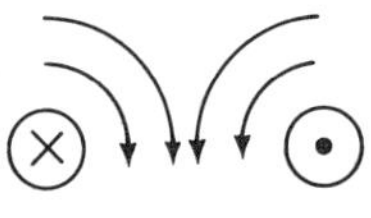

Fig. 24.11

that the symbol ⊗ means that the current flow is into the page, whereas the symbol ⊙ means that the current flow is out of the page.

24.7. The Force on a Charge in a Magnetic Field

If the charge q, the velocity v, and the flux density B (at right angles to the velocity) are known, then the force on the charge is given by the equation

$$f = qvB \qquad \left\{ \begin{array}{l} q = \text{coul} \\ v = \text{m/sec} \\ B = \text{wb/m}^2 \\ f = \text{nt} \end{array} \right\}$$

In Fig. 24.12(a), the velocity of a charge q_+ is seen to be at an angle θ with the field B. *The velocity component at right angles to the field* is $v \sin \theta$. The field B exerts a force f on this velocity component of $qvB \sin \theta$ at right angles to the velocity component.

The velocity component, the field, and the force on the *positive* charge

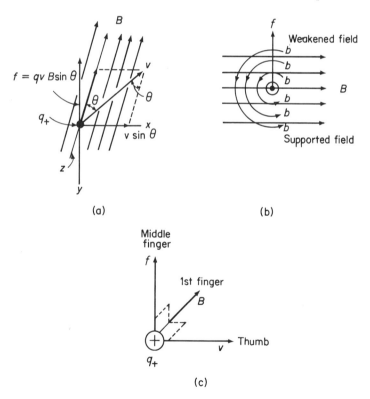

Fig. 24.12

Sec. 24.8   The Force on a Straight Wire   367

[Fig. 24.12(a)] may be found with the right-hand rule. If it is assumed that the moving charge sets up a flux of its own (Fig. 24.11), the right-hand rule may be used to determine the direction of the force created on the velocity component of the charge. In Fig. 24.12(b), the end view of the velocity-component direction of the positive charge q_+ is shown. This moving charge sets up its own field B, which is circular and which supports the external field B at the bottom of the charge and cancels the effect of some of the external flux at the top of the charge. If the weakened field is thought of as a vacuum region and the strengthened field as a pressure region, then the charge will have a force placed on it in an upward direction as shown in Fig. 24.12(b)—toward the weakened region and away from the strengthened region.

Another method is to apply the "three-fingered right-hand rule," Fig. 24.12(c). If the thumb is pointed in the direction of the *velocity component* and the first finger in the direction of the field, the middle finger will point in the direction of the force exerted on the *positive charge* by the *external* field B. The field, velocity component, and the force are at 90° to each other.

24.8. The Force on a Straight Wire Placed into a Magnetic Field

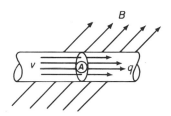

Fig. 24.13

A wire with a uniform cross-sectional area A has a length l. Assume N charges across the area, Fig. 24.13. If the charges cross the area with a velocity v, then the current will be

$$I = NvAq$$

It is also true that the number of charges n in a given volume is

$$n = NAl$$

The total force on all the charges (the entire wire) is

$$f = nqvB = NAlqvB$$

Bracket the expressions:

$$f = (NvAq)(lB)$$

Substitute I for $NvAq$:

$$f = IlB$$

The general equation is

$$f = IlB \sin \theta$$

$\left\{\begin{array}{l} A = \text{area: m}^2 \\ N = \text{no. of charges crossing area} \\ n = \text{no. of charges per volume} \\ q = \text{charge: coul} \\ v = \text{velocity of charges: m/sec} \\ I = \text{current: amp} \\ L = \text{length of wire: m} \\ B = \text{external field: wb/m}^2 \\ f = \text{force: nt} \\ \theta = \text{angle wire makes with field} \end{array}\right\}$

Example 5

A wire 50 cm long is placed in a magnetic field, flux density 2×10^{-6} wb/m². Calculate the force on the wire if 3 amp of current flow in the wire.

Solution

The force on the wire is

$$f = IlB \sin \theta = 3 \times 0.50 \times (2 \times 10^{-6}) \times \sin 90°$$
$$= 3 \times 10^{-6} \text{ nt}$$

$$\left\{ \begin{array}{l} I = 3 \text{ amp} \\ l = 0.50 \text{ m} \\ B = 2 \times 10^{-6} \text{ wb/m}^2 \\ \theta = 90° \end{array} \right\}$$

24.9. The Force and Torque on a Coil Placed into a Magnetic Field

Figure 24.14(a) shows a coil which rotates about a central axis *x–x*. The angle this central *axis* makes with the external field B is θ. The angle the *coil* makes with the field B is α. Let us assume that the current direction in the coil is as shown and that the axis of the coil is at 90° to B. Therefore, the force on wire *a* will be

$$f_a = IAB \sin \theta = Iab$$

The force on wire *a* will always be a maximum if the axis angle θ is a right angle to the direction of the flux. This is the case in electric motors.

The force on wire *b* will change as α changes and will be

$$f_b = IbB \sin \alpha$$

But the force f_b must act along the axis of rotation, since the field and the current direction are perpendicular to each other, as shown in Fig. 24.14(a). Since the *force* acts along the axis of rotation, this *torque* contributes nothing to the rotation of the coil. The resultant of all the *forces* must be zero. The resultant of all the *torques* is not zero.

Figure 24.14(b), (c), (d) shows the coil from Fig. 24.14(a) cut through the middle with the front half removed. If the coil angle with the field is α, then the torque will be a function of cos α. That is, if $\alpha = 0$, the torque is a maximum; if α is 90°, the torque is zero.

In Fig. 24.14(c) and (d), the angle $\theta = 90°$. The torque is the force times $b \cos \alpha$, where $b \cos \alpha$ is the distance component perpendicular to the force direction. Thus the torque on *a* is

$$L_a = BabI \sin \theta \cos \alpha$$

Sec. 24.9   The Force and Torque on a Coil   369

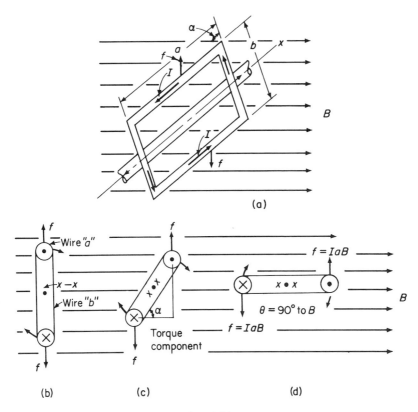

Fig. 24.14

Since ab is the area A of the coil of wire,
$$L_a = BAI \sin \theta \cos \alpha$$
But θ is usually 90° to the direction of the field; therefore, the torque is
$$L = IAB \cos \alpha$$
If the coil has N number of turns of wire, then
$$L = IABN \cos \alpha$$

Example 6

A coil, with 400 turns of wire and dimensions of 3 cm × 8 cm is suspended in a field, flux density 0.5 wb/m². If the current in the coil is 0.04 amp, calculate the torque when the coil is rotated into a 30° position with the flux direction. The axis of rotation of the coil is at 90° with the flux direction.

Solution

The torque is

$$L = IABN \cos \alpha$$
$$= 0.04 \times (24 \times 10^{-4}) \times 0.5 \times 400 \times \cos 30°$$
$$= 1.66 \times 10^{-2} \text{ nt-m}$$

$$\left\{ \begin{array}{l} I = 0.04 \text{ amp} \\ A = 3 \times 8 = 24 \text{ cm}^2 \\ = 24 \times 10^{-4} \text{ m}^2 \\ B = 0.05 \text{ wb/m}^2 \\ N = 400 \text{ turns} \\ \alpha = 30° \end{array} \right.$$

Problems

1. A long magnet (pole strength 850×10^{-8} wb) is placed 12 cm from another long magnet (pole strength 400×10^{-8} wb). If the poles facing each other are both S-poles, calculate the force of repulsion between the two magnets.
2. Two magnets are placed as shown in Fig. 24.15. The pole strength of the 24-cm magnet is 850×10^{-8} wb and the 10-cm magnet is 400×10^{-8} wb. Calculate the magnitude and the direction of the net forces on the 10-cm magnet.

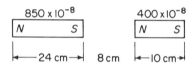

Fig. 24.15

3. A long magnet, pole strength 600×10^{-8} wb, has its south pole placed between two other magnets as shown in Fig. 24.16. If the 10-cm magnet has a pole strength of 800×10^{-8} wb and the 24-cm magnet has a pole strength of 120×10^{-8} wb, find the magnitude and direction of the force on the central magnet.

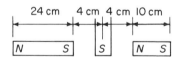

Fig. 24.16

4. A thin bar magnet has a pole strength of 850×10^{-8} wb. Assuming the magnet is 12 cm long, calculate the field strength at a point 10 cm from both poles of the magnet (see Fig. 24.17).

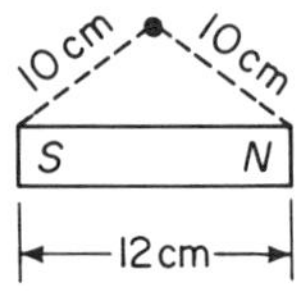

Fig. 24.17

Problems

5. Given the same magnet as shown in Fig. 24.17, calculate the field strength at a point 5 cm directly under the N-pole.
6. The line connecting a point $\boxed{N}$ in a field makes an angle of 30° with the north pole of a magnet 12 cm long. If the distance from the point $\boxed{N}$ to the south pole of the magnet is 60°, and the pole strength of the magnet is 10^{-7} wb, calculate the field strength at $\boxed{N}$.
7. A magnet has a pole strength of 60×10^{-4} wb. If the magnet is 12 cm long, find the flux density and the magnetic intensity at the point P. The line connecting the north pole and the point P forms a 60° angle, and the line connecting the south pole with the point P forms a 30° angle with the axis of the magnet.
8. Calculate the flux density for Prob. 4.
9. Calculate the flux density for Prob. 5.
10. Calculate the flux density for Prob. 6.
11. A bar magnet 10 cm long and with a pole strength of 60×10^{-7} wb is placed into an external field (density 3×10^{-3} wb/m^2) at an angle of 60°. Calculate (a) the torque on the magnet; (b) the magnetic-field intensity; (c) the intensity of magnification (area 8 cm^2); (d) the magnetic moment; (e) the force on the magnetic poles.
12. A magnet 15 cm long has a pole strength of 750×10^{-8} wb. If the magnet is placed at 45° in an external field (flux density 600×10^{-7} wb/m^2), calculate the torque on the magnet.
13. A wire 10 cm long is suspended in a magnetic field, intensity 40×10^2 amp-turn wb/m^2. If the wire carries 30 amp of current, calculate the force on the wire.
14. A coil of wire is placed into a magnetic field (flux density 6×10^{-4} wb/m^2). The coil is circular, has a mean diameter of 8 cm, and carries 20 amp of current. Calculate the torque on the coil when the plane of the coil is in a 40° position with the external field.
15. A coil of wire carrying 25 amp is placed into a magnetic field (flux density 3 wb/m^2). The coil is rectangular, having dimensions of 6 cm $\times$ 14 cm, and rotates with its axis parallel to the 14-cm side and at 90° to the flux direction. What is the torque, in the mks system of units, on the coil when the plane of the coil makes an angle of 40° with the flux?
16. A galvanometer coil measures 2 cm $\times$ 4 cm and has 80 turns of wire. Calculate the current which will cause the mirror to rotate through 10° when a magnetic field of 0.475 wb/m^2 creates a torque of 12×10^{-8} nt-m.

25

Magnetic Fields and Circuits

25.1. Ampère's Theorem

In Fig. 25.1, a segment of a conductor Δl carrying a current I, creates a field of a given intensity at P. This field is perpendicular to the plane containing the point P, a distance s from Δl. The field strength at P is directly proportional to the current, to the segment length Δl, and to the sine of θ. It is inversely proportional to the square of the distance s. Thus

$$\Delta H \propto \frac{I \Delta l \sin \theta}{s^2}$$

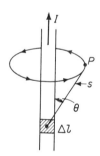

Fig. 25.1

and

$$H = k \frac{I \Delta l \sin \theta}{s^2}, \quad \text{where } k = \frac{1}{4\pi}$$

Therefore, if $H = B/\mu_0$,

$$\Delta B = \frac{\mu_0}{4\pi} \sum \frac{I \Delta l \sin \theta}{s^2}$$

I = current: amp
Δl = length of conductor: m
s = distance of segment to point in field: m
B = flux density: wb/m²
H = magnetic intensity: amp-turns/m
μ_0 = permeability (vacuum): wb/amp-m

This equation, known as Ampere's theorem, may best be handled with the calculus.

25.2. A Loop of Wire

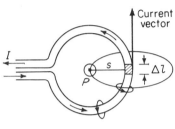

Fig. 25.2

In Fig. 25.2, it is desired to determine the magnetic flux density at point P, at the center of the loop, due to the current in the loop of wire. The direction of the flux density B (created by the current in the wire) acts out from the center of the loop at P. This may be verified using the right-hand rule. If the wire is grasped with the right hand so that the thumb is in the direction of the current, the fingers will give the direction of the flux.

The general equation from Sec. 25.1 may be written as

$$\Delta B = \frac{\mu_0}{4\pi} \frac{I}{s^2} \sum \Delta l \sin \theta$$

The sum of the Δl's equals $2\pi s$, and θ is always $90°$. Thus

$$B = \frac{\mu_0}{4\pi} \frac{I}{s^2} (2\pi s) \quad (1)$$

If more than one turn of wire is concentrated into a coil larger in diameter than in length, the total field strength becomes

$$B = \frac{\mu_0 NI}{2s} \qquad \left\{ \begin{array}{l} I = \text{current: amp} \\ N = \text{number of turns} \\ s = \text{radius of loop: m} \\ B = \text{flux density: wb/m}^2 \\ \mu_0 = 4\pi \times 10^{-7} \text{ wb/amp-m} \end{array} \right\}$$

Example 1

A circular coil of wire which has 200 turns produces flux density of 5×10^{-5} wb/m². The diameter of the coil is 20 cm and is wrapped so that the length of the coil may be disregarded. Calculate the current in the wire.

Solution

Solving for I,

$$I = \frac{2sB}{\mu_0 N} = \frac{2 \times 0.1 \times (5 \times 10^{-5})}{(4\pi \times 10^{-7})200} \qquad \left\{ \begin{array}{l} s = 0.1 \text{ m} \\ B = 5 \times 10^{-5} \text{ wb/m}^2 \\ N = 200 \end{array} \right\}$$

$$= 0.4 \times 10^{-1} \text{ amp}$$

25.3. The Solenoid and the Toroid

Figure 25.3(a) shows a *solenoid*, which consists of a series of windings on a nonmagnetic core. The field set up by the current in the winding is shown and may be verified by the right-hand rule. If one of the wires is grasped so that the thumb points in the direction of the current and the other fingers of the right hand are *inside* the solenoid, the fingers will give the direction of the flux inside the coil. Note that, should the current in the wire be reversed, the flux direction will also be reversed. (Verify this with the right-hand rule.)

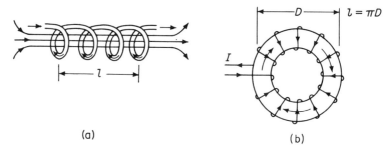

(a)   (b)

Fig. 25.3

The equation which yields the flux density at the center of the coil is as follows:

$$B = \frac{\mu_0 N I}{l}$$

A *toroid* is a solenoid which has been wound about a circular mandrel as shown in Fig. 25.3(b). The flux is concentrated wholly within the windings and the solenoid equation for flux density applies. The quantity l is the mean circumference of the coil as shown.

Example 2

A solenoid has a length of 30 cm and has 1500 coils. If the flux density is 190×10^{-4} wb/m², calculate the amperage in the solenoid.

Solution

Solve the solenoid equation for I:

$$I = \frac{Bl}{\mu_0 N} = \frac{(190 \times 10^{-4})0.3}{(4\pi \times 10^{-7})1500} = 3 \text{ amp} \qquad \left\{ \begin{array}{l} B = 190 \times 10^{-4} \text{ wb/m}^2 \\ l = 0.3 \text{ m} \\ N = 1500 \end{array} \right\}$$

25.4. The Long Straight Wire

The long wire in Fig. 25.4 creates a field when the current I flows as shown. The flux density P is given by the equation

$$B = \frac{\mu_0 I}{2\pi s}$$

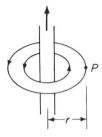

Fig. 25.4

Example 3

Assuming 15 amp in the wire in Fig. 25.4, calculate the flux density 4 cm from the wire.

Solution

The flux density is

$$B = \frac{\mu_0 I}{2\pi s} = \frac{(4\pi \times 10^{-7})15}{2\pi \times 0.04} = 7.5 \times 10^{-5} \text{ wb/m}^2 \qquad \left\{ \begin{array}{l} I = 15 \text{ amp} \\ s = 0.04 \text{ m} \end{array} \right\}$$

25.5. The Force Between Two Current-Carrying Wires

In Fig. 25.5(a), it is assumed that wire 2 is placed into the field of wire 1. Since both wires are carrying current, each will have its own radial flux. In Fig. 25.5(b), the left wire (current into the page) generates flux in a clockwise direction. The right wire generates its flux in a counterclockwise direction. Thus the flux from wire 2 supports the flux from wire 1 *between* the wires. This increase of flux density between the wires puts a force on both wires which tends to *separate* them.

The reverse takes place if the current in both wires flows in the same direction. This is shown in Fig. 25.5(c). The forces tend to cause the wires to attract each other.

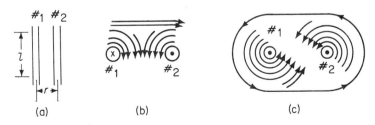

Fig. 25.5

From equations

$$B_1 = \frac{\mu_0 I_1}{2\pi s} \quad \text{and} \quad f = I_2 B_1 l_2 \sin \theta$$

substituting,

$$f = \frac{\mu_0 I_1 I_2 l_2}{2\pi s} \sin \theta$$

θ = angle of second wire to the field of the first wire = 90°
l_2 = length of wire 2
I_2 = current in wire 2
I_1 = current in wire 1
s = distance between wires

Example 4

Two parallel wires carry currents of 20 and 30 amp. Calculate (1) the force of attraction per unit length between them if they are placed 50 cm apart; (2) the total force between the two wires if they are each 2 mi long.

Solution

1. The force per unit length between the two wires is

$$f = \frac{\mu_0 I_1 I_2}{2\pi s} = \frac{(4\pi \times 10^{-7}) \times 20 \times 30}{2\pi \times 0.5} = 2.4 \times 10^{-4} \text{ nt/m}$$

$I_1 = 20$ amp
$I_2 = 30$ amp
$s = 0.5$ m
$l = 3218.7$ m

2. The force between the wires for 2 mi of length is

$$f_t = fl = (2.4 \times 10^{-4})3218.7 = 0.77 \text{ nt}$$

25.6. Ferromagnetism and the Hysteresis Loop

If a material placed into an external magnetic field aligns itself at right angles to the field, the material is said to be *diamagnetic*. If it aligns itself parallel to the field, it is said to be *paramagnetic*. Most materials react only slightly when placed into an external magnetic field. A few react rather strongly when placed into an external magnetic field; these materials are said to be *ferromagnetic*.

We have seen that the equation for the flux density of a coil is

$$B = \mu_0 \frac{NI}{l}$$

where μ_0 is the permeability of a vacuum (air).

If the same coil is wrapped around an iron core, it will be found that the flux density generated by the coil is much greater than when there was no core. This results from a considerable contribution of magnetic flux from the electron spins in the iron core to the magnetic flux generated by the current-carrying coil. The new *permeability constant* reflects this phenomenon. Thus the equation becomes

Sec. 25.6 Ferromagnetism and the Hysteresis Loop

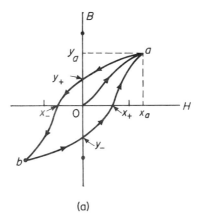

(a)

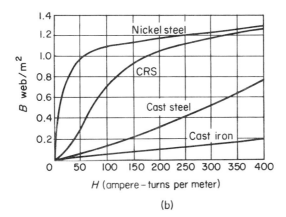

(b)

Fig. 25.6

$$B = \mu \frac{NI}{l} \quad \text{when} \quad \mu = \text{permeability of core material}$$

Figure 25.6(a) shows a *hysteresis curve* when a ferromagnetic material is magnetized and brought slowly from zero magnetization to maximum magnetization at a. That is, if the field strength H is increased slowly to x_a along the x-axis, the flux density B will increase to y_a. But if the field strength H is reduced from x_a to zero along the x-axis, the flux density will drop from y_a to y_+ and be given by $0y_+$, the *retentivity*.

Note carefully that, although no magnetic-field strength is present, the magnet retains its magnetism at y_+. To lose all its flux density, the field strength H must be reversed and reduced to x_Δ along the x-axis. This is given by $0x_\Delta$ and is called the *coercivity*.

The portion of the curve $0a$ in Fig. 25.6(a) is shown for several materials in Fig. 25.6(b). These curves show that μ is not constant for all values of B and H. The permeability curve is a straight line for air only. Values for μ may be calculated from values taken from the curves, Fig. 25.6(b), or from tables such as Table 25.1 compiled from these curves.

Table 25.1

(1)	(2)	(3)	(4)	(5)
Flux Density	Magnetic Intensity	Flux Density of Windings	Flux Density Due to Spin	Permeability
B	H	$\mu_0 H$	$B - \mu_0 H$	$\mu = \dfrac{B}{H}$
(wb/m²)	(amp-turns/m)	(wb/m²)	(wb/m²)	(wb/amp-m)
0.75	100	0.0001256	0.7499	0.0075

From the cold-rolled curve (Fig. 25.6b), the ordinate axis reads $B = 0.75$ wb/m². Trace the B-value to the right from 0.75 until it cuts the soft-steel curve. Then trace to the division at H equal to 100 amp-turns/m. Divide the B-reading by the H-reading to get the permeability μ for the cold-rolled steel (CRS) curve. Thus

$$\mu = \frac{B}{H} = \frac{0.75}{100} = 0.0075$$

In column (3), $H \times \mu_0$ (the latter value being $4\pi \times 10^{-7}$) will give the flux density $\mu_0 H$ of the *external field*. Therefore, in column (4), $B - \mu_0 H$ gives the density of the flux contributed by the spins.

In terms of scientific notation, and so the student may compare this value with $\mu_0 = 4\pi \times 10^{-7}$, the value of μ may be written as $75,000 \times 10^{-7}$ wb/amp-m.

25.7. Magnetic Circuits: mmf

The *flux loop*, created by a coil of wire, in a core (the toroid, Fig. 25.7) may be considered a magnetic circuit. It should be noted that the windings in Fig. 25.7(a) and (b) will yield the same flux density B provided N, I, l, and μ remain unchanged. The shape of the core will not affect the flux density. The flux-density equation for a toroid with a core is the same as for a toroid without a core, except that μ_0 is replaced by μ. Thus

$$B = \mu \frac{NI}{l}$$

Sec. 25.7 *Magnetic Circuits: mmf* 379

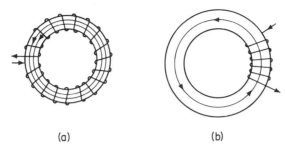

(a)   (b)

Fig. 25.7

It is important to remember that μ must be calculated for values B and H, which have been taken from the permeability curves or from tables. Thus

$$\mu = \frac{B}{H}$$

The ratio μ_R of the permeability μ of a material to the permeability of air μ_0 is called *relative permeability* and is sometimes useful:

$$\mu_R = \frac{\mu}{\mu_0}$$

Example 5

A toroid has an area of 6 cm² and a value of l equal to 14.6 cm. The flux density of its field is 0.5 wb/m² when 5 amp are flowing in the wire coil. Calculate (1) the number of turns of wire needed if the core is cast steel; (2) the relative permeability of the core; (3) the total flux in the core.

Solution

1. The turns of wire needed are found as follows:

From Fig. 25.6(b), if $B = 0.5$ wb/m², then $H = 275$ amp-turns/m. Therefore, the permeability of cast steel in this case is

$$\mu = \frac{B}{H} = \frac{0.5}{275} = 1.82 \times 10^{-3}$$

The ampere-turns required are

$$NI = \frac{Bl}{\mu} = \frac{0.5 \times 0.146}{1.82 \times 10^{-3}} = 40 \text{ amp-turns}$$

Thus, if 5 amp flow in the coil, the number of coils is

$$\left\{\begin{array}{l} B = 0.5 \text{ wb/m}^2 \\ H = 275 \text{ amp-turns/m} \\ l = 14.6 \text{ cm} = 0.146 \text{ m} \\ I = 5 \text{ amp} \\ A = 6 \text{ cm}^2 = 6 \times 10^{-4} \text{ m}^2 \end{array}\right\}$$

$$N = \frac{40 \text{ amp-turns}}{5 \text{ amp}} = 8 \text{ turns of wire}$$

380 Magentic Fields and Circuits Chap. 25

2. The relative permeability is

$$\mu_R = \frac{\mu}{\mu_0} = \frac{1.82 \times 10^{-3}}{4\pi \times 10^{-7}} = 1449 \quad (\text{no units})$$

3. The total flux in the core is

$$\phi = BA = 0.5(6 \times 10^{-4}) = 3 \times 10^{-4} \text{ wb}$$

If the core is cut so that an airgap is created as shown in Fig. 25.8(a), the total flux in the core and in the gap is the same, although the flux density is more concentrated in the core than in the airgap. The flux in the core and the flux in the airgap may be considered in series. This is shown in Fig. 25.8(a). A parallel magnetic circuit is shown in Fig. 25.8(b).

Now, let us consider a closed coil as shown in Fig. 25.7. Substituting $\phi = BA$ in $B = \mu NI/l$ and solving for ϕ, the equation for total flux is

$$\phi = \frac{\mu NIA}{l}$$

Rearranging the terms, $\quad \left\{\begin{array}{c} \text{mmf} = \text{amp-turns} \\ \phi = \text{wb} \\ \mathscr{R} = \text{rowlands} \end{array}\right\}$

$$\phi = \frac{NI}{l/\mu A} = \frac{\text{mmf}}{\mathscr{R}}$$

The quantity $l/\mu A$, called the *reluctance* of a magnetic circuit, is symbolized by a script $\mathscr{R}$, and is expressed in *rowlands*. The quantity NI is called the *magnetomotive force* (mmf).

Note the similiarity between

$$\phi = \frac{\text{mmf}}{\mathscr{R}} \quad \text{and} \quad I = \frac{V}{R}$$

A close analogy may be drawn between ϕ (the flux) and the current I; between mmf (the magnetomotive force) and V (the voltage); and between $\mathscr{R}$ (the reluctance) and R (the resistance).

The method for adding reluctances is the same as adding resistances.

Example 6

If the coil in Example 5 has a 2-mm gap cut into the core, calculate the number of turns of wire needed for the 5 amp.

Solution

The length of the cast-steel core is

$$l_c = l - l_a = 0.146 - 0.002 = 0.144 \text{ m}$$

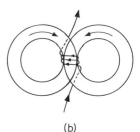

(a)

(b)

Fig. 25.8

If $B = 0.5$, then $H = 275$ amp-turns/m. Therefore,

$$NI = \frac{Bl_c}{\mu} = \frac{0.5 \times 0.144}{1.82 \times 10^{-3}} = 39.6 \text{ amp-turns}$$

For the airgap,

$$NI = \frac{Bl_a}{\mu_0} = \frac{0.5 \times 0.002}{4\pi \times 10^{-7}} = 800 \text{ amp-turns}$$

$\left\{\begin{array}{l} H = 275 \text{ amp-turns/m} \\ B = 0.5 \text{ wb/m}^2 \\ A = 6 \times 10^{-4} \text{ m}^2 \\ l = 0.146 \text{ m} \\ l_a = 0.002 \text{ m} \\ l_c = \text{length of core} \\ \mu = 1.82 \times 10^{-3} \\ I = 5 \text{ amp} \end{array}\right\}$

The total ampere-turns needed is

$$N_t I = 800 + 39.6 = 839.6 \text{ amp-turns}$$

The number of turns of wire needed when an airgap is present is

$$N_t = \frac{839.6}{5} = 168 \text{ turns}$$

The student should note the difference in the number of turns of wire needed when a small airgap is present as contrasted to Example 5, where only 8 turns were needed.

Alternate Solution

$$\mathcal{R}_c = \frac{l_c}{\mu A} = \frac{0.144}{(1.82 \times 10^{-3})(6 \times 10^{-4})} = 0.013 \times 10^7 \text{ rowlands}$$

$$\mathcal{R}_a = \frac{l_a}{\mu_0 A} = \frac{0.002}{(4\pi \times 10^{-7})(6 \times 10^{-4})} = 0.26 \times 10^7 \text{ rowlands}$$

The total reluctance in series is

$$\mathcal{R} = \mathcal{R}_c + \mathcal{R}_a = (0.013 \times 10^7) + (0.260 \times 10^7)$$
$$= 0.273 \times 10^7 \text{ rowlands}$$

The total flux is

$$\phi = BA = 0.5(6 \times 10^{-4}) = 3 \times 10^{-4} \text{ wb}$$

The mmf (amp-turns) is

$$\text{mmf} = \phi\mathcal{R} = (3 \times 10^{-4})(0.273 \times 10^7)$$
$$= 819 \text{ amp-turns}$$

The number of turns needed when 5 amp flow is

$$N = \frac{\text{mmf}}{I} = \frac{819}{5} = 164 \text{ turns}$$

25.8. The Earth's Magnetic Field

The Earth's *geographic axis* points toward the North Star, whereas its *magnetic axis* is displaced approximately 15° as shown in Fig. 25.9. Flux enters the magnetic north and leaves at the magnetic south. A compass needle will set itself parallel to the flux direction. Thus it will be drawn toward the Earth in the northern hemisphere and point away from the Earth in the southern hemisphere. This is called *dip*.

A coil of wire with a compass mounted at its center (*tangent galvanometer*), when oriented parallel to the Earth's field, has its needle oriented in the direction of the Earth's magnetic flux. If current flows in the coil, a deflection takes place, with $\tan \theta = H_I/H_e$. The field strength is

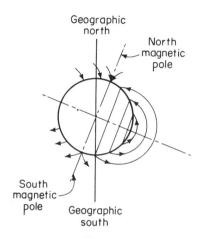

Fig. 25.9

$$H_I = \frac{NI}{2rs} \quad \text{(mks)}$$

If a bar magnet is suspended in the Earth's field, displaced a small amount, and released, its pole strength may be determined from its period of oscillation:

$$T = 2\sqrt{\frac{I_0}{mlH_e}}$$

I_0 = moment of inertia of the bar magnet; kg-m²
m = unit pole: wb
l = distance between poles: m
H_e = Earth's field strength: amp-turns/m
H_I = field created by current in coil
T = period of oscillation: sec

Problems

1. A coil of wire has 10 turns and produces a flux density of 5×10^{-6} wb/m². If the diameter of the coil is 20 cm and is wrapped so that the length of the coil may be disregarded, calculate the current in the wire.
2. A circular coil of wire has 50 turns closely wound into a loop (diameter 12 cm). A second coil has 20 turns of wire closely wound into a circular diameter of 10 cm. Calculate (a) the total field strength if the fluxes support each other when the current flow is 30 amp; (b) the total field strength when the current flows in the opposite direction through the two loops; (c) the flux densities in (a) and (b).
3. A solenoid is 12 cm long and has 600 turns of wire. Calculate the flux density at the center of the solenoid when the current is 30 amp.
4. A solenoid is 25 cm long and has 940 turns of wire. Calculate the flux density at the center of the solenoid when the current is 20 amp.
5. The flux density produced by a solenoid is 1540×10^{-4} wb/m² when the current is 15 amp. How many turns must the solenoid have, if its length is 8 cm, to produce this flux density?

Problems

6. A toroid has a mean diameter of 24 cm and 180 turns of wire. Calculate (a) the field strength inside the toroid, when the current is 10 amp; (b) the flux density.
7. Calculate the force exerted in a north pole of a magnet (strength 2×10^{-4} wb) if the magnet is placed at the center of the toroid in Prob. 6.
8. Calculate the current flow if the coil of wire in Prob. 1 is stretched to a 75-cm length.
9. When the current in a long straight wire is 10 amp and the wire is suspended in air, calculate (a) the flux density 2 cm from the wire; (b) the field strength 2 cm from the wire.
10. Two parallel wires are separated by a distance of 10 cm. If the current in one wire is 30 amp and the force per unit length between the wires is 2×10^{-4} nt, calculate the current in the other wire.
11. Two parallel wires carry currents of 15 and 25 amp. Calculate (a) the force per unit length between the wires if they are separated by a distance of 30 cm; (b) the total force for a $\frac{3}{4}$-mi length.
12. Two wires, each 10 m long and 60 cm apart, are subject to a total force of 8×10^{-4} nt. Calculate the current in each wire if the currents in the wires are equal.
13. A toroid, which has an area of 4 cm^2 and a value of l equal to 20 cm, has a flux density of 0.45 wb/m^2. (a) How many turns of wire are needed if the core is cast steel and the coil of wire carries 5 amp? (b) Calculate the relative permeability of the core. (c) Calculate the total flux in the core.
14. The mean length of a Rowland ring, Fig. 25.7(a), is 60 cm and its cross-sectional area is 8 cm^2. The core is made of cast steel and has 1800 windings. Calculate (a) the current in the windings necessary to produce a field strength of 600 amp-turns/m; (b) the total flux.
15. If the coil in Prob. 13 has a 2-mm airgap cut into it, calculate the number of turns of wire needed for 5-amp flow.
16. A toroid has a mean circumference of 45 cm with a cross-sectional area of 10 cm^2 and 600 windings. The core is of cast iron and has a permeability of 1800×10^{-7}. If 2 amp flow in the coil, find (a) the flux density; (b) the magnetic intensity; (c) the total flux involved; (d) the relative permeability.
17. The toroid in Prob. 16 has a 1-cm airgap cut into it. Find (a) the flux densities in the gap and in the core; (b) the mmf; (c) the total reluctance; (d) the total flux in the system; (e) the magnetic intensities in the core and in the airgap.

26

Induction

26.1. Induced Voltage in a Straight Wire

During the early part and middle of the nineteenth century, scientists such as Faraday, Henry, and Oersted did extensive research in electric current. They found that a magnetic field may *induce a voltage* in a conductor and that a voltage has a magnetic field associated with it.

The research showed that a wire in a magnetic field may have a voltage induced in it by either moving the wire in that field, or by varying the field, or both. Thus, if a length of wire is caused to move with a velocity v through a magnetic field which has a flux density B, the wire will "cut" a given number of lines of flux per unit time and an emf will be induced in the wire.

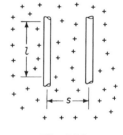

Fig. 26.1

The wire in Fig. 26.1 is caused to move from left to right, cutting the lines of flux* of an external field. We have already seen that the flux density is equal to the number of lines of flux concentrated in a unit area of space. Thus

$$B = \frac{\phi}{A} \quad \left\{ \begin{array}{l} \phi = \text{total flux} \\ A = \text{area} \\ B = \text{flux density} \end{array} \right\}$$

*The crosses ($\otimes$) indicate that the field is *into* the page of the text. Dots ($\odot$) indicate a field *out* of the page.

Induced Voltage in a Moving Coil

If the wire in Fig. 26.1 has a length l and moves a distance Δs, it will "cut" the lines of flux present in the area $l\,\Delta s$. Thus

$$B = \frac{\phi}{A} = \frac{\Delta\phi}{l\,\Delta s}$$

Since the induced emf is a function of the number of lines of flux cut per unit time, the general equation for induced emf may be written

$$E = \frac{N\,\Delta\phi}{\Delta t}$$

Solving the flux-density equation for $\Delta\phi$ and substituting into the emf equation,

$\begin{cases} N = \text{number of wires} \\ \phi = \text{flux: wb} \\ B = \text{flux density: wb/m}^2 \\ l = \text{length: m} \\ v = \text{velocity: m/sec} \\ E = \text{emf: volts} \\ \Delta s = \text{distance moved} \end{cases}$

$$E = \frac{NBl\,\Delta s}{\Delta t}$$

Since $\Delta s/\Delta t = v$,

$$E = NBl(v)$$

If the wire moves through the field at an angle δ with the direction of the flux, the equation becomes

$$E = NBlv \sin \delta$$

Example 1

A wire 30 cm long moves through an external magnetic field at an angle of 60° with the flux direction and with a velocity of 20 cm/sec. The flux density of the field is 5×10^{-2} wb/m². Calculate the induced emf in the wire.

Solution

The induced emf is

$E = NBlv \sin \delta$
$= 1 \times (5 \times 10^{-2}) \times 0.3 \times 0.2 \times \sin 60°$
$= 2.6 \times 10^{-3}$ volt

$\begin{cases} v = 20 \text{ cm/sec} = 0.2 \text{ m/sec} \\ B = 5 \times 10^{-2} \text{ wb/m}^2 \\ l = 0.30 \text{ m} \\ \delta = 60° \\ N = 1 \end{cases}$

26.2. Induced Voltage in a Moving Coil

In Fig. 26.2(a) and (b), an external field is present as a result of the two magnetic poles. The coil $xyzw$ is placed into this field and caused to rotate. Let us consider, for the moment, only the wire xy in Fig. 26.2(a). The slightest movement of the wire will cause lines of flux from the external field to be "cut." In Fig. 26.2(b), however, the wire xy is moving in the same direction as the

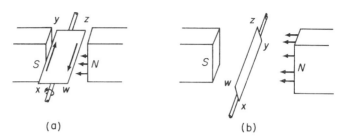

Fig. 26.2

flux lines. That is, at the instant that the coil is in the upright position, there will be no lines of flux cut, and consequently no induced emf generated in the coil. It will be seen that this system will generate an emf in the coil which, if plotted, will yield a sine curve.

Therefore, both wires, xy and zw, will contribute to the emf generated in the wire. Wires xw and yz will contribute nothing to the induced emf because they are rotating parallel to the lines of flux which make up the external magnetic field.

Figure 26.3 shows the end view of the coil, rotating in a counterclockwise direction. Angle θ is taken to be the angle made by the *normal to the plane of the coil* and the flux direction. This angle θ is also the angle made by the instantaneous velocity vector v and the flux direction.

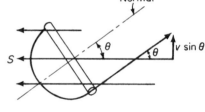

Fig. 26.3

The general equation for a rotating coil is, therefore,

$$E_{inst} = E_{max} \sin \theta$$

The quantity E_{inst} is the *induced voltage* at a particular position of the coil. The induced voltage is E_{max} when $\theta = 90°$ and $\sin \theta = 1$.

When the plane of the coil is at an angle with the field, the flux which threads the coil is θ. The equation for maximum induced emf is

$$E_{max} = NBlv$$

The velocity component perpendicular to the field is $v \sin \theta$ and*

---

*The function $\sin \theta$ refers to the radial position of the wire in the field. To indicate the position of the central axis with the lines of flux, a function, $\sin \delta$, must be included. It would be as if the central axis of a motor coil were at an angle δ with the field. The coil equation would become

$$E_{inst} = E_{max} NBlv \sin \theta \sin \delta \quad (\delta = 90° \text{ for a motor})$$

Sec. 26.2   *Induced Voltage in a Moving Coil*   387

$$E_{\text{inst}} = E_{\text{max}} \, NBlv \sin \theta$$

Since $v = \omega r$ and $\theta = \omega t$, the equation for instantaneous emf may be written as

$$E_{\text{inst}} = NBl\omega r \sin \omega t$$

But $lr = A$ and, therefore,

$$E_{\text{inst}} = NBA\omega \sin \omega t$$
$$= NBA 2\pi F \sin 2\pi F t$$

$$\left\{\begin{array}{l} E = \text{emf} \\ \omega = \text{angular velocity} \\ r = \text{radius of rotation} \\ \theta = \text{angular displacement} \\ t = \text{time} \\ F = \text{frequency} \\ 2\pi F = \text{rad/sec} \\ N = \text{number of lines} \\ B = \text{flux density} \\ A = \text{area of coil} \end{array}\right.$$

Example 2

A coil is made of 400 turns of wire and has dimensions of 6 cm × 10 cm. It rotates at a rate of 1200 rpm in a magnetic field of 200×10^{-4} wb/m². (1) Calculate the induced emf when the coil makes an angle of 60° with the field. (2) Calculate the average value of the induced voltage in $\frac{1}{2}$ rev, starting with a vertical coil.

Solution

The angular velocity is

$$\omega = 2\pi F = 2\pi\left(1200 \times \frac{1}{60}\right) = 40\pi \text{ rad/sec}$$

$$\left\{\begin{array}{l} v = 1200 \text{ rpm} \\ N = 400 \text{ turns} \\ B = 200 \times 10^{-4} \text{ wb/m}^2 \\ A = 6 \times 10 = 60 \text{ cm}^2 \\ = 60 \times 10^{-4} \text{ m}^2 \\ \theta = 30° \text{ with normal} \end{array}\right.$$

1. The instantaneous emf is

$$E_{\text{inst}} = NBA\omega \sin \theta$$
$$= 400(200 \times 10^{-4})(60 \times 10^{-4})40\pi \sin 30°$$
$$= 3 \text{ volts}$$

2. If the coil rotates at a rate of 20 rps, it will take $\frac{1}{20}$ sec for 1 revolution, and $\frac{1}{40}$ sec for $\frac{1}{2}$ rev.

The average voltage induced in the coil is

$$E_{\text{av}} = N \frac{\Delta \phi}{\Delta t}$$

If

$$\Delta \phi = BA = (200 \times 10^{-4})(60 \times 10^{-4})$$
$$= 1.2 \times 10^{-4} \text{ wb}$$

Then

$$E_{\text{av}} = 400 \times \frac{1.2 \times 10^{-4}}{\frac{1}{40}} = 1.9 \text{ volts}$$

26.3. Lenz's Law

Figure 26.4(a) shows a coil of wire and a bar magnet. The first requirement for induced emfs is that there be motion of the conductor or field. Here the magnet, polarized north and south as shown, is given a velocity *toward* the coil. The coil, according to *Lenz's law*, should become polarized to *oppose* this motion. Since the magnetic north pole is close to the coil, the coil must become polarized north in order to oppose the approaching magnet. This repulsion between the magnetic north pole and the induced north pole of the coil follows Lenz's law.

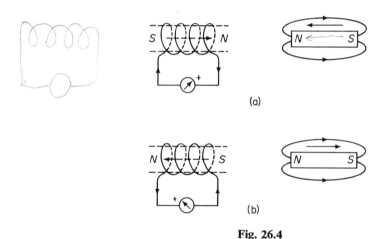

Fig. 26.4

According to our conventions, the flux direction *inside* a magnet (or a coil) is from S to N and the flux direction *outside* the magnet (or coil) is N to S. Once the coil is polarized, it is possible to determine the direction of the emf induced in the coil. If the coil is placed on the palm of the *right* hand and grasped in such a manner that the extended thumb points in the direction of the induced flux *inside* the coil, the fingers will point in the direction of the induced emf in the wire. Thus in Fig. 26.4(a), the thumb of the right hand will point toward the right (S to N) and the fingers of the right hand will yield the direction of the induced emf in the wire, as shown.

In Fig. 26.4(b), the motion of the magnet is to the right. The coil will polarize to oppose this motion. Since like poles attract, the coil will polarize as shown. This induced polarization of the coil establishes the direction of the internal flux. If the coil is grasped in the right hand so that the thumb points to the left, the induced emf will be as shown in Fig. 26.4(b).

Sec. 26.3   Lenz's Law

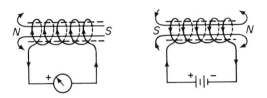

Fig. 26.5

Figure 26.5 shows two coils, one connected to a battery and the other connected to a meter. According to our conventions, current flows from + to − from the battery. If the battery-operated coil is grasped so that the fingers are in the direction of this conventional current flow, the thumb will point in the direction of the internal flux, which indicates a S-to-N polarization as shown.

If the coils are moved toward each other, Lenz's law says that the left coil will polarize in such a manner that the coils will repel. To do this, the poles next to each other in Fig. 26.5 must be the same. This establishes the polarization of the induced flux. Using the right-hand rule, the induced emf is also established.

Also, a change in current flow will cause a build-up or decrease in the number of flux lines. Two stationary coils, Fig. 26.6, will induce in the meter coil an emf whose direction depends on whether the switch is closed or opened.

Note that the current and flux directions in the battery coils are in the same direction in both Fig. 26.6(a) and Fig. 26.6(b).

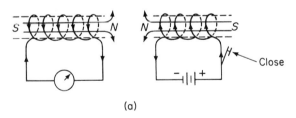

(a)

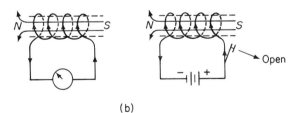

(b)

Fig. 26.6

390 Induction Chap. 26

If the switch is closed, the increased current and flux in the battery coil will cause an increased flux in the meter coil and consequently a repulsion will result (Lenz's Law). Since the battery coil is polarized as shown (N–S), a repulsion can take place only if the meter coil is polarized N–S as shown in Fig. 26.6(a).

If the switch is opened, Fig. 26.6(b), a decreased flux results and polarizes the meter coil in such a manner as to create an attraction to overcome the breakdown of the battery-coil field. The meter coil will polarize as shown in Fig. 26.6(b).

26.4. Generators and Motors

In Sec. 26.1, a discussion of the effect of an external magnetic field on a wire moving took place. Figure 26.7(a) shows a wire moving to the right with a velocity v in a field into the page of the text. If one were to think of this wire as "piling up" flux lines on the right side (velocity side) of the wire, this pile-up would put a force f on the wire to oppose the motion. This force is shown to act toward the left (Lenz's law). If the wire is grasped with the right hand so that the fingers are pointing in the direction of the *external* flux on the pile-up side of the wire, the extended thumb will point in the direction of the induced emf and current. The student should verify the direction of I in Fig. 26.7(b), (c), and (d).

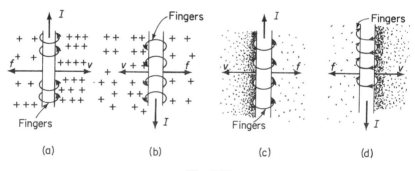

Fig. 26.7

The counterclockwise rotation of a coil of wire in a magnetic field is shown in the sequence in Fig. 26.8. In Fig. 26.8(a), the instantaneous velocity of the wires A and B is such that the velocity vector is parallel to the magnetic-flux direction. Therefore, the induced voltage is zero. Assume the coil turns counterclockwise; then Fig. 26.8(b) shows the coil rotated through 45°, etc.

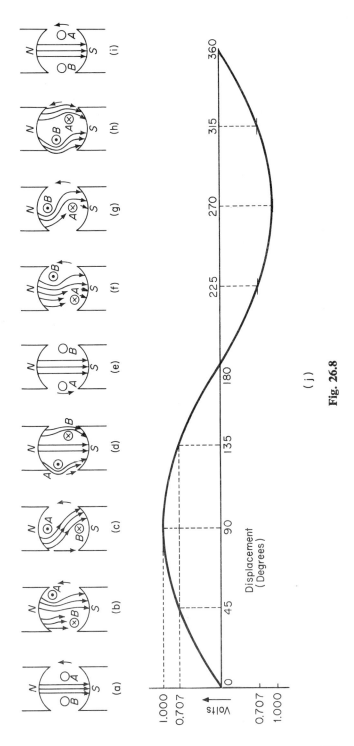

Fig. 26.8

It can be seen that the induced-current direction in the coil reverses itself twice during 1 rev.

The plot—volts against displacement in degrees—is the sine curve, shown in Fig. 26.8(j). The induced emf oscillates between zero, to maximum positive, to zero, to maximum negative, to zero (see Fig. 26.9 and Table 26.1).

Table 26.1

$\theta°$	$\sin \theta$	$E_i = E_m \sin \theta$
0	0	0
90	1	E_m
180	0	0
270	−1	−E_m
360	0	0

This, of course, is the principle of the *generator*. If the two ends of the coil, wires *A* and *B* (Fig. 26.9) are connected to two solid rings, as shown in Fig. 26.10(a), an *alternating-current* (ac) *generator* will result.

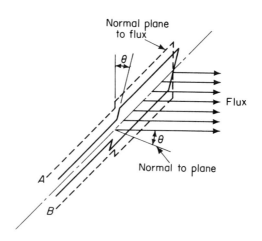

Fig. 26.9

Assume the coil rotates in a counterclockwise direction, cutting lines of flux created by the magnet. Wires *A* and *B* are both "piling up" flux in front of the coil as it rotates. If the wire near the magnetic north pole is grasped with the right hand so that the fingers are in the direction of the flux and on the

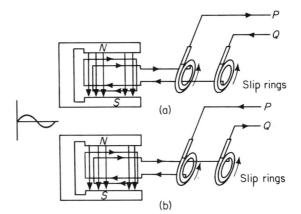

Fig. 26.10

flux-dense side of the coil, the thumb will point in the direction of the induced emf, which will flow toward the slip rings in wire A, out the brushes and the wire P to the right. Notice that, in Fig. 26.10(b), the emf in wire P has reversed its direction when the coil has rotated through 180°. This is an ac generator.

Figure 26.11(a) shows the same coil, except that wires A and B are connected to two halves of a ring separated by an insulating material. The brushes contact first one half of the ring, then another as the coil and the *commutator* revolve. Thus in Fig. 26.11(a), the upper part of the coil, wire A, has an emf induced toward brush P because of its counterclockwise rotation. In Fig. 26.11(b), the emf is induced in the same direction even though wire B is now in the up position; brush P receives the emf in the same direction as before. Thus the emf in wires P and Q is always the same. This is a *direct-current* (dc) *generator*.

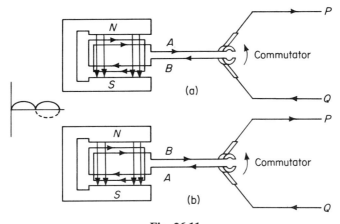

Fig. 26.11

The generator should deliver an almost continuous current. To do this, the commutator is made up of several segments, each with its own coil wound around the *armature*. Thus instead of two segments delivering one sine wave for every revolution of the armature as shown in Fig. 26.10(a), the several coils connected in series deliver several sine curves for one revolution of the armature. If these sine curves are added, the resultant sine curve appears as the upper curve in Fig. 26.12. This may become an almost straight line if many segments are used in the commutator, and the "ripple" become insignificant.

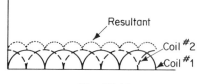

Fig. 26.12

In Fig. 26.13(a), we see a 2-pole generator with the armature and the windings shown as an end view. The flux circulates through the magnet as the commutator rotates, causing the magnetic lines of flux to be cut. Figure 26.13(b) shows the relationship between the number of coils, turns, and conductors.

We have seen that the average emf generated by a single wire is

$$E_{av} = \phi PN$$

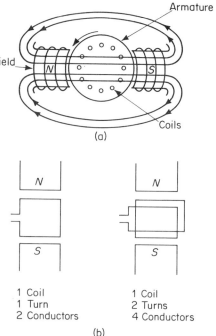

Fig. 26.13

Sec. 26.4   *Generators and Motors*   395

If Z equals the number of conductors [see Fig. 26.13(b)] wound on the armature, and q equals the number of paths involved, the equation for several coils becomes

$$E_{av} = \frac{\phi PNZ}{q}$$

For any given generator, PZ/q is a constant (K). The equation for any particular generator (dc) becomes

$$E_{av} = K\phi N$$

$\left\{ \begin{array}{l} \phi = \text{flux; wb/pole} \\ P = \text{number of poles} \\ N = \text{rps} \\ E_{av} = \text{average emf induced} \\ q = \text{number of paths} \\ Z = \text{number of conductors} \end{array} \right\}$

Notice very carefully that Z, the total number of conductors, may be arrived at by

$Z =$ number of turns $\times$ 2 $\times$ number of coils

Example 3

A dc generator is made with 8 poles, each pole emitting 2×10^{-2} wb/pole. The armature has 6 paths and 40 coils, each coil having 2 turns. If the armature revolves at the rate of 600 rpm, find the emf generated.

Solution

Each coil has 2 turns, each turn has 2 conductors; therefore, the total number of conductors is 4 for each coil:

$Z = 40$ coils $\times$ 4 conductors $= 160$ conductors

The average emf is

$$E_{av} = \frac{\phi PNZ}{q}$$

$$= \frac{(2 \times 10^{-2}) \times 8 \times 10 \times 160}{6}$$

$\left\{ \begin{array}{l} \phi/\text{pole} = 2 \times 10^{-2} \text{ wb/pole} \\ N = 600 \text{ rpm} = 10 \text{ rps} \\ P = \text{number of poles} = 8 \\ q = \text{number of paths} = 6 \end{array} \right\}$

$= 42.7$ volts

Whereas a generator coil (armature) revolves in a magnetic field and thus creates an emf, a *motor* uses the voltage supplied to it to convert electrical energy into mechanical energy. It was seen that there is a force put on a wire moving in an external field. Thus the current-carrying wire in Fig. 26.14 has a force put on it of

$$f = BIlZ$$

Because engineers and technicians usually work in lb-force, this equation is also given as

$$f = 0.225BIlZ$$

Induction

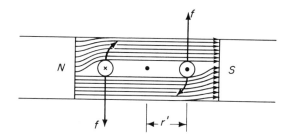

Fig. 26.14

The torque is (Fig. 26.14)

$$L = fr'$$

The horsepower is based on the work per revolution multiplied by the number of revolutions per second:

$$\text{hp} = \frac{2\pi r' fN}{550} = \frac{2\pi LN}{550}$$

$\left\{\begin{array}{l} B = \text{flux density} \\ I = \text{current} \\ l = \text{length of wire} \\ Z = \text{number of conductors} \\ L = \text{torque} \\ r' = \text{radius of coil} \\ N = \text{rps} \\ f = \text{force} \end{array}\right\}$

If the horsepower is based on rpm, the equation becomes

$$\text{hp} = \frac{2\pi r' fN}{33,000} = \frac{2\pi LN}{33,000}$$

Example 4

An armature has 80 conductors, each 10 cm long. The flux density of the field is 20×10^{-2} wb/m² and the total current is 30 amp. There are two paths in the armature whose radius is 18 cm. Calculate (1) the force developed; (2) the torque; (3) the horsepower developed if the coils rotate at 1800 rpm.

Solution

1. The total force is

$$f = 0.225\, BIlZ$$
$$= 0.225 \times (20 \times 10^{-2}) \times 15 \times 0.1 \times 80$$
$$= 5.4 \text{ lb}$$

$\left\{\begin{array}{l} B = 20 \times 10^{-2} \text{ wb/m}^2 \\ I = \frac{30}{2} = 15 \text{ amp each} \\ l = 0.1 \text{ m} \\ Z = 80 \text{ conductors} \\ r' = 18 \text{ cm} = 7.1 \text{ in.} \end{array}\right\}$

2. The torque is

$$L = fr' = 5.4 \times \frac{7.1}{12} = 3.2 \text{ ft-lb}$$

3. The horsepower developed is

$$\text{hp} = \frac{2\pi LN}{33{,}000} = \frac{2\pi \times 3.2 \times 1800}{33{,}000}$$
$$= 1.09$$

26.5. Mutual Inductance

In Sec. 26.4, it was seen that one coil may induce an emf in another coil. The unit of *mutual inductance* is the *henry*, designated by the letter $\mathscr{L}$. The inductance is said to be 1 henry when a change in current of 1 ampere per second in a primary coil induces an emf of 1 volt in the secondary coil. We have seen that inductance is in opposition to an increasing or decreasing current flow. It may be thought of as "electrical inertia".

$$E = \mathscr{L}_m \frac{\Delta I}{\Delta t}$$

Solving for $\mathscr{L}_m$,

$$\mathscr{L}_m = \frac{E \Delta t}{\Delta I}$$

The defining equation for induced emf is

$$E_m = N \frac{\Delta \phi}{\Delta t}$$

$\{$ E = emf in secondary: volts
ΔI = change in current in primary: amp
Δt = change in time in primary: sec
$\mathscr{L}_m$ = mutual inductance: henrys
$\Delta \phi$ = flux: wb
N = number of turns $\}$

Equating both equations,

$$\mathscr{L}_m = \frac{N \Delta \phi}{\Delta I}$$

Example 5

The direct current reaches a maximum value of 8 amp in 0.15 sec in a 60-turn coil. As a result, a 300-turn secondary coil produces 40×10^{-3} wb of flux. Calculate (1) the mutual inductance in the secondary coil; (2) the induced emf in the secondary coil.

Solution

1. Mutual inductance in the secondary:

$$\mathscr{L}_m = \frac{N_2 \phi_2}{I_1} = \frac{300 \times (40 \times 10^{-3})}{8} = 1.5 \text{ henrys} \quad \left\{ \begin{array}{l} N_2 = 300 \\ \phi_2 = 40 \times 10^{-3} \text{ wb} \\ I_1 = 8 \text{ amp} \\ t = 0.15 \text{ sec} \end{array} \right\}$$

2. Induced emf in the secondary:

$$E_2 = \mathscr{L}_m \frac{\Delta I}{\Delta t} = 1.5 \times \frac{8}{0.15} = 80 \text{ volts}$$

The induction coil, Fig. 26.15(a), is a device for making and breaking the primary coil, thus inducing a voltage in the secondary coil which is insulated from the primary. When the circuit is broken in the primary, the current decays very rapidly, creating a large induced voltage in the secondary. Therefore, with a large number of windings in the secondary and a rapid decay in the primary, a large emf is induced in the secondary. The capacitor in the primary circuit stores the charge when the circuit at the gap is broken so that a sudden decay takes place. The induced voltage for the secondary is small as the current is *growing* because the growth takes longer [see Fig. 26.15(b)].

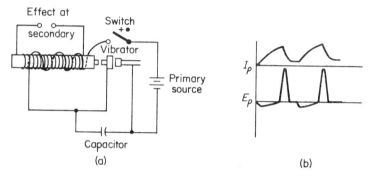

Fig. 26.15

The *transformer* uses the principle of mutual inductance. Two coils are wound on a common core so that the flux generated in the primary coil passes through, or links, the secondary coil when a changing current passes through the primary coil. Transformers may be used as step-up transformers, as in Fig. 26.16(a), or as step-down transformers, as in Fig. 26.16(b). If an alternating current is used in the primary, an alternating current will be induced in the secondary.

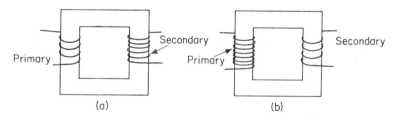

Fig. 26.16

The ratio of the voltages to the number of turns in the primary is equal to the ratio of the voltages in the secondary to the number of turns in the secondary:

Sec. 26.6 *Self-Inductance* 399

$$\frac{N_p}{E_p} = \frac{N_s}{E_s} \quad \text{or} \quad \frac{E_p}{E_s} = \frac{N_p}{N_s}$$

The current and voltage are inversely proportional to each other, so that

$$\frac{E_p}{E_s} = \frac{I_s}{I_p}$$

Example 6

A step-up transformer increases the voltage from 120 volts to 2800 volts. (1) If the primary coil has 240 turns, calculate the number of windings in the secondary coil. (2) Assuming the resistance in the primary coil to be 2 ohms, calculate the current and resistance in the secondary coil. (Assume no losses.)

Solution

1. The number of turns in the secondary:

$$\frac{N_p}{E_p} = \frac{N_s}{E_s}$$

Solving for N_s,

$$N_s = \frac{N_p E_s}{E_p} = \frac{240 \times 2800}{120} = 5600 \text{ turns}$$

$$\left\{ \begin{array}{l} E_p = 120 \text{ volts} \\ E_s = 2800 \text{ volts} \\ N_p = 240 \\ R_p = 240 \\ R_p = 2 \text{ ohms} \end{array} \right\}$$

2. The current and resistance in the secondary:

$$I_p = \frac{E_p}{R_p} = \frac{120}{2} = 60 \text{ amp}$$

$$I_s = \frac{E_p}{E_s} I_p = \frac{120}{2800} \times 60 - 2.6 \text{ amp}$$

$$R_s = \frac{E_s}{I_s} = \frac{2800}{2.6} = 1080 \text{ ohms}$$

26.6. Self-Inductance

If a wire is connected to a battery and the switch closed, the amperage rises according to Ohm's law, $I = V/r$, as shown in the plot in Fig. 26.17(a). If the same wire (same resistance) is wound into a coil and the switch closed, there will be a time lag before the current reaches a maximum value. This time lag is owing to the inductance created by the coil which opposes the rise of current from zero to maximum, as in Fig. 26.17(b). Notice that the same thing occurs when the switch is opened. The coil sets up a *counter-emf* (or back-emf) which opposes the current drop.

If a current is changing at the rate of 1 ampere per second in a coil and 1 volt of emf is induced in the coil, the *self-inductance* in the coil is said to be 1 henry. Therefore,

$$E_s = \mathscr{L}_s \frac{\Delta I}{\Delta t}$$

Also,

$$E_s = N \frac{\Delta \phi}{\Delta t}$$

Es = back-emf; volts
$\mathscr{L}s$ = self-inductance: henrys
ΔI = change in current: amp
Δt = change in time: sec
N = number of turns
$\Delta \phi$ = change in flux: wb

Equating,

$$\mathscr{L}_s = N \frac{\Delta \phi}{\Delta I}$$

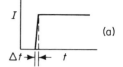

Fig. 26.17

Example 7

A coil is connected to a 110-volt source and a coil of resistance 5 ohms. The steady current output is 15 amp. When the switch is opened, the current drops to zero in 0.03 sec. Calculate (1) the self-inductance in the coil; (2) the change of flux in a 500-turn coil.

Solution

1. The induced voltage is

$$E_s = V - IR = 110 - (15 \times 5) = 35 \text{ volts}$$

The self-inductance is

$$\mathscr{L}_s = E_s \frac{\Delta t}{\Delta I} = 35 \times \frac{0.03}{15} = 0.07 \text{ henry}$$

V = 110 volts
R = 5 ohms
ΔI = 15 amp
Δt = 0.05 sec
N = 500

2. The change of flux is

$$\Delta \phi = \mathscr{L}_s \frac{\Delta I}{\Delta N} = 0.07 \times \frac{15}{500} = 21 \times 10^{-4} \text{ wb}$$

It was shown that the equation for flux density B for a solenoid within an air core is

$$B = \mu_0 \frac{NI}{l'}$$

Since $\phi = BA$, then

l' = length of solenoid: m
μ_0 = permeability ($4\pi \times 10^{-7}$): henry/m (*or*) wb/amp-m
B = flux density: wb/m²
A = area of coil: m²

$$\phi = \frac{\mu_0 NIA}{l'}$$

Substituting in the equation for self-inductance

$$\mathscr{L}_s = N \frac{\mu_0 NIA}{l' I} = \frac{\mu_0 N^2 A}{l'}$$

Sec. 26.7   *Growth and Decay: Inductance–Resistance*   401

If the core is a material other than air, the value of μ for that material should be used.

Example 8

One-hundred twenty turns of wire are wound into a solenoid 20 cm long. Calculate the inductance in the coil if the cross-sectional area of the coil is 12 cm² and the core is iron (permeability $\mu = 4 \times 10^{-4}$ henry/m).

Solution

$$\mathscr{L}_s = \frac{\mu N^2 A}{l'} = \frac{(4 \times 10^{-4}) \times 120^2 \times (12 \times 10^{-4})}{0.20} \qquad \left\{ \begin{array}{l} N = 120 \\ l' = 20 \text{ cm} = 0.20 \text{ m} \\ A = 12 \times 10^{-4} \text{ m}^2 \\ \mu = 4 \times 10^{-4} \text{ henry/m} \end{array} \right\}$$

$$= 3.5 \times 10^{-2} \text{ henry}$$

26.7. Growth and Decay: Inductance–Resistance

It was seen that a change of flux in a coil generates two opposing forces to free flow of current. One of these forces to the growth or decay of flux is resistance, the other is inductance. Also, there is a time lag between the closing of the switch and the growth of the flux to its maximum value. Thus

$$E = \mathscr{L}_s \frac{\Delta I}{\Delta t} \quad \text{and} \quad E = IR$$

In an inductance–resistance circuit, the general equation is

$$E = \mathscr{L}_s \frac{\Delta I}{\Delta t} + IR$$

Note that at the instant the switch is closed, before the resistance becomes effective,

$$IR = 0 \quad \text{and} \quad E = \mathscr{L}_s \frac{\Delta I}{\Delta t}$$

Once the current achieves steady state, there can be no inductance. Therefore,

$$E = IR, \quad \text{but} \quad \mathscr{L}_s \frac{\Delta I}{\Delta t} = 0$$

If the general equation is integrated using the calculus, the equation for *growth* in a resistance–capacitance circuit becomes

$$I_g = \frac{E}{R}(1 - e^{-Rt/\mathscr{L}})$$

If the switch is opened, so that decay takes place, the equation becomes

$$I_d = \frac{E}{R}(e^{-Rt/\mathscr{L}})$$

{ E = back-emf: volts
R = resistance: ohms
t = time: sec
l = inductance: henrys
e = natural log: 2.718
I_g = current growth: amp
I_d = current decay: amp }

The graph for growth is shown in Fig. 26.18(a) and the graph for decay is shown in Fig. 26.18(b).

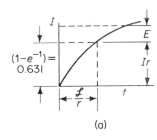
(a)

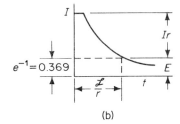
(b)

Fig. 26.18

Note that, in the growth equation, if

$$\frac{Rt}{\mathscr{L}} = 0$$

the growth equation reduces to $E = IR$.
If

$$\frac{Rt}{\mathscr{L}} = 1$$

then

$$e^{-Rt/\mathscr{L}} = \frac{1}{e^1} = 0.369$$

Therefore, to calculate the time required to reach 0.631, the final value for growth, and 0.369, the final value for decay, from

$$\frac{Rt}{\mathscr{L}} = 1$$

the time is

$$t = \frac{\mathscr{L}}{R}$$

Example 9

A battery supplies 6 volts to a coil whose inductance is 4 henrys and whose resistance is 60 ohms. Find (1) the steady-state current; (2) the initial current rate of change; (3) the time constant; (4) the current 0.02 sec after the current starts to flow. (5) After a period of time, the switch is opened: find the decay 0.04 sec after the switch is opened.

Solution

1. Steady-state current:

$$I = \frac{E}{R} = \frac{6}{60} = 0.1 \text{ amp} \quad \left\{ \begin{array}{l} E = 6 \text{ volts} \\ R = 60 \text{ ohms} \\ \mathscr{L}_s = 4 \text{ henrys} \end{array} \right\}$$

2. Initial current rate of change: from

$$E = \mathscr{L}_s \frac{\Delta I}{\Delta t}$$

Solve for $\Delta I/\Delta t$:

$$\frac{\Delta I}{\Delta t} = \frac{E}{\mathscr{L}_s} = \frac{6}{4} = 1.5 \text{ amp/sec}$$

3. The time constant:

$$t = \frac{\mathscr{L}}{R} = \frac{4}{60} = 0.067 \text{ sec}$$

4. Current after 0.02 sec of growth:

$$I_g = \frac{E}{R}(1 - e^{-Rt/\mathscr{L}}) = \frac{6}{60}[1 - e^{-(60 \times 0.02)/4}]$$

$$= 0.1(1 - e^{-0.3}) = 0.1\left(1 - \frac{1}{1.35}\right) = 0.026 \text{ amp}$$

5. Current after 0.04 sec of decay:

$$I_d = \frac{E}{R}(e^{-Rt/\mathscr{L}}) = \frac{6}{60}[e^{-(60 \times 0.04)/4}]$$

$$= 0.1 e^{-0.6} = 0.1\left(\frac{1}{1.822}\right) = 0.055 \text{ amp}$$

26.8. Growth and Decay: Resistance–Capacitance

In a resistance–capacitance circuit, the voltage output of a battery is balanced by an opposition to the emf build-up by the voltage drop due to the resistance and the back-emf. Thus

$$E = \frac{Q}{C} + R\frac{\Delta Q}{\Delta t}$$

Integrating, using calculus, the equation of growth for the charge on a capacitor becomes

$$Q_g = CE(1 - e^{t/RC}) \quad \left\{ \begin{array}{l} Q_g = \text{charge during growth} \\ Q_d = \text{charge during decay} \\ C = \text{capacitance} \end{array} \right\}$$

The decay equation for discharging a capacitor is

$$Q_d = CE(e^{-t/RC})$$

The time constant may be arrived at by setting

$$\frac{t}{RC} = 1$$

and solving for t:

$$t = RC$$

The graphs for charging and discharging a capacitor through a resistance take on the same appearance as Fig. 26.18(a) and (b).

Example 10

A 20,000-ohm resistor is connected to a 40 μf capacitor. Both are connected to a 60-volt dc power supply. Find (1) the steady-state current; (2) the time constant; (3) the charge on the capacitor after 0.2 sec; (4) the voltage across the capacitor after 0.2 sec; (5) the charge on the capacitor 0.6 sec after the switch is opened; (6) the voltage on the capacitor after 0.6 sec of decay.

Solution

1. Steady-state current:

$$I = \frac{V}{R} = \frac{60}{2 \times 10^4} = 3 \text{ ma}$$

2. Time constant:

$$t = RC = (2 \times 10^4)(40 \times 10^{-6}) = 0.8 \text{ sec}$$

3. Growth after 0.2 sec:

$$Q_g = CE(1 - e^{-t/RC})$$
$$= (40 \times 10^{-6}) \times 60 (1 - e^{-0.2/(2 \times 10^4)(40 \times 10^{-6})})$$
$$= (2.4 \times 10^{-3})(1 - e^{-0.25}) = 528 \text{ }\mu\text{coul}$$

$$\begin{cases} R = 2 \times 10^4 \text{ ohms} \\ V = 60 \text{ volts} \\ C = 40 \times 10^{-6} \text{ farad} \end{cases}$$

4. Voltage after 0.2 sec for growth:

$$V_g = \frac{Q_g}{C} = \frac{528 \times 10^{-6}}{40 \times 10^{-6}} = 13.2 \text{ volts}$$

5. Decay 0.6 sec after switch is opened:

$$Q_d = CE(e^{-t/RC}) = (40 \times 10^{-6}) \times 60 (e^{-0.6/(2 \times 10^4)(40 \times 10^{-6})})$$
$$= 2.4 \times 10^{-3}(e^{-0.75}) = 1.14 \times 10^{-3} \text{ coul}$$

6. Voltage after 0.6 sec of decay:

$$V_d = \frac{Q_d}{C} = \frac{1.14 \times 10^{-3}}{40 \times 10^{-6}} = 28.5 \text{ volts}$$

Sec. 26.10 Eddy Currents 405

26.9. Energy

The energy stored in a coil is given by

$$E = \frac{1}{2}\mathscr{L}I^2$$

The similarity between this equation and the equation for energy in a capacitor should be noted:

$$W = \frac{1}{2}CV^2$$

26.10. Eddy Currents

In Fig. 26.19(a), we see an iron core wound with a coil which creates flux that circles the core clockwise when the current is increased. These flux lines are themselves built up to a maximum. While increasing, they set up secondary flux which, according to Lenz's law, opposes the growth of the primary flux.

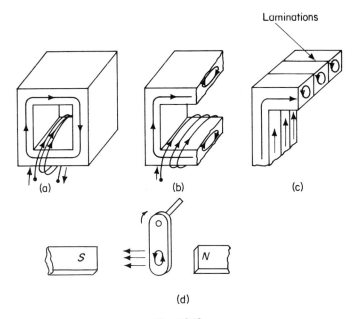

Fig. 26.19

These changing secondary flux lines (Fig. 26.19b) cause IR losses and therefore forces which oppose the operation of meters, generators, etc. They are called *eddy currents* and may be minimized by laminating the core out of very thin materials, each coated with a thin layer of shellac to form a small insulating barrier (see Fig. 26.19c).

Figure 26.19(d) shows a segment rotating in a clockwise direction. As it enters the field, eddy currents build up and create a force on the segment in opposition to the direction of motion of the disk, which creates a damping effect on the rotation.

The losses in motors and generators may be calculated according to the I^2R losses in the armature and the field.

Problems

1. A wire 40 cm long moves with a velocity of 30 cm/sec in an external field, flux density 5×10^{-2} wb/m². Calculate the induced voltage for the wire cutting the external field at an angle of 30° with the flux.
2. A conductor 10 cm long moves with a velocity of 40 cm/sec at an angle of 45° with the field. If the flux density of the external field is 8×10^{-2} wb/m², calculate the induced voltage in the wire.
3. A conductor 80 cm long is passed through an external field at an angle of 60° and at a velocity of 15 cm/sec. If the induced voltage generated is 2×10^{-2} volt, calculate the strength of the external field.
4. A coil rotates at a rate of 1200 rpm. The coil is 10 cm × 18 cm, has 600 turns, and cuts lines of force when rotated in a field of 200×10^{-4} wb/m². (a) Calculate the induced emf when the coil makes an angle of 60° with the field. (b) Calculate the average value of the induced voltage in $\frac{1}{2}$ rev, starting with the vertical coil.
5. A coil is rotated in an external magnetic field (flux density 44×10^{-2} wb/m²) in a time interval of 0.04 sec. The coil has 2000 turns of wire wound in a 6 cm × 8 cm rectangle. Calculate the induced emf.
6. Assuming the coil in Prob. 5 to have a resistance of 8 ohms, calculate (a) the current; (b) the charges induced in the coil.
7. A coil rotates at a rate of 1800 rpm in a field, flux density 10^{-3} wb/m². The coil has 5 turns and dimensions of 6 cm × 8 cm. Calculate (a) the instantaneous induced emf when the plane of the coil is 30° with the field; (b) the maximum emf; (c) the average emf induced in $\frac{1}{4}$ rev of the coil.
8. An armature revolves at a rate of 1200 rpm. The generator has 10 poles, and each pole strength is 9×10^{-2} wb. There are 8 coils and 4 paths and each coil has 12 turns. Calculate the emf generated.
9. A motor revolves at 500 rpm when the armature amperage is 40 amp. The line voltage is 220 volts and the resistance in the armature is 0.8 ohm. The armature has 320 conductors, 2 paths, a flux density of 0.5 wb/m², and a radius of 18 cm. Find (a) the force; (b) the torque developed when the length of each conductor is 16 cm. (c) What horsepower is developed at 1200 rpm?

Problems

10. An armature has 50 conductors, each 8 cm long. The flux density of the field is 2×10^{-2} wb/m^2 and the total current in all the conductors is 30 amp. Calculate (a) the total force if the armature has 3 paths; (b) the torque if the coil path radius is 15 cm; (c) the horsepower developed if the armature revolves at 1200 rpm.
11. Two coils—coil 1 and coil 2—have 50 and 200 turns, respectively. The 50-turn coil carries a direct current which reaches a maximum value of 5 amp in 0.2 sec. If the 200-turn coil produces 30×10^{-3} wb of flux, calculate (a) the mutual inductance in coil 2; (b) the induced voltage in coil 2.
12. The mutual inductance of two coils is 8 henrys. The primary coil is subject to a current change of 12 amp in 0.04 sec. Calculate (a) the induced emf in the secondary coil; (b) the induced flux in a secondary which has 1800 coils; (c) the energy stored in the coil if the maximum current is 12 amp.
13. A coil has 600 turns which produce 5×10^{-4} wb of flux when 3 amp flow in the wire. This produces 6×10^{-3} wb in a 1000-turn secondary coil when a switch is opened. The current drops in 0.2 sec. Find (a) the mutual inductance of the coil; (b) the induced emf in the secondary coil; (c) the self-inductance in the primary coil.
14. A transformer decreases the voltage from 2300 volts to 575 volts. (a) If the primary coil has 6900 turns, calculate the turns in the secondary. (b) Assuming a resistance in the primary coil of 40 ohms, calculate the current and resistance in the secondary. (Assume no losses.)
15. A switch is used to open and close a circuit which has a coil connected to a battery. The battery voltage is 120 volts and the coil resistance is 5 ohms. The steady current is 20 amp and there is a drop to zero amperage in 0.04 sec. (a) What is the self-inductance? (b) What is the value of the change in flux in a 400-turn coil?
16. An iron-core solenoid 10 cm long is wound with 60 turns of wire. What is the inductance in the coil if the cross-sectional area of the coil is 3.14 cm^2? ($\mu = 4 \times 10^{-4}$)
17. A solenoid has 1500 turns of wire wound into a diameter of 5 cm and into a length of 12 cm. Calculate (a) the self-inductance; (b) the self-inductance if the coil is wound on a core 70×10^{-4} wb/amp-m.
18. Using 5 henrys and 50 ohms resistance in a resistance–inductance circuit connected to a 100-volt source (dc), plot a curve with current as ordinate and time as the abscissa. Construct a table giving the current for time intervals of 0.05 sec, 0.1 sec, 0.15 sec, 0.2 sec, etc. to show that the current approaches steady state.
19. A coil whose inductance is 6 henrys has a resistance of 36 ohms. The battery supplies 12 volts to the coil. Find (a) the time constant; (b) the current 0.01 sec after the current starts to flow; (c) the current 0.024 sec after the current starts to flow; (d) the steady-state current; (e) the initial current rate of change.
20. Find the current 0.03 sec after the battery is short-circuited in Prob. 19.
21. A 12-volt battery is connected to a 24-μf capacitor and a 6000-ohm resistor. A switch is closed. Find (a) the charge on the capacitor after 0.1 sec; (b) the voltage on the capacitor after 0.1 sec; (c) the time constant; (d) the steady-state charge on the capacitor.
22. (a) In Prob. 21, find the decay 0.2 sec after a switch short-circuits the battery. (b) What is the voltage on the capacitor at this time?

27

Alternating Current

27.1. The Sine Wave: Instantaneous emf

In Chapter 26, the principle for generating an alternating current was discussed. A sine wave is generated as shown in Fig. 27.1. The curve at a is zero. At b, it is a maximum, E_{max}. At c, it is again zero. At 270° (d), it is negative maximum. At 360° (e), it is again zero. This oscillation is called a *cycle*. One cycle is called a *period*. The number of cycles per unit time (sec, min, etc.) is called the *frequency*.

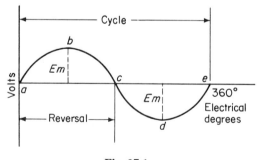

Fig. 27.1

It was seen in Chapter 26 that the frequency of an electric generator depends upon the number of poles and the revolutions per second of the armature. There are as many cycles per revolution as there are *pairs* of poles.

Sec. 27.1   The Sine Wave: Instantaneous emf   409

Example 1

An ac generator rotates at 1800 rpm. Calculate the cycles per second (cps) needed to operate the generator if it has 8 poles.

Solution

Cycles per revolution = 8/2 = 4 pairs of poles = 4 cycles/rev.

Cycles per second:

$$4 \times \frac{1800}{60} = 120 \text{ rps}$$

It was also shown that the equation for a coil revolving in a magnetic field generates an instantaneous emf of

$$E_{\text{inst}} = E_{\text{max}} \sin \theta = NBA\omega \sin \theta$$

When $\theta = 90°$,

$$E_{\text{inst}} = E_{\text{max}} = NBA\omega$$

In terms of angular velocity and time, if

$$\theta = \omega t$$

then

$$E_{\text{inst}} = NBA\omega \sin \omega t$$

$\left\{ \begin{array}{l} E_{\text{inst}} = \text{instantaneous voltage} \\ E_{\text{max}} = \text{voltage when } \theta \text{ is } 90° \\ \omega = \text{angular velocity: rad/sec} \\ F = \text{frequency: cps} \\ t = \text{time: sec} \\ A = \text{area of the coil} \\ B = \text{external-field density} \\ N = \text{number of turns} \\ \theta = \text{angular displacement with normal to field} \end{array} \right\}$

If $\omega = 2\pi F$, the equation in terms of frequency and time becomes

$$E_{\text{inst}} = NBA 2\pi F \sin 2\pi F t$$

In Fig. 27.2, the sine curve has a maximum value of 1 volt. As the coil rotates in the field, the instantaneous voltage is a function of the same θ from the equation

$$E_{\text{inst}} = E_{\text{max}} \sin \theta$$

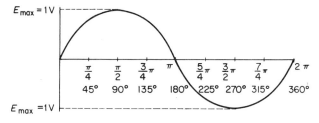

Fig. 27.2

Table 27.1 shows these values for all angles from 0° to 360° in increments of 30°.

Table 27.1

θ (degrees)	$\sin \theta$	$E_{max} \sin \theta$
0	0	0
30	0.500	0.500
45	0.707	0.707
60	0.866	0.866
90	1.000	1.000
120	0.866	0.866
135	0.707	0.707
150	0.500	0.500
180	0	0
210	−0.500	−0.500
225	−0.707	−0.707
240	−0.866	−0.866
270	−1.000	−1.000
300	−0.866	−0.866
315	−0.707	−0.707
330	−0.500	−0.500
360	0	0

27.2. Average emf

If, in Fig. 27.2, the average of the separate areas is taken, the sine curve, Fig. 27.3, and Table 27.2 result. Note that the dotted lines yield the average height of the area intervals.

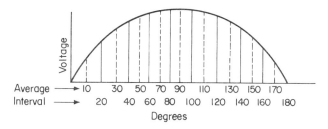

Fig. 27.3

If the arithmetic average of the sum of the nine readings in Table 27.2 is taken, the result is

$$E_{av} = 0.640 E_{max}$$

Sec. 27.2 *Average emf* 411

Table 27.2

Interval (degrees)	Midpoint (intervals 0)	Function (sin θ, base 1 volt)
0–20	10	0.174
20–40	30	0.500
40–60	50	0.766
60–80	70	0.940
80–100	90	1.000
100–120	110	0.940
120–140	130	0.766
140–160	150	0.500
160–180	170	0.174
	sum	5.760
	E-average	0.640

If smaller segments of the area under the sine curve are taken, the average emf approaches the true value of the average emf, or

$$E_{av} = \frac{2}{\pi} E_{max} = 0.637 E_{max}$$

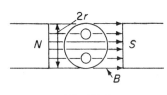

Fig. 27.4

Another method for determining this constant is to consider a rotating coil whose path diameter of rotation is equal to the width of a magnetic field, as shown in Fig. 27.4. In *one-half* revolution, the coil will cut flux ranging from zero lines cut (position of the coil as shown), to a maximum, to zero. The total flux cut will be

$$\Delta\phi = Bl2r$$

The time required for $\frac{1}{2}$ rev (πr) is

$$t = \frac{s}{v} = \frac{\pi r}{v}$$

But average emf is

$$E_{av} = \frac{\Delta\phi}{\Delta t} = \frac{Bl2r}{\pi r/v} = \frac{2Blv}{\pi}$$

From $Blv = E_{max}$ and $2/\pi = 0.637$, the average value is*

$$E_{av} = \frac{2E_{max}}{\pi} = 0.637 E_{max}$$

*The following relationship may be verified using the calculus.

27.3. Effective emf

Another value for current and voltage is the *effective value* related to the dc equivalent energy which an alternating current flow produces.

The direct current through a resistor is calculated using the power equation I^2R. If *one* ampere from an alternating current is passed through a resistor, it will produce the same heating effect; therefore, the *average* amperage is

$$I_1 = I_{max} \sin \theta_1, \quad I_2 = I_{max} \sin \theta_2, \quad \ldots$$

The power at any instant is

$$I_1^2 R, \quad I_2^2 R, \quad \ldots$$

The average power is

$$I_1^2 R + I_2^2 R + \cdots +$$

Factoring an R,

$$R(I_1^2 + I_2^2 + \cdots +)$$

The average power is $I_{eff}^2 R$; thus

$$I_{eff}^2 R = (I_1^2 + I_2^2 + \cdots +)R$$

Substituting for I_1, I_2, I_3, etc.,

$$I_{eff}^2 = I_{max}^2 \sin^2 \theta_1 + I_{max}^2 \sin^2 \theta_2 + \cdots +$$

Factoring an I_{max}^2 and taking the square root of both sides of the equation,

$$I_{eff} = I_{max} \sqrt{\sin^2 \theta_1 + \sin^2 \theta_2 + \cdots +}$$

Since the maximum value for $\theta = 90°$, then $\sin 90° = 1$. The average value is equal to $\frac{1}{2}$. Therefore,*

$$I_{eff} = I_{max} \sqrt{\tfrac{1}{2}} = 0.707 I_{max}$$

Example 2

An ac generator has 4 poles and generates a sine wave which shows a maximum value of 210 volts when it rotates at 1800 rpm. Calculate (1) the electrical degrees for $\frac{1}{180}$ sec; (2) the instantaneous emf at $\frac{1}{180}$ sec; (3) the effective emf; (4) the average emf.

---

*The following may be proved with the use of the calculus.

Solution

1. The frequency:

    2 poles × 30 rps = 60 cps

    $60 \times \frac{1}{180} = \frac{1}{3}$ cycle

    $\frac{1}{3} \times 360 = 120$ electrical degrees

2. The instantaneous emf:

    $E_{\text{inst}} = E_{\max} \sin \theta = 210 \times \sin 120°$
    $= 182$ volts

    $\left\{ \begin{array}{l} \text{rps} = 30 \\ \text{poles} = 4 \\ \text{voltage} = 210 \text{ volts} \end{array} \right\}$

3. The effective voltage:

    $E_{\text{eff}} = 0.707 E_{\max} = 0.707 \times 210$
    $= 149$ volts

4. The average voltage:

    $E_{\text{av}} = 0.636 E_{\max} = 0.636 \times 210$
    $= 134$ volts

27.4. Graphic Representation

When voltages or currents are to be added, the sine curves or their vector representations may be added. Two sine curves having the same frequencies may be in phase, or out of phase, and have their maximum or zero values occurring at the same or at different times. Thus in Fig. 27.5(a), we see two curves, current and voltage. Their vector representation shows the current vector acting along the x-axis and the voltage vector also acting along the x-axis and in the same direction. Since both curves have their zero points at the ordinate, reach a maximum at 90°, and return to zero, etc., at same time, they are said to be *in phase*.

In Fig. 27.5(b), the current wave reaches a maximum of 90° and the voltage curve reaches a maximum of 180°. The curves are said to be *out of phase* by $\theta = 90°$. In other words, it can be seen that the current wave is zero and is starting to "climb." Thus, if in the vector representation a *counterclockwise* rotation is assumed and if the I-vector is on the x-axis, as shown in Fig. 27.5(b), the E-vector is 90° behind the I-vector and the I-vector is considered to be *leading* the E-vector by 90°. This situation may be referred to as "E *lagging I*."

414 *Alternating Current* Chap. 27

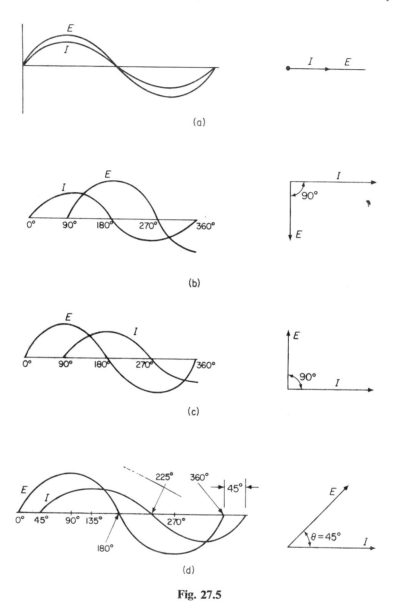

Fig. 27.5

Another possibility is that the voltage curve reaches a maximum of 90° before the current reaches a maximum (see Fig. 27.5c). Here, the voltage has reached a maximum before the current and the two peaks of both curves are displaced by 90°. In the vector representation, if the conventional counter-clockwise direction is considered, it can be seen that the voltage will always be

Sec. 27.6 *Pure Inductance* **415**

90° ahead of the current vector. Thus if E moves through 90°, I will be on the x-axis and I will be 90° behind E. The current is said to *lag* the voltage (or the student may want to consider this as a case where the voltage *leads* the current).

Figure 27.5(d) shows a curve and its corresponding vector diagram where E leads I by 45°. That is, the peak voltage is reached at 90° but the current curve reaches a maximum at $I = 135°$. This creates a current lag of $\theta = 45°$. The vector representation may also be seen in Fig. 27.5(d).

27.5. Pure Resistance

If an alternating current is sent through a resistor, as shown in Fig. 27.6(a), the current flow may be represented as shown in Fig. 27.6(b) and (c). The current flow will create an IR drop, which may be represented as a voltage-drop sine curve (Fig. 27.6b). At the same instant, a voltage E must be generated to overcome this IR drop. This voltage is shown as an equal and

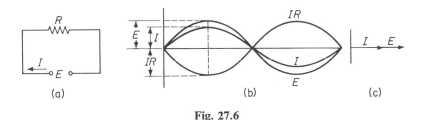

Fig. 27.6

opposite voltage curve E and is necessary in order to maintain the current flow. The voltage required to overcome this *pure resistance* to current flow is proportional to the current, because the resistance is constant. The current and voltage are said to be in phase.

27.6. Pure Inductance

Assume a circuit with a coil, having no resistance, attached to a battery as in Fig. 27.7(a). The current will generate an emf of self-inductance because of its rapidly changing values (ac). This emf of self-inductance will act in opposition to the current flow. The current curve I and the self-inductance curve are shown in Fig. 27.7(b). If there is no core in the coil and the coil has been assumed not to have any resistance, a voltage E must be impressed on the circuit to keep the current flowing.

In Fig. 27.7(b), note very carefully that, at 90°, the current curve is changing at a maximum rate and is increasing. Therefore, the inductance at 90° is a maximum and negative in opposition to this maximum change. At 180°, the current curve change is a minimum. Therefore, the inductance is at the zero point.

Figure 27.7(c) shows the vector representation of the current *lagging* the voltage by 90° for *pure inductance*.

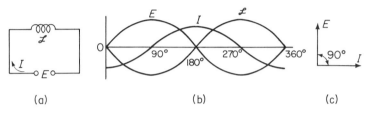

(a)   (b)   (c)

Fig. 27.7

The ratio of the induced voltage to the current is called the *inductive reactance*. The current increases from zero to maximum in one-quarter the time cycle, or period. If

$$T = \frac{1}{F}$$

$\{T =$ period, $F =$ frequency, $\mathscr{L} =$ inductance, $I_{max} =$ maximum current, $I_{av} =$ average current$\}$

the time when I is a maximum is

$$t = \frac{1}{4}T = \frac{1}{4F}$$

The equation for induced voltage is $E_{av} = \mathscr{L}_s(\Delta I/\Delta t)$. Making the substitutions,

$$E_{av} = \mathscr{L}\frac{I_{max}}{1/4F} = \mathscr{L}_s I_{max} 4F$$

If $I_{max} = (\pi/2) I_{av}$, then

$$E_{av} = \frac{\pi}{2} I_{av} \mathscr{L}_s 4F$$

$$= 2\pi F \mathscr{L}_s I_{av}$$

The expression $2\pi F \mathscr{L}_s$ is called the *inductive reactance*, and is designated by $X_\mathscr{L}$. The equation is

$\{X_\mathscr{L} =$ inductive reactance: ohms, $E =$ effective emf: volts, $I =$ effective current: amp, $\mathscr{L}_s =$ self-inductance: henrys, $F =$ frequency: cps, $Z =$ impedance: ohms$\}$

$$\frac{E_\mathscr{L}}{I_{eff}} = 2\pi F \mathscr{L}_s = X_\mathscr{L}$$

Sec. 27.7 Pure Capacitance 417

The ratio of effective voltage to effective current is also called *impedance* and is designed by the letter Z. Thus

$$X_{\mathscr{L}} = \frac{E}{I} = Z$$

Example 3

A circuit, negligible resistance, has a coil connected to a battery which produces 40 amp. If the coil has an inductance of 0.06 henry and a frequency of 35 cps, calculate (1) the inductive reactance; (2) the voltage required to force the 40 amp through the coil.

Solution

1. The inductive reactance is

$$X_{\mathscr{L}} = 2\pi F \mathscr{L}_s = 2\pi \times 35 \times 0.06 = 13.2 \text{ ohms} \qquad \left\{ \begin{array}{l} \mathscr{L}_s = 0.06 \text{ henry} \\ I = 40 \text{ amp} \\ F = 35 \text{ cps} \end{array} \right\}$$

2. The voltage is

$$E = IX_{\mathscr{L}} = 40 \times 13.2 = 528 \text{ volts}$$

27.7. Pure Capacitance

Figure 27.8(a) shows a capacitor connected to a voltage source with the resistance in the circuit negligible. At $0°$, the voltage is a minimum and is changing at the slowest rate. The current is zero. At the $90°$ part of the cycle, the voltage is changing at its fastest rate and the capacitor is charged to its maximum value. When the capacitor has discharged to its equilibrium state, the voltage is at its highest point and not changing. This is shown at $180°$. Thus when the current curve is at a maximum, the voltage curve is at zero or the current *leads* the voltage. If

$$E = \frac{q}{C} \quad \text{and} \quad q = It$$

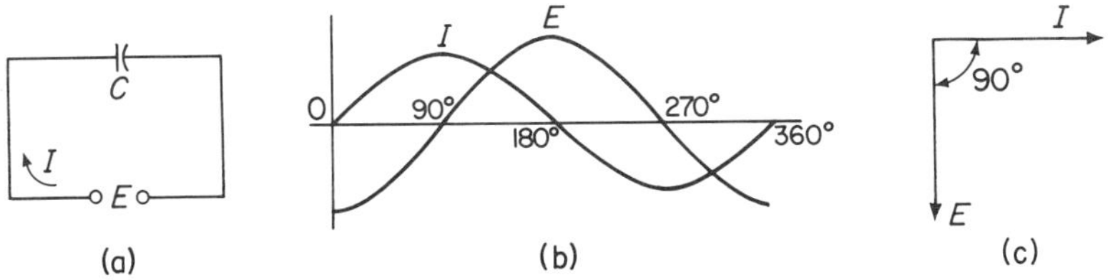

(a) (b) (c)

Fig. 27.8

then

$$EC = It$$

Again, if $t = 1/4F$, then

$$EC = I \times \frac{1}{4F}$$

$$\begin{cases} q = \text{charge} \\ I = \text{current} \\ t = \text{time} \\ E = \text{voltage} \\ F = \text{frequency} \end{cases}$$

Solving for E,

$$E = \frac{1}{4FC}.$$

But $E = E_{av} \times \pi/2$; therefore,

$$E_{av} = \frac{1}{2\pi FC} \times I_{eff}$$

This pure capacitance, or *capacitive reactance*, is designated by X_C and the impedance by Z:

$$X_C = \frac{E}{I} = \frac{1}{2\pi FC} = Z$$

Example 4

A 8-μf capacitor is connected to a 215-volt source at a frequency of 60 cps. Calculate (1) the capacitive reactance; (2) the current in the circuit.

Solution

1. The capacitive reactance is

$$X_C = \frac{1}{2\pi FC} = \frac{1}{2\pi 60 \times (8 \times 10^{-6})} = 332 \text{ ohms}$$

$$\begin{cases} C = 8\ \mu f \\ = 8 \times 10^{-6}\ \text{farad} \\ F = 60\ \text{cps} \\ E = 220\ \text{volts} \end{cases}$$

2. The current in the circuit is

$$I = \frac{E}{X_C} = \frac{220}{332} = 0.66 \text{ amp}$$

27.8. Resistance and Inductance

If a circuit has both resistance and inductance (Fig. 27.9a), the vector summation of the blocking effect may be represented as vectors and added vectorially. It has been shown that the inductance is

$$X_{\mathscr{L}} = \frac{E}{I} \quad \text{or} \quad E_{\mathscr{L}} = IX_{\mathscr{L}}$$

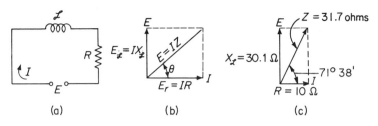

Fig. 27.9

The blocking effect due to the resistance is

$$E_r = IR$$

Figure 27.9(b) is the vector representation of the voltage *leading* the current. These vectors may be added, using the Pythagorean equation, or

$$E^2 = E_{\mathscr{L}}^2 + E_R^2$$

Substituting,

$$(IZ)^2 = (IR)^2 + (IX_{\mathscr{L}})^2$$

If the current I is factored, the equation for impedance becomes

$$Z = \sqrt{R^2 + X_{\mathscr{L}}^2}$$

Since the inductive reactance is $X_{\mathscr{L}} = 2\pi F \mathscr{L}$, the impedance becomes

$$Z = \sqrt{R^2 + (2\pi F \mathscr{L})^2}$$

Multiplying both sides by I,

$$E = IZ = I\sqrt{R^2 + (2\pi F \mathscr{L})^2}$$

or

$$I_{\text{eff}} = \frac{E}{Z} = \frac{E}{\sqrt{R^2 + (2\pi F \mathscr{L})^2}}$$

The phase angle becomes (from 27.9b)

$$\tan \theta = \frac{X_{\mathscr{L}}}{R} = \frac{2\pi F \mathscr{L}}{R}$$

or the phase angle can also be found from

$$\cos \theta = \frac{R}{Z}$$

the current *lags* the emf by the angle θ.

Example 5

A circuit consists of a 40-ohm resistor and a coil whose value is 0.08 henry, and a 60-cps 220-volt source. Calculate (1) the inductance reactance; (2) the impedance; (3) the current in the coil; (4) the voltage across each component; (5) the phase angle. (6) Draw the vector diagram.

Solution

1. The inductive reactance is

$$X_{\mathscr{L}} = 2\pi F \mathscr{L} = 2\pi 60 \times 0.08 = 30.1 \text{ ohms}$$

2. The impedance is

$$Z = \sqrt{R^2 + X_{\mathscr{L}}^2} = \sqrt{(10^2) + (30.1)^2}$$
$$= 31.7 \text{ ohms}$$

3. The current in the coil is

$$I = \frac{E}{R} = \frac{220}{10} = 22 \text{ amp}$$

$$\left\{\begin{array}{l} F = 60 \text{ cps} \\ \mathscr{L} = 0.08 \text{ henry} \\ R = 10 \text{ ohms} \\ E = 220 \text{ volts} \end{array}\right\}$$

4. The voltage across each component is

$$E_{\mathscr{L}} = X_{\mathscr{L}}I = 30.1 \times 22 = 662 \text{ volts}$$
$$E_R = RI = 10 \times 22 = 220 \text{ volts}$$

5. The phase angle is

$$\cos \theta = \frac{R}{Z} = \frac{10}{31.7} = 0.315$$
$$\theta = 71°38'$$

6. The phase diagram is as shown in Fig. 27.9(c).

27.9. Resistance and Capacitance

If the circuit has resistance and capacitance, the vector summation of the blocking effect of the resistance and the lag of the voltage to the current may be represented as shown in Fig. 27.10(b).

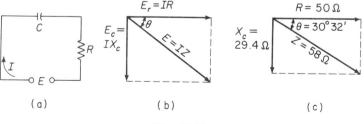

Fig. 27.10

Sec. 27.9   *Resistance and Capacitance*   **421**

The capacitive reactance is

$$X_C = \frac{E_C}{I}$$

In Fig. 27.10(b), the vector summation is

$$E^2 = E_R^2 + E_C^2$$

Substituting,

$$(IZ)^2 = (IR)^2 + (IX_C)^2$$

The impedance becomes

$$Z = \sqrt{R^2 + X_C^2}$$

If the capacitance reactance is $X_C = \frac{1}{2\pi FC}$, the impedance is

$$Z = \sqrt{R^2 + \left(\frac{1}{2\pi FC}\right)^2}$$

Since $E = IZ$

$$E = I\sqrt{R^2 + \left(\frac{1}{2\pi FC}\right)^2}$$

or

$$I = \frac{E}{Z} = \frac{E}{\sqrt{R^2 + \left(\frac{1}{2\pi FC}\right)^2}}$$

The phase angle θ may be calculated from Fig. 27.10(b), or

$$\tan \theta = \frac{X_C}{R} = \frac{\left(\frac{1}{2\pi FC}\right)}{R}$$

or from

$$\cos \theta = \frac{R}{Z}$$

The current *leads* the voltage by the phase angle θ.

Example 6

A circuit contains a 90-μf capacitor and a 50-ohm resistor which are connected to a 120-volt 60-cps line. Calculate (1) the capacitive reactance; (2) the impedance; (3) the current in the capacity; (4) the phase angle. (5) Draw the vector diagram.

Solution

1. Capacitive reactance:

$$X_C = \frac{1}{2\pi FC} = \frac{1}{2\pi 60 \times (90 \times 10^{-6})}$$
$$= 29.4 \text{ ohms}$$

2. Impedance:

$$Z = \sqrt{R^2 + X_C^2} = \sqrt{(50)^2 + (29.4)^2}$$
$$= 58 \text{ ohms}$$

$\begin{cases} R = 50 \text{ ohms} \\ C = 90 \times 10^{-6} \text{ farad} \\ E = 120 \text{ volts} \\ F = 60 \text{ cps} \end{cases}$

3. Current:

$$I = \frac{E}{Z} = \frac{120}{58} = 2.07 \text{ amp}$$

4. Phase angle:

$$\cos\theta = \frac{R}{Z} = \frac{50}{58} = 0.862$$
$$\theta = 30°32'$$

5. The current *leads* the voltage (Fig. 27.10c).

27.10. Resistance, Inductance, and Capacitance

Figure 27.11(a) shows an ac circuit which contains a capacitor, a coil, and a resistor. The vector diagram is shown in Fig. 27.11(b) and the resultant diagram of the vector addition is shown in Fig. 27.11(c). From Fig. 27.11(c),

$$(IZ)^2 = (IR)^2 + (IX_{\mathscr{L}} - IX_C)^2$$

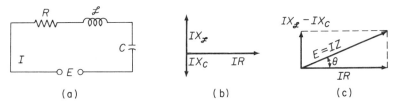

Fig. 27.11

Solving for impedance,

$$Z = \sqrt{R^2 + (X_{\mathscr{L}} - X_C)^2}$$

Sec. 27.10 Resistance, Inductance, and Capacitance

If $X_\mathscr{L} = 2\pi F \mathscr{L}$ and $X_C = \dfrac{1}{2\pi FC}$, the impedance becomes

$$Z = \sqrt{R^2 + \left[2\pi F \mathscr{L} - \left(\dfrac{1}{2\pi FC}\right)\right]^2}$$

The voltage is

$$E = IZ = I\sqrt{R^2 + \left[2\pi F \mathscr{L} - \left(\dfrac{1}{2\pi FC}\right)\right]^2}$$

and the current is

$$I = \dfrac{E}{\sqrt{R^2 + \left[2\pi F \mathscr{L} - \left(\dfrac{1}{2\pi FC}\right)\right]^2}}$$

Note that if $X_\mathscr{L} - X_C$ is positive, the resultant y-vector *leads* the current vector. If the resultant y-vector is negative, it will *lag* the current vector. The phase angle in Fig. 27.11(c) is

$$\tan \theta = \dfrac{X_\mathscr{L} - X_C}{R}$$

or

$$\cos \theta = \dfrac{R}{Z}$$

Example 7

Calculate (1) the inductive reactance; (2) the capacitive reactance; (3) the impedance; (4) the current; (5) the phase angle in Fig. 27.12(a). (6) Draw the phase diagram.

Solution

1. To calculate the inductive resistance, the total induction is

$$\mathscr{L} = \mathscr{L}_1 + \mathscr{L}_2 = 0.6 + 0.8 = 1.4 \text{ henrys}$$

and the inductive reactance is

$$X_\mathscr{L} = 2\pi F \mathscr{L} = 2\pi 60 \times 1.4 = 528 \text{ ohms}$$

2. The capacitive reactance is calculated from the total capacitance, which is

$$\dfrac{1}{C} = \dfrac{1}{C_1} + \dfrac{1}{C_2} = \dfrac{1}{20 \times 10^{-6}} + \dfrac{1}{30 \times 10^{-6}}$$

$$C = 12 \times 10^{-6} \text{ farad}$$

and the capacitive reactance is

$$X_C = \dfrac{1}{2\pi FC} = \dfrac{1}{2\pi 60 \times 12 \times 10^{-6}} = 220 \text{ ohms}$$

3. To calculate the impedance, the sum of the resistance is calculated first:

$$R = 50 + 8 + 4 = 62 \text{ ohms}$$

The impedance is

$$Z = \sqrt{R^2 + (X_\mathscr{L} - X_C)^2} = \sqrt{(62)^2 + (528 - 220)^2}$$
$$= 314 \text{ ohms}$$

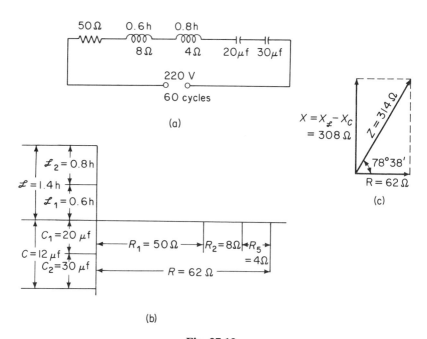

Fig. 27.12

4. The current is

$$I = \frac{E}{Z} = \frac{220}{314} = 0.701 \text{ amp}$$

5. The phase angle is

$$\cos\theta = \frac{R}{Z} = \frac{62}{314} = 0.197$$
$$\theta = 78°38'$$

6. The voltage *leads* the current (Fig. 27.12b and c).

27.11. The Power in an ac Circuit

If two curves are plotted, one for current and the other for voltage, as in Fig. 27.13(a), they may be combined into a power curve. In the case of a pure resistance circuit,

$$P = I^2R = EI$$

The power curve remains above the x-axis because E and I are either both $+$ or both $-$ and, therefore, the power P is always $+$. The pure resistance circuit dissipates energy in the form of heat which is never recovered.

Suppose the current either leads or lags the voltage by 90°, as is the case in capacitive–inductive reactance. Figure 27.13(b) shows the plot of the voltage, amperage, and power in an inductive circuit. Notice that in a pure inductive circuit, the power area under the curve, which indicates power removed from the circuit, is put back into the circuit, making the average power zero. The same is true of a pure capacitive-reactance circuit.

As soon as the circuit is not a pure circuit, the major portion of the resultant power curve will be above the x-axis, as shown in Fig. 27.13(c).

The average-power equation is the product of the effective current and the effective voltage, or

$$P = EI \cos \theta$$

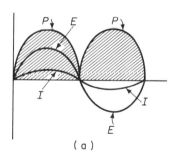
(a)

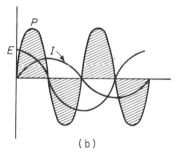
(b)

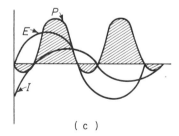
(c)

Fig. 27.13

426 Alternating Current Chap. 27

The power factor is cos θ and may be any value between zero and 1. The value of cos θ may be found from

$$\cos \theta = \frac{R}{Z}$$

Example 8

Find (1) the power factor in Example 7; (2) the power consumed by the circuit. (3) Show that this power consumption is all in the resistance.

Solution

1. Power factor:

$$\cos \theta = \frac{R}{Z} = \frac{62}{314} = 0.197$$

2. Power consumed:

$$P = EI \cos \theta = 220 \times 0.701 \times 0.197$$
$$= 30.38 \text{ watts}$$

3. Power from the resistance equation:

$$P = I^2 R = 0.701^2 \times 62$$
$$= 30.38 \text{ watts} \quad (check)$$

27.12. Resonance

In a series circuit containing resistance, inductance, and capacitance, the general equation is

$$Z = \sqrt{R^2 + (X_\mathscr{L} - X_C)^2} = \sqrt{R^2 + \left[2\pi F \mathscr{L} - \left(\frac{1}{2\pi FC}\right)\right]^2}$$

If the voltage drop, or reactance, of one balances the other, there is a point where $X_\mathscr{L} = X_C$ and $Z = R$. This is called *resonance*. Thus if

$$X_\mathscr{L} = X_C$$

then

$$2\pi F \mathscr{L} = \frac{1}{2\pi FC} \quad \left\{ \begin{array}{l} F_r = \text{resonant frequency} \\ \mathscr{L} = \text{inductance} \\ C = \text{capacitance} \end{array} \right\}$$

Solving for F_r,

$$F_r = \frac{1}{2\pi\sqrt{\mathscr{L}C}}$$

Example 9

A circuit has an inductance of 0.5 henry and a capacitance of 30 µf. If the circuit is a resonant circuit, what is the frequency?

Solution

$$F_r = \frac{1}{2\pi\sqrt{\mathscr{L}C}} = \frac{1}{2\pi\sqrt{0.5(32\times 10^{-6})}} \qquad \left\{ \begin{array}{l} \mathscr{L} = 0.5 \text{ henry} \\ C = 32\times 10^{-6} \text{ farad} \end{array} \right\}$$

$$= 40 \text{ cps}$$

Problems

1. Calculate the number of poles in a generator if it operates at 1725 rpm and 60 cps.
2. An ac 440-volt line at 60 cps carries 80 amp. Calculate (a) the angular velocity; (b) the maximum emf; (c) the maximum current; (d) the instantaneous voltage 0.004 sec after the zero point is reached.
3. Assume an ac generator has 10 poles and that the effective voltage is 120 volts when it rotates at 4200 rpm. (a) How many electrical degrees has the generator passed in 0.003 sec? Calculate (b) the maximum voltage; (c) the instantaneous voltage; (d) the average voltage.
4. A coil has an inductance of 0.006 henry. It is inserted into a 120-volt, 60-cps ac circuit. Calculate (a) the inductive reactance in the coil; (b) the impedance; (c) the current in the coil. (Assume no resistance.)
5. A coil whose inductance is 0.02 henry is connected to a battery which produces 30 amp. If the resistance is negligible: (a) What is the inductive reactance? (b) What is the voltage required to force the current through the coil?
6. An 8-µf capacitor is inserted into a 120-volt 60-cps ac circuit. Calculate (a) the capacitive reactance; (b) the impedance; (c) the current. (Assume no resistance.)
7. A capacitor has a voltage of 220 volts applied at a frequency of 60 cps. If the capacitance is 6 µf, find (a) the capacitive reactance; (b) the current in the circuit.
8. A resistance of 30 ohms is inserted into the circuit in Prob. 4. Calculate (a) the inductive reactance; (b) the impedance; (c) the current; (d) the angle made by the resultant with the x-axis. (e) Does the current lag or lead the voltage? Draw the diagram.
9. A coil has an inductance of 0.05 henry and a resistance of 10 ohms. The voltage at 60 cps is 110 volts ac. Find (a) the inductance reactance; (b) the impedance; (c) the current in the coil; (d) the voltage across each component; (e) the phase angle. (f) Draw the vector diagram.
10. If a 30-ohm resistance is inserted into the circuit in Prob. 6, calculate (a) the

capacitive reactance; (b) the impedance; (c) the current; (d) the angle made by the resultant with the x-axis. (e) Does the current lag or lead the voltage? Draw diagrams.

11. A capacitor and a resistor are connected to a 100-volt 50-cps ac line. The capacitor has a capacitance of 40 μf and the resistance is 20 ohms in the resistor. Find (a) the capacitive reactance; (b) the impedance; (c) the current in the capacitor; (d) the phase angle. (e) Draw the diagram.

12. A coil has an inductance 0.8 henry and a resistance of 50 ohms. The coil is inserted in series with a 60-μf capacitor. The voltage in the line is 220 volts at 50 cps. Calculate (a) the inductive reactance; (b) the capacitive reactance; (c) the impedance; (d) the current; (e) the phase angle. Draw the diagram.

13. A coil having a resistance of 40 ohms and an inductance of 0.5 henry is connected in series to a capacitor of 30 μf. The voltage in the line is 110 volts at 50 cps. Find (a) the inductive reactance; (b) the capacitive reactance; (c) the impedance; (d) the current; (e) the phase angle.

14. Find (a) the impedance; (b) the current; (c) the phase angle in Fig. 27.14. Draw the phase diagram.

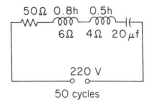

Fig. 27.14

15. In Prob. 8, calculate (a) the power factor; (b) the power consumed by the circuit.
16. In Prob. 9, calculate the (a) power factor; (b) the power consumed by the circuit.
17. In Prob. 12, calculate (a) the power factor; (b) the power consumed by the circuit.
18. In Prob. 13, calculate (a) the power factor; (b) the power consumed by the circuit.
19. In Prob. 14, calculate (a) the power factor; (b) the power consumed by the circuit.
20. Calculate the resonant frequency of an ac circuit if the inductance is 0.8 henry and the capacitance is 50 μf.
21. If the frequency is 56 cps and the inductance is 0.4 henry, calculate the capacitance in a resonant circuit.

28

Wave Motion and Sound

28.1. Transverse and Longitudinal Waves

With respect to energy propagation as a result of particle vibration, it is the moving particle which generates kinetic energy. The particle itself oscillates and transmits this energy from one particle to another. Thus it is the energy which propagates and not the particle.

Consider a rope attached at one end to a wall. If the loose end is snapped, a wave is generated in the rope toward the wall. The particles oscillate up and down, yet the energy wave propagates toward the wall. This type of particle vibration, where the particle vibrates at right angles to the wave propagation, is called *transverse* and the wave is called a *transverse wave*.

The energy may also be propagated because of particle oscillation in the same direction, or parallel to the propagation direction of the wave. This type of motion is called *longitudinal* or *compressional* and the wave is called a *longitudinal* or *compressional wave*.

Waves which are a combination of both are also possible. Figure 28.1 shows a vibrating string. The particles of the string oscillate up and down in their own path, each particle pulling its neighbor up and down, as the case may be. This creates a transverse sine wave which, in Fig. 28.1, propagates from right to left, shown by the cross-hatched circle.

Thus in Fig. 28.2, at the very instant the rope is snapped, the particle next

430 Wave Motion and Sound Chap. 28

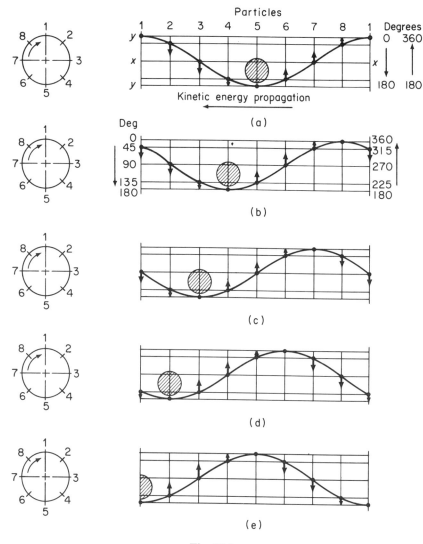

Fig. 28.1

to the hand doing the snapping moves first, and this causes the next particle to move, etc. In Fig. 28.2(a), particle a has moved to the position shown and is ready to move particle b in an upward direction. In Fig. 28.2(b), a short time interval later, particle a is on the way down; it has caused b to move upward. Particle c remains unaffected to this point.

Notice that particles must be close together to have an effect on one

Sec. 28.1 Transverse and Longitudinal Waves 431

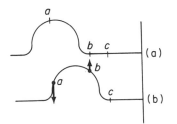

Fig. 28.2

another. Thus there can be no transverse transmission of energy in a gas in this sense. It is also important to recognize that in a transverse wave the particles never leave their horizontal positions, yet the energy is propagated at right angles to the particle motion.

A compressional or longitudinal wave propagates kinetic energy along the same axis and in the same direction in which it vibrates. Figure 28.3(c) shows the neutral position of the particles. Figure 28.3(b) shows the position of the vibrating particles at a particular instant of time, and Fig. 28.3(d) shows the vibrating particles at a later time interval. The condensed region can be seen to have moved from right to left.

These regions of *condensation* and *rarefaction* may be started by a vibrating diaphragm. The diaphragm will cause the particles immediately in front of it to compress and the particles immediately in back of it to separate. This

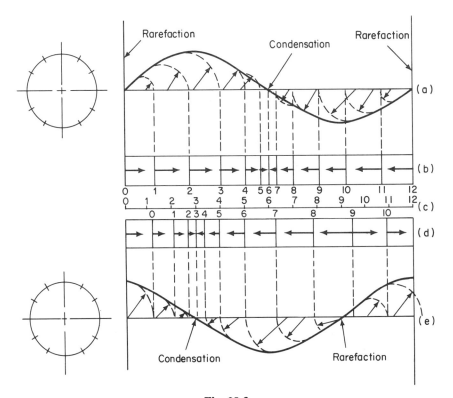

Fig. 28.3

432 Wave Motion and Sound Chap. 28

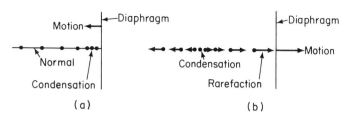

Fig. 28.4

creates a crowded condition in front of the diaphragm with the particles being compressed opposing the motion of the diaphragm.

But once these particles have been set into motion, their momentum carries them forward to create a region of condensation (see Fig. 28.4a). As the diaphragm is pulled to the right (Fig. 28.4b), a region of rarefaction is created because the diaphragm is now moving to the right while the particles are moving to the left. This creates an evacuated region immediately to the left of the diaphragm, causing the particles to reverse their direction in an attempt to fill this vacuum. But these particles, even though they reverse their direction, have already started the same process moving from particle to particle toward the left. As a result, even though the energy is moving to the left, the particles are being first pushed to the left and then drawn back

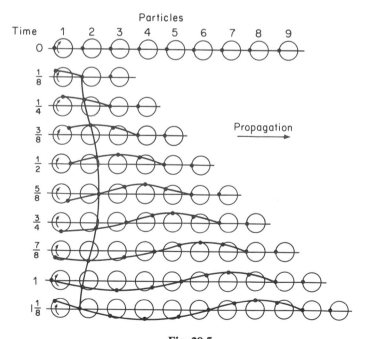

Fig. 28.5

Sec. 28.2 *Particle Displacement: Velocity* 433

into the evacuated region, only to run head-on into another high-energy region.

Water waves are a combination of longitudinal and transverse waves (see Fig. 28.5). The transverse characteristics of a water wave will cause a wooden block to move up and down at the same time that it is moving to the right and to the left. Thus, if the wave is propagated to the right, the energy is propagated to the right, and the block will possess transverse motion perpendicular to the direction of energy propagation and longitudinal motion parallel to the direction of energy propagation. Each row in Fig. 28.5 represents a time interval of $\frac{1}{8}t$.

Recall that, in *purely transverse motion*, a particle traces a straight-line oscillating path perpendicular to the direction of propagation of energy. In *purely longitudinal motion*, a particle traces a straight-line oscillating path parallel to the direction of propagation of motion. In Fig. 28.5, each particle (the path of particle 2 is shown) traces an elliptical path.

28.2. Particle Displacement: Velocity

Figure 28.6 shows three particles of a wave, the wave being propagated to the right. Particles 1 and 3 are said to be *in phase*. They are oscillating in the same direction and represent 360° of separation on a reference circle. Particle 2 is out of phase with particles 1 and 3; it is moving in the opposite direction from them.

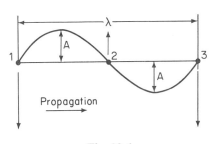

Fig. 28.6

The *wavelength* is defined as the shortest distance between two particles vibrating in phase. The *amplitude* of a wave is the maximum displacement of any particle from its neutral position. It is one-half the distance from the top of the crest to the bottom of the trough. In Fig. 28.6, it is designated as A.

The *period* is the *time* required for a particle to complete one cycle from neutral position to the top of the crest, to the bottom of the trough, and back to the neutral position. Note that during this one cycle, the wave has moved forward one wavelength. The number of cycles completed in 1 second, the reciprocal of the period, is called the *frequency*:

$$F = \frac{1}{P} \quad \left\{ \begin{array}{l} F = \text{frequency: vib/sec} \\ P = \text{period: sec/vib} \end{array} \right\}$$

434 *Wave Motion and Sound* Chap. 28

Thus P represents the time it takes one wave to complete its vibration and F represents the number of waves that pass a fixed point in 1 sec. If we know how long it will take one wave (λ) to pass a given point, the velocity v may be written as

$$v = \frac{\lambda}{P}$$

Substituting for $1/P = F$, the velocity of a wave becomes $\quad \left\{ \begin{array}{l} v = \text{velocity: ft/sec } or \text{ cm/sec} \\ \lambda = \text{wavelength: ft } or \text{ cm} \end{array} \right\}$

$$v = F\lambda$$

Some values for velocity of sound in various materials are given in Appendix Table A.18.

Example 1

The velocity of sound in water is taken to be 4800 ft/sec. If the period of a sound wave in water is 0.004 sec, calculate (1) the wavelength; (2) the frequency of vibration.

Solution

1. The wavelength is

$$\lambda = vP = 4800 \times 0.004 = 19.2 \text{ ft}$$

2. The frequency is $\quad \left\{ \begin{array}{l} v = 4800 \text{ ft/sec} \\ P = 0.004 \text{ sec/vib} \end{array} \right\}$

$$F = \frac{v}{\lambda} = \frac{4800}{19.2} = 250 \text{ vib/sec}$$

The velocity of a *transverse* wave depends upon the elasticity of the carrier medium and the inertia characteristics of the medium. The equation for velocity of a transverse wave in a wire or string is

$$v = \sqrt{\frac{f}{m/l}} \quad \left\{ \begin{array}{ll} \qquad\quad British & \quad Metric \\ f \text{ (tension)} = \text{lb} & \text{load} \times 980 \\ m \text{ (mass)} = \frac{w}{g} = \text{slugs} & \quad\text{g} \\ l \text{ (length)} = \text{ft} & \quad\text{cm} \\ v \text{ (velocity)} = \text{ft/sec} & \quad\text{cm/sec} \end{array} \right\}$$

Example 2

A copper wire is 4 ft long and weighs 2 oz. If the tension in the wire is 12 lb, calculate the velocity of a wave in the vibrating wire.

Sec. 28.2   Particle Displacement: Velocity   **435**

Solution

The velocity is

$$v = \sqrt{\frac{f}{m/l}} = \sqrt{\frac{12 \times 32 \times 4}{\frac{1}{8}}} \qquad \left\{\begin{array}{l} f = 12 \text{ lb} \\ l = 4 \text{ ft} \\ m = 2 \text{ oz} = \frac{1}{8}/32 \text{ slug} \end{array}\right\}$$

$$= 111 \text{ ft/sec}$$

Example 3

A string, mass 0.8 g and 50 cm long, is stretched by a load of 2000 g. Calculate the velocity of a transverse wave in the string.

Solution

The velocity is

$$v = \sqrt{\frac{f}{m/l}} = \sqrt{\frac{2000 \times 980}{0.8/50}} \qquad \left\{\begin{array}{l} f = (2000 \times 980) \text{ dynes} \\ l = 50 \text{ cm} \\ m = 0.8 \text{ g} \end{array}\right\}$$

$$= 1.11 \times 10^4 \text{ cm/sec}$$

It has been shown that, for a longitudinal wave, when a force is applied to a material, the material is distorted and a force opposing this distortion is created in an attempt to restore the status quo. This restoring force is a function of the elasticity of the material. The velocity depends upon the elasticity and the density of the material in which the wave is being propagated.

The general equation for the velocity of a *longitudinal wave* is

$$v = \sqrt{\frac{E}{d}} \qquad \left\{\begin{array}{lll} & \textit{British} & \textit{Metric} \\ (v) \text{ velocity} & = \text{ft/sec} & \text{cm/sec} \\ (E) \text{ elasticity} & = \text{lb/ft}^2 & \text{dynes/cm}^2 \\ (d) \text{ density} & = \text{lb/ft}^3 & \text{g/cm}^3 \\ (Y) \text{ Young's mod} \\ (B) \text{ bulk mod} & = \text{dynes/cm}^2 & \text{psi} \times 144 \\ (P) \text{ abs pressure} & = \text{dynes/cm}^2 & \text{psfa} \\ & \gamma = \frac{c_p}{c_v} \end{array}\right\}$$

where $E = Y$ *for solids*
         $E = B$ *for liquids*
         $E = \gamma P$ *for gases*

Example 4

The specific weight of brass is 535 lb/ft³. Young's modulus is 16×10^6 psi. Calculate the velocity of a longitudinal wave in a brass rod.

Solution

The velocity of a compressional wave is

$$v = \sqrt{\frac{Y}{d}} = \sqrt{\frac{(16 \times 10^6) \times 144}{535/32}} \qquad \left\{\begin{array}{l} Y = \text{Young's mod} \\ \quad = (16 \times 10^6 \text{ psi} \times 144) \text{ psf} \\ d = \text{density} \\ \quad = 535/32 \text{ slugs} \end{array}\right\}$$

$$= 1.18 \times 10^4 \text{ ft/sec}$$

Example 5

Calculate the velocity of a longitudinal wave in mercury.

Solution

The velocity is

$$v = \sqrt{\frac{B}{d}} = \sqrt{\frac{(3.8 \times 10^6) \times 144}{845/32}} \qquad \left\{ \begin{array}{l} B = \text{bulk mod} \\ = (3.8 \times 10^6 \text{ psi} \times 144) \text{ psf} \\ d = 845 \text{ lb/ft}^3 \end{array} \right\}$$
$$= 4.6 \times 10^3 \text{ ft/sec}$$

Example 6

Calculate the velocity of a longitudinal wave in air at STP (standard temperature and pressure) in the British system of units if the specific weight of air is 0.0806 lb/ft^3, and if c_p(constant pressure) = 0.2375 and c_v(constant volume) = 0.1689.

Solution

The ratio of specific heat at constant pressure (c_p) to specific heat at constant volume (c_v) is

$$\gamma = \frac{c_p}{c_v} = \frac{0.2375}{0.1689} = 1.406$$

The velocity of the longitudinal wave in air is

$$v = \sqrt{\frac{\gamma P}{d}} = \sqrt{\frac{1.406 \times 14.7 \times 144}{0.0806/32}} \qquad \left\{ \begin{array}{l} P = 14.7 \times 144 \text{ lb/ft}^2 \\ W = 0.0806 \text{ lb/ft}^3 \\ c_p = 0.2375 \\ c_v = 0.1689 \end{array} \right\}$$
$$= 1085 \text{ ft/sec}$$

The ratio of velocities to the ratio of the absolute temperatures by

$$\frac{v_1}{v_2} = \sqrt{\frac{T_1}{T_2}}$$

Example 7

Calculate the velocity of sound at 70°F if the velocity at 32°F is 1088 ft/sec.

Solution

The ratio of the velocities is

$$\frac{v_1}{v_2} = \sqrt{\frac{T_1}{T_2}}$$

Sec. 28.3 *Stationary Waves* 437

Substituting,

$$\frac{1088}{v_2} = \sqrt{\frac{492}{530}} \qquad \left\{ \begin{array}{l} T_1 = 32 + 460 = 492°R \\ T_2 = 70 + 460 = 530°R \\ v_1 = 1088 \text{ ft/sec} \end{array} \right\}$$

Solving for v_2,

$$v_2 = 1129 \text{ ft/sec}$$

28.3. Stationary Waves

When waves with the same frequency and amplitude, but going in the *opposite* directions, are superimposed one upon the other, the point of *destructive interference* is called a *node*. The wave at that point has a zero amplitude. The point at which one wave reinforces the other is called the *antinode*. At this point, the amplitude is a maximum and the *interference* is *constructive*. A vibrating string, one end fastened, will yield nodes and antinodes as shown in Fig. 28.7.

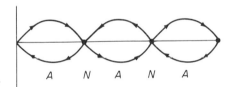

Fig. 28.7

The string (Fig. 28.8a) vibrates up and down as the disturbance moves toward the left, and strikes the wall; then, not being able to move the wall, the reaction force of the wall forces the first antinode next to the wall to whip up, as shown in Fig. 28.8(b), and start back toward the right. As a result, the incoming wave is a trough and the reflected wave starts out as a crest. Thus the reflected wave at the fixed end is 180° out of phase with the incoming wave.

If the disturbances were to continue as a series of incoming connected waves, there would be complete interference (nodes) between the incoming and reflected waves at every half-wavelength from the wall and complete reinforcement (antinodes) at 90° and 270°, etc., from the wall. Both ends of the string must be nodes.

If the wall end of the string were looped loosely about a smooth rod, Fig. 28.8(c), this loop would start to move down the rod as the disturbance approached the rod (see Fig. 28.8d). Since there is no wall to create a reaction force, the loop moves down the rod until the disturbance itself stops the movement of the loop. But the downward inertia of the loop has started the

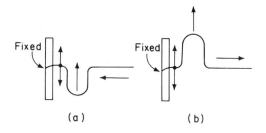

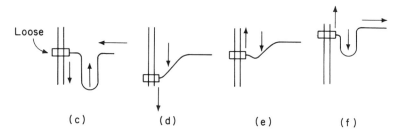

Fig. 28.8

rope moving down again, at which time the loop starts back up the rod [Fig. 28.8(e)]. In so doing, it starts a disturbance back along the string as in Fig. 28.8(f). The incoming wave is a trough and the reflected wave starts out as a trough. There is no phase change involved and the loop end of the rope whips back and forth along the rod like the end of a snake's tail.

28.4. Vibrating String

Figure 28.9(a) shows a string supported at both ends in such a manner that the distance between the supports is equal to one-half wavelength. Thus

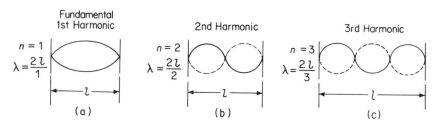

Fig. 28.9

Sec. 28.4   Vibrating String   439

$$\frac{\lambda}{2} = l \quad \text{or} \quad \lambda = 2l$$

This type of vibration is called the *fundamental* or *first harmonic* and generates one *loop* with nodes at each end of the string.

When the distance between the supports (Fig. 28.9b) is equal to *one* full wavelength, the string is said to vibrate according to its *second harmonic*, or *first overtone*, and

$$\frac{2\lambda}{2} = l \quad \text{or} \quad \lambda = l$$

If the string vibrates according to its third harmonic (or second overtone), as shown in Fig. 28.9(c), then the string generates three full loops:

$$\frac{3\lambda}{2} = l \quad \text{or} \quad \lambda = \frac{2}{3}l$$

Thus if the order of the harmonic, n, is related to the number of loops, the general equation may be written as

$$\frac{n\lambda}{2} = l$$

Since $v = F\lambda$, the equation for frequency may be written as

$$F = \frac{v}{\lambda} = \frac{v}{2l/n} = \frac{nv}{2l}$$

Also,

$$v = \sqrt{\frac{f}{m/l}}$$

$\begin{cases} n = \text{order of harmonic} \\ l = \text{length} \\ \lambda = \text{wavelength} \\ F = \text{frequency} \\ v = \text{velocity} \\ f = \text{force} \end{cases}$

The equation for frequency in terms of tension and mass per unit length becomes

$$F = \frac{n}{2l}\sqrt{\frac{f}{m/l}}$$

Note very carefully that n is the number corresponding to the number of loops generated for all nodes of vibration in this chapter.

Example 8

A steel wire weighs 0.24 oz and is 6 ft long. If it is stretched by a 12-lb force, calculate (1) the frequency of the fundamental; (2) the wavelength of the first overtone; (3) the frequency of the second harmonic.

Solution

1. Both ends of the wire vibrate as nodes, thus creating one loop for the fundamental, or $n = 1$:

$$F_1 = \frac{n}{2}\sqrt{\frac{f}{m/l}} = \frac{1}{2 \times 6}\sqrt{\frac{12 \times 6}{4.66 \times 10^{-4}}}$$
$$= 33 \text{ vib/sec}$$

2. The wavelength of the first overtone, or second harmonic, is

$$\lambda = \frac{2l}{n} = \frac{2}{2} \times 6 = 6 \text{ ft}$$

$$\left\{ \begin{array}{l} n = 1 \\ f = 12 \text{ lb} \\ w = 0.24/16 = 0.015 \text{ lb} \\ m = 0.0015/32 \\ = 4.66 \times 10^{-4} \text{ slug} \\ n = 2 \\ l = 6 \text{ ft} \end{array} \right\}$$

3. The frequency is

$$F_2 = n \times \text{ frequency of the fundamental}$$
$$= 2 \times 33 = 66 \text{ vib/sec}$$

Example 9

A vibrator causes an 0.8-g string to vibrate at 140 vib/sec. If the string is 150 cm long, calculate the tension which will cause the string to vibrate according to (1) its fundamental; (2) its fourth overtone.

Solution

1. The tension when vibrations according to the fundamental: occur

$$F = \frac{2l}{n}\sqrt{\frac{f}{m/l}}$$

$$140 = \frac{1}{2 \times 150}\sqrt{\frac{f \times 980}{0.8/150}} \qquad \left\{ \begin{array}{l} m = 0.8 \text{ g} \\ F = 140 \text{ vib/sec} \\ l = 150 \text{ cm} \\ n = 1 \end{array} \right\}$$

$$f = 9.6 \times 10^3 \text{ g}$$

2. The tension when vibrations occur according to the 4th overtone ($n = 5$):

$$140 = \frac{5}{2 \times 150}\sqrt{\frac{f \times 980}{0.8/150}} \qquad \{n = 5\}$$

$$f = 384 \text{ g}$$

28.5. The Vibrating Rod

A rod may be caused to vibrate longitudinally if stroked with a resined cloth. Since the rod may be held at the end or anywhere between the ends, it can be made to vibrate with nodes at the restricted points, or with antinodes at the nonrestricted ends. Several examples are shown in Fig. 28.10(a) and (b).

Sec. 28.5 The Vibrating Rod

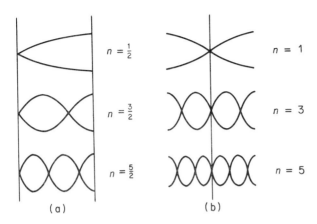

Fig. 28.10

The wavelength relationship is

$$\lambda = \frac{2l}{n}$$

The frequency equation is

$$F = \frac{nv}{2l}$$

The velocity equation is $v = \sqrt{Y/d}$; therefore,

$$F = \frac{n}{2l}\sqrt{\frac{Y}{d}} \qquad \left\{ \begin{array}{l} Y = \text{Young's mod} \\ d = \text{density: slugs} \\ l = \text{length} \\ n = \text{number of loops} \end{array} \right\}$$

Example 10

A brass rod is 48 in. long and is clamped at the 12-in. point. The specific weight of brass is 535 lb/ft³ and Young's modulus is 16 × 10⁶ psi. Calculate (1) the fundamental frequency for the rod; (2) the wavelength for the fundamental frequency; (3) the velocity of sound for this rod.

Solution

1. The fundamental frequency is $n = 2$ (both ends free). This is the smallest possible number of loops (half-wavelengths) because there must be an antinode at each end of the rod and a node where the rod is being restricted. This is shown in Fig. 28.11.

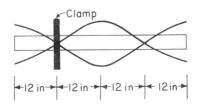

Fig. 28.11

$$F = \frac{n}{2l}\sqrt{\frac{Y}{d}} = \frac{2}{2 \times 4}\sqrt{\frac{16 \times 10^6 \times 144}{535/32}}$$
$$= 2940 \text{ vib/sec}$$

2. The wavelength for the fundamental frequency is

$$\left\{ \begin{array}{l} m = 2 \\ l = 48 \text{ in.} = 4 \text{ ft} \\ Y = (16 \times 10^6 \times 144) \text{ psf} \\ d = 535/32 \text{ slugs} \end{array} \right\}$$

$$\lambda = \frac{2l}{n} = \frac{2 \times 4}{2} = 4 \text{ ft}$$

3. The velocity of sound in this brass rod is

$$v = F\lambda = 2940 \times 4 = 11{,}760 \text{ ft/sec}$$

The *Kundt's tube*, Fig. 28.12, is used to determine the velocity of sound in a metal (or gas). It is essentially a long glass tube with an adjustable tightfitting plunger P_1 and a movable rod equipped with a plunger P_2 at the other end of the tube. The rod has a length l and is supported in the middle. The chamber inside the glass contains air (or some other gas) and either cork dust or Lycopodium powder which will respond to the vibrations in the gas.

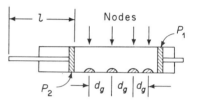

Fig. 28.12

It has been shown that a rod clamped in the middle will vibrate with a node at the point of clamping and antinodes at each end. Thus when this rod is stroked with a resined cloth, it vibrates longitudinally, giving off its fundamental. If piston P_1 is adjusted to put the reflected wave in the gas in phase with the incident wave, the powder will be unaffected at the nodes and form into ridges, perpendicular to the length of the tube, at the antinodes. The wavelength of the sound in the tube will be twice the distance between two nodes, or $\lambda = 2dg$ (Fig. 28.12).

However, the frequency in the rod and in the tube is the same, even though the velocities and the wavelengths are not the same; therefore,

$$v_r = F\lambda_r \quad \text{and} \quad v_g = F\lambda_g$$

Solving both equations for F,

$$F = \frac{v_r}{\lambda_r} = \frac{v_g}{\lambda_g}$$

$$\left\{ \begin{array}{l} v_r = \text{velocity in the rod} \\ v_g = \text{velocity in the glass} \\ \lambda_r = \text{wavelength (rod)} \\ \lambda_g = \text{wavelength (glass)} \\ F = \text{frequency} \end{array} \right\}$$

But $\lambda_r = 2l_r$ for the fundamental, and $\lambda_g = 2d_g$. Substituting,

$$\frac{v_r}{l_r} = \frac{v_g}{d_g} \quad \left\{ \begin{array}{l} d_g = \text{distance between the nodes (gas)} \\ l_r = \text{length (rod)} \end{array} \right\}$$

Sec. 28.6　　　　　　　　　*Organ Pipes*　　　　　　　　　443

Example 11

A brass rod 4 ft long is clamped in the center in a Kundt's tube which contains air. The velocity of sound in air is 1090 ft/sec and in the rod 11,500 ft/sec. Calculate the distance between two nodes in the tube when the rod is stroked with a resined cloth.

Solution

The ratio of velocity to length is

$$\frac{v_r}{l_r} = \frac{v_g}{d_g}$$

Solving for d_g and substituting,

$$\left\{ \begin{array}{l} v_r = 11{,}500 \text{ ft/sec} \\ v_g = 1090 \text{ ft/sec} \\ l_r = 4 \text{ ft} \end{array} \right\}$$

$$d_g = l_r \frac{v_g}{v_r} = 4 \times \frac{1090}{11{,}500} = 0.376 \text{ ft} = 4.51 \text{ in.}$$

28.6. Organ Pipes

Somewhat the same conditions prevail in organ pipes as are present in a vibrating rod. If the vibrations take place in an organ pipe open at one end, the open end vibrates as an antinode, the closed end a node (see Fig. 28.13). If the pipe is open at both ends, then both ends vibrate as antinodes, as shown in Fig. 28.14.

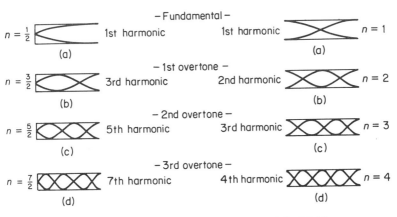

Fig. 28.13　　　　　　　　　Fig. 28.14

The equation for the wavelength is

$$\lambda = \frac{2l}{n}$$

The frequency is

$$F = \frac{nv}{2l}$$

Note that in Fig. 28.13(a), $n = \frac{1}{2}$ and the wavelength and frequency become

$\{ \begin{array}{l} l = \text{length of column} \\ \lambda = \text{wavelength} \\ v = \text{velocity} \\ n = \text{number of loops} \end{array} \}$

$$\lambda = \frac{2}{\frac{1}{2}} = 4l$$

$$F = \frac{\frac{1}{2}v}{2l} = \frac{v}{4l}$$

Example 12

Calculate the length of a closed-end pipe which will yield the fundamental when the velocity is 1088 ft/sec and the frequency is 512 vib/sec at 32°F.

Solution

This wavelength for a closed-end pipe is

$$\lambda = \frac{v}{F} = \frac{1088}{512} = 2.125 \text{ ft}$$

The length of the pipe is

$\{ \begin{array}{l} v = 1088 \text{ ft/sec} \\ n = 1/2 \\ F = 512 \text{ vib/sec} \end{array} \}$

$$l = \frac{n\lambda}{2} = \frac{\frac{1}{2} \times 2.125}{2} = 0.531 \text{ ft}$$

Example 13

Assume the conditions are the same as in Example 12 except that both ends of the pipe are open. (1) Calculate the length of the tube. (2) How much must the length of the tube change if the vibrations were taking place at 70°F?

Solution

1. The wavelength for a pipe open at both ends is

$$\lambda = \frac{1088}{512} = 2.125 \text{ ft}$$

The length of the pipe is

$\{ \begin{array}{l} n = 1 \\ v_1 = 1088 \text{ ft/sec} \\ F = 512 \text{ vib/sec} \\ T_1 = 460 + 32 = 492°R \\ T_2 = 460 + 70 = 530°F \end{array} \}$

$$l = \frac{n\lambda}{2} = \frac{1 \times 2.125}{2} = 1.0625 \text{ ft}$$

Sec. 28.7   *Resonance*   **445**

2. The ratio of velocities is

$$\frac{v_1}{v_2} = \sqrt{\frac{T_1}{T_2}}$$

Solving for v_2 and substituting,

$$v_2 = v_1 \sqrt{\frac{T_2}{T_1}} = 1088\sqrt{\frac{530}{492}}$$

$$= 1129 \text{ ft/sec}$$

Therefore, the wavelength is

$$\lambda = \frac{v}{F} = \frac{1129}{512} = 2.205 \text{ ft}$$

The length of the tube is 70°F is

$$l = \frac{n\lambda}{2} = \frac{1 \times 2.205}{2} = 1.1025 \text{ ft}$$

The change in length is

$$\Delta l = 1.1025 - 1.0625 = 0.04 \text{ ft} = 0.48 \text{ in.}$$

The *siren disk* is an instrument used to illustrate the relationship of pitch to frequency. This disk is constructed so that it has evenly spaced holes on several different radii. If the wheel is rotating at a fixed number of revolutions per second, the more holes in a circle, the more times per second an air jet, directed at that particular circle of holes, will be separated from the main column of air, thus creating "puffs" of air which are really regions of rarefaction and condensation, or sound. The more of these separations per second, the higher the frequency, and the higher the pitch.

28.7. Resonance

Resonance takes place in a sympathetic vibratory system where the frequency set up by the vibrating body causes another body to respond with one of its own natural frequencies of vibration. Thus, one tuning fork will cause another tuning fork to vibrate, if the natural frequencies of both tuning forks are the same or if one is a multiple of the other. Bridges have collapsed because some disturbance has caused the structure to vibrate in sympathy with it, thereby increasing the amplitude of the structure beyond the safe point.

Figure 28.15 shows a long glass tube connected to a reservoir. The water level in the tube may be controlled by raising or lowering the container. Therefore, starting with the tube full of water, striking the tuning fork with a rubber mallet will set the column vibrating. The first resonant point may be found by very slowly lowering the water level in the glass tube. A point will be reached where the sound will suddenly become loud. This is attributed to the natural frequency of the vibrating column of air in the tube and the natural frequency of the tuning fork supporting each other.

If more water is let out of the glass tube, the sound will diminish until a point is reached—where the length of the column is $\frac{3}{4}\lambda$ and again where the length of the column is $\frac{5}{4}\lambda$—where the sound will become louder. It is presumed that at these lengths, a standing wave is created in the tube which can support, on rebound from the water surface, the frequency of the tuning fork.

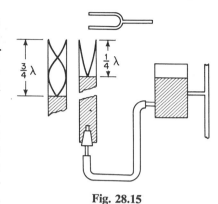

Fig. 28.15

The determination of the velocity of sound in air may be accomplished with this method if a tuning fork of known frequency is available.

Example 14

A tuning fork which vibrates at 720 vib/sec is struck over an adjustable resonating column. As the water level is dropped, the first point of resonance is found to be at 4.56 in. Calculate (1) the velocity of sound in air; (2) the length of the column for the second resonant point; (3) the wavelength of the sound.

Solution

1. For the first harmonic of this closed-end tube, $n = \frac{1}{2}$. The velocity of sound in air for this column from

$$F = \frac{nv}{2l_1}$$

is
$\left\{ \begin{array}{l} n = \frac{1}{2} \\ l_1 = 4.56 \text{ in.} \\ F = 720 \text{ vib/sec} \end{array} \right\}$

$$v = \frac{2Fl_1}{n} = \frac{2 \times 720 \times 4.56/12}{\frac{1}{2}}$$
$$= 1094.4 \text{ ft/sec}$$

2. The second resonant point ($n = \frac{3}{2}$) gives a column length of

$$l_2 = \frac{nv}{2F} = \frac{\frac{3}{2} \times 1094.4}{2 \times 720} = 1.14 \text{ ft} = 13.68 \text{ in.}$$

3. From the first resonant point ($n = \frac{1}{2}$), the wavelength is

$$\lambda = \frac{2l_1}{n} = \frac{2 \times 4.56/12}{\frac{1}{2}} = 1.52 \text{ ft}$$

Sec. 28.8   *Doppler Effect for Sound*   447

From the second resonant point ($n = \frac{3}{2}$), the wavelength is

$$\lambda = \frac{2 \times 1.14}{\frac{3}{2}} = 1.52 \text{ ft} \quad (check)$$

28.8. Doppler Effect for Sound

An extraordinary effect is created when a train, blowing its whistle, approaches and passes an observer. As the train approaches the observer, the frequency seems to increase. It falls off rapidly as the train passes and moves away from the observer. This rise and fall of frequency is called the *Doppler effect.*

There are three situations which can exist. They are (1) where the source generating the waves is stationary and the observer is moving either toward or away from the source; (2) where the observer is stationary and the source generating the waves is moving either toward or away from the observer; (3) where both the observer and the source are in motion.

If the source and observer are stationary, the observer will pick up the frequencies F emitted by the source such that

$$F = \frac{v}{\lambda}$$

If the observer is moving *toward* the source, he will pick up a greater number of frequencies per unit time than the source is emitting as a result of the observer's velocity. If the observer is moving *away* from the source, the observer will pick up fewer of the emitted frequencies per unit time.

If the source is moving *toward* a stationary observer, the effect is to shorten the wavelength. Thus in Fig. 28.16, λ' is the lengthened wavelength and λ is the shortened wavelength. The observer picks up these shortened wavelengths.

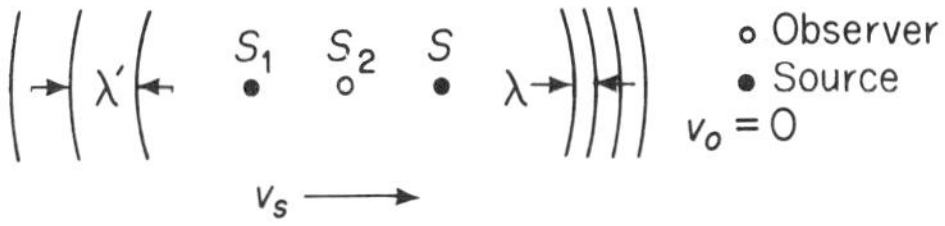

Fig. 28.16

If we state several rules to be followed, the application of the general equation will become routine:

448   *Wave Motion and Sound*   Chap. 28

1. The general relationship between the wavelengths of the source and the observer is

$$\frac{v - V_o}{F_o} = \frac{v - V_s}{F_s}$$

If this relationship is solved for F_o,

$$F_o = F_s\left(\frac{v - V_o}{v - V_s}\right)$$

$\begin{cases} v = \text{velocity of sound} \\ V_0 = \text{velocity of observer} \\ V_s = \text{velocity of source} \\ F_0 = \text{frequency to observer} \\ F_s = \text{frequency from source} \end{cases}$

2. The quantity v is the velocity of sound and is always $+$.
3. The quantities V_o and V_s are $+$ if they act to the right and $-$ if they act to the left.
4. When applying rule 3, the object should be shown to the *right* of the source as shown in all the drawings in Fig. 28.17.

Figure 28.17(a) through (h) show how the rule is applied.

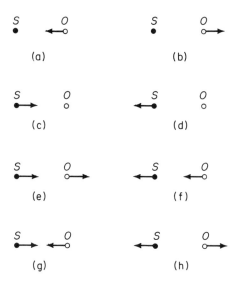

Fig. 28.17

Case 1. SOURCE STATIONARY, OBSERVER IN MOTION

1. Observer moving toward source [Fig. 28.17(a)]:

$$F_0 = F_s\left[\frac{v - (-V_0)}{v}\right] = F_s\left(\frac{v + V_0}{v}\right)$$

2. Observer moving away from source [Fig. 28.17(b)]:

$$F_0 = F_s\left(\frac{v - V_0}{v}\right)$$

$\begin{cases} v = \text{velocity of sound} \\ V_0 = \text{velocity of observer} \\ V_s = \text{velocity of source} \\ F_0 = \text{frequency to observer} \\ F_s = \text{frequency from source} \end{cases}$

Sec. 28.8   Doppler Effect for Sound   449

Case 2. SOURCE IN MOTION, OBSERVER STATIONARY

1. Source moving toward observer [Fig. 28.17(c)]:

$$F_0 = F_s\left(\frac{v}{v - V_s}\right)$$

2. Source moving away from observer [Fig. 28.17(d)]:

$$F_0 = F_s\left[\frac{v}{v - (-V_s)}\right] = F_s\left(\frac{v}{v + V_s}\right)$$

Case 3. SOURCE AND OBSERVER IN MOTION

1. Both source and observer moving to the right [Fig. 28.17(e)]:

$$F_0 = F_s\left(\frac{v - V_0}{v - V_s}\right)$$

2. Both source and observer moving to the left [Fig. 28.17(f)]:

$$F_0 = F_s\left[\frac{v - (-V_0)}{v - (-V_s)}\right] = F_s\left(\frac{v + V_0}{v + V_s}\right)$$

3. Source and observer moving toward each other [Fig. 28.17(g)]:

$$F_0 = F_s\left[\frac{v - (-V_0)}{v - V_s}\right] = F_s\left(\frac{v + V_0}{v - V_s}\right)$$

4. Source and observer moving away from each other [Fig. 28.17(h)]:

$$F_0 = F_s\left[\frac{v - V_0}{v - (-V_s)}\right] = F_s\left(\frac{v - V_0}{v + V_s}\right)$$

Example 15

Assume two trains approach each other. One train is blowing its whistle with a frequency of 800 cps (cycles per second) while its speed is 100 mph. The engineer on the second train hears the whistle. His train is traveling with a speed of 60 mph. Calculate the frequency heard by the engineer on the second train.

Solution

Since the trains are approaching each other [Fig. 28.17(g)],

$$F_o = F_s\left[\frac{v - (-V_o)}{v - V_s}\right] = F_s\left(\frac{v + V_o}{v - V_s}\right)$$

$$= 800\left(\frac{1088 + 88}{1088 - 147}\right) \quad \left\{\begin{array}{l} v = 1088 \text{ ft/sec} \\ V_o = 60 \text{ mph} = 88 \text{ ft/sec} \\ V_s = 100 \text{ mph} = 147 \text{ ft/sec} \\ F_s = 800 \text{ cps} \end{array}\right\}$$

$$= 999.7 \text{ vib/sec}$$

Problems

1. Explain the mechanism by which transverse and longitudinal waves propagate energy.
2. Assume the velocity of sound in water to be 4800 ft/sec. If the period of vibration of a sound wave in water is 0.0035 sec, find (a) the wavelength; (b) the frequency of vibration.
3. The period of vibration of sound in alcohol is 2.5 milliseconds and its wavelength is 9.75 ft. Calculate (a) its velocity in alcohol; (b) the frequency.
4. The tension in a copper wire is 12 lb. The wire is 3 ft long and weighs 1 oz. Calculate the velocity of a wave in the wire.
5. A 0.7-g string is 30 cm long. If it is stretched by a load of 4000 g, what is the velocity of a transverse wave in the string?
6. A steel wire is under a tension of 50 lb. If the wire is 24 in. long and weighs 6 oz, calculate the speed of a wave set up in the wire.
7. The velocity of a transverse wave along a wire is 3500 cm/sec when a load of 3000 g stretches the wire. If the wire is 100 cm long, calculate the mass of the wire for these conditions.
8. Calculate the speed of sound in air at STP in the cgs system of units. Note that the pressure must be in dynes. (Assume $\lambda = 1.4$.)
9. Calculate the wavelength of a compressional wave in benzene if the frequency is 512 vib/sec.
10. Young's modulus for aluminum is 10×10^6 lb/in.^{2}. What is the velocity of a compressional wave if the specific weight of aluminum is 170 lb/ft?
11. Calculate the specific weight of cast iron if the velocity of a longitudinal wave in the rod is 1.74×10^4 ft/sec and Young's modulus is 29.5×10^6 psi.
12. Calculate the velocity of a longitudinal wave in alcohol (bulk mod = 0.13×10^6 psi).
13. Find the wavelength of a compressional wave in helium if the wave vibrates at 800 vib/sec at STP, constant pressure = 1.25, constant volume = 0.755, and the specific weight of helium is 0.0111.
14. A steel wire weighs 0.24 oz and is 20 ft long. If it is stretched by a 12-lb force, find (a) the frequency of the fundamental; (b) the wavelength of the first overtone; (c) the frequency of the first overtone.
15. A wire weighs 0.04 lb and is 4 ft long. If it is stretched by a 5-lb force, calculate (a) the fundamental frequency; (b) the wavelength of the fourth overtone; (c) the frequency of the third harmonic.
16. An electric vibrator oscillates at the rate of 120 vib/sec. A string of mass 0.5 g is 100 cm long. What tension must be hung on the end of the string to cause it to vibrate (a) according to its fundamental; (b) with its fourth overtone?
17. Calculate the velocity of sound in oxygen at $-10°$F.
18. A 0.8-oz string attached to a vibrator displays 8 loops when it vibrates at 440 vib/sec. The string is 56 in. long. Calculate (a) the tension in the string; (b) the wavelength exhibited; (c) the velocity of the wave.
19. Assume an aluminum rod 60 in. long clamped at the 15-in. point. The specific

gravity of the rod is 2.7 and it has a modulus of elasticity of 10×10^6 psi. (a) What is the fundamental frequency of vibration for this rod? (b) What is the wavelength for this fundamental frequency? (c) Find the velocity of sound for the aluminum rod.

20. A brass rod 24 in. long is held at one end and caused to vibrate by stroking with a resined cloth. The specific weight of brass is 540 lb/ft^3 and Young's modulus is 16×10^6 psi. Calculate (a) the fundamental frequency; (b) the velocity of the sound wave in the rod; (c) the velocity of the sound wave caused in the air at 90°F if velocity in air is 1090 ft/sec.
21. The same rod as in Prob. 20 is held in the middle. Calculate (a) the fundamental; (b) the velocity of the sound wave in the rod; (c) the frequency for the second overtone.
22. A glass rod is placed into a Kundt's tube and clamped in the center. The velocity of sound in the glass rod when it is stroked with a resined cloth is 16,400 ft/sec. The tube is filled with air. If the velocity of sound in air is 1088 ft/sec and the rod is 4 ft long, what is the distance between two nodes?
23. Solve Prob. 22, given the same data except that the tube is filled with hydrogen ($v = 4165$ ft/sec at 32°F) instead of air.
24. Calculate (a) the fundamental frequency of a closed pipe 3 ft long if the velocity of sound is 1088 ft/sec; (b) the wavelength of the fifth harmonic.
25. Solve Prob. 24 using the same data, except that the pipe is open at both ends.
26. A pipe open at both ends vibrates with a frequency of 512 vib/sec. Calculate (a) the length of the pipe if the velocity of sound is 1196 ft/sec; (b) the change in length of the pipe if the temperature goes from 32°F to 90°F.
27. Solve Prob. 26 if the pipe is open at one end.
28. An organ pipe open at one end vibrates with a frequency of 220 vib/sec when air is blown into it. If hydrogen is blown into it, find (a) the wavelength; (b) the frequency; (c) the length of the pipe.
29. What is the frequency and wavelength of the sound created when a disk having a circle of 60 evenly spaced holes rotates at the rate of 600 rpm? (Assume STP.)
30. Assume that the conditions in Prob. 29 prevail at 80°F. Find (a) the frequency; (b) the wavelength of the sound.
31. A tuning fork is struck over an adjustable resonating column. As the water level is dropped, the first point of resonance is found to be at 14 cm. (a) If the tuning fork vibrates at 600 vib/sec, find the velocity of sound in air. (b) What is the length of the column for the second resonant point? (c) What is the wavelength of the sound?
32. A tuning fork vibrates at a rate of 512 vib/sec. Resonance occurs when the air column is 0.54 ft long. Calculate (a) the velocity of sound in air; (b) the wavelength for the first resonant point; (c) the difference in wavelengths for the first and fourth resonant points.
33. A motorcycle and a train approach each other as the train whistle is blown. The frequency of the sound is 1000 vib/sec. The velocity of the train is 80 mph, and that of the motorcycle is 100 mph. Calculate the apparent frequency heard by the man on the motorcycle.
34. A man riding in an automobile approaches a siren which is emitting 1200 vib/sec. The automobile is traveling at the rate of 60 mph. (a) What is the apparent

frequency that the man hears? (b) What is the frequency that he hears after he has passed the siren?

35. An airplane is approaching an observation tower with a velocity of 500 mph. It is emitting a dominant wave frequency of 2000 vib/sec. (a) What is the apparent frequency as heard by the man in the tower? (b) What is the apparent frequency as the plane passes over the tower? (c) What is the apparent frequency after the plane has passed the tower?

36. Two pilots are approaching each other at a rate of 300 mph and 400 mph, respectively. A dominant sound frequency of 1800 vib/sec is being emitted by the second plane. (a) What is the apparent-frequency pickup by the pilot of the first plane? (b) What is the apparent-frequency pickup after the planes pass each other?

29

Light: The Point Source

29.1. Development of Light Theory

From the beginning of time, man thought of light as a directed beam from his eye to the object being viewed. Pythagoras and his followers thought light to be a stream of particles coming from an object to the eye. Others thought of light as a disturbance rather than as a particle.

Leonardo da Vinci suggested that light could be a wave disturbance. In the latter part of the seventeenth century, Christian Huygens developed the wave theory suggested by da Vinci. At about the same time, Sir Isaac Newton developed his particle theory.

Newton's experiments regarding the attraction of one body for another led him to the conclusion that light travels faster in the denser of two materials. Huygens believed that light, as a wave, must travel more slowly in the denser of two materials. If Newton were correct, the light which refracts, or bends, when entering a denser medium must bend *away* from the normal. If Huygens were correct, light must bend *toward* the normal on entering a denser medium. We know now that light bends *toward* the normal when entering a dense medium.

Very early in the nineteenth century, Thomas Young performed his experiments in interference which shifted the weight of experimental evidence in favor of Huygen's wave theory. That is, if light is made of particles, then a beam of particles directed at a sharp edge, Fig. 29.1, will be absorbed by

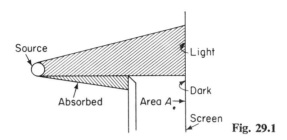

Fig. 29.1

that sharp edge and there will be a sharp line of demarcation between the bright and dark portions of the screen. As it turns out, this is *not* what happens. Light somehow actually bends around this sharp edge and predictable patterns appear on the screen at area A, Fig. 29.1. This can only happen if a wave is involved. Also, if light needs a vibrating particle, how is it that it will pass through an evacuated region where there are no vibrating particles?

It was Young's series of experiments and the verification of James Maxwell's electromagnetic-wave theory which seemed to indicate that light is indeed a wave phenomena. Maxwell's theory viewed light as being constructed of two waves in phase and at right angles to each other, as shown in Fig. 29.2. The combination propagates to the right.

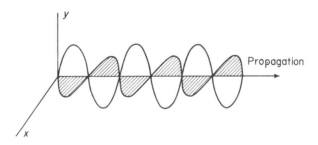

Fig. 29.2

Experimental evidence and reason seemed to support Maxwell's theory until it was observed that, in the photoelectric effect, when light strikes a photoelectric surface, the small amount of light energy striking the surface cannot possibly supply the amount of energy necessary to release electrons from that surface.

To explain this, Albert Einstein, using the particle theory proposed by Max Planck, theorized that light is a discrete *photon*—a wavelet, a small packet of energy. This particle is quite different from Newton's particle. Planck's particle is a packet, or quantum, of energy, called a photon; it is an energy "particle."

Neither the quantum theory nor the wave theory explains all the known phenomena of light. At present, both theories, aided by the mathematics of wave mechanics, are used to account for the existing known phenomena of light. Propagation is explained by the wave theory, energy transfer by the particle theory.

29.2. The Velocity of Light

It will be seen in subsequent discussions that atoms are composed of orbital electrons which are in equilibrium (ground-state) orbits. When energy is supplied to these electrons, they will absorb energy in discrete amounts and jump to higher-energy orbits. When the opportunity and conditions are just right, these electrons will jump back to an appropriate ground state with a prescribed release of energy. This energy is called "light," whether it comes off as a wave or a particle; this constitutes the dual theory of light.

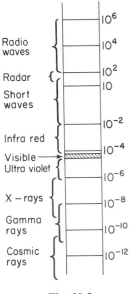

Fig. 29.3

The level of the energy released determines the type of wave with which we are dealing and its position in the electromagnetic spectrum, Fig. 29.3. With the discovery of radio waves, radar waves, X rays, gamma rays, and cosmic rays in addition to the infrared, visible-light, and ultraviolet rays, the theory fits the proportionality

$$\lambda \propto \frac{1}{F}$$

or, solving for the velocity of light,

$$c = F\lambda \quad \left\{ \begin{array}{l} \lambda = \text{wavelength} \\ F = \text{frequency} \\ c = \text{velocity of light} \end{array} \right\}$$

All electromagnetic waves, once released from the atom, travel *in a vacuum* at a speed of 2.998×10^8 m/sec (3×10^8 m/sec), or 186,000 mi/sec. It is interesting to note that, if we consider that electromagnetic waves propagate in a vacuum, the velocity of these electromagnetic waves is a function of the product of the permeability and permittivity of free space; or

$$c^2 = \frac{1}{\varepsilon_0 \mu_0}$$

$\left\{ \begin{array}{l} c = \text{velocity of light} \\ \varepsilon_0 = 8.85 \times 10^{-12} \text{ permeability} \\ \quad\quad \text{in vacuum} \\ \mu_0 = 4\pi \times 10^{-7} \text{ permittivity} \\ \quad\quad \text{in vacuum} \end{array} \right.$

or

$$c = \frac{1}{\sqrt{\varepsilon_0 \mu_0}} = \frac{1}{\sqrt{(8.85 \times 10^{-12}) \times (4\pi \times 10^{-7})}}$$

$$= 2.9982 \times 10^8 \text{ m/sec}$$

As late as the seventeenth century, the arguments about the velocity of light revolved about whether the speed of light was infinite or finite.

Galileo attempted to "clock" the speed of light by placing a lantern atop each of two mountains. One group of assistants, atop one mountain, uncovered its lantern. When the second group saw the light from the first lantern, they uncovered the second lantern. The expired time interval was clocked. It soon became evident that they were clocking human-reaction time and not the velocity of light. Most of the difficulty encountered in the measurement of the speed of light grew out of the extremely high velocity of light; methods for proving this speed to be finite were not sufficiently refined.

As early as 1675, Olaf Roemer, an astonomer, demonstrated that the speed of light is finite. Figure 29.4 shows the orbit of Earth around the Sun and the orbit of one of the moons of Jupiter about the planet Jupiter. When Earth is in the position shown in Figure 29.4(a) and the moon of Jupiter is eclipsed, Roemer should have been able to calculate the time when the eclipse of the moon of Jupiter was to take place as the time that the Earth is in the position shown in Figure 29.4(b), some time later. But each time Roemer

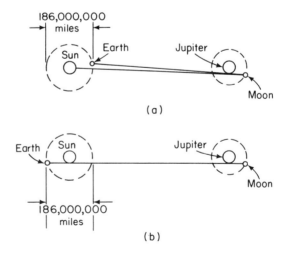

Fig. 29.4

Sec. 29.2   The Velocity of Light

made his calculations for the half-cycle rotation of the Earth about the Sun, he was approximately 1000 sec off. Using a calculated value for the diameter of the Earth's orbit of 186,000,000 mi, Roemer concluded that the additional time, which he considered an error, was due to the additional distance the light had to travel across the Earth's orbit. Therefore, he calculated the velocity of light as 186,000 mi/sec. On the basis of the calculated value of the Earth's orbit, Roemer's value was actually 192,000 mi/sec. The importance of Roemmer's work is not its accuracy, but his having shown that the speed of light is finite.

In the middle of the nineteenth century, Armand Fizeau, a French scientist, used a revolving spoked wheel to measure the velocity of light. (The apparatus is shown in Fig. 29.5.) Figure 29.5 shows two *select* light rays from a source. The two rays pass through a converging lens and are focused onto a glass plate M_1. Some of the light is reflected so that the two rays cross at the spoked wheel N and proceed to the collimating lens L_2. They then pass through the lens L_3, which focuses them on the mirror M_2. The central ray reflects off the mirror and returns to the spoked wheel over its original path. The outer ray reflects at the mirror and is shown returning to the spoked wheel at the bottom of the column. The two rays then pass through the tooth opening, glass plate, and lens L_4 to the eye.

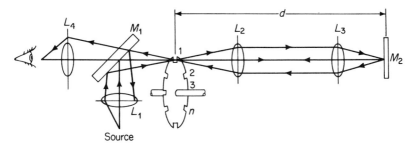

Fig. 29.5

If the revolutions per second of the spoked wheel are such that the light passes through slot 1 and, on returning from M_2, passes through slot 2, it is possible to calculate the velocity of light over the distance d.

If the distance traveled by the light is $2d$ and the velocity is c, then from $s = vt$, the time is

$$t = \frac{s}{v} = \frac{2d}{c}$$

Also, the time it takes the wheel to revolve a distance equal to one tooth and one space is

$$t = \frac{1}{2n} \times \frac{1}{F} = \frac{1}{2Nn}$$

Equating both equations for t,

$$\frac{2d}{c} = \frac{1}{2Nn}$$

$\begin{cases} t = \text{time: sec} \\ F = \text{frequency: vib/sec} \\ N = \text{rps} \\ 2n = \text{number of teeth and spaces} \\ d = \text{distance from wheel to mirror } M_2 \\ c = \text{velocity of light} \end{cases}$

Solving for c,

$$c = 4dNn$$

Example 1

A spoked wheel has 720 teeth and rotates at 720 rpm. What must be the distance between the spoked wheel and the reflecting mirror if the calculations show the speed of light to be 186,350 mi/sec?

Solution

Solve the preceding equation for d:

$$d = \frac{c}{4Nn} = \frac{186,350}{4 \times 12 \times 720} \qquad \begin{cases} n = 720 \text{ spokes} \\ N = (720 \text{ rpm} \times \frac{1}{60}) \text{ rps} \end{cases}$$
$$= 5.392 \text{ mi}$$

Jean Foucault modified the Fizeau apparatus by replacing the toothed wheel with a rotating mirror. His calculations were rather inaccurate, but by placing water between the rotating mirror and the source, he was able to prove

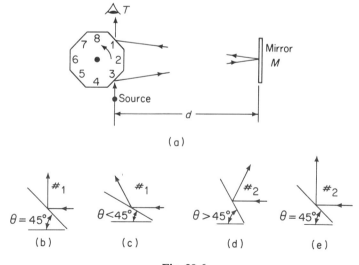

Fig. 29.6

Sec. 29.3   *Luminous Flux and Radiant Flux*   459

that light travels more slowly in materials denser than air. This was a significant contribution to the study of light.

Albert Michelson, using Fizeau's method, measured the velocity of light with extreme accuracy. Fundamentally, Michelson's apparatus consisted of a flashing light source, an octagonal mirrored wheel, a mirror placed about 22 miles from the wheel, and a sensitive scope for picking up the reflected beam. Figure 29.6(a) shows the rest position at which the light reflects off face 3, proceeds to mirror M, reflects back to face 1, and splits the cross-hairs in the telescope T.

As the wheel starts to rotate, the reflected ray takes on the positions shown in Fig. 29.6(b), (c), (d), and (e) and all the intermediate positions. As the mirror rotates faster and faster, face 2 reaches the position occupied by face 1, and the reflected ray once again splits the cross-hairs.

Since it takes one-eighth of a revolution of the wheel for the light to traverse the distance $2d$, the time in seconds is

$$t = \frac{1}{8} \times \frac{1}{N} = \frac{1}{8N}$$

Therefore, $\{ N = \text{speed of mirror: rps} \atop d = \text{distance} \}$

$$c = \frac{s}{t} = \frac{2d}{1/8N} = 16Nd$$

Example 2

The speed of light in the Michelson apparatus is calculated to be 186,340 mi/sec. If the wheel is to rotate at 530 rps, how far is the reflecting mirror from the rotating wheel?

Solution

From the preceding equation,

$$d = \frac{c}{16N} = \frac{186,340}{16 \times 530} \qquad \{ c = 186{,}340 \text{ mi/sec} \atop N = 530 \text{ rps} \}$$
$$= 21.974 \text{ mi}$$

Recent observations would seem to confirm the theory that the speed of light is independent of the velocity of the light source, of the color of the light, and of the intensity of the light.

29.3. Luminous Flux and Radiant Flux

Luminous flux represents the flux emerging from a sphere, 1 ft in radius, through a 1-ft^2 opening. To establish a unit of luminous flux in the British

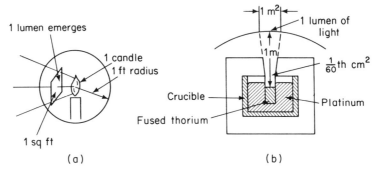

Fig. 29.7

system, a spermaceti candle was used as a light source burning at the rate of 120 grains/hr (Fig. 29.7a). If the candle radiates 4π lumens in all directions, then only 1 lumen of light can emerge from the 1-ft² opening. Thus the unit of light flux became defined, and the intensity of the source is said to be 1 *candle*.

Figure 29.7(b) shows the method used by the International Committee on Weights and Standards in 1948 to define the lumen. The platinum is brought to its melting point (2042°K), which causes the fused thorium to glow, emitting radiation. If the opening is $\frac{1}{60}$ cm², it will emit flux when permitted to fall upon 1 m² of surface of a sphere which has a radius of 1 m. The light emitted through the opening is defined as the *lumen*. Thus *luminous flux* represents the capacity of a light source to produce a visible sensation.

Whereas luminous flux is a subjective phenomenon, *radiant flux* is an energy concept. It represents the energy per unit time radiated by an electromagnetic wave. It is possible that waves outside the visible spectrum may radiate the same amount of *radiant* energy but emit no luminous flux at all. Radiant flux is measured with instruments such as lightmeters, pyrometers, etc. With the help of the *luminosity curve* (Fig. 29.8) one type of flux can be converted into the other.

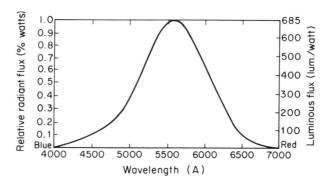

Fig. 29.8

Sec. 29.4 The Point Source 461

The luminosity curve is based on the fact that the eye is most sensitive to bright yellow-green light of wavelength 555 mμ. If the radiant flux for maximum conditions is chosen as 1 watt and the emission is 685 lumens/watt, it is possible to establish the relationship between the number of lumens evoked by 1 watt of radiant flux from the curve (Fig. 29.8) for any visible wavelength.

Example 3

The radiant flux of a certain light is 1 watt and its wavelength is 620 mμ. Calculate (1) the lumens evoked by this light; (2) the watts of radiant flux required to evoke 120 (lumens) that would be required for 555-mμ wavelength.

Solution

1. From the curve, Fig. 29.8, 1 watt of radiant flux will evoke only 40% of the lumens evoked by the light, wavelength 555 mμ. Therefore, the radiant flux

$$685 \times 0.40 = 274 \text{ lm/watt}$$

required to evoke 120 lm of 625 mμ is

$$\frac{120}{274} = 0.438 \text{ watt}$$

2. The radiant flux required for 120 lm of 555 mμ is

$$\frac{120}{685} = 0.175 \text{ watt}$$

The wattage difference is

$$0.438 - 0.175 \text{ watt} = 263 \text{ watts}$$

The *luminous efficiency*, in watts, is the ratio of the lumen output to the input power.

$$\text{eff} = \frac{\text{luminous flux}}{\text{radiant flux}} = \text{lm/watt}$$

Some typical luminous-efficiency values, arrived at empirically, are shown in appendix table A. 19.

29.4. The Point Source

A *point source* of light radiates flux in all directions equally over the entire surface of a sphere, radius r, as shown in Fig. 29.9(a). If another sphere, concentric about the source s, with a radius r_2 encircles the first sphere, then the same number of light rays (flux) fall on both spheres. Thus, if F is the luminous flux and E is the illumination per unit area on the surface of both spheres, the equations representing the total flux on the two spheres are

462   Light: The Point Source   Chap. 29

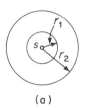
(a)

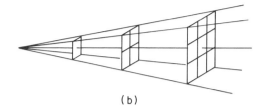
(b)

Fig. 29.9

and
$$F = E_1 A_1 = E_1 4\pi r_1^2$$

$$F = E_2 A_2 = E_2 4\pi r_2^2$$

If the flux is constant, the illumination per unit area must be inversely proportional to the square of the radius of the sphere:

$$E \propto \frac{1}{r^2}$$

This is shown in Fig. 29.9(b). As the distance r increases by increments of 1 cm, the number of squares increase as the square of this distance. That is, the same flux illuminating the first square with a given intensity must spread itself over 9 times the surface of a square 3 times removed from the source. The intensity is reduced by a factor of 9.

The total flux radiated through a solid angle ω by a light source is 4π *steradians*.* Therefore, the *lumen* is the flux radiated per steradian from a point source (Fig. 29.10a). If the luminous intensity I is defined as the flux per unit solid angle, since there are 4π steradians in a sphere, the total flux emitted by a point source is

$$F = 4\pi I = \omega I \quad \left\{ \begin{array}{l} I = \text{candle} \\ F = \text{lm} \end{array} \right\}$$

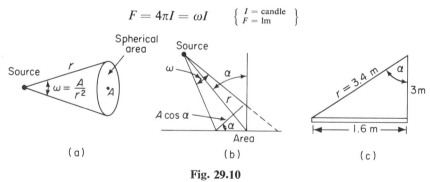

Fig. 29.10

---

*One steradian is defined as the solid angle generated from the center of a sphere which intercepts at the surface of the sphere, an area exactly equal to the square of the radius.

Sec. 29.4   *The Point Source*   463

Since the solid angle equals

$$\omega = \frac{A}{r^2}$$

then

$$F = \frac{A}{r^2} I$$

The flux per unit area reaching a surface is called the *illuminance E*. That is,

$$E = \frac{F}{A}$$

In Fig. 29.10(b), the solid angle ω subtended at the source by the area A is

$$\omega = \frac{A \cos \alpha}{r^2}$$

Since $F = I\omega$, the general equation for total flux generated by a point source is

$$F = \frac{IA \cos \alpha}{r^2}$$

The illumination is

$$E = \frac{F}{A} = \frac{IA \cos \alpha}{r^2 A} = \frac{I}{r^2} \cos \alpha$$

When the illuminated surface is normal to the source $\alpha = 0$ and $\cos \alpha = 1$, the illumination becomes

$$E = \frac{I}{r^2}$$

Example 4

A 60-candle lamp (75 watts) is 3 m above a circular table, radius 80 cm. The lamp is directly over the center of the table. Calculate (1) the solid angle subtended by the table; (2) the flux striking the table; (3) the illumination of the surface of the table; (4) the total flux emitted by the source; (5) the power necessary to operate this lamp if the efficiency is 14 lm/watt. (6) Assuming the lamp is moved so that it is directly over the edge of the table, calculate the illumination at the farthest edge of the table. (7) Calculate the luminous efficiency of the lamp.

Solution

1. The solid angle subtended by the table is

$$\omega = \frac{A \cos \alpha}{r^2} = \frac{\pi (0.80)^2 \cos 0°}{(3)^2}$$

$$= 0.22 \text{ steradians}$$

2. The flux striking the table or emitted by the source is

$$F = I\omega = 60 \times 0.22 = 13.2 \text{ lm}$$

3. The illumination of the surface of the table is

$$E = \frac{I \cos \alpha}{r^2} = \frac{60 \times \cos 0°}{(3)^2} = 6.7 \text{ lm/m}^2$$

$$\left\{ \begin{array}{l} r = 3 \text{ m} \\ I = 60 \text{ candles} \\ \alpha = 0° \\ \text{radius} = 0.080 \text{ m} \\ \text{radiant flux} = 75 \text{ watts} \end{array} \right\}$$

4. The total flux emitted by the source is

$$F = 4\pi I = 4\pi(60) = 754 \text{ lm}$$

5. The power input is

$$\text{radiant flux} = \frac{754}{14} = 53.9 \text{ watts}$$

6. The illumination at the farthest edge of the table is (Fig. 29.10c)

$$E = \frac{I}{r^2} \cos \alpha = \frac{60}{(3.4)^2} \times \frac{3}{3.4} = 4.6 \text{ lm/m}^2$$

7. The luminous efficiency is

$$\text{eff} = \frac{\text{luminous flux}}{\text{radiant flux}} = \frac{754}{75}$$

$$= 10.05 \text{ lm/watt}$$

29.5. Photometry

Figure 29.11(a) shows an optical-bench setup with a standard lamp P mounted at a known distance from either a Bunsen photometer, or a Lummer-Brodhun photometer (Fig. 29.11b). The Bunsen photometer has a grease spot on the screen which will transmit some light. An unknown lamp P_x is placed as ⋯, so that it will reflect light from a focusing mirror upon the grease spot. Since the brighter lamp causes the grease spot to transmit more light, the spot will appear darker on its side than on the side of the grease spot facing the weaker lamp. The unknown lamp is moved closer and closer until the grease spot disappears. The illumination on both sides of the screen will then be equal, and

$$E_k = E_x$$

or

$$\frac{I_k}{r_k^2} = \frac{I_x}{r_x^2}$$

$$\left\{ \begin{array}{l} E_k = \text{known illumination} \\ E_x = \text{unknown illumination} \\ I_k = \text{known intensity} \\ I_x = \text{unknown intensity} \\ r_k = \text{known distance} \\ r_x = \text{unknown distance} \end{array} \right\}$$

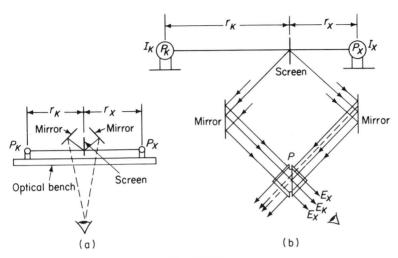

Fig. 29.11

Example 5

A standard 40-candle lamp is placed 45 in. from a Bunsen photometer when the grease spot disappears. This occurs when the unknown lamp is placed 30 in. from the photometer. Calculate the candlepower (i.e., the luminous intensity) of the unknown lamp.

Solution

Since

$$\frac{I_k}{r_k^2} = \frac{I_x}{r_x^2}$$

$$\left\{\begin{array}{l} I_k = 40 \text{ candles} \\ r_x = 30 \text{ in.} \\ r_k = 45 \text{ in.} \end{array}\right\}$$

Then

$$I_x = \frac{I_k r_x^2}{r_k^2} = \frac{40(30)^2}{(45)^2} = 17.8 \text{ candles}$$

The student should understand the following facts:

1. The luminous intensity (I) applies to the source of the flux. The units are

    BRITISH   *METRIC*

    $I = \text{candles} = \dfrac{\text{lumens}}{\text{steradian}}$ candles

2. The illumination (E) applies to the area illuminated.

    $E = \text{ft-candles} = \dfrac{\text{lumens}}{\text{ft}^2}$ meter-candles; $\dfrac{\text{lumens}}{\text{m}^2}$; lux

3. The total flux applies to the source or the area illuminated.

F = lumens   lumens

4. The following are defined equations:

$$\omega = 4\pi = A/r^2$$
$$I = F/\omega$$
$$E = F/A$$
$$\text{efficiency} = F/\text{watts}$$

Problems

1. A revolving spoked wheel is used to determine the velocity of light. The velocity of light is calculated at 186,300 mi/sec when the spoked wheel revolves at 900 rpm and is placed 5 mi from the source of light. Calculate the number of teeth in the wheel.
2. Michelson's apparatus calculates the velocity of light using an 8-sided mirror rotating at a speed of 500 rps. If the velocity of light is calculated at 186,300 mi/sec, calculate the distance from the mirror to the source.
3. The velocity of light is calculated at 186,300 mi/sec using the Michelson apparatus. If the mirrored wheel rotates with a speed of 18,000 rpm and is 20 mi from the source, how many mirrors must be in the wheel?
4. Given 1 watt of light, wavelength of 600 mμ, calculate (a) the lumens evoked; (b) the watts of radiant flux evoked for 300 lumens of the light in part (a); (c) the wattage required to produce the same brightness for 555-mμ wavelength.
5. (a) Find the number of lumens evoked by a light if its wavelength is 625 mμ when the radiant flux is 1 watt. (b) How many more watts of radiant flux are required to evoke 80 lumens than would be required for 555-mμ wavelength?
6. Given a mixture of three monochromatic wavelengths—450 mμ, 480 mμ, and 590 mμ—which contribute 10 watts, 15 watts, and 20 watts, respectively, what is: (a) the radiant flux? (b) the luminous flux? (c) the luminous efficiency?
7. A 250-watt lamp generates 3500 lumens of flux. Calculate (a) the luminous intensity; (b) the luminous efficiency of the lamp.
8. The illuminance at a point directly under a lamp 15 ft above a floor is 3.2 foot candles. Calculate the candlepower of the lamp.
9. A drawing table is illuminated by a 40-watt fluorescent lamp 3 ft off the center of the surface of the table. Calculate (a) the illuminance directly under the lamp; (b) the illuminance directly under the lamp if the table is inclined at 30°.
10. A 150-candle lamp is 3 ft above a table. How much closer must the lamp be moved to give one-third more lighting?
11. A 200-candle lamp is 2ft above a table. How much must the lamp be moved to give one-half less light?
12. (a) How many 40-watt bulbs are needed to light up a swimming pool, area 100 × 30 ft, with an average illumination of 10 lm/ft², if the efficiency is 40%? The

luminous efficiency of the lamp is 11 lm/watt. (b) Assume the lamp to be a fluorescent 40-watt lamp with a luminous efficiency of 60 lm/watt. How many flourescent lamps are needed? (c) Compare the numbers of lamps needed in both cases.

13. A lamp (luminous intensity 40 candles) is 240 cm from a circular table (radius 45 cm) whose surface is normal to the direction of the light ray. Find (a) the solid angle subtended by the table; (b) the flux falling on the table; (c) the illumination of the surface of the table; (d) the total flux emitted by the source. (e) If the luminous efficiency is 15 lm/watt, what is the power necessary to operate this lamp? (f) Assuming the light to be moved so that it makes an angle of 45° with the normal to the table, find the illumination on the table.

14. A lamp is 160 cm over the top of the center of a round table 120 cm in diameter. If illumination at the farthest edge of the table is to be 50 lm/m², calculate (a) the intensity of the lamp; (b) the illumination normal to the top; (c) the solid angle subtended by the tabletop; (d) the total flux reaching the tabletop; (e) the total flux emitted by the source; (f) the efficiency of the lamp if it takes 20 watts to operate the lamp.

15. A lamp is connected to the edge of a 40-cm diameter round table. The lamp is 50 cm off the tabletop and normal to it. If the illumination at the farthest edge of the table is to be 50 lm/m², find (a) the intensity of the lamp; (b) the illumination normal to the lamp; (c) the solid angle, in steradians, subtended by the tabletop; (d) the total flux on the tabletop; (e) the total flux emitted by the source. (f) If it takes 20 watts to operate this lamp, what is its luminous efficiency?

16. A room is 16 ft × 12 ft and has a ceiling of 8 ft. A ceiling light of 60 lm/watt is placed in the ceiling directly in the center of the room. The minimum requirement directly under the lamp is 50 candlepower. Calculate (a) the luminous intensity; (b) the total flux output; (c) the wattage needed; (d) the illumination in one corner of the room.

17. Two street lamps are 15 ft apart and atop 12-ft poles. Each lamp has a luminous intensity of 3000 candles. Calculate the illumination (a) on the street midway between both lamps; (b) directly underneath one lamp.

18. Two lamps, 600 candles each, are placed atop 9-ft and 12-ft posts, 8 ft apart. Calculate the illumination under the 9-ft lamppost.

19. Repeat Prob. 18 for the illumination under the 12-ft lamp.

20. Repeat Prob. 18 for illumination halfway between the two posts.

21. A 100-watt standard lamp rated at 80 candles is placed 4 ft from a photometer screen. An unknown lamp is 2.6 ft from the photometer screen when the grease spot disappears. Calculate the rating of the unknown lamp.

22. A 15-watt fluorescent lamp has an efficiency of 50 lm/watt. (a) What is the luminous intensity of the lamp? (b) What is the illumination 4 ft under the lamp?

23. Two lamps are rated at 25 candles and 100 candles. They are mounted on an optical bench 6 ft apart. (a) At what point between the two lamps will the illumination be equal? (b) Calculate another position where the illumination will be equal.

30

Mirrors and Thin Lenses

30.1. Plane Mirrors

Objects can reflect, transmit or absorb light. This chapter will deal with reflection and refraction by plane and spherical surfaces. According to modern light theory, reflection and refraction demand a wave theory.

Huygen's theory of wavelets is used to explain reflection and refraction. Figure 30.1(a) shows a *plane mirror MM* and a *point source S* which emits

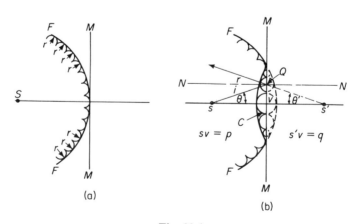

Fig. 30.1

Sec. 30.1 **Plane Mirrors** 469

light in all directions. If equal time intervals are chosen, then the various radii are equal. A curve FF drawn tangent to all these wavelets, at any instant, is called a *wave front*. In Fig. 30.1(a), this wave front is shown at the very instant that it touches the plane mirror. The radius of this wave front FF is generated from S.

Figure 30.1(b), shows the wave front from Fig. 30.1(a) at some short time interval later. If there had been no reflecting surface intervening, the wave front would have propagated according to the dotted lines. Reflection is a result of the ability of the reflecting surface to turn back the wavelets (wave front). Since these waves are traveling in the same medium before and after being turned back, their velocities are the same after impact as before impact. It is therefore possible to complete the spherical wavelets c as shown in Fig. 30.1(b).

In Fig. 30.1(b), it is also true that had the wavelet c been allowed to continue unobstructed by MM, it would have reached point S' in the same time interval as it took wavelet c to get to surface MM from S. If the distance SV is taken to be the *object distance p*, and the distance $S'V$ is taken to be the *image distance q*, then

$$p = q$$

In triangles SQV and $S'QV$, since $SV = S'V$ and VQ is common to both triangles,

$$\theta = \theta'$$

Finally, the *incident angle i* equals θ, and the *angle of reflection r* equals θ'. Therefore,

$$i = r$$

The *law of reflection* states that the angle the incident ray makes with the normal equals the angle the reflected ray makes with the normal. The normal, incident, and reflected rays all are in the same plane.

The graphic method, Fig. 30.2, for locating an image is as follows:

1. The object is to the left of the plane mirror MM.
2. At the point where the beam OV strikes the mirror, a normal NN to the mirror MM is drawn.
3. Since the angle of incidence i is equal to the angle of reflection r, the reflected rays will propagate as shown in Fig. 30.2. It will appear to come from point. I.
4. The ray OX will reflect back on itself and appear to come from I.
5. It should be noted that ray OV' also appears to be reflected from I.

Therefore, from Fig. 30.2,

$$\angle i = \angle r, \qquad OX = XI, \qquad p = q$$

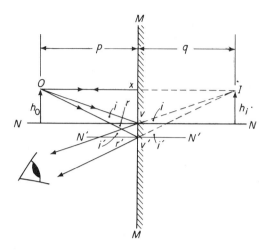

Fig. 30.2

Also, because the triangles $h_o p(\overline{OV})$ and $h_i q(\overline{IV})$ are equal, the magnification M is*

$$M = \frac{\text{image distance}}{\text{object distance}} = \left|\frac{q}{p}\right|$$

$$= \frac{\text{image height}}{\text{object height}} = \left|\frac{h_i}{h_o}\right|$$

Since the object distance is equal to the image distance, the magnification for a plane mirror is equal to 1.

30.2. Fundamental Concepts and Conventions

At this time, some concepts will be established which will be used throughout the presentation of mirrors and lenses.

1. The *principal focus* or *focal point* F is the point on the principal axis to which all rays converge [Fig. 30.3(a)], or appear to converge [Fig. 30.3(b)], when the rays come from an object at infinity. The *focal distance*, f, is the distance from the focal point F to the vertex V.
2. *Objects* are *real* if the light actually comes from them. *Objects* are taken to be *virtual* if light *appears* to come from them.†

---

*In the following, the symbol | | means "absolute value."
†This concept will be developed later in this chapter. The concept of a virtual object will be useful when considering lens combinations. They have no real basis in fact.

Sec. 30.2 Fundamental Concepts and Conventions 471

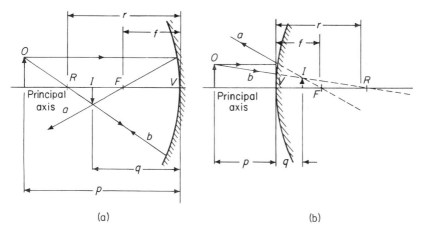

Fig. 30.3

3. *Images* are *real* if light actually passes through them and *virtual* if it appears that light passes through them.
4. In Fig. 30.3(a) and (b), light is being propagated in all directions by the object. Only two rays are needed to locate the image. A ray *a* from infinity will be parallel to the central axis; will reflect so that the angle of incidence is equal to the angle of reflection; and will pass, or appear to pass, through the focal point. Another ray, *b*, which propagates through the center of curvature, will strike the mirror and reflect back on itself. The point at which these two rays intersect is the *image position*.

The following conventions apply to mirrors and lenses.

1. Light will *always* be considered as coming from the left of the *vertex V*, and propagated from left to right.
2. If the object O is to the left of the vertex V, the object distance p is positive. If the object O is to the right of the vertex V, the object distance p is negative.
3. If the image I is real (to the left of the vertex for mirrors), the image distance q is positive. If the image I is virtual (to the right of the vertex for mirrors), the image distance q is negative.
4. If the mirror (or lens) causes reflected rays to converge on the focal point F (Fig. 30.3a), the mirror (or lens) is said to be positive and the focal distance f is positive. If the mirror (or lens) causes reflected rays to *appear* as though they converge on the focal point F (Fig. 30.3b), the mirror (or lens) is said to be negative and the focal distance f is negative.
5. If the radius of curvature r is to the right of the vertex V (Fig. 30.3b), it is positive. If the radius of curvature r is to the left of the vertex V, it is negative.
6. If the magnification M is positive, the image is on the side of the principal axis opposite the object. Thus if the object is above the principal axis, the image will be

below the central axis when M is positive. If the magnification is negative, the image will be on the same side of the principal axis as the object.

A summary of these conventions is shown in Table 30.1

Table 30.1

Left	Mirrors Vertex V	Right
	$+p$	$p-$
(real)	$+q$	$q-$ (virtual)
(converging)	$+f$	$f-$ (diverging)
	$-r$	$r+$

M is $+$, image inverted
M is $-$, image erect

30.3. Convex and Concave Mirrors

According to the conventions established in Sec. 30.2, the following equation will apply to all mirrors:

$$\frac{1}{p}+\frac{1}{q}=-\frac{2}{r} \quad \left\{\begin{array}{l}p = \text{object distance} \\ q = \text{image distance} \\ r = \text{radius of curvature}\end{array}\right\}$$

If an object is at infinity, the image is at the focal point. Thus

$$\frac{1}{\infty}+\frac{1}{q}=-\frac{2}{r}$$

and

$$q=-\frac{r}{2}$$

Since the focal point is one-half the distance from the vertex V to R,

$$-\frac{r}{2}=f$$

and, therefore,

$$-\frac{2}{r}=\frac{1}{f}$$

or

$$\frac{1}{p}+\frac{1}{q}=-\frac{2}{r}=\frac{1}{f}$$

Sec. 30.3   Convex and Concave Mirrors   473

The magnification is

$$M = \frac{q}{p}$$

Example 1

A concave mirror has a focal length of 3 in. Calculate the image position and size when a 1.5-inch object is (1) 9 in. to the left of the vertex; (2) 2 in. to the left of the vertex. (3) Draw the ray diagrams. (4) Is the image real or virtual, inverted or erect?

Solution

1. Image distance:

$$\frac{1}{p} + \frac{1}{q} = \frac{1}{f}$$

Substituting,

$$\frac{1}{9} + \frac{1}{q} = \frac{1}{3}$$

$$q = 4.5 \text{ in.} \quad (\textit{real image})$$

The magnification is
$$\left\{ \begin{array}{l} f = +3 \text{ in.} \\ p = +9 \text{ in.} \\ h_o = 1.5 \text{ in.} \end{array} \right\}$$

$$M = \frac{q}{p} = \frac{4.5}{9} = 0.5 \times \quad (\textit{inverted image})$$

The size of the image is

$$h_i = h_o M = 1.5(0.5) = 0.75 \text{ in.}$$

The ray diagram is shown in Fig. 30.4(a).

2. Image distance:

$$\frac{1}{2} + \frac{1}{q} = \frac{1}{3}$$

$$q = -6 \text{ in.} \quad (\textit{virtual image})$$

The magnification is
$$\left\{ \begin{array}{l} f = +3 \text{ in.} \\ p = +2 \text{ in.} \\ h_o = 1.5 \text{ in.} \end{array} \right\}$$

$$M = \frac{q}{p} = \frac{-6}{2} = -3 \times \quad (\textit{erect image})$$

The size of the image is

$$h_i = h_o M = 1.5(3) = 4.5 \text{ in.}$$

The ray diagram is shown in Fig. 30.4(b).

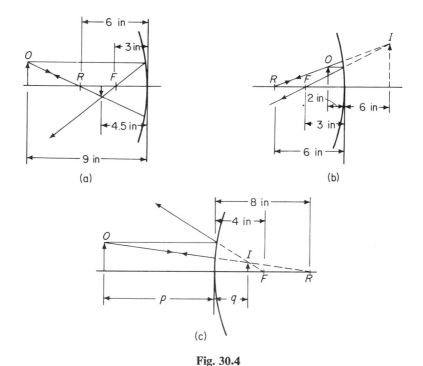

Fig. 30.4

Example 2

A convex mirror has a radius of curvature of 8 in. Calculate the image position and magnification when a 2-in. object is 12 in. to the left of the vertex. Draw the ray diagram. Is the image real or virtual, inverted or erect?

Solution

The image distance is

$$\frac{1}{p} + \frac{1}{q} = -\frac{2}{r}$$

Substituting and solving for q,

$$\frac{1}{12} + \frac{1}{q} = -\frac{2}{+8}$$

$$q = -3 \quad (virtual\ image)$$

$$\left\{ \begin{array}{l} r = +8 \\ p = +12 \text{ in.} \\ h_0 = 2 \text{ in.} \end{array} \right\}$$

The magnification is

$$M = \frac{q}{p} = -\frac{-3}{12} = -\frac{1}{4} \times \quad (erect\ image)$$

Sec. 30.4 Refraction and the Index of Refraction

The image size is

$$h_i = h_o M = 2\left(\frac{1}{4}\right) = 0.5 \text{ in.}$$

The ray diagram is shown in Fig. 30.4(c).

30.4. Refraction and the Index of Refraction

Light rays which enter a denser medium are transmitted at a reduced velocity. In Fig. 30.5(a), a light ray traveling in air with a velocity v enters a denser medium. The new velocity in the denser medium is v_1. If the velocity of light in the new medium is less than its velocity in the first medium, the ray is bent toward the normal NN. If the velocity of light in the second medium is greater than its velocity in the first medium, the ray is bent away from the normal.

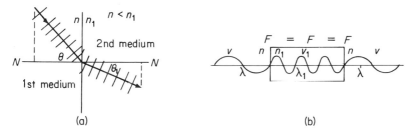

Fig. 30.5

The ratio of the speed of light in a vacuum (air) to the speed of light in a denser medium may be represented by a proportionality constant n, called the *index of refraction*, so that

$$n = \frac{c}{v} \quad (\textit{first medium})$$

$$n_1 = \frac{c}{v_1} \quad (\textit{second medium})$$

The index of refraction will always be greater than 1, as shown in the Appendix Table A.20.

Solving both equations above for c and equating,

$$nv = n_1 v_1$$

or

$$\frac{n}{n_1} = \frac{v_1}{v}$$

$$\left\{\begin{array}{l} n = \text{index of refraction 1st medium} \\ n_1 = \text{index of refraction 2nd medium} \\ v = \text{velocity of light of incident ray} \\ v_1 = \text{velocity of light of refracted ray} \end{array}\right\}$$

476   *Mirrors and Thin Lenses*   Chap. 30

Referring to the two triangles in Fig. 30.5(a), the ratio of the sine of the angle of incidence to the velocity of the incident ray is equal to the ratio of the sine of the angle of refraction to the velocity of the refracted ray.

$$\frac{\sin \theta}{v} = \frac{\sin \theta_1}{v_1}$$

Solving for the ratio v_1/v,

$$\frac{v_1}{v} = \frac{\sin \theta_1}{\sin \theta}$$

Since $n/n_1 = v_1/v$, then
$\left\{ \begin{array}{l} \theta = \text{incident angle} \\ \theta_1 = \text{refracting angle} \end{array} \right\}$

$$\frac{n}{n_1} = \frac{\sin \theta_1}{\sin \theta}$$

From this, *Snell's law* is

$$n \sin \theta = n_1 \sin \theta_1$$

Example 3

A light ray strikes a piece of flint glass (index of refraction 1.63) at an angle of incidence of 30°. Calculate the refraction in the flint glass.

Solution

The angle of refraction is

$$\sin \theta_1 = \frac{n}{n_1} \sin \theta = \frac{1}{1.63} \sin 30° \qquad \left\{ \begin{array}{l} n_1 = 1.63 \\ \theta = 30° \\ n = 1 \end{array} \right\}$$

$$\theta_1 = 17°53'$$

It has also been shown that when light travels in different materials, the wavelengths are proportional to the velocities in the respective material. The frequencies, however, are constant in both materials. Therefore, as the velocity of the light reduces, it does so because the wavelength is shortened. The frequency, or number of vibrations per second, remains constant as shown in Fig. 30.5(b). Thus

$$v = F\lambda, \qquad v_1 = F\lambda_1$$

Solving both equations for f and equating,

$$\frac{v}{\lambda} = \frac{v_1}{\lambda_1}$$

Solving for v_1/v,

$$\frac{v_1}{v} = \frac{\lambda_1}{\lambda} \qquad \left\{ \begin{array}{l} \lambda = \text{wavelength in incident medium} \\ \lambda_1 = \text{wavelength in refracting medium} \end{array} \right\}$$

Sec. 30.4   Refraction and the Index of Refraction   477

A summation of the preceding equation is

$$\frac{n}{n_1} = \frac{\sin\theta_1}{\sin\theta} = \frac{v_1}{v} = \frac{\lambda_1}{\lambda}$$

Example 4

A glass rod 5 cm long is placed between a light source and a screen. The light source emits light of wavelength 550 mμ. The distance from the light source to the screen is 12 cm. If the index of refraction of the glass is 1.6, calculate the number of waves between the source and the screen.

Solution

The total distance traveled in air is

$$12 \text{ cm} - 5 \text{ cm} = 7 \text{ cm}$$

The equivalent optical distance while the light is in the glass is

$$5 \times 1.6 = 8 \text{ cm}$$

The total equivalent distance is

$$7 + 8 = 15 \text{ cm}$$

Since each wave is 550×10^{-7} cm long, the number of waves is

$$N = \frac{15}{550 \times 10^{-7}} = 2.73 \times 10^5 \text{ waves} \qquad \{N = \text{number of waves}\}$$

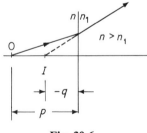

Fig. 30.6

Apparent depth may be treated as a refracting plane surface. If the conventional sequence for n and n_1 is observed and if the student remembers that light comes from the left, the drawing becomes oriented as shown in Fig. 30.6. Note that the image is virtual (no light actually passes through I) and therefore q will be negative when substituted from a problem.

The ratio of indices of refraction to the distances for apparent depth is

$$\frac{n}{p} = \frac{n_1}{-q}$$

The magnification for apparent depth is

$$M = -\frac{nq}{n_1 p}$$

Mirrors and Thin Lenses — Chap. 30

Example 5

An object is 12 ft below the surface of a pool of water (index of refraction 1.33). How far below the surface of the pool does the image appear to be?

Solution

$$q = -\frac{n_1}{n}p = -\frac{1}{1.33} \times 12 = -9.02 \text{ ft} \qquad \left\{\begin{array}{l} n = 1.33 \\ n_1 = 1 \\ p = 12 \text{ ft} \end{array}\right\}$$

Another phenomenon, refraction-reflection at a plane surface, results when light crosses an interface $n > n_1$, as shown in Fig. 30.7.

According to Snell's law, an incident angle θ will have a corresponding angle of refraction θ_1. But when n is greater than n_1, angle θ_1 will be greater than angle θ. Therefore, for some angle θ_c, the angle θ_1 will be equal to 90°. When $\theta_1 = 90°$, θ_c is called the *critical angle* (see Fig. 30.7).

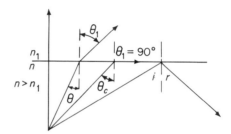

Fig. 30.7

If the incident angle is greater than the critical angle, the ray will be reflected internally and the angle of incidence i will be equal to the angle of reflection r.

From Snell's law, $n \sin \theta = n_1 \sin \theta_1$. If $\sin \theta_1 = 90°$,

$$\sin \theta_c = \frac{n_1}{n}$$

When n_1 is for air, $n_1 = 1$ and

$$\sin \theta_c = \frac{1}{n}$$

Example 6

Calculate the critical angle for Canada balsam (index of refraction 1.53 referenced to air).

Sec. 30.5   Thin Lenses   479

Solution

$$\sin \theta_c = \frac{1}{1.53} = 0.653$$

$$\theta_c = 40°46'$$

Note that at an incident angle of 40°, the ray will refract; at an incident angle of 41°, the ray will reflect.

30.5. Thin Lenses

As with mirrors, a lens is labeled positive if it causes rays from an object at infinity to converge, and negative if it causes the rays to diverge. Figure 30.8(a) shows a lens which causes the incident parallel rays to refract and pass through the focal point F. Note that there are three select rays, which may be used to locate an image. Only two of these rays need be used. The ray parallel to the axis will converge through F. Another ray will go through the point where the principal axis is normal to and crosses the central axis of the lens. These rays will intersect at the arrowhead image I. The third ray which could have been used passes through F' and converges into a parallel path after passing through the lens. It also crosses the arrowhead at I.

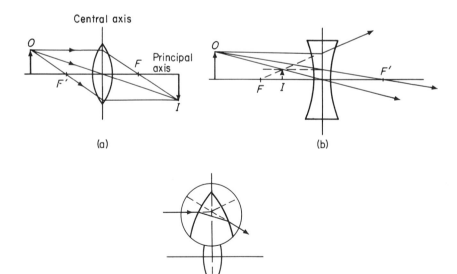

Fig. 30.8

Figure 30.8(b) shows a diverging lens. Since the rays coming from the left will diverge after leaving the lens, a parallel ray will diverge after leaving the lens in such a manner that the ray, when projected back (dotted line), will pass through the second focal point F. Another ray will cross the intersection of the principal axis and the lens axis normal to the principal axis. It will intersect the first ray described and determine the position of the image I. A third ray may be drawn toward the first focal point F'. This ray will diverge on passing through the lens so that it will emerge parallel to the principal axis. If this ray is projected back (dotted line), it will pass through the intersection point of the first two rays just described.

Note that there are actually two refractions taking place, as shown in Fig. 30.8(c). For convenience, when drawing ray diagrams for a thin lens, the vertical centerline of the lens may be used.

It should also be noted at this point that lenses which are thicker in the center than in the outer periphery will cause rays to converge, Fig. 30.9(a). Those which are thicker at the outer periphery than at the center will cause rays to diverge, Fig. 30.9(b). Lenses are characterized according to the curvature of their surfaces *when viewed from the right*. Thus, the lens in Fig. 30.8(b) is called a concave-convex lens.

Converging lenses, thicker in the center, slow down a wave front in the thicker portion of the lens for a longer period of time than in the outer, thinner periphery (see Fig. 30.9c). Thus the center of the wave front lags behind the outer edges of the wave fronts. The perpendicular to these wave fronts converges at F. The same reasoning may be used for a diverging lens, as shown in Fig. 30.9(d).

This discussion establishes the principle that a lens which causes rays to converge is a *positive lens* and a lens which causes a ray to diverge is a *negative*

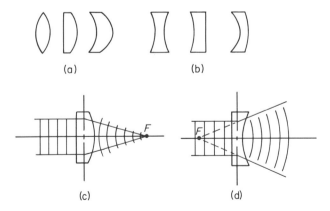

Fig. 30.9

Sec. 30.5   Thin Lenses   481

lens. Positive lenses have positive focal distances (f) and negative lenses have negative focal distances (f). All other conventions established for mirrors in Sec. 30.2 are to be used for lenses. Figure 30.10(a) and (b) and Table 30.2, illustrate the sign conventions.

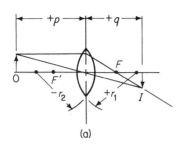 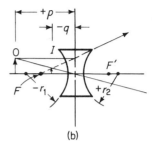

(a)   (b)

Fig. 30.10

Table 30.2

	Lenses Vertex	
Left	V	Right
(virtual) $-q$ (converge) $+f$ $-r$	$+p$	$p-$ $q+$ (real) $f-$ (diverge) $r+$

M is $+$, image inverted
M is $-$, image erect

The equation used for a thin lens, which has two radii r_1 and r_2 as shown in Fig. 30.10(a) and (b) is

$$\frac{1}{p} + \frac{1}{q} = \left(\frac{n_1}{n} - 1\right)\left(\frac{1}{r_1} - \frac{1}{r_2}\right)$$

It has been shown that if an object is at infinity, the image will be at F. Therefore, substituting $p = \infty$, $q = f$, and $n = 1$,

$$\frac{1}{\infty} + \frac{1}{f} = (n_1 - 1)\left(\frac{1}{r_1} - \frac{1}{r_2}\right)$$

The thin-lens equation is

$$\frac{1}{f} = \frac{1}{p} + \frac{1}{q}$$

$\left\{\begin{array}{l} r_1 = \text{1st surface light strikes} \\ r_2 = \text{2nd surface light strikes} \\ n = \text{1st medium} \\ n_1 = \text{2nd medium} \\ p = \text{object distance} \\ q = \text{image distance} \end{array}\right\}$

The focal length of thin lenses is sometimes given in terms of the *diopters*, which is defined as the inverse of the focal length when the focal length is given in meters; or

$$\text{diopter} = \frac{1}{f\text{(meters)}}$$

The magnification is

$$M = \frac{q}{p} = \frac{h_i}{h_o}$$

Example 7

A converging lens has a focal length of 5 in. A 2-in. object is placed on the principal axis in the various positions shown (Fig. 30.11). Calculate the position of the image and image size, and whether the image is real or virtual, and inverted or erect when the object is (1) at infinity; (2) 12 in. to the left of the vertex; (3) 2 in. to the left of the vertex. Draw the ray diagrams.

Solution

1. Object at infinity (Fig. 30.11a).

    (a) Image position:

    $$\frac{1}{p} + \frac{1}{q} = \frac{1}{f}$$

    $$\frac{1}{\infty} + \frac{1}{q} = \frac{1}{5} \qquad \left\{ \begin{array}{l} f = +5 \text{ in.} \\ p = \infty \end{array} \right\}$$

    $q = 5$ in.   (real and to the right of the vertex)

    (b) Magnification:

    $$M = \frac{q}{p} = \frac{5}{\infty} = 0 \times \qquad (a \text{ point at } F)$$

    $$h_i = h_o M = 0(2) = 0 \text{ in.}$$

2. Object at 12 in. (Fig. 30.11b).

    (a) Image position:

    $$\frac{1}{12} + \frac{1}{q} = \frac{1}{5}$$

    $q = 8.6$ in.   (real and to the right of the vertex)

    (b) Magnification:

    $$M = \frac{q}{p} = \frac{8.6}{12} = 0.72 \times \qquad (inverted\ image)$$

    $$h_i = h_o M = 2(0.72) = 1.44 \text{ in.}$$

Sec. 30.5   *Thin Lenses*   **483**

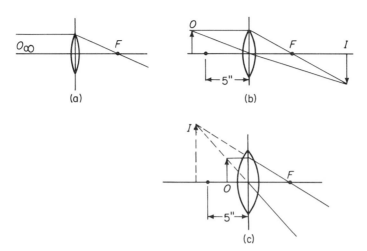

Fig. 30.11

3. Object at 2 in. (Fig. 30.11c).

    (a) Image position:
    $$\frac{1}{2} + \frac{1}{q} = \frac{1}{5}$$

    $q = -3.3$ (*virtual and to the left of the vertex*)

    (b) Magnification:
    $$M = \frac{q}{p} = \frac{-3.3}{2} = -1.65 \times \quad (\textit{erect image})$$
    $$h_i = h_o M = 2(1.65) = 3.30 \text{ in.}$$

Example 8

A diverging lens has a focal length of 8 in. A 2-in. object is placed on the principal axis in the positions shown (Fig. 30.12). Find the position of the image and image size, and whether the image is real or virtual, and erect or inverted. The object is at (1) infinity; (2) 16 in. to the left of the vertex; (3) 6 in. to the left of the vertex. Draw the diagrams.

Solution

1. Object at infinity (Fig. 30.12a).

    (a) Image position:
    $$\frac{1}{\infty} + \frac{1}{q} = \frac{1}{-8}$$

    $q = -8$ in. (*virtual and to the left of the vertex*)

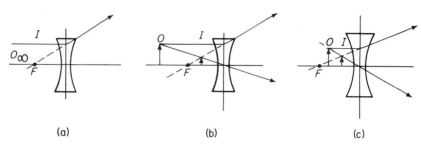

(a) (b) (c)

Fig. 30.12

(b) Magnification:

$$M = \frac{q}{p} = \frac{-8}{\infty} = 0 \times \quad (a\ point\ at\ F)$$

$$h_i = h_o M = 2(0) = 0$$

2. The object at 16 in. (Fig. 30.12b).

   (a) Image position:

   $$\frac{1}{16} + \frac{1}{q} = \frac{1}{-8}$$

   $q = -5.3$ (*virtual and to the left of the vertex*)

   (b) Magnification:

   $$M = \frac{q}{p} = \frac{-5.3}{16} = -0.33 \times \quad (erect\ image)$$

   $$h_i = h_o M = 2(0.33) = 0.66\ \text{in.}$$

3. Object at 6 in. (Fig. 30.12c).

   (a) Image position:

   $$\frac{1}{6} + \frac{1}{q} = \frac{1}{-8}$$

   $q = -3.43$ in.   (*virtual and to the left of the vertex*)

   (b) Magnification:

   $$M = \frac{q}{p} = \frac{-3.43}{6} = -0.57 \times \quad (erect\ image)$$

   $$h_i = h_o M = 2(0.57) = 1.14\ \text{in.}$$

The following equation for thin lenses is used when the lens has two radii, r_1 and r_2, as shown in Fig. 30.13(a) and (b), and when approximate results are required. We have seen that this equation for thin lenses is

$$\frac{1}{f} = (n - 1)\left(\frac{1}{r_1} - \frac{1}{r_2}\right)$$

Sec. 30.5 Thin Lenses

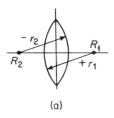
(a)

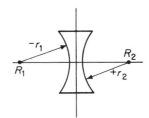
(b)

Fig. 30.13

Example 9

An object is placed 20 in. to the left of a convex–concave thin lens (index of refraction 1.3) which has radii of 12 in. and 5 in. respectively. Calculate (1) the focal distance; (2) the image distance; (3) the magnification. (4) Is the image real or virtual, inverted or erect? (5) Draw the ray diagram.

Solution

1. The focal distance is

$$\frac{1}{f} = (n-1)\left(\frac{1}{r_1} - \frac{1}{r_2}\right)$$

$$= (1.3 - 1)\left(\frac{1}{12} - \frac{1}{-5}\right)$$

$$f = 11.8 \text{ in.}$$

2. The image distance is

$$\frac{1}{p} + \frac{1}{q} = \frac{1}{f} \qquad \left\{\begin{array}{l} n = 1.3 \\ p = +20 \text{ in.} \\ r_1 = +12 \text{ in.} \\ r_2 = -5 \text{ in.} \end{array}\right\}$$

$$\frac{1}{20} + \frac{1}{q} = \frac{1}{11.8}$$

$$q = 28.8 \text{ in.} \qquad (real\ image)$$

3. The magnification is

$$M = \frac{q}{p} = \frac{28.8}{20} = 1.44 \times \qquad (inverted\ image)$$

4. The image is real and inverted.

5. The ray diagram is shown in Fig. 30.14.

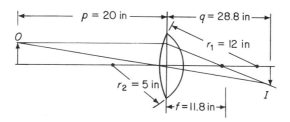

Fig. 30.14

If more accurate results are required, the single-surface equation should be used. This equation is

$$\frac{n}{p} + \frac{n_1}{q} = \frac{n_1 - n}{r}$$

$\{$ n = index of refraction of 1st medium; n_1 = index of refraction of 2nd medium; p = object; q = image; r = radius of spherical surface $\}$

The magnification is

$$M = \frac{nq}{n_1 p} = \frac{h_i}{h_o}$$

If the focal length f_1 in air ($n_1 - 1$) is known, then

$$\frac{1}{f_1} = \left(\frac{n}{n_1} - 1\right)\left(\frac{1}{r_1} - \frac{1}{r_2}\right)$$

$\{$ n = index of refraction of lens; n_1 = 1st medium; n_2 = 2nd medium $\}$

If the same lens is placed into a second fluid (index of refraction n_2), then

$$\frac{1}{f_2} = \left(\frac{n}{n_2} - 1\right)\left(\frac{1}{r_1} - \frac{1}{r_2}\right)$$

Solving both equations for $1/r_1 - 1/r_2$ and equating,

$$\frac{n_1}{f_1(n - n_1)} = \frac{n_2}{f_2(n - n_2)}$$

Example 10

The focal length of a thin lens (index of refraction 1.6) is -8 in. Calculate its focal length in glycerin (index of refraction 1.48).

Solution

$$\frac{n_1}{f_1(n - n_1)} = \frac{n_2}{f_2(n - n_2)}$$

$$\frac{1}{-8(1.6 - 1)} = \frac{1.48}{f_2(1.6 - 1.48)}$$

$\{$ $f_1 = -8$ in.; $n = 1.6$; $n_1 = 1$; $n_2 = 1.48$ $\}$

$$f_2 = -59.2 \quad (diverging)$$

Example 11

A spherical convex surface has a radius of curvature of 2 in. This spherical surface is at one end of a hollow rod which is then filled with ethyl alcohol (index

Sec. 30.5   *Thin Lenses*   **487**

of refraction 1.36). An object is placed 20 in. to the left of the vertex of the sphere. If the object is 0.5 in. high, find (1) the position of the image; (2) the final size of the image. (Neglect the thickness of the glass.)

Solution

1. The position of the image:

$$\frac{n}{p} + \frac{n_1}{q} = \frac{n_1 - n}{r}$$

Substituting,

$$\frac{1}{20} + \frac{1.36}{q} = \frac{1.36 - 1}{2}$$

$$q = 10.5 \text{ in.} \quad (real)$$

$$\begin{Bmatrix} r = +2 \text{ in.} \\ n = 1 \\ n_1 = 1.36 \\ p = 20 \text{ in.} \\ h_0 = 0.5 \text{ in.} \end{Bmatrix}$$

2. The magnification and final size of the image:

$$M = \frac{nq}{n_1 p} = \frac{1 \times 10.5}{1.36 \times 20} = 0.39 \times \quad (inverted)$$

$$h_i = h_o M = 0.5(0.39) = 0.195 \text{ in.}$$

It is also possible to have a combination of two spherical surfaces far enough apart so that it is not possible to consider the combination a thin lens. Under these conditions, the original object may yield a real image. This real image may be either to the right or to the left of the second spherical surface. This image is treated as an object by the second surface. Thus it is theoretically possible to have a real or virtual object with respect to the second surface. This second object may yield either a real or virtual final image.

Example 12

A 10-in. glass rod (index of refraction 1.5) has hemispherical radii ground on each end as shown in Fig. 30.15(a). The radius at the left end of the rod is 4 in. At the right end of the rod, the radius is 6 in. The object is 32 in. to the left of the first surface. Calculate the position of the final image. Draw the ray diagram.

Solution

1. Refraction at the first surface [Fig. 30.15(a)]:

$$\frac{n}{p_1} + \frac{n_1}{q_1} = \frac{n_1 - n}{r_1}$$

Substituting,

$$\frac{1}{32} + \frac{1.5}{q_1} = \frac{1.5 - 1}{4}$$

$$\begin{Bmatrix} n = 1 \\ n_1 = 1.5 \\ r_1 = +4 \\ p_1 = 32 \text{ in.} \end{Bmatrix}$$

$$q_1 = 16 \text{ in.} \quad (real)$$

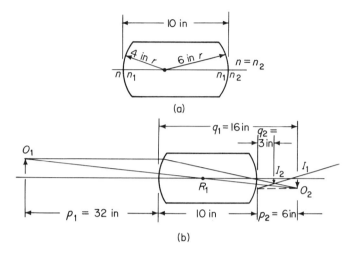

Fig. 30.15

2. Refraction at the second surface (Fig. 30.15b):

$$\frac{n_1}{p_2} + \frac{n_2}{q_2} = \frac{n_2 - n_1}{r_2}$$

Substituting,

$$\frac{1.5}{-6} + \frac{1}{q_2} = \frac{1 - 1.5}{-6} \qquad \left\{ \begin{array}{l} n_1 = 1.5 \\ n_2 = 1 \\ r_2 = -6 \text{ in.} \\ p_2 = -6 \text{ in. (virtual)} \end{array} \right\}$$

$q_2 = 3$ in.   (*real and to right of rod*)

30.6. Thin Lenses in Contact

Two thin lenses in contact may be treated as one thin lens having a combined focal effect equal to the sum of the focal lengths of each of the lenses. Thus

$$\frac{1}{F_{\text{comb}}} = \frac{1}{f_1} + \frac{1}{f_2}$$

It is important that the focal length of the lens combination be kept small with respect to the focal length of each lens.

The proof for the preceding equation revolves about the principle that the image from the first lens is positive and becomes the negative object for the second lens. Therefore, for the first lens,

$$\frac{1}{p_1} + \frac{1}{q_1} = \frac{1}{f_1}$$

Sec. 30.6   Thin Lenses in Contact

If the combination is considered a thin lens and if
$$q_1 = -p_2$$
then $p_2 = -q_1$, and for the second lens,
$$\frac{1}{-q_1} + \frac{1}{q_2} = \frac{1}{f_2}$$
Solving both equations for $1/q_1$ and equating,
$$\frac{1}{F_{comb}} = \frac{1}{p_1} + \frac{1}{q_2} = \frac{1}{f_1} + \frac{1}{f_2}$$
This shows that the *initial* object position and the final *image* position are functions of the sum of the focal lengths of the separate lenses.

We have seen that the diopter is the inverse of the focal length expressed in meters. Therefore, it is also true that
$$D = D_1 + D_2$$

Example 13

Three thin lenses are placed in contact to form a single thin lens. The focal lengths of the lenses are -1.0 cm, $+4.0$ cm, and $+3.0$ cm. (1) What is the focal length of the combination? (2) What is the diopter power of the combination? (3) Find the image position, if the object is 160 cm to the left of the center of the lens combination.

Solution

1. Focal length of the combination:
$$\frac{1}{F_{comb}} = \frac{1}{-1.0} + \frac{1}{4.0} + \frac{1}{3.0} = -\frac{5}{12}$$
$$F_{comb} = -2.4 \text{ cm} \quad (\textit{diverging lens})$$

2. Diopter power:
$$D = \frac{1}{-2.4 \times 10^{-2} \text{ m}}$$
$$= -41.67 \text{ diopters} \quad (\textit{diverging lens})$$

3. Image position:
$$\frac{1}{F_{comb}} = \frac{1}{p} + \frac{1}{q}$$
$$\frac{1}{-2.4} = \frac{1}{160} + \frac{1}{q}$$
Then
$$q = -2.36 \text{ cm} \quad (\textit{to the left of the lens})$$

30.7. Aberrations

Chromatic aberration occurs because the index of refraction differs for different wavelengths of light. The lens separates light according to wavelength and several color images result instead of a single predicted image. This is shown in Fig. 30.16 and is called *axial* chromatic aberration. Aberration normal to the principal axis is called *lateral* chromatic aberration. It results from differences in the height of the image.

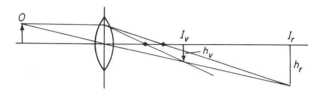

Fig. 30.16

Other aberrations which may take place result in the distortion of images even if the light is monochromatic. They are the following:

1. *spherical* aberration, which results when parallel rays approach a spherical lens and the outer rays are refracted more than the centrally located rays;
2. *astigmatism*, which results when rays from a distant object fail to focus at a point, but converge in an elliptical cross-section, the minor axis degenerating into a line image;
3. *curvature-of-field* aberration, when primary and secondary images have different degrees of flatness;
4. *coma*, when the differences in magnification for different parts of a lens throw the image off the principal axis;
5. *distortion* (lateral and pincushion), which results from differences in magnification normal to the principal axis.

Problems

1. A convex mirror has a radius of curvature of 20 cm. An object is placed 50 cm to the left of the vertex. Calculate (a) the position of the image; (b) the image size if the object is 2 cm high. (c) What is the nature of the image? (d) Draw the ray diagram.
2. A convex mirror has a radius of curvature of 10 in. A $1\frac{1}{2}$-in. object is placed 5

in. to the left of the vertex. Calculate (a) the position of the image; (b) the image size. (c) Is the image real or virtual, inverted or erect? (d) Draw the ray diagram.
3. A concave mirror has a radius of curvature of 10 in. A 2-in. object is placed 8 in. to the right of the vertex. Calculate (a) the image position; (b) the size of the image. (c) Characterize the image as virtual or real, inverted or erect. (d) Draw the ray diagram.
4. A concave mirror has a focal length of 5 in. The magnification is $5\times$. (a) Calculate the object and image distances from the vertex. (b) Draw the ray diagram.
5. An object is placed 20 in. to the left of the vertex of a mirror. If the image is 5 in. to the right of the mirror, (a) calculate the focal length of the mirror. (b) Is the mirror concave or convex? (c) Is the image erect or inverted, virtual or real? (d) Calculate the radius of curvature of the mirror.
6. An object is 5 in. to the left of a concave mirror. The focal length of the mirror is 12 in. Calculate (a) the image position; (b) the image size if the object is 2 in. long. (c) Is the image real or virtual, erect or inverted? (d) Draw the ray diagram.
7. A light ray strikes a piece of quartz at an angle of incidence of $30°$. What is the angle of refraction in the quartz?
8. The index of refraction of water is 1.33. Light, wavelength 6000 A units (6000 $\times$ 10^{-8} cm), strikes the water normal to its surface. Calculate (a) the velocity of this light in the water; (b) the wavelength of the light in the water.
9. A plate glass 6 mm thick is placed so that it intercepts a point source of light of 500-mμ wavelength. If the distance from a point source to a screen is 4 cm, find the number of waves between the source and the screen. The index of refraction of the glass is 1.5.
10. An object is 10 ft below the surface of a tank of glycerin (index of refraction 1.48). How far below the surface of the glycerin does the image appear to be if viewed normal to the surface?
11. Calculate the critical angle for quartz.
12. A bulb, propagating a narrow ray of light, is immersed into a tank of glycerin. (a) What is the critical angle for the glycerin? (b) What is the diameter of the largest bright spot possible if the bulb is 4 in. below the surface of the glycerin ($n = 1.48$)?
13. Repeat Prob. 12 if the liquid is water.
14. An object is placed 10 in. to the left of a positive lens, focal length 5 in. Calculate the image position, nature, orientation, and magnification. Draw the diagram.
15. An object is 30 in. to the left of a positive lens, focal length 12 in. Calculate the image position, nature, orientation, and magnification. Draw the diagram.
16. Repeat Prob. 15 if the object is 8 in. to the left of the vertex.
17. A diverging lens has a 2-in. object placed 20 in. to the left of the vertex. The focal length of the lens is 8 in. Calculate (a) the image position; (b) the image size. (c) Is the image real or virtual, inverted or erect? (d) Draw the ray diagram.
18. Repeat Prob. 17 when the object is 6 in. to the left of the vertex.
19. A thin flint-glass, double-concave lens is constructed so that the first radius is 5 in. and the second is 10 in. Calculate the focal length of the lens.
20. Work Prob. 19 if the first surface is concave and the second surface is convex.
21. In Prob. 19, an object is placed 30 in. to the left of the lens. Calculate (a) the

492   *Mirrors and Thin Lenses*   Chap. 30

position of the image; (b) the magnification. (c) Is the image real or virtual, inverted or erect? (d) Draw the ray diagram.

22. In Prob. 20, an object is placed 8 in. to the left of the lens. Calculate (a) the image position; (b) the magnification. (c) Is the image real or virtual, inverted or erect? (d) Draw the ray diagram.

23. An object 20 in. to the left of a thin lens casts an image 8 in. to the left of the lens. If the first radius is 10 in. concave and the second radius is 20 in. convex, calculate the index of refraction of the lens.

24. Calculate the focal length in Prob. 23 in diopters.

25. A rod of thin glass is hollow and has a 4-in. spherical convex radius at its left end. The rod is filled with a liquid (index of refraction 1.4). An object is placed 30 in. to the left of the vertex. If the object is 0.5 in. long, calculate (a) the position of the image; (b) the image size. (c) Is the image real or virtual, inverted or erect?

26. Work Prob. 25 if the radius in the end of the rod is spherical concave.

27. A spherical object, diameter of 12 in. and made of glass, has a small object embedded 2 in. to the left of the center of curvature. If the object is viewed from the left, how far from the center of the sphere does the object appear to be? What is the magnification?

28. A quartz rod is ground with a 6-in. concave spherical surface at one end. An object is placed 18 in. to the left of the spherical surface. If the object is $\frac{1}{2}$-in. high, calculate (a) the position of the image; (b) the size of the image. (c) Characterize the image as real or virtual, inverted or erect.

29. A crown-glass rod is ground to a 2-in. convex spherical surface and then immersed into a carbon tetrachloride solution. A $\frac{1}{4}$-in. object is placed 4 in. to the left of the vertex. Both object and rod are submerged. Where is the image?

30. A rod (index of refraction 1.6) is ground at both ends to hemispherical surfaces of 1.2-in. and 6-in. radii, respectively, as shown in Fig. 30.17(a). If the rod is 6 in. long and the object is 8 in. to the left of the rod, calculate the final image distance.

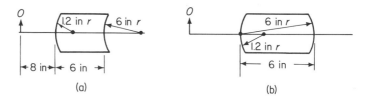

Fig. 30.17

31. Solve Prob. 30 if the hemispherical surfaces are ground as shown in Fig. 30.17(b).
32. A rod (index of refraction 1.4) is ground as shown in Fig. 30.18(a) and has an object as shown. Calculate the position of the final image.
33. Repeat Prob. 32 for Fig. 30.18(b).
34. Repeat Prob. 32 for Fig. 30.18(c).

Problems

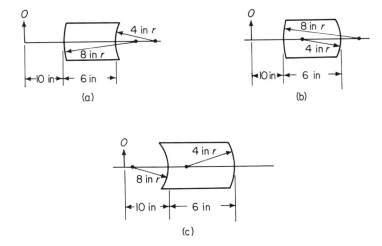

Fig. 30.18

35. Two thin lenses in contact have focal lengths of 2.8 diopters and 20 cm. Calculate (a) focal length; (b) the diopter power of the combination.
36. The focal length of a thin lens (index of refraction 1.5) is 18 cm. The lens is immersed in alcohol (index of refraction 1.3). Find the focal length of this lens in the alcohol.
37. A thin lens having radii of 15 cm and 10 cm, respectively, has an index of refraction of 1.6. What is its focal length in water (index of refraction 1.3)?

31

Multiple Lenses

31.1. The Eye

The eye, as an optical instrument, is very flexible in its ability to accommodate to illuminance or brightness, to form a sharp image of distant or near objects, to receive very small quantities or very large quantities of light, and to respond to small changes in wavelength (which makes it possible to detect color differences).

The eye, approximately 1.8 cm in diameter, is nearly spherical in shape (Fig. 31.1). Its lens is made up of layers of transparent tissue. Its density increases toward the center and its average index of refraction is 1.437. Immediately in front of the lens is a cavity filled with fluid known as the *aqueous humor* (index of refraction 1.336). Confining and protecting this fluid is the cornea which, of course, is also transparent (index of refraction 1.376). The iris, with its central opening, the pupil, contracts or dilates to control the amount of light permitted to pass through the lens.

Behind the lens is another fluid (index of refraction 1.336), called the *vitreous humor*. This humor is confined by the retina, which is a layer of nerve fibers sensitive to light. The retina is composed of rods and cones, which are nerve ends coming from the optic nerve. The rods are sensitive to light, but seem not to be sensitive to color. The cones are sensitive to color. They are most numerous at the fovea, which has the highest sensitivity of vision. Because of this high sensitivity, the eye has "learned" to focus the image it receives at the fovea. Since that part of the retina where the nerve fibers leave the eye has neither rods nor cones, there is a complete absence of vision at this point. This is called the *blind spot*.

Sec. 31.1   *The Eye*   495

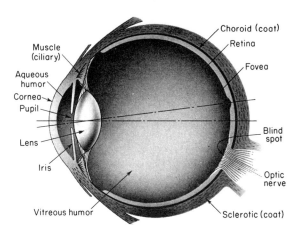

Fig. 31.1

The refraction of a light ray is accomplished by the cornea and the lens. The lens is completely relaxed for light rays from infinity and the focal point of the lens is at the fovea. As the object approaches the eye, sets of muscles cause the lens to become more spherical, so that the image will once again be focused on the fovea for clear vision. This accommodation may take place for images at infinity and for images as near as approximately 7 cm in young people. When a clear image is formed at infinity, that point is called the *far point*. As a person's age increases, it becomes more difficult to make the near-point accommodation. The average normal distance for clear vision is taken as 25 cm (10 in.). The normal eye, the farsighted eye, and the nearsighted eye are shown in Fig. 31.2 (a), (b), and (c).

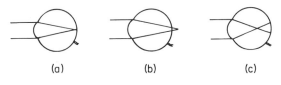

(a)   (b)   (c)

Fig. 31.2

The *near point* is that distance from the eye within which the eye cannot see an object clearly. That is, if the near point of an eye is at 50 cm, an object placed at 30 cm cannot be seen clearly. It therefore becomes necessary to project an image to the 50-cm distance for clear vision. A person who cannot see objects closer than his near point is said to be *farsighted*.

If the *far point* of an eye is at 300 cm in front of that eye, this means that the eye cannot clearly see an object *beyond* the 300-cm point. Thus if an object is at 500 cm, its image must be projected to 300 cm for clear vision. Since objects farther from the eye than the far point cannot be seen clearly, this person is said to be *nearsighted*.

Thus nearsighted people have a far point beyond which they cannot see objects clearly. Farsighted people have a near point closer than which they cannot see objects clearly.

Example 1

The far point of an eye is at 30 cm. Calculate the focal length of a lens needed to see an object clearly at 80 cm [see Fig. 31.3(a)].

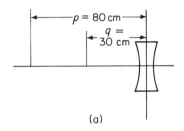
(a)

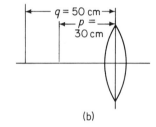
(b)

Fig. 31.3

Solution

$$\frac{1}{f} = \frac{1}{p} + \frac{1}{q} = \frac{1}{80} + \frac{1}{-30} \qquad \left\{ \begin{array}{l} p = 80 \text{ cm} \\ q = -30 \text{ cm} \end{array} \right\}$$

$$f = -48 \text{ cm} \qquad (diverging)$$

Example 2

The near point of an eye is at 50 cm. Calculate the focal length of a lens needed to see an object clearly at 30 cm.

Solution

$$\frac{1}{f} = \frac{1}{30} + \frac{1}{-50} \qquad \left\{ \begin{array}{l} p = 30 \text{ cm} \\ q = -50 \text{ cm} \end{array} \right\}$$

$$f = 75 \text{ cm} \qquad (converging)$$

31.2. Lens Systems

Lens combinations *not* in contact with each other are calculated using the principle that the image from the first lens is used as the object for the

Lens Systems

second lens. The methods to be used may best be demonstrated by the use of illustrated examples.

Example 3

An object is 1.5 in. long and is 30 in. in front of a converging lens, focal length 10 in. A second converging lens, focal length 15 in., is 35 in. to the right of the first lens. Calculate (1) the position of the final image; (2) the size and character of the final image. (3) Draw the ray diagram.

Solution

1. Image position.

    (a) Position of the first image:

    $$\frac{1}{p_1} + \frac{1}{q_1} = \frac{1}{f_1}.$$

    $$\frac{1}{q_1} = \frac{1}{f_1} - \frac{1}{p_1} = \frac{1}{10} - \frac{1}{30} \qquad \left\{ \begin{array}{l} p = +30 \text{ in.} \\ f_1 = +10 \text{ in.} \end{array} \right\}$$

    $$q_1 = +15 \text{ in.}$$

    (b) Second-object distance:

    $$p_2 = 35 - 15 = 20 \text{ in.}$$

    (c) Position of the second image:

    $$\frac{1}{p_2} + \frac{1}{q_2} = \frac{1}{f_2}$$

    $$\frac{1}{q_2} = \frac{1}{f_2} - \frac{1}{p_2} = \frac{1}{15} - \frac{1}{20} \qquad \left\{ \begin{array}{l} f_2 = +15 \text{ in.} \\ p_2 = 20 \text{ in.} \end{array} \right\}$$

    $$q_2 = +60 \text{ in.} \qquad \text{(right of the second lens)}$$

2. Size and character of the final image:

    (a) Magnification:

    $$M = \left(\frac{q_1}{p_1}\right)\left(\frac{q_2}{p_2}\right) = \frac{15}{30} \times \frac{60}{20} = 1.5 \times$$

    (b) Image length:

    $$h_i = h_o M = 1.5(1.5) = 2.250 \text{ in.}$$

    (c) Characterization:
    Real final image. The first image is inverted with respect to the first object. The second (final) image is erect with respect to the initial object.

3. The ray diagram is as shown in Fig. 31.4.

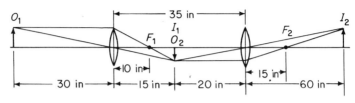

Fig. 31.4

Example 4

Two lenses are 16 in. apart. The focal length of the first lens is +15 in. The focal length of the second lens is −10 in. A 2-in. object is placed 40 in. to the left of the first lens. Calculate (1) the position of the final image; (2) the magnification and character of the final image. (3) Draw the ray diagram.

Solution

1. The final image.

   (a) Position of the first image:

   $$\frac{1}{p_1} + \frac{1}{q_1} = \frac{1}{f_1}$$

   $$\frac{1}{q_1} = \frac{1}{f_1} - \frac{1}{p_1} = \frac{1}{15} - \frac{1}{40} \qquad \left\{ \begin{array}{l} p_1 = +40 \text{ in.} \\ f_1 = +15 \text{ in.} \end{array} \right\}$$

   $$q_1 = 24 \text{ in.}$$

   (b) Second-object distance:

   $$p_2 = 16 - 24 = -8 \text{ in.} \qquad \left\{ \begin{array}{l} P_2 = -8 \text{ in.} \\ f_2 = -10 \text{ in.} \end{array} \right\}$$

   (c) Position of the second image:

   $$\frac{1}{p_2} + \frac{1}{q_2} = \frac{1}{f_2}$$

   $$\frac{1}{q_2} = \frac{1}{f_2} - \frac{1}{p_2} = \frac{1}{-10} - \frac{1}{-8} \qquad \left\{ \begin{array}{l} p_2 = -8 \text{ in.} \\ f_2 = -10 \text{ in.} \end{array} \right\}$$

   $$q_2 = +40 \text{ in.}$$

2. Magnification and character of the image.

   (a) Magnification:

   $$M = \left(\frac{q_1}{p_1}\right)\left(\frac{q_2}{p_2}\right) = \left(\frac{24}{40}\right) \times \left(\frac{40}{-8}\right) = -3 \times$$

   (b) Image length:

   $$h_i = h_o M = 2(3) = 6 \text{ in.}$$

(c) Characterization:

Real final image. The first image is inverted with respect to the first object. The second image is erect with respect to the second object. Therefore, the final image is inverted with respect to the initial object.

3. The ray diagram is shown in Fig. 31.5

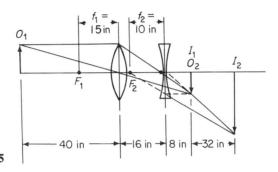

Fig. 31.5

Example 5

A positive lens, focal length 8 in., is placed 30 in. to the left of a concave mirror, focal length 5 in. An object is placed 40 in. to the left of the lens. Calculate (1) the position of the final image with respect to the original object; (2) the magnification and characterization of the final image. (3) Draw the ray diagram.

Solution

1. Image position.

    (a) Position of the first image:

    $$\frac{1}{p_1} + \frac{1}{q_1} = \frac{1}{f_1}$$

    $$\frac{1}{q_1} = \frac{1}{f_1} - \frac{1}{p_1} = \frac{1}{8} - \frac{1}{40} = \frac{40-8}{320} = \frac{32}{320} \qquad \left\{ \begin{array}{l} p_1 = 40 \text{ in.} \\ f_1 = 8 \text{ in.} \end{array} \right\}$$

    $$q_1 = 10 \text{ in.}$$

    (b) Second-object distance:

    $$p_2 = 30 - 10 = 20 \text{ in.}$$

    (c) Position of the final image from the initial object:

    $$\frac{1}{p_2} + \frac{1}{q_2} = \frac{1}{f_2}$$

$$\frac{1}{q_2} = \frac{1}{f_2} - \frac{1}{p_2} = \frac{1}{5} - \frac{1}{20} = \frac{20-5}{100} = \frac{3}{20} \qquad \left\{ \begin{array}{l} p_2 = 20 \text{ in.} \\ f_2 = 5 \text{ in.} \end{array} \right\}$$

$$q_2 = 6.7 \text{ in.} \qquad \text{(to left of the mirror)}$$

The position of the image with respect to the initial object is

$$40 \text{ in.} + 30 \text{ in.} - 6.7 \text{ in.} = 63.3 \text{ in.}$$

2. Magnification and characterization.

(a) $M = \left(\dfrac{q_1}{p_1}\right)\left(\dfrac{q_2}{p_2}\right) = \dfrac{10}{40} \times \dfrac{6.7}{20} = 0.08 \times$

(b) The final image is erect with respect to the initial object.

3. The ray diagram is shown in Fig. 31.6

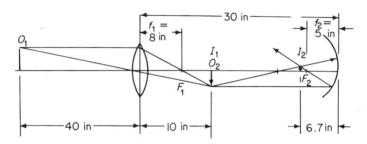

Fig. 31.6

31.3. The Magnifying Glass

The *magnifying glass* is a simple positive lens which creates a virtual erect image when the object is just inside the focal point, as shown in Fig. 31.7(a). Since the most distinct vision occurs at 25 cm (10 in.) in front of the eye, the image should be formed at that position. To calculate the magnification (M_e), p is positive, q is negative, and f is positive.

Substituting these values into the lens equation,

$$\frac{1}{p} + \frac{1}{-25} = \frac{1}{f}$$

Solving for $1/p$,

$$\frac{1}{p} = \frac{1}{f} + \frac{1}{25}$$

Multiplying through by 25,

$$\frac{25}{p} = \frac{25}{f} + 1$$

Sec. 31.3   The Magnifying Glass   501

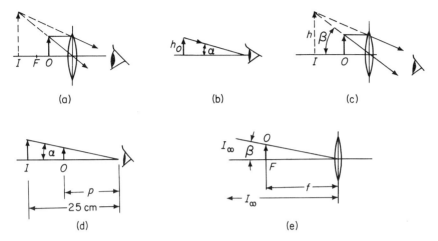

Fig. 31.7

Since $25/p$ is the magnification, then

$$M_e = \frac{25}{f} + 1$$

The magnification may also be expressed as an angular relationship by considering the ratio of the visual angle α created by the object when viewed by the naked eye [Fig. 31.7(b)] and the angle β made by the apparent image with the lens interposed [Fig. 31.7(c)].

The angular magnification is

$$M_e = \frac{\beta}{\alpha} \qquad \left\{ \begin{array}{l} \beta = \text{angle when lens is inserted} \\ \alpha = \text{angle with naked eye} \end{array} \right\}$$

Example 6

An image is created at 25 cm when a lens, 10-cm focal length, is used as a magnifier. (1) How large is the image if the object is 4 cm long? (2) Where is the object? (3) Verify the magnification using the object—image-distance ratio.

Solution

1. Image magnification:

$$M = \frac{25}{f} + 1 = \frac{25}{10} + 1 = 3.5 \times$$

Image size:

$$h_i = h_o M = 4\,(3.5) = 14 \text{ cm}$$

2. Object distance: $\{f = 10 \text{ cm}, q = -25 \text{ cm}\}$

$$\frac{1}{p} = \frac{1}{f} - \frac{1}{q} = \frac{1}{10} - \frac{1}{-25}$$

$$p = 7.1 \text{ cm}$$

3. Verification of the magnification:

$$M = \frac{q}{p} = \frac{-25}{7.1} = 3.5\times \quad (check)$$

When a virtual image is caused to form at infinity, the angles α and β are very small and an approximation may be used, such that the magnification is a function of the angles as shown in Fig. 31.7(d) and (e).

This is made possible when the image is at infinity, because the angle subtended by O in Fig. 31.7(d) and the angle subtended by I_∞ in Fig. 31.7(e) are equal. The equation is

$$M_e = \frac{\beta}{\alpha} = \frac{O/f}{O/25} = \frac{25}{f}$$

31.4. The Compound Microscope

The fundamental feature of the lens system in a microscope is a system of two lenses. The lens nearer the eye is the *eyepiece*. The lens nearer the object is the *objective*. In a *compound microscope*, the eyepiece has a normal focal length and an objective with a very short focal length.

If the object is close to the focal point of the objective lens, it will form an image I_o inside the focal point of the eyepiece, as shown in Fig. 31.8(a). This image I_o becomes the objective O_e for the eyepiece which forms a large virtual inverted image.

The magnification for the objective lens is

$$M_o = \frac{q_o}{p_o}$$

The magnification of the eyepiece is

$$M_e = \frac{25}{f_o} + 1$$

The total magnification is

$$M = M_o M_e = \frac{q_o}{p_o}\left(\frac{25}{f_e} + 1\right)$$

Sec. 31.5 Telescopes

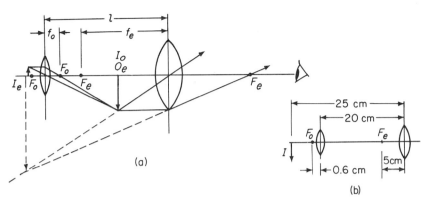

Fig. 31.8

Example 7

The distance between two converging lenses in a microscope is 20 cm. The focal length of the eyepiece is 5 cm; the focal length of the objective is 0.6 cm. A clear image is to be formed at 25 cm to the left of the eyepiece. Calculate (1) the position of the object viewed; (2) the magnification [see Fig. 31.8(b)].

Solution

1. The object distance for the eyepiece lens is

$$\frac{1}{p_e} = \frac{1}{f_e} - \frac{1}{q_e} = \frac{1}{5} - \frac{1}{-25}$$

$$p = 4.17 \text{ cm}$$

The image distance for the objective is

$$q_o = 20 - 4.17 = 15.83 \text{ cm}$$

$\left\{ \begin{array}{l} l = 20 \text{ cm} \\ f_e = 5 \text{ cm} \\ f_0 = 0.6 \text{ cm} \\ q_e = -25 \end{array} \right\}$

The object distance for the objective lens is

$$\frac{1}{p_o} = \frac{1}{f_o} - \frac{1}{q_o} = \frac{1}{0.6} - \frac{1}{15.83}$$

$$p_o = 0.62 \text{ cm to the left of the objective}$$

2. The magnification is

$$M = \frac{q_o}{p_o}\left(\frac{25}{f_e} + 1\right) = \frac{15.83}{0.62}\left(\frac{25}{5} + 1\right) = 153.2 \times \quad (inverted)$$

31.5. Telescopes

The *astronomical telescope* is shown in Fig. 31.9(a). The object to be viewed is a long way off. Note that the focal point of the objective F_o is shown to be *just* inside the focal point of the eyepiece F_e. For our purposes, the two focal points are close enough to each other to be considered as one. Since the object O_o is a long way off, the image I_o will be very close to F_o. This image I_o is taken to be the object O_e for the eyepiece and it will form a very large image I_e.

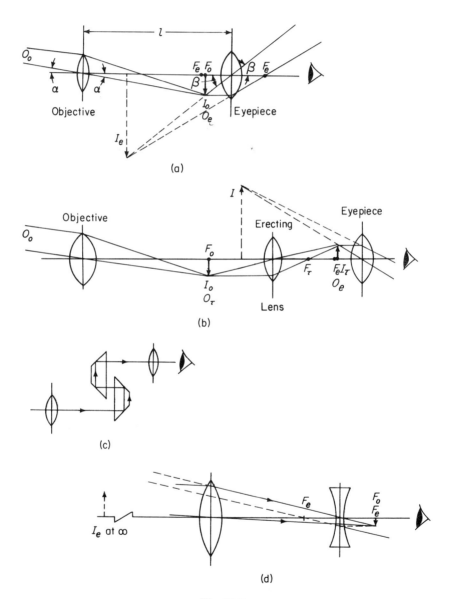

Fig. 31.9

The length of the telescope is
$$L = f_o + f_e$$
The angular magnification will be
$$M = \frac{\text{angle subtended at eye by } I_e}{\text{angle subtended by the unaided eye } O_o} = \frac{\beta}{\alpha}$$

If the object O_o is at infinity (far away), then f_e and f_o will, for all practical purposes, coincide. Thus
$$M = \frac{\beta}{\alpha} = \frac{f_o}{f_e} \quad \text{or} \quad M = \frac{f_o}{f_e}$$

This inverted image may be corrected by inserting a correcting (erecting) lens between the objective and the eyepiece. This lens does not affect the magnification. It increases the length of the telescope by four times the focal length of the erecting lens (see Fig. 31.9b). The proof of the additional length follows:

Since the magnification is not affected,
$$m = \frac{q}{p} = 1$$
Therefore, the object distance equals the image distance; or
$$p_t = q_t$$
Substituting into the lens equation,
$$\frac{1}{f_t} = \frac{1}{p_t} + \frac{1}{p_t} \quad \text{or} \quad \frac{1}{f_t} = \frac{2}{p_t}$$
Therefore,
$$2f_t = p_t \quad \text{and} \quad 2f_t = q_t$$
Adding,
$$p_t + q_t = 2f_t + 2f_t = 4f_t$$
Therefore, the length of the telescope with erecting lens is
$$L_t = f_o + f_e + 4f_t$$

Since this makes the telescope rather long, the path in prism binoculars is erected by reflecting prisms as shown in Fig. 31.9(c).

Example 8

A telescope has an objective lens (focal length $+30$ cm) and an eyepiece (focal length $+2$ cm). If an erecting lens (focal length 8 cm) is inserted, calculate (1) the magnification; (2) the length of the telescope; (3) the length of the telescope with the erecting lens.

Solution

1. The magnification is

$$M = \frac{f_o}{f_e} = \frac{30}{2} = 15\times$$

2. The length of the telescope is

$$\left\{\begin{array}{l} f_o = 30 \text{ cm} \\ f_e = 2 \text{ cm} \\ f_t = 8 \text{ cm} \end{array}\right\}$$

$$L = 30 + 2 = 32 \text{ cm}$$

3. The length of the telescope with the erecting lens is

$$L_t = f_o + f_e + 4f_t = 30 + 2 + 4(8) = 64 \text{ cm}$$

By substituting a diverging eyepiece, as in Fig. 31.9(d), the final image remains erect without increasing the overall length of the telescope. This is the *Galilean telescope*.

The magnification is

$$M = \frac{f_o}{f_e}$$

Since f_e is negative, the magnification will become positive, showing that the final image is erect. It is better to leave the equation as shown and apply the conventions as needed.

The length of the telescope becomes

$$L_g = f_o + (-f_e) = f_o - f_e$$

The reflecting telescope is used to eliminate chromatic and spherical aberration. This is possible because a parabolic mirror is used instead of a lens (Fig. 31.10). These telescopes have high *light-gathering power*. The light-gathering power of a reflecting telescope is a function of the area of the objective lens, or the square of the diameter. Thus a 100-in. telescope will gather 4 times the light of a 50-in. telescope.

The magnification is the product of the focal length of the mirror divided by the focal length of the eyepiece, or

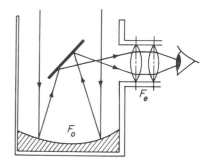

Fig. 31.10

$$M = \frac{f_o}{f_e}$$

Example 9

A concave mirror (assumed to be spherical) has a radius of curvature of 1200 cm. The focal length of the eyepiece is 1.5 cm. What is the magnifying power of the reflecting telescope?

Solution

$$M = \frac{f_o}{f_e} = \frac{1200/2}{1.5} = 400 \times \qquad \begin{Bmatrix} r = 1200 \text{ cm} \\ f_o = 1200/2 \text{ cm} \\ f_e = 1.5 \text{ cm} \end{Bmatrix}$$

31.6. The Camera and the Projector

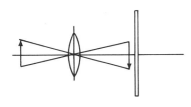

Fig. 31.11

The *camera* lens is essentially a converging lens, which will form a real inverted image on a photosensitive plate, as shown in Fig. 31.11. A shutter controls the exposure time and the diaphragm keeps light away from the periphery of the lens, thus eliminating distortion. Short exposure time, large aperture, and wide field coverage are desirable.

The *F-number* is the light-gathering power of a lens acting in conjunction with the aperture. The focal length and the aperture opening are combined to indicate the relative aperture so that

$$\text{relative aperture} = \frac{d}{f}$$

The *F*-number is the reciprocal of the relative aperture and is

$$F\text{-number} = \frac{f}{d}$$

That is, a lens designed as $f/6$ means that the diameter of the aperture is reduced to one-sixth its focal length:

$$\frac{f}{6} = \frac{1}{6} \times f = \text{effective diameter of aperture opening}$$

A lens designated as $f/12$ has an effective opening one-half smaller and will collect one-fourth the light gathered by the $f/6$ lens. The exposure time increases as the square of the *F*-number.

Thus the relationship between the exposure time and the *F*-number is

$$\frac{t_1}{(f/d_1)^2} = \frac{t_2}{(f/d_2)^2} \qquad \begin{Bmatrix} t_1 = \text{exposure time for } t/d_1 \\ t_2 = \text{exposure time for } t/d_2 \end{Bmatrix}$$

Example 10

A camera uses a speed of $\frac{1}{40}$ second when the *F*-number is $f/2$. Calculate the exposure time if the *F*-number is $f/8$.

Solution

$$t_2 = t_1\left(\frac{d_1}{d_2}\right)^2 = \frac{1}{40}\left(\frac{f/2}{f/8}\right)^2 \quad \left\{\begin{array}{l} d_1 = f/2 \\ d_2 = f/8 \\ t_1 = 1/40 \text{ sec} \end{array}\right\}$$

$$= \frac{2}{5} \text{ sec}$$

The *projector* (Fig. 31.12) uses the reverse procedure of the camera. In this case, the instrument illuminates a small inverted object and projects a large erect imate. A condensing lens collects light from a source and concentrates it upon the slide. The slide is slightly to the left of the focal point of the objective. This forms a large real image on a screen.

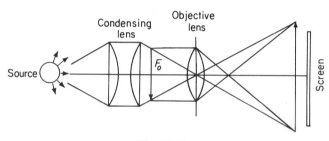

Fig. 31.12

Example 11

A projection lens illuminates an object slide 2 cm long and throws an image upon a screen 20 m from the lens. If the lens has a focal length of 40 cm, what is the size of image?

Solution

The object distance is

$$\frac{1}{p} = \frac{1}{f} - \frac{1}{q} = \frac{1}{40} - \frac{1}{2000} = 40.8 \text{ cm}$$

The magnification is $\quad\quad\quad\quad\quad\quad\quad\quad\quad\quad \left\{\begin{array}{l} f = 40 \text{ cm} \\ q = 2000 \text{ cm} \end{array}\right\}$

$$M = \frac{q}{p} = \frac{2000}{40.8} = 49 \times$$

The image size is

$$49 \times 2 = 98 \text{ cm} \quad\quad (inverted)$$

Problems

1. Calculate the focal length of a lens needed to correct farsighted vision if the near point is 60 cm.
2. Calculate the focal length of a lens needed to correct nearsighted vision to see an object at infinity if the far point is at 60 cm.
3. A farsighted person has a near point of 40 cm. Calculate the focal length needed for normal vision.
4. A certain eye is to view an object at 90 cm. The correction is made with a diverging lens, focal length 50 cm. Where is the far point?
5. An object is to be seen clearly at 50 cm. The near point of the eye is at 75 cm. Calculate the focal length of the lens.
6. An object is to be seen clearly at 120 cm. The far point of the eye is at 40 cm. Calculate the focal length of the lens.
7. An object is 2 in. long and is 50 in. in front of a converging lens of focal length 10 in. A second converging lens (focal length 15 in.) is 37.5 in. to the right of the first lens. (a) Where is the final image? (b) What is the size and character of the final image? (c) Draw the ray diagram.
8. Two positive lenses are used as in combination to project an image. A 2-cm object is placed 16 cm to the left of the first lens (focal length 8 cm). The second lens (focal length 5 cm) is placed 30 cm to the right of the second lens. Calculate the final image (a) position; (b) size; (c) characterization. (d) Draw the ray diagram.
9. Repeat Prob. 8 if the second lens is placed 21 cm to the right of the first lens.
10. Repeat Prob. 8 if the second lens is placed 18 cm to the right of the first lens.
11. Repeat Prob. 8 if the second lens is at 10 cm to the right of the first lens.
12. A diverging lens (focal length 12 cm) is placed 60 cm to the right of a converging lens (focal length 4 cm). A 3-cm object is placed 10 cm to the left of the system. Calculate the final image (a) position; (b) magnification; size; (c) the characterization. (d) Draw the ray diagram of the system.
13. Repeat Prob. 12 if the distance between the lenses is 15 cm.
14. A converging lens (focal length 5 cm) is placed 40 cm to the left of a concave mirror (radius of curvature 18 cm). An object is placed 88 cm to the left of the lens. (a) Where is the final image? (b) What is its magnification and the nature of the final image? (d) Draw the ray diagram.
15. Assuming the data given in Prob. 14, except that the mirror is convex, solve the problem.
16. Calculate the focal length of a 20 × magnifying glass.
17. The magnifying power of a lens is 15 ×. If the screen is placed 20 cm to the right of the lens. (a) Calculate the position of the object which will produce a clear image at the screen. (b) Assume the image to be 36 cm long. How long is the object?
18. A 12 × magnifying glass produces a virtual image at 18 cm. If the object is 6 cm long, calculate (a) the focal length of the lens; (b) the size of the image; (c) the position of an object for clear vision.

19. A man is to read a newspaper with a four-power magnifier. If the man wishes to hold the magnifier 3 cm from the newspaper, (a) calculate the position of the image. Calculate the focal length of the lens for clear vision (b) at 25 cm; (c) at infinity.
20. A compound microscope is to view an object at 0.7 cm from the objective lens (focal length 0.5 cm). The eyepiece is 14 cm from the objective. A clear final image is formed 35 cm to the left of the eyepiece. Calculate (a) the focal length of the eyepiece; (b) the lateral magnification.
21. The distance between two converging lenses in a microscope is 12 cm. The focal length of the eyepiece is 8 cm; the focal length of the objective is 0.6 cm. A clear image is to be formed 25 cm to the left of the eyepiece. Calculate (a) the initial position of the object; (b) the magnification.
22. (a) What is the magnifying power of a telescope having an objective of focal length $+50$ cm and an eyepiece of focal length $+2$ cm? (b) What is the length of the telescope? (c) What is the length of the telescope if an erecting lens (focal length 10 cm) is added?
23. An astronomical telescope has two lenses: 2 diopters, and 12 diopters. Calculate (a) the magnification; (b) the length of the telescope tube.
24. An erecting lens (focal length 5 diopters) is inserted into the telescope in Prob. 23. Calculate the length of the telescope.
25. Calculate the length of the tube if the telescope in Prob. 23 were a Galilean telescope.
26. Calculate the magnification power of a reflecting telescope if a concave mirror has a radius of curvature (assumed to be spherical) of 800 cm and a focal length of 2 cm for the eyepiece.
27. A camera has an F-number of $f/6$ and its shutter is set at $\frac{1}{30}$ sec. Calculate the equivalent exposure time if the F-number is $f/11$.
28. A camera lens has a focal length of 18 cm. The free diameter is 2 cm. (a) Calculate the F-number. (b) Calculate the F-number if the free diameter is 1 cm (c) If the exposure time for the F-number (part a) is $\frac{1}{200}$ sec, calculate the exposure time required for the free diameter on part (b).
29. The focal length of a camera is 12 cm for an $f/1.9$ lens. (a) If the correct exposure is $\frac{1}{100}$ sec for $f/1.9$, what is the exposure time at $f/8$? (b) What is the effective diameter for both stops?
30. A projection lens throws an image on a screen 30 ft away. The lens illuminates an object slide 2 in. high. If the final image must be 24 in. long, calculate the focal length of the lens.
31. Two objectives (focal lengths 16 mm and 1.8 mm) and two eyepieces (angular magnifications $4 \times$ and $10 \times$ are furnished with a microscope. The image is formed at 25 cm from the eypiece. What magnifications are possible? (The distance between the lenses is 15 cm.)

32

A Further Study of Refraction

32.1. Displacement

It has been shown that light refracted at an interface between two media will refract toward the normal if the second medium is denser than the first. It will refract away from the normal if the second medium is less dense than the first. Thus in Fig. 32.1, at interface No. 1 the ray refracts toward the normal and at interface No. 2 the ray refracts away from the normal.

If the two interfaces, Fig. 32.1, are parallel, the entering and emerging rays will be parallel. The proof follows:

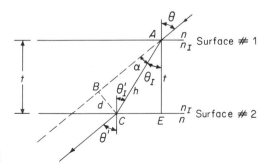

Fig. 32.1

A Further Study of Refraction

1. The refraction at the first interface is

$$n \sin \theta = n_1 \sin \theta_1$$

2. The refraction at the second interface is

$$n_1 \sin \theta_1' = n \sin \theta'$$

3. The alternate interior angle $\theta_1' = \theta_1$. Therefore, substituting θ_1 for θ_1' in the previous equation and equating with the equation in step 1,

$$n \sin \theta = n \sin \theta'$$

4. Dividing both sides of the equation by n,

$$\theta = \theta'$$

The entering ray, Fig. 32.1, and the emerging ray are therefore equal. The following steps are used to calculate the *displacement:*

1. From triangle AEC,

$$\cos \theta_1 = \frac{t}{h}$$

2. Solving for h,

$$h = \frac{t}{\cos \theta_1}$$

3. Since $\alpha + \theta_1 = \theta$,

$$\alpha = \theta - \theta_1$$

4. From triangle ABC,

$$\sin \alpha = \frac{d}{h}$$

5. Solving for d, the displacement is

$$d = h \sin \alpha$$

Example 1

A flat glass plate (index of refraction 1.4), 3 in. thick, is used to displace a light ray. Assume the incident ray enters the first interface at an angle of 40°. Calculate the displacement of the emerging ray.

Solution

1. The refraction angle is

$$n \sin \theta = n_1 \sin \theta_1$$

$$\sin \theta_1 = \frac{n}{n_1} \sin \theta = \frac{1}{1.4} \sin 40° \qquad \left\{ \begin{array}{l} t = 3 \text{ in.} \\ n = 1 \\ n_1 = 1.4 \\ \theta = 40° \end{array} \right\}$$

$$\theta_1 = 27°23'$$

2. Side h is

$$h = \frac{t}{\cos \theta_1} = \frac{3}{\cos 27°23'} = 3.378 \text{ in.}$$

3. The displacement is

$$d = h \sin \alpha = 3.378 \sin (40 - 27°23') \quad \text{where } \alpha = \theta - \theta_1$$
$$= 0.738 \text{ in.}$$

32.2. Deviation

Figure 32.2(a) shows a prism which has an apex angle A. The light enters the left face of the prism at an angle θ with the normal N_1 and is refracted toward the normal N_1, thus creating an angle of θ_1. This ray approaches the right side of the prism at an angle θ_2 with the normal N_2. It emerges at an angle θ_3. If the entering ray is projected (dotted line), it will form an angle D with the emerging ray. This angle D is called the *deviation angle*. Snell's law is used to solve this kind of problem.

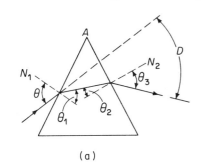

(a)

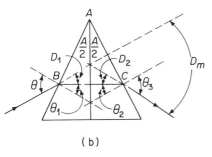

(b)

Fig. 32.2

It can be shown that for a particular entering angle θ, the angle D will be a minimum. This angle will be called the *angle of mimimum deviation*, designated by D_m. For any angle greater than or less than angle θ, the deviation angle will be greater than D_m. In Fig. 32.2(b), the deviation angle will have a minimum value when the entering angle θ is equal to the emerging angle θ_3.

Figure 32.2(b) show a prism where the conditions of refraction create an angle of minimum deviation D_m as just described. The incident ray makes an angle θ with the normal and passes through the prism so that it is perpendicular to the bisector of angle A. This forms an isosceles triangle ABC. Note that the base of the prism does not affect the problem.

The ray refracts at the second surface so that

$$\theta = \theta_3$$

if
$$\theta = \theta_1 + D_1 \quad \text{and} \quad \theta_3 = \theta_2 + D_2$$
Then solving for D_1 and D_2,
$$D_1 = \theta - \theta_1 \quad \text{and} \quad D_2 = \theta_3 - \theta_2$$
But, since D_1 and D_2 are exterior angles, then
$$D_m = D_1 + D_2$$
Substituting,
$$D_m = (\theta - \theta_1) + (\theta_3 - \theta_2) = (\theta + \theta_3) - (\theta_1 + \theta_2)$$
But since
$$\theta_1 + \theta_2 = A \quad \text{and} \quad \theta = \theta_3 \quad (\textit{minimum deviation})$$
then
$$D_m = 2\theta - A$$
Solving for θ,
$$\theta = \frac{D_m + A}{2}$$
Also,
$$\theta_1 = \frac{A}{2}$$
Applying Snell's law,
$$n_1 \sin \theta = n \sin \theta_1 \quad \left\{ \begin{array}{l} A = \text{apex angle} \\ D_m = \text{angle of minimum deviation} \\ n = \text{index of refraction of prism material} \end{array} \right\}$$
Since $n_1 = 1$,
$$n = \frac{\sin \theta}{\sin \theta_1}$$
Substituting for θ and θ_1,
$$n = \frac{\sin [(D_m + A)/2]}{\sin A/2}$$
$$= \frac{\sin \tfrac{1}{2}(D_m + A)}{\sin \tfrac{1}{2} A}$$
When the apex angle is small, a very important approximation is
$$n = \frac{A + D_m}{A}$$

Example 2

Calculate the index of refraction of a 60° glass prism when the angle of minimum deviation is 46°16′.

Solution

$$n = \frac{\sin \frac{1}{2}(A + D_m)}{\sin \frac{1}{2}A} = \frac{\sin \frac{1}{2}(60° + 46°16')}{\sin \frac{1}{2}(60°)} = 1.9 \quad \left\{ \begin{array}{l} D_m = 46°16' \\ A = 60° \end{array} \right\}$$

Example 3

A prism has an apex angle of 60° and an index of refraction of 1.7. Calculate the angle of minimum deviation.

Solution

$$n = \frac{\sin \frac{1}{2}(A + D_m)}{\sin \frac{1}{2}A}$$

$$1.7 = \frac{\sin \frac{1}{2}(60° + D_m)}{\sin \frac{1}{2}(60°)}$$

$$1.7 \times \sin 30° = \sin\left(\frac{60° + D_m}{2}\right)$$

Therefore,

$$\sin^{-1} 0.850 = \sin^{-1}\left(\frac{60° + D_m}{2}\right)$$

or

$$58°13' \times 2 = 60° + D_m$$

Therefore, the angle of minimum deviation is

$$D_m = 56°26'$$

The *spectrometer* shown in Fig. 32.3(a) is used to measure the angle of minimum deviation. The collimator has an adjustable slit placed at the focal point of the lens. If the slit is considered the source of light, rays passing through the lens will converge into parallel rays on leaving the lens. This light is refracted through the prism, which may be rotated and locked into position. The refracted light is then picked up by the telescope and the angular graduations read off a vernier. By manipulating the prism table and the telescope, the angle of minimum deviation may be found.

The apex angle A may be found by reflecting light from both sides of the prism when the apex angle is pointed in the direction of the collimator [Fig. 32.2(b)]. The prism and table are clamped and the angles of reflection are read from both sides; one-half the difference of the two readings yields the apex angle. That is, since

$$\angle i = \angle r \quad \text{and} \quad \delta = \delta_1 = \tfrac{1}{2} A$$

by the law of reflection,

$$\delta_1 = \delta_2$$

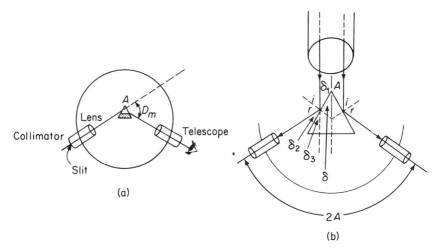

Fig. 32.3

and by alternate interior angles,

$$\delta_1 = \delta_3$$

Therefore,

$$\delta_2 = \delta_3$$

And since $\delta_2 = \delta_3$ is one-half the angular reading, then

$$\delta_2 + \delta_3 = A$$

32.3. Spectra

Before we can proceed with the study of dispersion, we must know something about spectra. As the wavelength of light (color) changes, the velocity of light, and consequently the index of refraction, changes for a given medium. The longer the wavelength, the smaller the index of refraction. The shorter the wavelength, the larger the index of refraction. Thus, as can be seen in Fig. 32.4, the shorter wavelengths refract more than the longer wavelengths: n is greater for short wavelengths than for long wavelengths.

In Fig. 32.4, a prism is used to separate the incident white light into its component colors. Following the reasoning just given, each wavelength will be refracted differently, thus separating the various wavelengths—from red, which has the longest wavelength and smallest refracting index, to violet, which has the shortest wavelength and largest refracting index. This array of color is called a *spectrum*.

The spectrum shown in Fig. 32.5(a) is *continuous* since the colors resolve

Sec. 32.3 *Spectra* 517

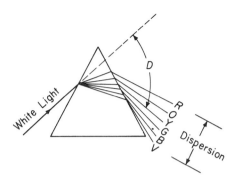

Fig. 32.4

one into the other with no break between the various colors. Such spectra result from incandescent solids or liquids and from gases under high pressure. The range of colors is continuous in the visible range from red to violet.

When the spectrum of a vapor in a gas flame or a low-pressure discharge tube is analyzed, there appears a series of bright lines characteristic of the material used. This type of spectrum [Fig. 32.5(b)] is called a *bright-line spectrum* and seems to occur because of the ability of the atom to absorb energy and emit characteristic wavelengths when the energy causes the atom to vibrate.

If a compound is placed in a carbon electrode cavity and the spectrum viewed, the lines appear very close together as bands attributed to molecular vibration. This spectrum is seen in Fig. 32.5(c) and is called a *band spectrum*.

Bright-line and band spectra may be classified as *emission spectra* because they result from the emission of energy waves as a result of the vibrations of atoms or molecules.

Late in the eighteenth century, a German physicist, Joseph Fraunhofer, studied the dark lines which he noticed in the continuous spectrum of the Sun. These dark lines (or cluster of lines) in a continuous spectrum—called *Fraunhofer lines*—are caused when the wavelengths pass through the cooler atmospheres of the Earth or of the Sun. If the atmosphere contains a vapor or gas which has the same wavelength as the wavelength trying to pass through the vapor, it is absorbed. This type of spectrum is called a *dark-line spectrum*, or *absorption spectrum*, and is shown in Fig. 32.5(d). Thus when a continuous spectrum from a hot source passes through a cooler gas, the wavelengths from the cooler gas will be missing in the continuous spectrum. This may be demonstrated by passing light from a carbon arc through a vapor (sodium) and then permitting the light to refract through a prism.

In general, emission spectra result from cold solids, liquids, or gases emitting wavelengths when heated; absorption spectra result from hot objects whose light passes through a cooler medium.

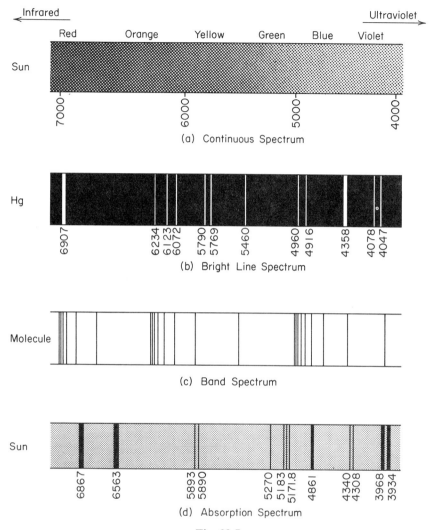

Fig. 32.5

Fraunhofer lines have been assigned the letters listed in Appendix Table A.21, which also shows the corresponding wavelengths, colors, and elements.

32.4. Dispersion

Figure 32.4 shows that when white light is passed through a prism, it will emerge in an array of colors and form a continuous spectrum. If the

Sec. 32.4   *Dispersion*   519

colors in the spectrum are arbitrarily grouped into six primary colors, each will be deviated according to its own wavelength, the *mean* deviation being that of yellow *D*-light, wavelength 5893 A (*angstrom units:* 5893 A = 5893 × 10^{-10} m = 5893 × 10^{-8} cm).

Dispersion is the angular separation between any two colors. For convenience, we shall develop the equations for red *C* (6563 A) and violet *H* (3968 A), as shown in Fig. 32.6. Since these prisms usually have small apex angles, the equations are developed using the approximation from Sec. 32.2:

$$n = \frac{D_m + A}{A}$$

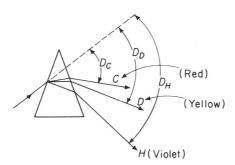

Fig. 32.6

The deviation of the *C*-light is

$$D_C = n_C A - A = A(n_C - 1)$$

The deviation for *H*-light is

$$D_H = A(n_H - 1)$$

From the definition of dispersion (where *D* represents deviation and D_{C-H} represents dispersion from *C*- to *H*-light),

$$D_{C-H} = D_H - D_C$$
$$= A(n_H - 1) - A(n_C - 1)$$

or

$$D_{C-H} = A(n_H - n_C)$$

$\left\{ \begin{array}{l} H_{C-H} = \text{dispersion} \\ D_H = \text{deviation of } H\text{-light} \\ D_C = \text{deviation of } C\text{-light} \end{array} \right\}$

The *dispersive power* ω_{C-H} is related to the mean deviation of yellow *D*-light. Thus

$$\omega_{C-H} = \frac{A(n_H - n_C)}{A(n_D - 1)} = \frac{n_H - n_C}{n_D - 1} \quad \begin{cases} \omega = \text{dispersive power} \\ D_H = \text{deviation of } H\text{-light} \\ D_C = \text{deviation of } C\text{-light} \\ D_D = \text{deviation of } D\text{-light} \\ n_H = \text{index of refraction of } H\text{-light (} violet \text{)} \\ n_C = \text{index of refraction of } C\text{-light (} red \text{)} \\ n_D = \text{index of refraction of } D\text{-light (} yellow \text{)} \end{cases}$$

The dispersive power is sometimes given as the reciprocal. When this is the case, the Greek letter *nu* (v) is used.

Since the index of refraction is a function of the refracting material and the wavelength, some representative values for glass are given in the table in Appendix A.22.

Example 4

The dispersive power of a barium–flint-glass prism is 0.0259. Its apex angle is 7°. Calculate the angular separation between the H-light and the G-light.

Solution

$$\omega_{F-H} = \frac{n_H - n_F}{n_D - 1}$$

Solve for

$$n_H - n_F = \omega(n_D - 1) = 0.0259(1.5682 - 1)$$
$$= 0.0147$$

$$\begin{cases} \text{Barium–flint glass} \\ n_D = 1.5682 \\ n_H = 1.5912 \\ n_F = 1.5765 \\ \omega = 0.0259 \\ A = 7° \end{cases}$$

The dispersion is

$$D_{F-H} = A(n_H - n_F) = 7°(1.5912 - 1.5765)$$
$$= 0.103° \text{ separation}$$

32.5. Direct-Vision Prisms: Angular Separation and Achromatization

Prisms may be combined to produce several effects. One effect is called *direct vision*. Two prisms may be combined so that the D-ray is not deviated at all; the net deviation of the remaining rays is zero, but the dispersion of all the rays is as shown (exaggerated) in Fig. 32.7(a). Thus the D-ray is not deviated and, although there is no *net* deviation, there is dispersion.

The angle of the second prism, in combination with the first, may be found as follows:

Since the D-line is not deviated,

$$D_D - D_{D_1} = 0$$

Sec. 32.5   *Direct-Vision Prisms*   521

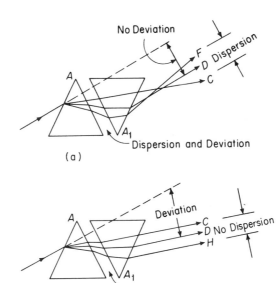

Fig. 32.7

Since
$$D_D = A(n_D - 1) \quad \text{and} \quad D_{D_1} = A_1(n_{D_1} - 1)$$

then
$$A(n_D - 1) - A_1(n_{D_1} - 1) = 0$$

Therefore, A_1 may be found from

$$A_1 = \frac{A(n_D - 1)}{n_{D_1} - 1}$$

$\left\{\begin{array}{l}\text{First prism}\\ D_D = \text{deviation of } D\text{-line}\\ A = \text{apex angle}\\ n_D = \text{index refraction of } D\text{-light}\\ \text{Second prism}\\ D_{D_1} = \text{deviation of } D\text{-line}\\ A_1 = \text{apex angle}\\ n_{D_1} = \text{index refraction of } D_1 \text{ light}\end{array}\right\}$

Example 5

An ordinary crown-glass prism (angle 8°) is combined with a dense flint-glass prism. What must be the apex angle of the dense fling-glass prism so that the combination is "direct vision"?

Solution

The apex angle of the dense flint is

$$A_1 = \frac{A(n_D - 1)}{n_{D_1} - 1} = \frac{8(1.5171 - 1)}{1.6555 - 1} = 6.31°$$

$\left\{\begin{array}{l}\text{Ordinary crown glass}\\ A = 8°\\ n_D = 1.5171\\ \text{Dense flint glass}\\ n_{D_1} = 1.6555\end{array}\right\}$

The angular separation between the *C*- and *H*-rays for two prisms would be a function of the separation between the C- and H-lights as a result of passing through the first prism and the separation on passing through the second prism:

The angular separation due to the first prism is

$$D_{C-H} = A(n_H - n_C)$$

The angular separation due to the second prism is

$$D_{C_1-H_1} = A_1(n_{H_1} - n_{C_1})$$

Therefore, the overall angular separation between the C- and H-lights is

$$D_{C_1-H_1} - D_{C-H} = A_1(n_{H_1} - n_{C_1}) - A(n_H - n_C)$$

Example 6

What would the angular separation between the C- and H-lights be for the combination in Example 5?

Solution

The angular separation between the C- and H-lights is

$$\begin{aligned} D_{C_1-H_1} - D_{C-H} &= A_1(n_{H_1} - n_{C_1}) - A(n_H - n_C) \\ &= 6.31(1.6940 - 1.6501) \\ &\quad - 8(1.5325 - 1.5146) \\ &= 0.1338° \end{aligned}$$

$$\left\{ \begin{array}{l} \text{Ordinary crown glass} \\ A = 8° \\ n_H = 1.5325 \\ n_C = 1.5146 \\ \text{Dense flint glass} \\ A_1 = 6.31° \\ n_{H_1} = 1.6940 \\ n_{C_1} = 1.6501 \end{array} \right.$$

Achromatization is a condition where any dispersion caused in the first prism is corrected by the refraction in the second prism [Fig. 32.7(b)]. Since there is no dispersion, all the rays emerge from the second prism as parallel rays. All the rays, however, are deviated by the same amount:
Since there is no net dispersion,

$$D_{H_1-C_1} = D_{H-C}$$

Therefore,

$$A_1(n_{H_1} - 1) - A_1(n_{C_1} - 1) = A(n_H - 1) - A(n_C - 1)$$

and

$$A_1(n_{H_1} - n_{C_1}) = A(n_H - n_C)$$

Since it is usually desirable to find the combination angle A_1, the preceding equation is solved for A_1; therefore,

Sec. 32.5   Direct-Vision Prisms   523

$$A_1 = \frac{A(n_H - n_C)}{n_{H_1} - n_{C_1}} \left\{ \begin{array}{l} \text{First prism} \\ A = \text{apex angle} \\ n_H = \text{index of refraction of } H\text{-light} \\ n_C = \text{index of refraction of } C\text{-light} \\ \text{Second prism} \\ A_1 = \text{apex angle} \\ n_{H_1} = \text{index of refraction of } H\text{-light} \\ n_{C_1} = \text{index of refraction of } C\text{-light} \end{array} \right\}$$

Example 7

(1) What must be the apex angle of a dense flint-glass prism for achromatization of the 8° ordinary crown-glass prism in Example 5? (2) What is the deviation of the light?

Solution

1. The apex angle for achromatization:

$$A_1 = \frac{A(n_H - n_C)}{n_{H_1} - n_{C_1}}$$
$$= \frac{8(1.5325 - 1.5146)}{1.6940 - 1.6501}$$
$$= 3.26°$$

$\left\{ \begin{array}{l} \text{Ordinary crown glass} \\ A = 8° \\ n_H = 1.5325 \\ n_C = 1.5146 \\ \text{Dense flint glass} \\ n_{H_1} = 1.6940 \\ n_{C_1} = 1.6501 \end{array} \right\}$

2. The deviation of the D-line:

$$D_D - D_{D_1} = A(n_D - 1) - A_1(n_{D_1} - 1)$$
$$= 8(1.5171 - 1) - 3.26(1.6555 - 1) \quad \left\{ \begin{array}{l} n_D = 1.5171 \\ n_{D_1} = 1.6555 \end{array} \right\}$$
$$= 2°$$

Note that chromatic aberration in lenses, Fig. 32.8(a), may be corrected in the same manner as in prisms. Figure 32.8(b) shows the correction made with a negative lens.

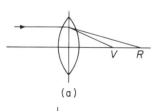
(a)

(b)

Fig. 32.8

Problems

1. A flat piece of glass (index of refraction 1.58) $1\frac{1}{2}$ in. thick is used to displace a ray of light incident to the normal by an angle of 30°. Calculate the displacement of the ray.
2. A block of glass is used to refract a beam of light. The index of refraction of the glass is 1.7. The refracted ray makes an angle of 22°35′ with the normal and in so doing displaces the ray 1.650 in. The length of the light path through the glass is 4.333 in. Calculate (a) the angle of incidence at the first interface; (b) the thickness of the glass.
3. Given a stack of three layers of flat glass, all indices of refraction different, show that the entering ray and the emerging ray are parallel and displaced.
4. A ray is incident to an equilateral-prism surface at an angle of 20°. If the index of refraction of the prism is 1.6, what happens to the ray?
5. In Prob. 4, find the incident angle when the ray leaves the prism *parallel* to the right side.
6. The index of refraction of a prism (apex angle 50°) is 1.4. Calculate the angle of minimum deviation.
7. A prism has an apex angle of 60° and an index of refraction of 1.5. Calculate the angle of minimum deviation.
8. Show that, if the angle of minimum deviation is 33°46′ for a quartz prism ($n = 1.46$), then for any incident angle greater than or less than the incident angle which yields minimum deviation, the deviation angle will be greater than the angle of minimum deviation. (The apex angle is 60°.)
9. Calculate the dispersion between the *C*- and *F*-lines for a light–flint-glass prism (apex angle 4°).
10. Calculate, for Prob. 9, (a) the dispersive power; (b) the *v*-value.
11. When the apex angle of a medium flint prism is 5°, calculate (a) the dispersion between the *F*- and *H*-lines; (b) the dispersive power of the prism.
12. A barium–flint-glass prism (apex angle 7°) is combined with a medium flint-glass prism. Calculate (a) the apex angle of the medium flint-glass prism so that the combination creates a "direct-vision" combination; (b) the angular separation between the *F*- and *H*-lines.
13. (a) What must be the apex angle of the medium flint-glass prism in Prob. 12 for achromatization of the barium–flint-glass prism? (b) What is the deviation of the light?
14. A prism (apex angle 9°) made of light flint glass is to be achromatized with another prism of dense flint glass for *C*- and *G*′-light. (a) What apex angle must the second prism have? (b) What is the deviation for this pair?

33

Interference and Diffraction

33.1. The Double Slit

If (monochromatic) single-wavelength light is used, waves may be algebraically added [Fig. 33.1(a)] or subtracted [Fig. 33.1(b)]. When two waves are added, they are said to *interfere constructively*. When two waves subtract, they are said to *interfere destructively*. Monochromatic light will produce white rings (or *fringes*) when waves interfere constructively, and dark rings (or fringes) when they interfere destructively.

Thomas Young, in the early nineteenth century, performed an experiment using a point, or slit, as a source of monochromatic light to illuminate a double slit as shown in Fig. 33.2. Thus a single wave coming from a light source, Fig. 33.2(a), will "split" into two bands, propagating toward slits S and S_1. Each

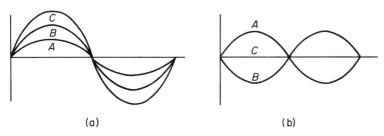

Fig. 33.1

526 *Interference and Diffraction* Chap. 33

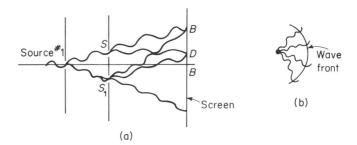

Fig. 33.2

of these bands will, in turn, act as a source for new bands which will propagate toward the screen.

If the two waves, one from slit S and one from slit S_1, reinforce each other at the screen, a bright fringe will appear. *Reinforcement* takes place whenever the distance traveled from the double slits to the screen is the same for both waves or when one wave travels any multiple of full wavelengths farther than the other.

If the distance difference is one half-wavelength greater, or any multiple of half-wavelengths greater, there will be total destruction at D.

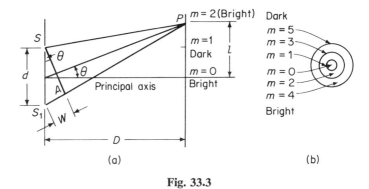

Fig. 33.3

Figure 33.3(a) shows two waves propagating from S and S_1 which interfere constructively at point P, a distance l from the principal axis. Assume this to be the first constructive-interference point removed from $m = 0$ so that W is equal to one full wavelength λ or two half-wavelengths; that is, $S_1 P$ is 2 half-wavelengths longer than SP. The line SA is then drawn as shown, and

$$W = d \sin \theta$$

But $\left\{ \begin{array}{l} d = \text{distance between slits} \\ W = \text{wavelength difference} \end{array} \right\}$

$$W = \lambda$$

Sec. 33.1 The Double Slit

If m (the *order*) equals a half-number of wavelengths—counting the central fringe as $m = 0$ [see Fig. 33.3(b)]—then

$$W = m\frac{\lambda}{2}$$

and

$$d \sin \theta = m\frac{\lambda}{2}$$

or

$$\sin \theta = \frac{m\lambda}{2d}$$

From Fig. 33.3(a),

$$\tan \theta = \frac{l}{D}$$

However, since the angle θ is usually small, the tangent function may be replaced with the sine function. Thus

$$\sin \theta = \frac{l}{D}$$

Equating,

$$\frac{l}{D} = \frac{m\lambda}{2d}$$

$\left\{\begin{array}{l} D = \text{distance from slits to screen} \\ d = \text{distance between slits} \\ \lambda = \text{wavelength of light} \\ m = \text{integers, or orders} \\ l = \text{distance from central bright band to fringe} \end{array}\right\}$

Solving for l, the distance from the central bright band to the fringe considered is

$$l = \frac{m\lambda}{2} \times \frac{D}{d} \qquad \left\{\begin{array}{c} \text{Bright} \\ m = 2, 4, 6, \ldots \text{ (even)} \\ \text{Dark} \\ m = 1, 3, 5, \ldots \text{ (odd)} \end{array}\right\}$$

Example 1

Show that (1) if $m = 0, 2, 4, \ldots$ (even integers), l yields constructive interference; (2) If $m = 1, 3, 5, \ldots$ (odd integers), l yields destructive interference.

Solution

1. The bright orders; m is equal to the even integers.

    (a) When $m = 0$:

    $$l = \frac{m\lambda}{2} = \frac{0\lambda}{2} = 0 \qquad (bright)$$

    Since the distance $SP = SP_1$, there is constructive interference and the central fringe will be a bright.

(b) When $m = 2$:
$$l = \frac{m\lambda}{2} = \frac{2\lambda}{2} = \lambda \quad (bright)$$

(c) When $m = 4$:
$$l = \frac{m\lambda}{2} = \frac{4\lambda}{2} = 2\lambda \quad (bright)$$

2. The dark orders; m is equal to the odd integers.

(a) When $m = 1$:
$$l = \frac{m\lambda}{2} = \frac{1}{2}\lambda \quad (dark)$$

(b) When $m = 3$:
$$l = \frac{m\lambda}{2} = \frac{3}{2}\lambda \quad (dark)$$

(c) When $m = 5$:
$$l = \frac{m\lambda}{2} = \frac{5}{2}\lambda \quad (dark)$$

Example 2

Two slits are 2 mm apart and 3 m from a screen. Light (wavelength 6500 A) illuminates the two slits. Calculate (1) the distance l of the first-order fringe from the center (zero order); (2) the distance l of the fourth-order fringe from the center; (3) the separation of the first-and fourth-order fringes.

Solution

1. The first-order fringe:
$$l = \frac{m\lambda}{2} \times \frac{D}{d} = \frac{1(6500 \times 10^{-8})}{2} \times \frac{300}{0.2}$$
$$= 4.8 \times 10^{-2} = 0.048 \text{ cm}$$

2. The fourth-order fringe:
$$l = \frac{m\lambda}{2} \times \frac{D}{d} = \frac{4(6500 \times 10^{-8})}{2} \times \frac{300}{0.2}$$
$$= 0.195 \text{ cm}$$

$$\begin{cases} \text{Dark fringe} \\ m = 1 \\ D = 6500 \times 10^{-8} \text{ cm} \\ \lambda = 300 \text{ cm} \\ d = 2 \text{ mm} = 0.2 \text{ cm} \\ \text{Light fringe} \\ m = 4 \end{cases}$$

3. The distance between the first and fourth orders is
$$\Delta l = 0.195 - 0.048 = 0.147 \text{ cm}$$

The *Michelson interferometer* is an instrument which makes use of the principle of interference of light waves to measure the wavelength of unknown light. In Fig. 33.4, a light beam from the source is split at a 45° half-silvered

Sec. 33.1 *The Double Slit* 529

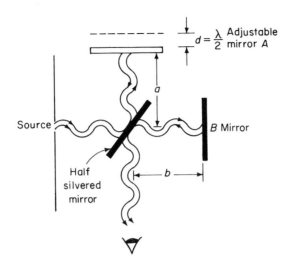

Fig. 33.4

mirror. Part of the ray propagates off this mirror a distance a to an adjustable mirror A, reflects, and passes through the 45° mirror to the eye. The other part of the split ray proceeds through the 45° mirror a distance b to mirror B, reflects off mirror B, reflects off the 45° mirror, and proceeds to the eye.

If paths a and b are equal, reinforcement occurs. If the adjustable mirror A is moved a distance $d = \lambda/2$, path a will be increased by twice $\lambda/2$, or λ. The waves from mirror A and mirror B will be in phase. If mirror A is moved a distance $\lambda/4$, there will be destructive interference, because the path a is increased by $2 \times \lambda/4$, or $\lambda/2$. Therefore,

$$d = \frac{\lambda}{2}m \qquad \left\{ \begin{array}{l} d = \text{distance moved by adjustable mirror} \\ \lambda = \text{wavelength} \\ m = \text{number of bands} \end{array} \right\}$$

Example 3

The adjustable mirror on the Michelson interferometer moves 0.020 mm when 80 dark bands are counted. What is the wavelength of the light causing the bands?

Solution

$$\lambda = \frac{2d}{m} = \frac{2 \times 0.002}{80} = 5000 \text{ A}$$

It should be evident, from all that has been said, that if more than single-wavelength light is used, there will be reinforcement at different distances

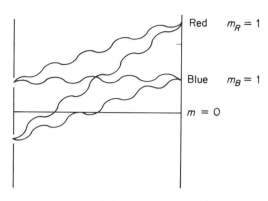

Fig. 33.5

from the central bright band. Assume a light is used which contains blue and red waves. Since the blue wavelength is shorter than the red wavelength, there will be constructive interference closer to the central bright band than for the red reinforcement (see Fig. 33.5). Thus as the wavelength decreases, the distance between the bands decreases.

33.2. Interference in a Thin Film; the Half-Wavelength Shift

Figure 33.6(a) shows the upper and lower surfaces of a thin film. Most of the light incident to the upper surface will pass through the film. Part of the light (x) will reflect off the upper surface of the film and part of the light (y) will reflect off the lower surface of the film. If the path difference of ray y and ray x is λ, there should be reinforcement. But ray x reflects from a surface whose index of refraction is greater than the incident medium (air); whereas ray y reflects from a surface of smaller index of refraction (lower surface) than the incident medium (n). Under these conditions, a *half-wavelength shift* takes place, and instead of reinforcement, destructive interference takes place between the x- and y-rays. Thus a half-wavelength loss (or gain) takes place at the upper surface, where the index of refraction of the medium of the incident ray is less than the index of refraction of the refracting (second) medium.

For normal incidence, the equivalent light path for reinforcement is 2 times the thickness of the film multiplied by the index of refraction ($\frac{1}{2}\lambda$ into the film in both directions):

$$2nt$$

Adding $\lambda/2$ to take care of the half-wavelength shift, the equation for constructive interference becomes

Sec. 33.2 Interference in a Thin Film 531

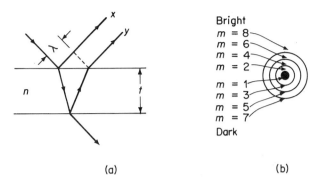

Fig. 33.6

$$2nt + \frac{\lambda}{2} \qquad \left\{ \begin{array}{l} n = \text{index of refraction of film} \\ t = \text{thickness of film} \\ \lambda = \text{wavelength} \\ m = \text{order} \end{array} \right\}$$

Equating with $m(\lambda/2)$,

$$2nt + \frac{\lambda}{2} = m\frac{\lambda}{2}$$

Solving this equation for the thickness of the film,

$$t = \frac{(m-1)\lambda}{4n} \qquad \left\{ \begin{array}{l} \text{Bright} \\ m = 2, 4, 6, \ldots \text{ (even)} \\ \text{Dark} \\ m = 1, 3, 5, \ldots \text{ (odd)} \end{array} \right\}$$

Note that in Fig. 33.6(b), the central fringe is dark.

Example 4

Show that (1) if $m = 2, 4, 6, \ldots$ (even), the fringes are bright; (2) if $m = 1, 3, 5, \ldots$ (odd), the fringes are dark.

Solution

1. For bright-order fringes.

    (a) When $m = 2$, the thickness of the film is

    $$t = \frac{(m-1)\lambda}{4n} = \frac{(2-1)\lambda}{4n} = \frac{1\lambda}{4n}$$

    If the thickness is $\frac{1}{4}\lambda$, the light must travel to the lower surface and back. Since the thickness is $\frac{1}{4}\lambda$, the light will travel $\frac{1}{2}\lambda$. Considering the half-wavelength shift, this adds to yield 1λ and a bright fringe.

    (b) When $m = 4$, the thickness is

    $$t = \frac{(4-1)\lambda}{4n} = \frac{3\lambda}{4n}$$

The wave path is

$$2\left(\frac{3\lambda}{4n}\right) = 1\tfrac{1}{2}\lambda + \tfrac{1}{2}\lambda \ (shift)$$

which yields

$$2\lambda \ (\text{a bright fringe})$$

2. For dark-order fringes.

(a) When $m = 1$, the thickness is

$$t = \frac{(m-1)\lambda}{4n} = \frac{(1-1)\lambda}{4n} = 0$$

The wave path should reinforce constructively. Considering the half-wavelength shift, destruction occurs, causing a central dark band.

(b) When $m = 3$, the thickness is

$$t = \frac{(m-1)\lambda}{4n} = \frac{(3-1)\lambda}{4n} = \frac{1\lambda}{2n}$$

The wave path is twice the thickness, or

$$2\left(\frac{1\lambda}{2n}\right) = \frac{\lambda}{n}$$

Add the half-wavelength shift and the fringe will be dark, or

$$\frac{\lambda}{n} + \frac{\lambda}{2n} = \frac{3\lambda}{2n}$$

33.3. Newton's Rings

An accurate lens (as shown in Fig. 33.7), if highly polished and placed on a flat surface, will create an *air wedge*. Thus light from the source S will reflect

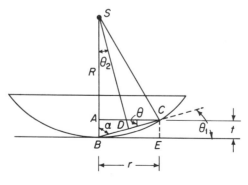

Fig. 33.7

off the lower surface of the lens and off the flat surface. These two rays will either reinforce or destroy each other. The point at which the lens touches the flat surface upon which it is resting will appear dark, because of the half-wavelength shift. Concentric light and dark rings, called *Newton's rings*, will appear as the distance increases from the center outward.

Thus in the triangles SDB and ABC,

$$\angle \alpha = \angle \alpha \quad \text{and} \quad \theta = \theta_2$$

Since $\theta = \theta_1$,

$$\theta_1 = \theta_2$$

Also,

$$\sin \theta_2 = \frac{BC/2}{R} = \frac{BC}{2R}$$

If t is small, BC may be replaced by r, the radius of the ring, and

$$\sin \theta_2 = \frac{r}{2R} = \sin \theta_1$$

From triangle BCE, if θ_1 is very small,

$$\tan \theta_1 = \frac{t}{r} = \sin \theta_1$$

$\left\{\begin{array}{l} t = \text{thickness of air wedge} \\ r = \text{radius of interference ring} \\ R = \text{radius of lens} \\ m = \text{order} \\ \lambda = \text{wavelength} \end{array}\right\}$

Equating,

$$\frac{t}{r} = \frac{r}{2R}$$

Solving for t,

$$t = \frac{r^2}{2R}$$

However, t is also equal to ($n = 1$ for the air wedge)

$$t = \frac{(m-1)\lambda}{4}$$

Equating the two values for t,

$$\frac{r^2}{2R} = \frac{(m-1)\lambda}{4}$$

$\left\{\begin{array}{c} \text{Bright rings} \\ m = 2, 4, 6, \ldots \text{ (even)} \\ \text{Dark rings} \\ m = 1, 3, 5, \ldots \text{ (odd)} \end{array}\right\}$

Example 5

The wavelength of light used to produce Newton's rings is 6000 A. Calculate (1) the thickness of the air wedge for the second-*order* ring; (2) the second bright fringe; (3) the fifth-*order* ring; (4) the fifth dark fringe; (5) the diameter of the fifth *order*, if the diameter of the lens used is 30 cm.

Solution

1. The thickness of the air wedge for the second-*order* ring is

$$t = \frac{(m-1)\lambda}{4} = \frac{(2-1)(6000 \times 10^{-8})}{4} \quad \left\{ \begin{array}{l} \lambda = 6000 \times 10^{-8} \text{ cm} \\ m = 2 \end{array} \right\}$$

$$= 1.5 \times 10^{-5} \text{ cm}$$

2. The order *m* of the *second bright fringe* as distinguished from the second order, may be found using the equation

$$m = 2N = 2(2) = 4$$

Thus the thickness of the air wedge is

$$t = \frac{(m-1)\lambda}{4} = \frac{(4-1)(6000 \times 10^{-8})}{4}$$

$$= 4.5 \times 10^{-5} \text{ cm}$$

3. The thickness of the air wedge, $m = 5$, is

$$t = \frac{(m-1)\lambda}{4} = \frac{(5-1)(6000 \times 10^{-5})}{4}$$

$$= 6 \times 10^{-5} \text{ cm}$$

4. *The fifth dark ring*, as distinguished from the fifth-*order* ring, may be calculated using the equation

$$m = 2N - 1 = 2(5) - 1 = 9$$

Thus the thickness of the air wedge is

$$t = \frac{(m-1)\lambda}{4} = \frac{(9-1)(6000 \times 10^{-8})}{4}$$

$$= 1.2 \times 10^{-4} \text{ cm}$$

5. The diameter of the fifth *order* is

$$r = \sqrt{\frac{2R\lambda(m-1)}{4}} = \sqrt{\frac{2 \times 15 \times (6000 \times 10^{-8})(5-1)}{4}}$$

$$= 4.24 \times 10^{-2} \text{ cm} \quad \left\{ \begin{array}{l} R = 30/2 = 15 \text{ cm} \\ m = 5 \end{array} \right\}$$

The diameter of the ring is

$$2 \times 4.24 \times 10^{-2} = 8.48 \times 10^{-2} \text{ cm.}$$

33.4. Single-Slit Diffraction

Single slits cause light to bend slightly around the corner of a sharp edge, if that edge is of the order of the wavelength of the light used. Figure 33.8(a) shows a slit and the principal axis *AB* from the slit to the screen. Two rays,

Sec. 33.4 *Single-Slit Diffraction* 535

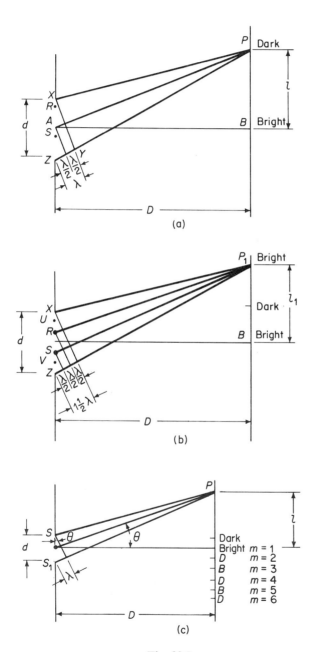

Fig. 33.8

one from each corner of the slit, are caused to converge by a lens (not shown) at point P. Another ray is drawn from A to point P. The perpendicular yx is drawn as shown.

Assume the distance zy is equal to one wavelength of the light used. Then distances xP and AP will be one half-wavelength out of phase and destructive interference will take place. The same will take place for distances zP and AP, with the same results. It is also true that any point in segment Ax of the slit and any corresponding point in segment Az—e.g., R and S—will also be one half-wavelength out of phase with each other. Thus point P will be dark.

In Fig. 33.8(b), if the edge of the slits from x and z, project waves to p_1, which create a path *difference* of $1\frac{1}{2}$ wavelengths, then the fringe will project as a bright fringe. If the $1\frac{1}{2}$-wavelength path difference is divided equally into three segments—xR, RS, and Sz—then the wave from point x will support the wave from point S, because it will have a path difference of one wavelength. The wave from point R will support the wave from point z in the same manner. Any point—say, u—from segment xR will support, constructively, any point—say, v—from segment Sz. The destructive effect from segment RS tends to reduce the intensity of the bright fringe at P_1. Since the constructive interference from segments xR and Sz is stronger than the destructive interference from segment RS, the point P_1 is bright, but with a reduced intensity. Note very carefully that the dark fringes are really the limits of the bright intensity.

Figure 33.8(c) shows the central bright fringe and the succeeding orders of m.

For maximum intensity,

$$d \sin \theta = m\frac{\lambda}{2}$$

The equation for the distance l is derived as in Sec. 33.1. Thus

$$l = \frac{m\lambda D}{2d}$$

$\begin{cases} \text{Bright order} \\ m = 1, 3, 5, \ldots \text{(odd)} \\ \text{Dark order} \\ m = 2, 4, 6, \ldots \text{(even)} \end{cases}$ $\begin{cases} D = \text{distance from screen to slit} \\ l = \text{distance from central band to band } P \\ \lambda = \text{wavelength} \\ m = \text{order} \end{cases}$

Example 6

Light of 6500 A passes through a single slit 0.05 cm wide and is diffracted on a screen 1.8 m away. Calculate the distance from the central band to the third-order fringe.

Solution

The distance from the central band to the third-order fringe is

$$l = \frac{m\lambda D}{2d} = \frac{3(6500 \times 10^{-8})(1.8 \times 10^2)}{2 \times 0.05}$$

$\begin{cases} \lambda = 6500 \times 10^{-8} \text{ cm} \\ d = 0.05 \text{ cm} \\ D = 1.8 \times 10^2 \text{ cm} \\ m = 3 \end{cases}$

$$= 0.35 \text{ cm}$$

33.5. Diffraction Gratings

The *diffraction grating* is an optically flat and transparent material with many closely ruled and equally spaced lines etched into its surface. Incident light passes through the space between these ruled lines. These transparent sections act like small, closely spaced slits and subject a beam of light to the combined phenomena of interference and diffraction. The interference of the rays causes interference patterns of maximum- and minimum-intensity bands. The diffraction effect reduces the width of the maximum-intensity band, thus making the first order bands distinguishable from the other maximum-intensity bands.

Under the conditions shown in Fig. 33.9, the diffraction of each wavelength inherent in the white light will occur at a different angle with the centerline. The shorter the wavelength, the less the diffraction; the longer the wavelength, the greater the diffraction. Therefore, a first-order *spectrum*, rather than a first-order single band, will appear (see Fig. 33.9). At some angle θ, this spectrum will repeat itself, but with much less intensity and greater spacing between the respective colors.

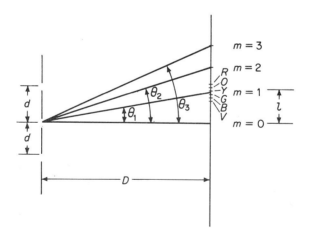

Fig. 33.9

If the grating has N number of lines per centimeter (or per inch), the distance d between two successive lines is

$$d = \frac{1}{N}$$

For interference, where the central band is bright and a mixture of all colors, the orders are $m = 1, 2, 3, \ldots$ and the equation is

$$m\lambda = d \sin \theta$$

Since

$$l = D \tan \theta \simeq D \sin \theta$$

solving both equations for $\sin \theta$ and equating,

$$\frac{l}{D} = \frac{m\lambda}{d}$$

$\left\{\begin{array}{l}\lambda = \text{wavelength} \\ m = \text{order of bright band} \\ d = \text{distance between two successive slits} \\ \theta = \text{angle the ray makes with central axis} \\ l = \text{distance from center axis to } m\text{-order} \\ D = \text{distance from grating to screen}\end{array}\right\}$

The equation for the wavelength is

$$\lambda = \frac{ld}{mD}$$

If a lens (focal length f) is used to focus the beam, then

$$D = f$$

Example 7

A grating has 6000 lines/cm etched into its surface. The distance from the grating to the screen is 60 cm. Calculate (1) the wavelength of the light used, for the first-order band, when the angle θ is $37°58'$; (2) the largest possible order for this grating and this wavelength of light.

Solution

1. The distance between the first-order band and the central ray is

$$l = D \tan \theta = 60 \tan 21°18' = 23.4 \text{ cm}$$

To calculate the wavelength,

$$\lambda = \frac{ld}{mD} = \frac{23.4}{1 \times 60} \times \frac{1}{6000} = 6500 \times 10^{-8}$$

$\left\{\begin{array}{l}d = 1/6000 \\ m = 1 \\ D = 60 \text{ cm} \\ \theta = 21°18'\end{array}\right\}$

$$= 6500 \text{ A}$$

2. From equation $m\lambda = d \sin \theta$, it should be evident that the magnitude of $\sin \theta$ cannot exceed 1.000. The largest order m which is possible is a function of the $\sin \theta$. Thus from the equation $m = d \sin \theta$, solve for m and set $\theta = 90°$:

$$m = \frac{d}{\lambda} \sin 90°$$

Substituting and solving for m,

$$m = \frac{1}{6000} \times \frac{\sin 90°}{6500 \times 10^{-8}} = 2.6$$

The largest possible order is

$$m = 2$$

33.6. Crystal Diffraction

Since X-rays are wave forms of very short wavelength, it is necessary to have a grating where the lines are very closely spaced if diffraction is to take place. As a matter of fact, it is necessary that the spacing be of the order of the wavelength of the X-ray. The spacing must be closer than it is possible to scribe lines. Certain crystals have ordered layers of atoms, and therefore may be used as gratings for very short X-rays. This has been supported experimentally. *Laue patterns* are the visual effects of crystal diffraction observed with an electron microscope. These patterns are an important means for studying crystal structure and the characteristics of short-wavelength rays. In Fig. 33.10 is shown a section through a crystal, showing the ordered layers.

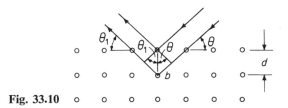

Fig. 33.10

If the angle θ is small, a selective scattering takes place, such that the incident angle θ and the maximum-intensity scattering angle θ_1 are equal. When these angles are equal, distances a and b are also equal. If the sum of a and b is a multiple of the wavelength of the X-rays, there will be reinforcement of the reflected rays. Therefore,

$$\sin \theta = \frac{b}{d} \quad \text{and} \quad \sin \theta_1 = \frac{a}{d}$$

Solving both equations for a and b,

$$b = d \sin \theta \quad \text{and} \quad a = d \sin \theta_1$$

If $\theta = \theta_1$ and there is reinforcement, then $a + b = \lambda$, and adding,

$$\lambda = a + b = 2d \sin \theta$$

For all orders m,

$$m\lambda = 2d \sin \theta$$

$\left\{ \begin{array}{l} d = \text{distance between layers of atoms} \\ \lambda = \text{wavelength} \\ m = \text{order of reinforcement} \\ \theta = \text{incident angle} \end{array} \right\}$

The calculation for d for sodium chloride might be done as follows:

1. Each mole contains 2 atoms. The atomic weight of sodium equals 23, and that of chlorine 35.46.

2. From Avogadro's number,

$$23.00 + 35.46 = 58.46 \text{ g of sodium chloride contains}$$
$$6.02 \times 10^{23} \text{ molecules}$$

3. Therefore, 1 g contains

$$\frac{6.02 \times 10^{23}}{58.46} \text{ molecules/g}$$

4. Each molecule contains 2 atoms; the number of atoms per gram is thus

$$2 \times \frac{6.02 \times 10^{23}}{58.46} = 2.06 \times 10^{22} \text{ atoms/g}$$

5. The density of rock salt is 2.164 g/cm³. Therefore, there are

$$2.06 \times 10^{22} \text{ atoms/g} \times 2.164 \text{ g/cm}^3 = 4.46 \times 10^{22} \text{ atoms/cm}^3$$

Since the structure is an ordered cubic, the distance between the atom layers becomes

$$d = \frac{1}{\sqrt[3]{4.46 \times 10^{22}}} = 2.82 \times 10^{-8} \text{ cm}$$

Example 8

A beam of X-rays is deflected by a calcite crystal with a first-order incident angle of 7°35'. The distance between the atom layers is 3.029×10^{-8} cm. What is the wavelength of the X-ray?

Solution

$$\lambda = \frac{2d \sin \theta}{m}$$

$$= \frac{2(3.029 \times 10^{-8}) \sin 7°35'}{1} \quad \left\{ \begin{array}{l} \theta = 7°35' \\ d = 3.029 \times 10^{-8} \text{ cm} \\ m = 1 \end{array} \right\}$$

$$= 0.8 \times 10^{-8} \text{ cm} = 0.8 \text{ A}$$

33.7. Resolving Power

A lens, because of the diffraction of two light rays passing though it, forms disks (circular diffraction patterns) instead of solid images. This lack of *resolution* is a function of the wavelength of the light used and the diameter of the lens.

As the diameter of the lens increases, it has greater ability to distinguish the two images. The *resolving power* of a lens is its ability to separate the overlapping images formed by two sources.

Sec. 33.7   Resolving Power

The equation for calculating the resolving power of a lens is

$$R = \frac{r}{f} = \frac{1.22\lambda}{D}$$

The distance between the image spots is

$$d = RS$$

The radius of the central spot is

$$r = \frac{1.22\lambda}{D}f$$

λ = wavelength
D = diameter of lens
R = resolving power: rad
f = focal length of lens
r = radius of central image spot
S = object distance
d = distance between two objects

Fig. 33.11

In Fig. 33.11, the two sources O and O_1 are separated by one radius and are similarly separated in the image pattern. Thus the objects become distinguishable—i.e., are resolved. This is the bare minimum and yields a central radius of the central bright spot which serves for resolution.

Example 9

A lens has a diameter of 4 cm and a focal length of 12.5 cm. Two light sources (wavelength 6000 A) are placed 1 km (kilometer) from a lens. (1) What is the closest to each other that the two spots can be and still be distinguished? (2) What is the radius of the central spot if a screen is placed at the focal point of the lens?

Solution

1. Distance between the objects:

$$R = \frac{1.22\lambda}{D} = \frac{1.22(6000 \times 10^{-8})}{4}$$

$$= 1.83 \times 10^{-5} \text{ rad}$$

$$d = (1 \times 10^5 \text{ cm})(1.83 \times 10^{-5})$$

$$= 1.83 \text{ cm apart}$$

$\lambda = 6000 \times 10^{-8}$ cm
$D = 4$ cm
R = resolving power
$S = 10^5$ cm
$f = 12.5$ cm

2. Radius of the central spot:

$$r = \frac{1.22\lambda}{D}f = (1.83 \times 10^{-5})12.5$$

$$= 22.875 \times 10^{-5} \text{ cm}$$

The resolution of a *spectrometer* is a function of the smallest wavelength.

It is found to be the ratio of the average wavelength λ_{av} to the change in wavelength $\Delta\lambda$, or

$$R_s = \frac{\lambda_{av}}{\Delta\lambda}$$

The resolving power for a grating for clear separation is

$$R_g = mN \quad \begin{cases} R_s = \text{resolving power of spectrometer} \\ R_g = \text{resolving power of grating} \\ N = \text{number of lines times the ruled width} \\ m = \text{order} \\ \lambda_{av} = \text{average wavelength} \\ \Delta\lambda = \text{change in wavelength} \end{cases}$$

Example 10

(1) What spectrometer resolving power is needed to show clearly two adjacent green neon lines (4708.86 A and 4704.39 A)? (2) What is the theoretical resolving power of a grating having 10,000 lines/cm, width of rulings 2.5 cm? (The order is the second.)

Solution

1. Resolving power:

$$R_s = \frac{\lambda_{av}}{\Delta\lambda} = \frac{4708.86 + 4704.39}{2} \times \frac{1}{4.47} \quad \{\Delta\lambda = 4708.86 - 4704.39 = 4.47\}$$

$$= 1053 \quad (\textit{resolving power to show both lines})$$

2. Capabilities of the grating:

$$R_g = mN = 2 \times 10,000 \times 2.5 = 50,000$$

Problems

1. The distance from a two-slit grating to a screen is 2 m. The distance between the two slits is 0.05 mm. If the wavelength of the light diffracted by the two slits is 5400 A, calculate the separation of adjacent bright fringes for two orders.
2. The separation of two slits in a double-slit grating is 0.8 mm. The distance from the central axis to the third-order fringe is 1.5 mm. If the screen is 1.5 m from the grating, calculate the wavelength of the light used.
3. In Prob. 1, determine the separation between two dark fringes.
4. In Prob. 1, calculate the distance between the third-order dark fringe and the sixth-order bright fringe.
5. Calculate the angular separation in Prob. 4 for the tenth bright fringe and the twelfth bright order.
6. The wavelength of light used with a Michelson interferometer is 8000 A. The adjustment d is equal to 0.05 mm. Calculate the number of bands.

Problems

7. Calculate the least thickness of a dark soap film when incident sodium light (5893 A) is normal to the surface and the index of refraction of the soap is 1.6.
8. A soap film has an index of refraction of 1.4. Calculate the wavelength of the light which will interfere constructively when the film is 1.2×10^{-5} cm thick.
9. The wavelength of light is 6500 A. What is the thickness of the air wedge (a) at the second bright fringe? (b) at the second dark fringe? (c) What is the diameter of the second bright fringe if the diameter of the lens is 20 cm?
10. The wavelength of light is 5400 A. Using a 40-cm diameter lens to produce Newton's rings, calculate (a) the thickness of the air wedge at the eleventh-order fringe; (b) the diameter of the ring. (c) Is the ring dark or bright?
11. The radius of a lens used to produce Newton's rings is 80 cm. Calculate the diameter of the ring produced by the nineteenth order when sodium light (5893 A) is used.
12. Two flat glass plates are separated at one end by a paper tissue 0.05 mm thick. Assume the wavelength of 6500 A is used to produce Newton's interference bands. (a) Calculate the thickness of the air wedge at the fifth dark band. (*Note:* the fifth dark band has an order of $2N-1$.) (b) How many dark lines are counted on the surface of the upper plate glass?
13. A slit is 200 cm from a screen. If the light used to illuminate the slit is 6000 A and the width of the slit is 0.5 mm wide, calculate (a) the distance from the central bright band to the sixth-order fringe; (b) the separation at the screen between the fifth and sixth order. (c) Is the fringe in part (a) bright or dark?
14. The distance between the central bright fringe and the seventh-*order* bright fringe is 0.5 cm. The distance from a single slit to the screen is 90 cm. If the slit is 0.04 cm wide, calculate (a) the wavelength of light used; (b) the angle between this beam and the central axis.
15. Light of 6000 A, on passing through a single slit, projects a bright fringe on a screen 1.6 m away. (a) What is the distance from the central bright fringe to the third-order bright fringe? (The slit is 0.04 cm wide.) (b) Find the distance from the central bright fringe to the third bright fringe. (*Note*: Observe the difference between third-order and third bright fringe.)
16. A diffraction grating, ruled with 6000 lines/cm, projects a spectrum on a screen 80 cm from the grating. What is the distance from the central bright band to the second-order bright band for light of 5400 A?
17. What is the maximum order possible in Prob. 16?
18. A grating has 15,000 lines/in. etched into its surface. The distance from the grating to the screen is 30 cm. The distance from the central axis to the first-order central axis is 9.5 cm. Calculate (a) the wavelength of the first-order light ray; (b) the largest possible order for this system.
19. A diffraction grating has 16,000 lines/in. What is the dispersion for 4000-A and 7000-A wavelengths of light for the first and second orders of light incident on the grating?
20. What is the angular separation for a grating with 6000 lines/cm for helium (a) between the red line (7065 A) and the yellow line (5876 A)? (b) between the red line (7065 A) and the blue line (4472 A)? (c) Find the distance from the central bright band to each of the lines above the first two orders, if the distance from the grating to the screen is 50 cm.

21. The distance between two layers of a sodium chloride crystal is 2.814×10^{-8} cm. What incident angle yields the first order when the wavelength of the X-rays used is 0.9 A?
22. It is desired to calculate the crystal spacing in a calcite crystal. An X-ray (wavelength 0.4 A) is used at an incident angle of $3°47'$. What is the spacing of the layers of atoms in the crystal?
23. A lens has a diameter of 10 cm and a focal length of 37.5 cm. Two light sources are used as objects 8 m from the lens. If the wavelengths of two lights are 5500 A, find (a) the closest the two light sources can be to each other and still be distinguished; (b) the radius of the central spot if the screen is at the focal point.
24. (a) What is the resolving power of the first order of a 7000-line/cm grating ruled in a 3-in. square? (b) The average wavelength of light used to illuminate this grating is 496 mμ. What is the smallest separation that can be detected by this grating in the first order? (c) What is the change in wavelength for the second-order spectrum lines?

34

The Polarization of Light

34.1. Polarization

Light has all the characteristics of a transverse wave and, as such, vibrations occur in a plane perpendicular to the direction of the propagation of the energy wave. The direction and magnitude of this vibration may be represented by an infinite number of vectors, as shown in Fig. 34.1. According to the

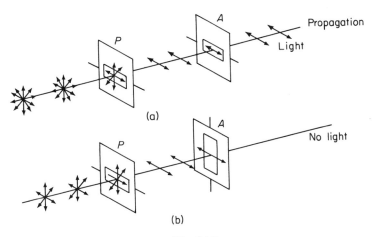

Fig. 34.1

electromagnetic-wave theory, the vibrations are both electronic and magnetic, in phase, and at right angles to each other. Only the electric vibrations will be considered.

If all the waves, except that group vibrating in one plane, are absorbed, the light is said to be *plane-polarized*. The mechanism used to isolate the single-plane vibrations is called a *polarizer* (*P* in Fig. 34.1). Actually, the components of the vectors which are blocked contribute to the intensity of the plane of vectors which get through the polarizer at full intensity.

A second polarizer is used to analyze this plane-polarized wave and is called an *analyzer* (*A* in Fig. 34.1). The analyzer may permit the plane-polarized wave to pass through it, or may block it out partially or wholly. In Fig. 34.1(a), all the rays are blocked out except the horizontal rays. These rays pass through the analyzer because its "slit" is parallel to the polarizer slit. If the analyzer slit is oriented perpendicular to the polarizer slit, as in Fig. 35.1(b), no light will appear to the right of the analyzer. Thus by manipulating the analyzer, it is possible to determine whether, and in which plane, the light transmitted by the polarizer has been polarized. Note that, however the polarizer is oriented, there will always be a plane of wave vibrations which will get through.

It is interesting to note that if the light waves were compressional instead of longitudinal, they could not be blocked out, since they would vibrate in the direction of propagation and would, therefore, simply pass through both the polarizer and the analyzer.

Polaroid is a lamination of plastic and long, thin iodosulfate crystals. Under controlled conditions, these crystals can be made to align themselves parallel to the face of the plastic sheet. Thus two sheets of this laminated material may be used as shown in Fig. 34.1(a) and (b). When the crystals in one sheet are parallel to the crystals in the other sheet, light will pass through both. If the axis of one sheet is 90° to the other, almost all the light is blocked out.

Thus, once the light is polarized and becomes unidirectional, the angle between the optical axis of the crystal of the polarizer and the analyzer becomes important to the control of the amplitude and the intensity of the light that eventually gets through both the polarizer and the analyzer. Therefore, in Fig. 34.2, P will pass through the polarizer; the component $P_1 \cos \theta_1$ will also pass through the polarizer; P_2 and the component $P_1 \sin \theta_1$ will be absorbed. Adding all the components into the intensity after polarization (shown as $\sum P$), the resultant vector may be resolved into its components, such that again only the vector $P \cos \theta$ will pass through the analyzer and the component $P \sin \theta$ will be absorbed. Thus when $\theta = 0°$ or $180°$, the amplitude and the intensity of the light passing through the analyzer are a maximum. When $\theta = 90°$ or $270°$, the amplitude and the intensity of the analyzed light are a minimum.

Sec. 34.1　　　　　　　　　　Polarization　　　　　　　　　　547

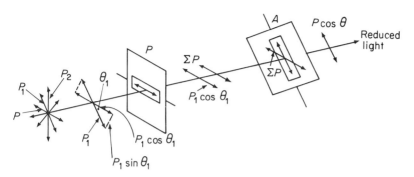

Fig. 34.2

If both the polarizer and the analyzer are perfect transmitters (no losses), about 50 per cent of the vibrations and components parallel to the optical axis of the polarizer will get through this polarizer. The resulting amplitude of the vibrations which get through the analyzer is $\cos \theta$. Since the intensity is a function of the square of the angle through which the analyzer is rotated, the intensity of the analyzed light is $\cos^2 \theta$.

For multiple plates, the intensity of the original *unpolarized* light transmitted is the product of $\frac{1}{2}$ (the polarizing intensity) multiplied by the $\cos^2 \theta$ which each optical axis makes with its neighbor. The total-intensity equation is

$$I = \tfrac{1}{2}(\cos^2 \theta)^{n-1} \qquad \{\, I = \text{total intensity}\,\}$$

Example 1

A polarizer transmits at a maximum value. The axis of an analyzer is rotated 30° with the polarizer axis. What percentage of the polarized light is transmitted by the analyzer?

Solution

The percentage of the *polarized* light transmitted by the polarizer is

$$\cos^2 \theta = \cos^2 30° = 0.750 = 75\%$$

Example 2

Assume three polarizing plates, so stacked that the first and last plates are at 90° to each other, with the central plate at 45° to the other two. What is the intensity of the light transmitted by the three plates?

Solution

1. The first polarizer transmits $\frac{1}{2}$ the incident light.

2. Each succeeding analyzer transmits

$$\cos^2 \theta = \cos^2 45° = 0.5$$

3. The total intensity transmitted is

$$I = \tfrac{1}{2}(0.5)^{n-1} = \tfrac{1}{2}(0.5)^2 = 12.5\% \qquad \{n = 3\}$$

Note this interesting point: if the two end polarizers remain at 90° to each other and the center plate is removed, the intensity of the original light reduces to zero, since

$$I = \tfrac{1}{2} \cos^2 90° = 0\%$$

34.2. Polarization by Reflection

Unpolarized light striking the surface of a flat plate of glass is partially refracted by the glass and partially reflected. If the reflected and refracted light is checked with an analyzer, both rays will be found to be partially polarized. *Brewster's law* states that if light is incident to the glass surface so that the angle between the reflected rays and the refracted rays is 90°, the reflected rays will be plane-polarized so that the vibrations are parallel to the reflecting surface and the majority of the vibrations of the refracted rays will be polarized in a plane parallel to the xy-plane (Fig. 34.3). The incident ray vibrates in all directions perpendicular to line AD.

Thus in Fig. 34.3, the reflected DC-ray is plane-polarized, the vibrations taking place perpendicular to the plane xy. The direction of vibration is shown as a dot, or end view of the vector arrow. The refracted ray BD will be found to have most of its vibrations taking place parallel to the xy-plane.

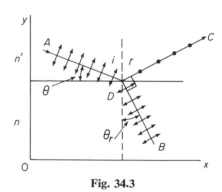

Fig. 34.3

Maximum polarization takes place when the rays CD and DB are at 90° to each other. Under this condition,

$$\theta = \theta_r \quad \text{and} \quad i + \theta_r = 90° \quad \text{or} \quad \theta_r = 90 - i$$

This makes the incident angle i the complementary angle of the refracting angle θ_r. From Snell's law,

$$n' \sin i = n \sin \theta_r = n \sin(90 - i)$$

Solving for n,

$$n = \frac{n' \sin i}{\sin(90 - i)}$$

But sin $(90 - i) = \cos i$; therefore,

$$n = \frac{n' \sin i}{\cos i} = n' \tan i$$

If $n' = 1$, Brewster's law becomes

$$n = \tan i$$

Example 3

The index of refraction of light (wavelength 5890 A) for water is 1.33. What is the polarizing angle?

Solution

The polarizing angle is

$$\tan i = n = 1.33$$
$$i = 53°4'$$

34.3. Polarization by Refraction

When a medium such as glass transmits waves with equal velocities in all directions, it is said to be *isotropic*. But stressed glass and many crystalline materials possess different properties in different directions—that is, the crystal can separate the transmitted light into two different kinds of wavelets, each having different velocities. One set of wavelets spreads out spherically; the other spreads out elliptically.

There is a point where the elliptical and spherical waves are tangent to each other. The line connecting the points of tangency is called the *optical axis* and is shown in Fig. 34.4(a).

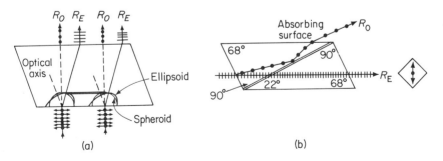

Fig. 34.4

The parallel lines, Fig. 34.4(a), drawn approaching the spheroids and the ellipsoids represent wave fronts of different velocities, the ellipsoid wave fronts having slower velocities than the spheroid wave fronts. The vibration of the wave front which passes through the crystal as a result of spheroid waves polarizes the light *perpendicular* to the plane of the page and is called the *ordinary ray* R_O. The ray which passes through the crystal according to the elliptical wavelets polarizes the light so that its vibrations are *parallel* to the page and is called the *extraordinary ray* R_E. If a ray is incident and normal to the surface of the crystal, the ordinary ray will pass through the crystal undeviated, but polarized. The extraordinary ray is refracted. The overall phenomenon is called *double refraction* and is shown in Fig. 34.4(a).

In Fig. 34.4(b), a calcite rhombohedron, with the natural-cleavage angle of 71° is altered, by polishing, to 68°. The crystal is then cut so that the cut plane forms an angle of 90° with the polished ends. The cut surfaces are polished and cemented with Canada balsam. Since Canada balsam has an index of refraction less than the ordinary ray but greater than the extraordinary ray, the ordinary ray, because of the cut angle, will be totally reflected. The extraordinary ray will be transmitted. The ordinary ray is usually absorbed as shown in Fig. 34.4(b), but the extraordinary ray is polarized and transmitted. This prism is called a *Nicol prism*.

Some of the values for the index of refraction for sodium light for both ordinary and extraordinary rays are shown in Table 34.1.

Table 34.1

Material	n_O	n_E
Canada balsam	1.540	—
calcite	1.658	1.486
ice	1.309	1.313
quartz	1.544	1.553
sodium nitrate	1.587	1.336
tourmaline	1.668	1.640

The indices of refraction for the ordinary and extraordinary rays are a function of the ratio of the velocity of light in the transmitting medium to the velocity of light. Thus

$$n_O = \frac{c}{v_O} \quad \text{and} \quad n_E = \frac{c}{v_E} \quad \left\{ \begin{array}{l} v_O = \text{velocity of ordinary ray} \\ v_E = \text{velocity of extraordinary ray} \\ n_O = \text{index of refraction of ordinary ray} \\ n_E = \text{index of refraction of extraordinary ray} \end{array} \right\}$$

Also, the wavelength of the incident ray divided by the wavelength of either the ordinary or the extraordinary ray gives the respective index of refrac-

Sec. 34.4 Polarization by Absorption and Scattering

tion of the transmitted light. Thus

$$n_O = \frac{\lambda}{\lambda_O} \quad \text{and} \quad n_E = \frac{\lambda}{\lambda_E} \quad \begin{cases} \lambda = \text{wavelength of incident ray} \\ \lambda_O = \text{wavelength of ordinary ray} \\ \lambda_E = \text{wavelength of extraordinary ray} \end{cases}$$

Example 4

Monochromatic light (wavelength 5893 A) is incident to a calcite crystal. (1) Find the velocities of the ordinary and extraordinary transmitted rays in the crystal. (2) What is the wavelength of the ordinary and extraordinary rays in the crystal?

Solution

1. The velocities are

$$v_O = \frac{c}{n_O} = \frac{3 \times 10^{10}}{1.658} = 1.81 \times 10^{10} \text{ cm/sec}$$

$$v_E = \frac{c}{n_E} = \frac{3 \times 10^{10}}{1.486} = 2.02 \times 10^{10} \text{ cm/sec}$$

2. The wavelengths are

$$\begin{cases} n_O = 1.658 \\ n_E = 1.486 \\ \lambda = 5893 \times 10^{-8} \text{ cm} \end{cases}$$

$$\lambda_O = \frac{\lambda}{n_O} = \frac{5893 \times 10^{-8}}{1.658} = 3554 \text{ A}$$

$$\lambda_E = \frac{\lambda}{n_E} = \frac{5893 \times 10^{-8}}{1.486} = 3965 \text{ A}$$

34.4. Polarization by Absorption and Scattering

Dichroic crystals will separate the vibrations of light into horizontal and vertical vibrations and absorb one set while transmitting the other. Tourmaline is such a material, but its transmitted ray is colored. (The process is shown in Fig. 34.5.)

Figure 34.5(a) is the top view of Fig. 34.5(b). In the top view, the vibrations parallel to the plane of the page are absorbed and never get through the

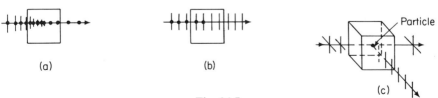

Fig. 34.5

crystal. The vibrations perpendicular to the page are transmitted with slightly reduced intensity. This may also be seen in Fig. 34.5(b).

Figure 34.5(c) shows the effect of *scattering* by fine particles. Incident light is scattered by the particles so that one ray will be transmitted through the body of dust partially polarized. The other will be transmitted at an angle with the incident ray such that only a univibrational direction may be observed with an analyzer.

Very small particles will transmit the short wavelengths of light (blue). The longer wavelengths are polarized as the particles get larger, until the transmitted light which is scattered becomes white. The transmission and polarization of sunlight through the atmosphere is an example of scattering.

34.5. Interference and Polarization

Section 34.3 showed how a *birefringent material* (capable of causing double refraction) separates a ray into electrical vibrations at 90° to each other. This separation is a function of the index of refraction, or difference in velocities, of the two rays after the light enters the material. If the velocities of the ordinary rays and the extraordinary rays are different, it should be possible to select a thickness of material such that the emerging rays will be one half-wavelength out of phase with each other. This should cause interference between the two rays. But observation of the rays shows this not to be the case. The two rays will be polarized perpendicular to each other.

If the two mutually perpendicular rays (vectors) are caused to pass through an analyzer, the vector components will emerge from the analyzer with their vibration planes parallel. These vibrations will interfere with each other and cause destruction of one wavelength and the transmission of the other. This transmitted ray is the complementary color of the wavelength destroyed.

This process is shown in Fig. 34.6(a). The light source enters the polarizer P which transmits the waves parallel to the optical axis. The slit in the polarizer is shown to be elliptical. The linear polarized light then passes through a double-refracting crystal R which is of the appropriate thickness so that the entering ray is divided into an ordinary and an extraordinary ray one half-wavelength out of phase. The optical axis of the crystal R is oriented at 45° to the polarizer P, so that only the components of the polarized vibrations are transmitted. Since the ordinary and extraordinary rays vibrate at 90° to each other, the vibrations between R and A are oriented at 45° to the polarizer and at 90° to each other, as shown in Fig. 34.6(a). They are also one half-wavelength out of phase with each other.

Sec. 34.5   *Interference and Polarization*   553

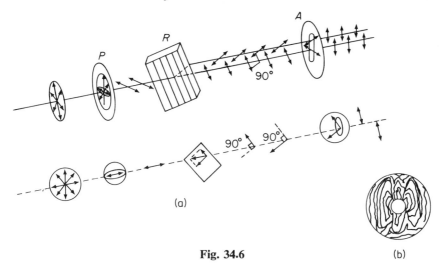

Fig. 34.6

The two vibrations do not occur in the same plane; hence they will not interfere. They must be passed through the analyzer A, which transmits the components parallel to the optical axis of the analyzer. These polarized vibrations are in the same plane and one half-wavelength out of phase and will therefore interfere.

Since it is possible to control the thickness of the polarizing material, some wavelengths will interfere constructively, others destructively. If all wavelengths are present, colored strain patterns will be visible. If the analyzer is now rotated through 90°, the patterns constructively interfered with will be interfered with destructively and the complementary colors will be visible. The reverse is true for colors subjected to destructive interference before the analyzer is rotated.

When a material such as glass is strained, it becomes double-refracting. When subjected to analysis between a polarizer and an analyzer, this material will set up strain patterns [Fig. 34.6(b)]. Machined parts which are to be subjected to severe strain during use may be studied by making a model of the part out of this material, subjecting the model to strain, and studying the strain patterns.

Some materials—quartz, for example—will rotate the ordinary and extraordinary rays and recombine them when they leave the crystal. Although the crystal itself is not rotated, the rays will be rotated through some angle. The vibrations may be out of phase by 90° or 180°. The entering vibration angle with the optical axis is the same as the emerging vibrating angle with the optical axis. If the two components are combined at some angle other than 45°, elliptical polarization will result. If the two components are combined at 45°, they will be equal, and the light will be circularly polarized.

Problems

1. Calculate how much of the light transmitted through the polarizer is transmitted through the analyzer if the analyzer is rotated through (a) 20°; (b) 40°; (c) 50°; (d) 75°.
2. How much of the original light is transmitted through the polarizer in Prob. 1?
3. Calculate the amplitude of the light after polarization in Prob. 1.
4. Four polarizing plates are stacked at equal angles when the first and last plate are at 90° to each other. How much of the original light is transmitted by the four plates?
5. Calculate the polarization angle for dense glass (index of refraction 1.65).
6. Repeat Prob. 5 for diamond (index of refraction 2.1).
7. What is the polarization angle if light (5890 A) is used for the following: (a) carbon tetrachloride ($n = 1.4607$)? (b) crown glass (1.534)? (c) flint glass (1.619)? (d) rock salt (1.544)?
8. What is the polarization angle for crown glass for (a) red B (oxygen) light (index of refraction 1.5301)? (b) red C (hydrogen) light (index of refraction 1.5311)? (c) yellow D (sodium) light (index of refraction 1.534)? (d) green F (hydrogen) light (index of refraction 1.541)? (e) ultraviolet H (calcium) (index of refraction 1.5509)?
9. What is the velocity of the ordinary and extraordinary rays for sodium light in sodium nitrate?
10. Find the wavelength of the light in Prob. 9.
11. What is the critical angle for—(a) the ordinary; (b) the extraordinary—rays passing through a calcite crystal to the Canada balsam in the Nicol prism?

35

Electron-Beam Acceleration and Deflection

35.1. Thermionic Emission

Charges may move or be caused to move in metals, liquids, or gases. The movement in a solid consists of the movement of negative charges. Movement in a gas is both ionic and electronic. It is obvious that if movement is to take place in a vacuum, the particles must be introduced into the vacuum.

An analogy may be drawn between a container of water and the particles closely bound in a solid. If the water is at room temperature, some of the particles in the water will pick up enough energy to break through the surface of the liquid. As more and more energy is added to the liquid, the particles vibrate faster and faster and more particles will leave the surface per unit time. In a metal, the "particle" which acquires this energy is the electron. If the electron has acquired enough energy, in the form of velocity, to break through the surface of the metal, it will leave the surface, and in so doing give up energy to the surface of the metal. Different metals have different escape-energy plateaus. The energy required for an electron to leave a metallic surface is called the *work function* of that metal.

If a potential difference (V volts) is created between two plates and a negatively charged particle is introduced between the plates, the particle will be accelerated toward the positive plate and away from the negative plate. The result is an increase in the velocity of the particle and consequently an increase in its kinetic energy.

The product of 1 volt of potential difference and the charge on an electron is defined as the *electron-volt* eV. Since the charge on an electron is 1.6×10^{-19} coulombs, the energy of 1 electron-volt is equal to 1.6×10^{-19} joules.

Heating a metal increases the electron kinetic energies, through increased frequency vibrations. This increase of kinetic energy may be enough to cause some of the electrons to break through the surface of the metal. Another method of increasing the energy of particles is by the transformation of light energy to kinetic energy of velocity. Still another method is the bombardment of electrons with other high-velocity particles.

Figure 35.1 shows an evacuated tube containing a loop of wire (*cathode*) and a collector plate (*anode*). The anode is connected to the + terminal of the battery. The cathode is used as a heater wire which causes electrons to be emitted between the cathode and the anode in a process called *thermionic emission*. As the cathode is heated to a higher and higher temperature, it emits more electrons. If the cathode releases more electrons than the anode can collect, the "excess" of electrons will drift toward the anode and create a negative space charge.

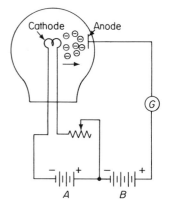

Fig. 35.1

If the heater voltage is held constant and the anode potential increased, the plate current will absorb greater quantities of electrons until it absorbs all the electrons which the cathode heater is capable of producing. It is possible to create a saturated space charge by this mechanism.

35.2. Rectifiers and the Triode

Thomas Edison was the first to observe that if the anode has a negative potential with respect to the cathode, all plate current disappears. A *diode* can then be made to act as a valve to control the quantity of electricity permitted to flow. If the plate is negative, no current flows. If the plate is positive and saturated, the flow is a maximum.

Thus if the plate is connected to an alternating current, the plate will permit current to flow when it is positive and will stop the current flow when it is negative. Since one-half of the cycle of an alternating current is positive, the input wave will be *half-wave rectified*. The hookup, termed a half-wave

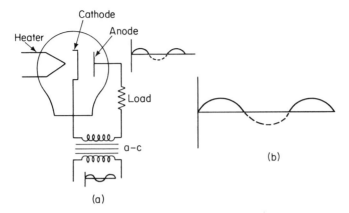

Fig. 35.2

rectification is shown in Fig. 35.2(a) and the rectified wave is shown in Fig. 35.2(b). The dotted line shows the input wave.

Figure 35.3(a) and (b) show *full-wave rectification*. Both plates are connected so that they are 180° out of phase.

The *triode*, a three-element tube, is shown in Fig. 35.4(a). The extra element is a screen called the *grid* which permits electrons to flow through from cathode to anode. DeForest discovered that by making the grid more positive or less positive with respect to the plate, he was able to exert better control of those electrons which are permitted to reach the plate. He found that a small potential change, when the grid is negative, creates a comparatively large plate-current increase. Thus, even though the plate current is still dependent upon the plate potential with respect to the cathode, the plate

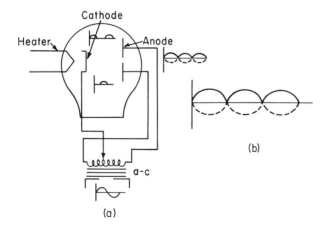

Fig. 35.3

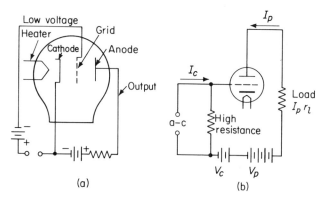

Fig. 35.4

current depends upon the grid potential with respect to the cathode to an even greater extent. Figure 35.4(b) shows how a triode tube can be used as an amplifier by virtue of the fact that small grid voltage produces a large plate current.

The triode may also be used as an oscillator. If the switch S_1 [Fig. 35.5(a)] is closed, the capacitor will charge as shown. When switch S_1 is opened and switch S_2 is closed, the capacitor discharges completely through the coil which, because of its inductance, supports the current until the capacitor has been oppositely charged. If the resistance in L is small, the capacitor will discharge opposite to its original sign and nearly to its original value. At this point, the discharge will take place in the opposite direction so that the capacitor is once again charged as shown. The process repeats itself until the losses (heat, resistance, etc.) cause the oscillations to stop. [A plot of the decrease in oscillations is given in Fig. 35.5(b).]

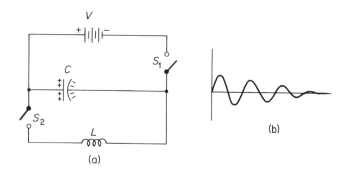

Fig. 35.5

Sec. 35.3   *Transistors*   559

When the resistance in the circuit is negligible, the oscillations will resonate according to

$$F = \frac{1}{2\pi\sqrt{LC}}$$

If energy is supplied to the circuit to replenish these losses, the circuit will continue to oscillate undiminished.

35.3. Transistors

Certain materials, such as metals, are good conductors; whereas other materials, such as ceramics or rubber, are poor conductors. Materials which lie in between the very good and the very poor conductors are called *semiconductors*. These materials conduct under certain conditions. If several of these semiconducting materials are used in combination to produce a desired result, the combination is called a transistor. Some semiconducting materials are aluminum oxide, copper oxide, germanium, and silicon.

Semiconductors contain relatively few free electrons. If a semiconductor is to conduct, it must get its electrons from some other source. This may be accomplished by introducing impurities (other materials) which have electrons, called *valence electrons*. Thus germanium will share its four valence electrons with other germanium atoms. This sharing of valence electrons is the source of energy which binds one atom to another. Thus in Fig. 35.6(a) (the lattice structure is shown flatteneed out), each of four electrons is tied to the center atom with a heavy line and linked to a neighboring atom with a thin line. Thus the center atom shares its four valence electrons with four neighboring electrons. They in turn share their valence electrons with their neighbors.

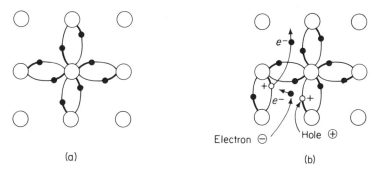

Fig. 35.6

At times, as a result of light-, heat-, or electrical-energy input, an electron is jarred loose, as shown in Fig. 35.6(b). One of the neighbors may also break loose and head for the spot vacated. If an electron vacates a position, that position is thought of as having lost a negative charge—we say that a *"hole"* is created. Since it loses one negative charge, the hole may be thought of as having a + charge.* Thus with respect to Fig. 35.6(b), we may say either that the electron is moving to the left or that the positive charge (hole) is moving toward the right.

Excess electrons may be "fed" into the system, as indicated, by introducing foreign materials which have, say, five valence electrons. Since, in our scheme, four valence electrons enter into the binding process, one valence electron is left to move in the materials and thus create holes. However, there are more free electrons moving than positive charges. This is called an *N-type* (negative) crystal.

Let us assume that a foreign material which has three valence electrons is introduced into our base material, which has four valence electrons. This deficiency of electrons generates holes, or positive charges. This is called a *P-type* (positive) crystal.

In its normal state, the P-type crystal has more holes than negative charges. Placing an N-type crystal in contact with a P-type crystal creates a barrier between the two crystals as shown in Fig. 35.7(a).

In its normal state, the N-type crystal has more negative charges than holes. The negative charges move from the N-crystal across the barrier, filling holes in the P-crystal. The migration of electrons from the N- to the P-crystal has the effect of making the P-crystal more negative and the N-crystal more positive. A potential difference is created and an ammeter will register a current.

In Fig. 35.7(b), when the alternating current is positive at the P-end, the electrons flow across and then *away* from the barrier because the alternating current is negative at the N-end. Note that the polarity of the alternating current will generate a large current flow in the P-N crystal because it is pulling negative charges from the N-crystal, where there are many negative charges. A large current results across the barrier.

Now let us assume that the polarity of the alternating current has changed

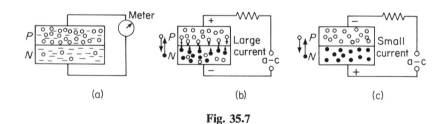

Fig. 35.7

*An excess of negative charges is represented by a solid dot ●. An excess of positive charges (holes) is represented by a circle ○.

as shown in Fig. 35.7(c). Now few electrons and few positive charges will cross the barrier and a very small current will result.

The effect is rectification. When the P-crystal is positive, a large current will flow. When the P-crystal is negative, a very small current will flow.

Figure 35.8(a) shows a *"junction type"* N-P-N *transistor* in which a thin positive wafer is sandwiched between two negative wafers. Three leads are taken off the wafers and thus the combination may be used as a triode to amplify or oscillate. It is also possible to construct P-N-P combinations as shown in Fig. 35.8(b).

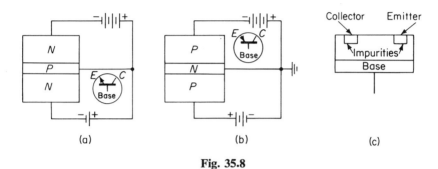

Fig. 35.8

In Fig. 35.8(c), we see a *"point-contact"* transistor, in which impurities are introduced into the crystal and then tapped.

35.4. The Thyratron

The fundamental concept of conduction in a gas tube is the acceleration of ions and electrons toward the appropriate electrode. That is, positive ions move toward the negative pole, and negative ions and electrons move toward the positive pole. These accelerated particles collide with other molecules in the tube and may knock off electrons, creating more electrons and charged ions. While this is going on, ions and electrons are being neutralized.

If a negatively charged ion or an electron should reach an electrode, the ion would give up its charge to the electrode and become neutral. The charge passes into the electrode and out to the battery. There is a general drift of positive and negative charges to the oppositively charged electrodes.

If a gas is added to the tube, more particles are made available for transporting these charges. Thus if the charged particles result from thermionic emission, the flow of electrons may be controlled by a grid. When ionization takes place, the grid has no control over the flow of current.

Removing an electron from the orbit of a neutral molecule to create an

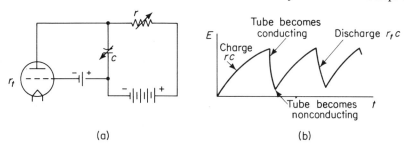

Fig. 35.9

ion requires a minimum potential, called the *ionization potential*. Ionization potentials of a few elements are: hydrogen, 13.53 volts; mercury, 10.4 volts; oxygen, 13.55 volts.

Figure 35.9(a) shows a circuit which uses a gas-filled tube called a *thyratron*. The tube is nonconducting until the variable capacitor is charged from the battery through the resistor. The variable resistor and capacitor are adjusted to the required time constant rc. Since the tube is nonconducting, the capacitor becomes charged until the charge on the capacitor reaches the breakdown value of the gas in the tube. At this instant, the capacitor discharges rapidly through the tube, which once again becomes nonconducting. The process is repeated.

During discharge, the time constant $r_t c$ is very small, because the resistance of the tube is very small during the decay part of the cycle. The charging time is longer because of the high resistance of the resistor in the circuit while the build-up is taking place. The number of cycles per second is controlled by varying either r or c. This type of charge–discharge functioning generates a sawtooth curve as shown in Fig. 35.9(b).

35.5. The Crookes Tube

The *Crookes tube* is a long glass tube with an electrode in each end connected to a high-voltage source and an opening which is connected to a vacuum pump. When a potential difference is established between the cathode and anode, a discharge takes place. If this discharge is at atmospheric pressure, sparks will jump between the two electrodes. As the pressure in the tube is reduced, a blue glow fills the tube from electrode to electrode. The voltage required to maintain this discharge drops. If the pressure is reduced farther, a blue-purple glow appears at the cathode. This glow moves away from the cathode as the pressure is lowered farther. When the pressure drops to about 0.1 mm, the regions separate, as shown in Fig. 35.10. Starting at the cathode,

Cathode Rays

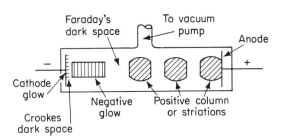

Fig. 35.10

a glow appears at the electrode, followed by Crookes dark space, the negative glow, the Faraday dark space, and the positive column.

As the pressure is dropped farther, the voltage required to support the discharge starts to increase. The Crookes dark space increases until no amount of voltage will support a discharge. Ions produced traverse the entire length of the tube without collision. Thus the discharge current decreases almost to zero and the voltage rises very slowly.

At high *pressures*, the energy required to cause ionization can be attained only by high voltages, because the distances between collision are very short. At intermediate pressures, the ionization energy (velocity) may be attained because of the space increase before collision. When the pressure is very low, the ions may travel the entire length of the tube without collision or ionization. But on collision with the cathode, electrons possessing very high velocities are released and cause the glass to glow green. This is a result of the fluorescence of the glass.

35.6. Cathode Rays

Electron beams, called *cathode rays*, possess certain characteristics:

1. Cathode rays may produce fluorescence or phosphorescence. If a cathode ray is caused to fall on a fluorescent material, some of the energy of the rays causes the molecules of this material to absorb energy from the rays and become excited. When the vibrating molecules return to their normal state, some of the energy is emitted in the form of light rays of longer wavelength than the exciting rays. When the excited molecules return to their nomal state almost immediately, the absorbing material is said to *fluoresce*. If there is an appreciable lag before the molecules return to their normal state, the material is said to *phosphoresce*.
2. Cathode rays are emitted perpendicular to the surface of the cathode. This may be demonstrated by predicting that the rays will focus at the geometric center of a concave mirror [see Fig. 35.11(a)].
3. Cathode rays travel in a straight line. This may be demonstrated with the Maltese

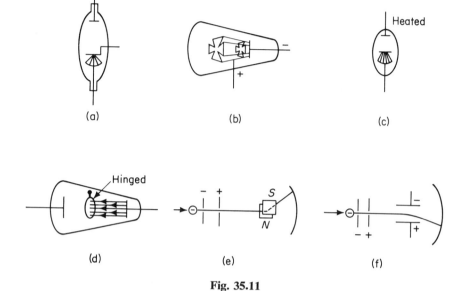

Fig. 35.11

cross equipment (Fig. 35.11b). When an object intercepts the rays, a sharp image of the object (Maltese cross) is seen on the fluorescent screen. The shadow will not fluoresce. The outline of the image is sharp, thus indicating that the rays travel in a straight line.

4. Cathode rays possess kinetic energy. This may be demonstrated by allowing the rays to strike a target. The target is heated by the rays as in Fig. 35.11 (c). Kinetic energy may be demonstrated as in Fig. 35.11(d), where an object is caused to "flop" over when struck by the rays.
5. Cathode rays may be deflected by a magnetic field. Figure 35. 11(e) show a small pencil of rays deflected according to the polarity of the plates.
6. Cathode rays may be deflected by an electric field. Figure 35.11(f) shows a small pencil of rays deflected according to the charge on the plates. This causes the deflection of the negative rays.
7. Cathode rays may produce X-rays. If a cathode ray has very high velocity (kinetic energy) and strikes a dense surface, such as tungsten, it will produce very short penetrating wavelengths called X-rays.
8. Cathode rays will penetrate thin sheets of metal. When cathode rays are directed at aluminum foil, the rays will penetrate the foil and produce luminosity on the opposite side of the foil from the source.

35.7. The Cathode-Ray Tube

The *cathode-ray tube*, Fig. 35.12(a), consists of a heater H, a cathode C, an anode A, accelerating plates, two vertical deflecting plates P_1, two horizontal

Sec. 35.7 The Cathode-Ray Tube 565

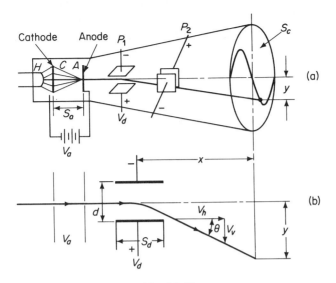

Fig. 35.12

deflecting plates P_2, and a fluorescent screen S_c. The accelerating voltage V_a is about 1000 volts. The deflecting plates P_2 cause the beam to sweep the screen in a horizontal direction at the same time that the plates P_1 cause the beam to deflect along the y-axis. The application of various voltages to P_1P_2 causes various configurations on the screen S_c.

If the voltage V_d is applied to the two plates, as shown in Fig. 35.12(b), a force will be put on the negative beam according to the plate polarity, and the beam will deflect toward the positive plate. Once the beam leaves these plates, it will travel in a straight line. It will travel a horizontal distance x and deflect a distance y on the screen. On leaving the plates, the horizontal velocity will be v_h and the vertical velocity will be v_v.

1. The horizontal velocity v_h.

    The horizontal velocity of the beam has kinetic energy

    $$\text{KE} = \tfrac{1}{2}mv_h^2$$

The kinetic energy is also a function of the charge on an electron and the potential difference of the two plates. Thus

$$\text{KE} = V_a e$$

Equating both kinetic energy equations,

$$\tfrac{1}{2}mv_h = V_a e$$

$\left\{\begin{array}{l} v_h = \text{horizontal velocity} \\ v_v = \text{vertical velocity} \\ V_a = \text{accelerating voltage} \\ e = \text{charge on an electron} \\ m = \text{mass of an electron} \end{array}\right\}$

Solving for v_h,

$$v_h = \sqrt{2V_a\left(\frac{e}{m}\right)}$$

2. The vertical velocity v_v.

The kinetic energy may be considered as force times the distance between the two plates. Thus*

$$KE = fd$$

Solving for f,

$$f = \frac{KE}{d}$$

Substituting for $KE = V_d e$

$$f = \frac{V_d e}{d}$$

From $v_v = at$ and $f = ma$,

$$v_v = \frac{f}{m} t$$

Substituting for $t = S_a/v_h$ and $f = V_d e/d$,

$$v_v = \frac{V_d e}{dm} \times \frac{S_d}{v_h} = \frac{V_d e S_d}{dm v_h}$$

3. The y-deflection.

From the small triangle in Fig. 35.12(b), it can be seen that

$$\tan \theta = \frac{v_v}{v_h}$$

Making the substitutions for v_v,

$$\tan \theta = \frac{v_v}{v_h} = \frac{V_d e S_d}{dm v_h^2}$$

Substituting the value for v_h,

$$\tan \theta = \frac{V_d e S_d}{dm \left(\sqrt{\frac{2V_a e}{m}}\right)^2}$$

Reducing the equation,

$$\tan \theta = \frac{V_d e S_d}{dm \left(\frac{2V_a e}{m}\right)} = \frac{V_d S_d}{2 d V_a}$$

v_v = vertical velocity
v_h = horizontal velocity
e = charge on an electron
 = 1.6×10^{-19} coul
V_d = deflecting voltage
V_a = accelerating voltage
S_d = length of deflecting plates
d = distance between plates
m = mass of an electron
 = 9.11×10^{31} kg
x = distance from centerline of deflecting plates to screen

Since $\tan \theta = y/x$, then

$$y = x \tan \theta = x \left(\frac{V_d S_d}{2 d V_a}\right)$$

*The quantity fd = work. Also, $f = ma$ and $v^2 = 2ad$. Solving the latter two equations for f and d and substituting into the work equation,

$$fd = (ma)\frac{v^2}{2a} = \frac{1}{2} mv^2$$

Example 1

A cathode-ray tube has a potential of 1100 volts applied to the accelerating plates and 120 volts to the deflecting plates. The length of the deflecting plates is 3 cm and the distance between the plates is 2 cm. If the distance from the center of the deflecting plates to the screen is 30 cm, find (1) the horizontal velocity of the electron beam, (2) the kinetic energy of the beam, (3) the vertical velocity, (4) the angle of deflection, and (5) the distance the beam deflects on the screen. (6) From the equation for vertical velocity, find the charge-to-mass ratio for the electron.

Solution

1. The horizontal velocity is

$$v_h = \sqrt{\frac{2eV_a}{m}} = \sqrt{\frac{2(1.6 \times 10^{-19})1100}{9.11 \times 10^{-31}}}$$

$$= 1.97 \times 10^7 \text{ m/sec}$$

2. The kinetic energy of the beam is

$$KE = \tfrac{1}{2}mv_h^2 = \tfrac{1}{2}(9.11 \times 10^{-31})(1.97 \times 10^7)^2$$

$$= 1.76 \times 10^{-16} \text{ joule}$$

3. The vertical velocity is

$$v_v = \frac{V_d e S_d}{dmv_h} = \frac{120(1.6 \times 10^{-19})0.03}{0.02(9.11 \times 10^{-31})(1.97 \times 10^7)}$$

$$= 1.6 \times 10^6 \text{ m/sec}$$

$$\left\{\begin{array}{l} V_a = 1100 \text{ volts} \\ V_d = 120 \text{ volts} \\ S_d = 3 \text{ cm} = 0.03 \text{ m} \\ d = 2 \text{ cm} = 0.02 \text{ m} \\ x = 30 \text{ cm} = 0.30 \text{ m} \end{array}\right\}$$

4. The angle of deflection θ is

$$\tan\theta = \frac{v_v}{v_h} = \frac{1.6 \times 10^6}{1.97 \times 10^7} = 0.0812$$

$$\theta = 4°40'$$

5. The deflection on the screen is

$$y = x \tan\theta = 0.30(8.12 \times 10^{-2})$$

$$= 0.0244 \text{ m}$$

6. The charge-to-mass ratio from the equation for vertical velocity is

$$\frac{e}{m} = \frac{v_h v_v d}{V_a S_d} = \frac{(1.97 \times 10^7)(1.6 \times 10^6)0.02}{120 \times 0.03}$$

$$= 1.750 \times 10^{11} \text{ coul/kg} = 1.750 \times 10^8 \text{ coul/g}*$$

*The accepted value is 1.7589×10^8 coul/g.

J. J. Thomson used a magnetic field to deflect an electron beam to calculate the charge-to-mass ratio of an electron.

A charged particle entering a magnetic field is subjected to a force

$$f = Bev \qquad \left\{ \begin{array}{l} B = \text{magnetic flux density: web/m}^2 \\ e = \text{charge on an electron: coul} \\ v = \text{velocity: m/sec} \end{array} \right\}$$

The centripetal force needed to hold a particle in a circular orbit is

$$f = \frac{mv^2}{r}$$

Equating the two force equations, solving for v, and substituting into the equation $Ve = \frac{1}{2}mv^2$ yields

$$Ve = \frac{1}{2}m\left(\frac{rBe}{m}\right)^2$$

Solving this equation for e/m,

$$\frac{e}{m} = \frac{2V}{B^2 r^2}$$

Example 2

If a cathode ray enters a magnetic field of 5×10^{-4} wb/m², what is the charge-to-mass ratio of the electron if the radius of curvature of the beam is 30.7 cm and the voltage is 2000 volts?

Solution

The charge-to-mass ratio is

$$\frac{e}{m} = \frac{2V}{B^2 r^2} = \frac{2 \times 2000}{(5 \times 10^{-4})^2 \times (30.7 \times 10^{-2})^2}$$
$$= 1.7 \times 10^{11} \text{ coul/kg} = 1.7 \times 10^8 \text{ coul/g}$$

The charge-to-mass ratio for any isotope, stable or unstable, may be found from

$$\frac{e}{m} = \frac{96{,}520 \times \text{ion charge} \times 10^3}{\text{at wt}} = \frac{\text{coul}}{\text{kg}}$$

Example 3

Calculate the charge-to-mass ratio of an alpha particle whose charge is $+2e$.

Solution

$$\frac{e}{m} = \frac{96{,}520 \times 2 \times 10^3}{4} = 4.826 \times 10^7 \text{ coul/kg}$$

35.8. The Mass Spectrograph

Several different kinds of *mass spectrographs* utilize the fact that a charged particle is affected by the force placed on it by either an electric field, a magnetic field, or both. Mass spectrographs are used to find the masses of positively charged ions or of isotopes.

Ions have their normal complement of protons and neutrons, but are not neutral, because they have either lost or gained electrons which affect the charge-balance between the nucleus and the electrons circling the nucleus. *Isotopes* have their normal complement of protons and electrons, but do not have the same number of neutrons. That is, they have the same atomic number, but different mass numbers. Whereas electrons affect the chemical properties of an element, the addition or loss of a neutron cannot be detected by chemical means. The loss or gain of neutrons, however, affects the mass of the element. The mass spectrograph can detect this mass difference.

The isotopes of hydrogen are

$$_1H^1 = \text{hydrogen}$$
$$_1H^2 = \text{deuterium}$$
$$_1H^3 = \text{tritium}$$

The symbols show the same atomic numbers, but different atomic masses. The additional mass is attributed to extra neutrons in the nucleus. They would appear on the plate of a mass spectrograph as darkened ribbons as shown in Fig. 35.13(a).

It should be evident that, if the various isotopes could all be accelerated to a single velocity and then caused to assume a circular path, those with the greater mass would traverse longer arc paths. That is, if the speeds of a heavy and a light object are the same, the radius of the path of the heavy object will be longer than the radius of the path of the light object. Thus a mass spectrograph accelerates a supply of isotopes, filters them to one speed, and then applies the same force to all the filtered isotopes so that the heavier isotopes will travel in a longer path than the lighter isotopes.

Figure 35.13(b) is a sketch of a Bainbridge mass spectrograph. A supply of positive ions enters the region between the positive and negative poles of the accelerating mechanism ($p_1 p_2$). These positive ions are accelerated and caused to pass into an area of a crossed magnetic and electric field which puts opposing forces on the ions. Balancing these forces for a given velocity permits these selected ions to pass through an opening in the ground plate p_3.

The force put on a positive charge, or charges, by an electric field is

$$f = Ve$$

And the force put on a positive charge or charges by a magnetic field is

$$f = evB_f \qquad \{B_f = \text{magnetic filter field}\}$$

Equating,

$$Ve = evB_f$$

Solving for velocity,

$$v = \frac{V}{B_f}$$

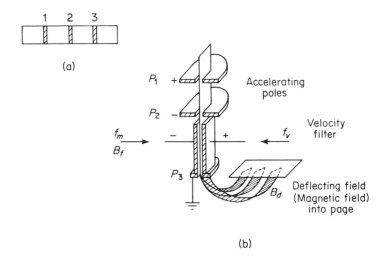

Fig. 35.13

The filtering effect of the crossed fields is to remove particles which do not have the proper velocities to get through the plates at P_3. Within desired limits, this insures a supply of ions which have the same velocities.

Below plate P_3 there exists only a constant magnetic flux density B_d which puts a force on those constant-velocity isotopes that can get through P_3. The radius of the path of a charged particle is

$$r = \frac{mv}{B_d e}$$

It should be pointed out that below P_3, the quantities v, B_d, and e are constant—because v has been filtered to a common velocity; B_d is adjusted and applies to all isotopes; and e has a single charge, or some multiple of the single charge. The mass m is proportional to the radius of curvature.

35.9. The Cyclotron and Particle Acceleration

Dr. Stanley Livingston and Dr. Ernest Lawrence developed the *cyclotron* for accelerating atomic particles. This instrument is essentially a pair of hollow semicircular chambers, Fig. 35.14(a), called *dees*, so assembled that there is a slit between the dees. When connected to a high-frequency alternating voltage, the dees become electrodes. Covering the dees is a chamber which is to be evacuated. Outside the dees are two very large magnets which generate the necessary magnitude flux to force into orbit the particle being accelerated as shown in Fig. 35.14(b).

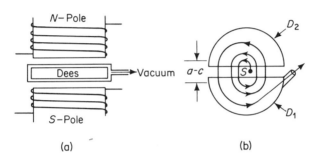

Fig. 35.14

In Fig. 35.14(b), ions are introduced at the source S in the evacuated chamber covering the dees. These ions may be hydrogen ions ($H+$), deuterons, or alpha particles. When the particle is in the space between the dees, D_1 is negative and D_2 is positive. The positive particle is given an acceleration toward D_1. On entering chamber D_1, only the magnetic field has any effect on the accelerated particle. The effect on the particle is to force it into a circular path of constant radius.

As the positive particle approaches D_2, the polarity of the dees changes: D_2 is now negative and D_1 is positive. Thus once the particle leaves D_1, it again comes under the influence of the electric field. The particle is given a new acceleration which is higher than the former acceleration. This new acceleration causes the particle to take a circular path having a greater radius in D_2 than it had before this acceleration because the time taken to make one journey through one dee is constant.

This process is repeated many times until the particle has acquired an energy somewhere in the neighborhood of 20 *million electron-volts* (MeV.) Note that if the voltage used to accelerate the single charged particle is 100,000 volts, each crossing of the space between the dees will cause the particle to pick up energy of 100,000 eV. Thus it will take 200 gap crossings to impart to the particle an energy of 20 MeV.

The radius of an orbiting particle in a magnetic field is

$$r = \frac{mv}{Bq}$$

If the angular velocity $\omega = v/r$, then solving the preceding equation for v/r,

$$\omega = \frac{v}{r} = \frac{Bq}{m}$$

The linear velocity is

$$v = \frac{Bqr}{m}$$

$\begin{Bmatrix} B = \text{flux density} \\ q = \text{charge} \\ m = \text{mass} \\ v = \text{velocity} \\ r = \text{radius} \\ \omega = \text{angular velocity} \end{Bmatrix}$

Since B, q, and m are constant, the *radius* of the orbiting particle is dependent upon its velocity only.

The distance covered in one dee (one-half of a circumference) is πr. The equation required to calculate the time for 1 half-cycle is

$$t = \frac{s}{v} = \frac{\pi r}{Bqr/m} = \frac{\pi m}{Bq}$$

In this instance, B, q, and m are constant. The *time* is constant and independent of the velocity and radius of the particle.

A controlling factor for the maximum velocity for a given cyclotron is the largest possible radius of the dees:

If

$$v_{max} = Br_{max}\frac{q}{m}$$

then the kinetic energy is

$$KE = \frac{1}{2}mv^2 = \frac{1}{2}m\left(\frac{Br_{max}q}{m}\right)^2$$

$$= \frac{B^2 r_{max}^2 q^2}{2m}$$

The voltage which will yield the same kinetic energy is

$$KE = qV$$

Equating the two kinetic-energy equations,

$$Vq = \frac{B^2 r_{max}^2 q^2}{2m}$$

Sec. 35.9   The Cyclotron and Particle Acceleration   573

Solving for V,

$$V = \frac{B^2 r_{max}^2}{2} \times \frac{q}{m}$$

The frequency may be found from $\omega = 2\pi F$. Thus

$$\omega = 2\pi F = B \times \frac{q}{m}$$

Therefore, the frequency is

$$F = \frac{B}{2\pi} \times \frac{q}{m}$$

Example 4

The flux density of a cyclotron is 2 wb/m². The particles being accelerated are protons, mass 1.67×10^{-27} kg. If the maximum orbital radius of the particle is 0.5 m, calculate (1) the voltage necessary to orbit the protons, (2) the maximum velocity of the protons, (3) the kinetic energy of the particle, (4) the period and the frequency, (5) the energy in electron-volts.

Solution

1. The voltage necessary to orbit the proton is

$$V = \frac{B^2 r_{max}^2}{2} \times \frac{q}{m} = \frac{2^2 \times 0.5^2}{2} \times \frac{1.6 \times 10^{-19}}{1.67 \times 10^{-27}}$$

$$= 4.8 \times 10^7 \text{ volts}$$

2. The maximum velocity of the proton is

$$v_{max} = B r_{max} \frac{q}{m} = 2 \times 0.5 \times \frac{1.6 \times 10^{-19}}{1.67 \times 10^{-27}}$$

$$= 9.58 \times 10^7 \text{ m/sec}$$

3. The kinetic energy of the particle is

$$\text{KE} = \frac{B^2 r_{max}^2 q^2}{2m} = \frac{2^2 \times 0.5^2 (1.6 \times 10^{-19})^2}{2(1.67 \times 10^{-27})} \quad \left\{ \begin{array}{l} m = 1.67 \times 10^{-27} \text{ kg} \\ B = 2 \text{ wb/m}^2 \\ r_{max} = 0.5 \text{ m} \\ q = +1.6 \times 10^{-19} \text{ coul} \end{array} \right\}$$

$$= 7.66 \times 10^{-12} \text{ joule}$$

4. The period for 1 half-cycle is

$$t = \frac{\pi m}{Bq} = \frac{\pi(1.67 \times 10^{-27})}{2(1.6 \times 10^{-19})}$$

$$= 1.64 \times 10^{-8} \text{ sec}$$

and the frequency is

$$F = \frac{Bq}{2\pi m} = \frac{2(1.6 \times 10^{-19})}{2\pi(1.67 \times 10^{-27})}$$

$$= 3.05 \times 10^7 \text{ cps}$$

5. The energy in electron-volt units is

$$E = \frac{7.66 \times 10^{-12}}{1.6 \times 10^{-19}} = 4.79 \times 10^7 \text{ eV}$$

The limitation to accelerating particles indefinitely is the relativistic mass increase which takes place as the velocity of the particle approaches the velocity of light. At high velocities, most of the energy goes into an increase in the mass, rather than into the increase in velocity. Thus the time required to make one trip through one dee is changed and the movement of the particle loses its phase relation to the frequency of oscillation. This happens when lightweight particles are accelerated in a cyclotron.

The *betatron*, Fig. 35.15, is used to accelerate electrons to very high velocities. As indicated above, the cyclotron is incapable of accelerating very light particles because the relativistic increase of a light particle changes the charge-to-mass ratio and makes it impossible to synchronize the frequency of rotation of the particle with the frequency oscillation of the accelerating fields.

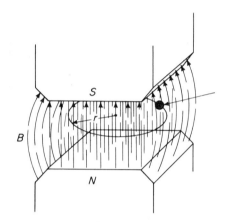

Fig. 35.15

The betatron uses a rapidly changing magnetic field which holds the electron in orbit while the resulting tangential force accelerates it. The electrons are injected into the magnetic field as the field is increasing. This speeds up the electron. Each time the field increases, it boosts the orbiting particle to a greater orbital velocity. This same magnetic field will hold the electron in orbit as its velocity increases. Thus very high velocities can be imparted to the electrons because their velocities are independent of the frequency of oscillation of an electric field.

The *synchrotron* operates with one dee and accelerates particles by applying the same principle as the betatron. Electrons are fired into the dee and a rapidly increasing magnetic-field strength increases their velocity in a fixed orbit. As their velocities approach the speed of light, the relativistic masses of the electrons increase rapidly. The magnetic induction is increased to hold the particle in a stable orbit. At the same time, an alternating potential difference is introduced to further speed up the electron.

Linear accelerators increase the velocity of electrons along a straight line by changing the potential in tubes. Each time the electron enters a section of tube, the electron is accelerated. The limitation to this type of accelerator is the necessary length of the tube at the point where the velocity of the particles is high.

Another type of linear accelerator uses an electron magnetic wave as a carrier. If charged particles are fired into the tube at the right instant, the magnetic wave will accelerate the particle toward the speed of light very rapidly.

35.10. The Photoelectric Effect

One of the theories used to explain the phenomena of light deals with light as discrete packets of energy which satisfy the proportion

$$E \propto F \quad \left\{ \begin{array}{l} F = \text{frequency} \\ E = \text{energy} \end{array} \right\}$$

Max Planck postulated that the absorption or emission of energy by a "black body" is accomplished through the movement of energy as whole-number units of energy; or,

$$E = hF \quad \{ h = \text{Planck's constant} \}$$

where h is the proportionality constant, called *Planck's constant*, and is equal to 6.63×10^{-34} joule-sec; F is the frequency of the radiation; and the entity hF, called the *photon*, must propagate as a unit. If

$$c = F\lambda$$

then the energy may also be equated as

$$E = \frac{hc}{\lambda}$$

The shorter the wavelength of the photon source, the higher the energy of the photon. Also, since wavelength is inversely proportional to the frequency, the higher the frequency, the greater the energy of each packet.

In the *photoelectric cell* (Fig. 35.16), an emission plate is curved so that emitted electrons will be focused upon a collector rod. This collector rod is connected to the $+$ terminal of a battery, making the collector the anode. The emission plate is connected to the $-$ terminal of the battery and becomes the cathode. Thus a ready supply of electrons exists at the cathode, just waiting to be released.

The surface of the metal cathode, however, is capable of holding these electrons within its structure. They may escape from the cathode only if their energy is greater than a characteristic energy value for that particular material. This value is called the *threshold frequency* of the surface.

If an electron can accumulate enough kinetic energy to break through the surface, it will leave that surface and be

Fig. 35.16

attracted to the positive collector rod. When a photon strikes an electron, some of the energy of the photon is transferred to the electron, which in turn acquires a velocity because of the energy transfer. If this velocity is great enough to allow the electron to escape from the surface of the material, it can escape only if it gives up the amount of energy required by the threshold frequency of the surface. After giving up this energy to the material surface, the electron leaves the surface, and whatever energy is left in the electron is kinetic energy of velocity.

Note that the transfer of energy from a photon to an electron is an "all-or-nothing" proposition. On collision, either the electron "takes" all the photon energy—in which case the photon ceases to exist;—or the electron "takes" none of the photon energy—in which case the photon remains in existence until it collides with another electron capable of taking all the photon energy.

The threshold frequency of a given material can be determined by making the collector more and more negative until all flow from the emitter stops. This stopping voltage is different for different materials. At this point, the threshold frequency is

$$hF_0 = hF$$

where hF_0 is the energy required to break through the surface. The stopping-voltage equation is

$$Ve = \tfrac{1}{2}mv^2$$

Each metal has its own critical wavelength, or threshold frequency. Therefore, the *velocity* of the emitted electron is a function of the frequency—not the intensity—of the light impinging on the cathode. *The intensity of the light controls the number of electrons emitted, but not their velocity.*

The velocity of an emitted electron may be calculated from

$$\tfrac{1}{2}mv^2 = hF - hF_0$$

where hF_0 = the work function.

In order to get electrons off a surface from which light is bounced,

$$W(\text{Work}) = hF > hF_0$$

Values for h
mks system = 6.63×10^{-34} joule-sec
cgs system = 6.63×10^{-27} erg-sec
electron-volts = 4.13×10^{-15} eV

If $hF = hF_0$, then the energy of a photon is

$$E = hF = h\frac{c}{\lambda}$$

Example 5

A retarding potential of 0.4 eV is required to block the movement of electrons from the cathode of a photoelectric cell when violet light of wavelength 4000

Sec. 35.10   The Photoelectric Effect   577

Angstrom units (A) strikes its surface. Find the following: (1) the frequency of this light, (2) the energy of this wavelength, (3) the work function of this surface, (4) the threshold frequency, (5) the wavelength for for this frequency, (6) the energy for wavelength of light 3000 A, (7) the net energy after the electron leaves the metal surface, (8) the energy in joule-seconds, (9) the velocity of this electron after it leaves the surface.

Solution

1. Frequency of this light:

$$F = \frac{c}{\lambda} = \frac{3 \times 10^8}{400 \times 10^{-9}}$$
$$= 7.5 \times 10^{14} \text{ vib/sec}$$

2. Energy of this wavelength:

$$E = hF = (4.13 \times 10^{-15})(7.5 \times 10^{14})$$
$$= 3.0975 \text{ eV}$$

3. Work function of this surface:

$$hF_0 = hF - \tfrac{1}{2}mv^2 = 3.0975 - 0.4$$
$$= 2.6975 \text{ eV}$$

4. Threshold frequency:

$$hF_0 = 2.6975$$

or

$$F_0 = \frac{2.6975}{4.13 \times 10^{-15}}$$
$$= 6.53 \times 10^{14} \text{ vib/sec}$$

5. Corresponding wavelength for this frequency:

$$\lambda = \frac{c}{F_0} = \frac{3 \times 10^8 \text{ m/sec}}{6.53 \times 10^{14} \text{ vib/sec}}$$
$$= 4594 \times 10^{-10} \text{ m}$$
$$= 4594 \text{ A}$$

Notice that, in order to release electrons from this surface, the incident light must possess a wavelength shorter than 4593 A.

6. Energy for light of wavelength 3000 A:

$$F = \frac{c}{\lambda} = \frac{3 \times 10^8}{300 \times 10^{-9}}$$
$$= 1 \times 10^{15} \text{ vib/sec}$$
$$hF = (4.13 \times 10^{-15})(1 \times 10^{15})$$
$$= 4.13 \text{ eV}$$

7. Energy after the electron leaves the surface:
$$E = hF - hF_0 = 4.1300 - 2.6975$$
$$= 1.4325 \text{ eV}$$

8. Energy in joules:
$$E = 1.4325(1.6 \times 10^{-19}) = 2.292 \times 10^{-19} \text{ joule}$$

9. Electron velocity after leaving the surface:
$$v = \sqrt{\frac{2E}{m}} = \sqrt{\frac{2(2.292 \times 10^{-19})}{9.11 \times 10^{-31}}}$$
$$= 7.1 \times 10^5 \text{ m/sec}$$

35.11. X-Rays

At this time our only interest in the production of X-rays is examining them as an inverse process to the photoelectric process. The process is the opposite in the sense that electrons, during collision with dense surfaces, cause very-short-wave photons to be emitted, as opposed to the transfer of energy from the photon to the electron during the photoelectric effect.

In 1897, Wilhelm Roentgen observed these very-highly-penetrating rays which are now called X-rays. These rays are electromagnetic in nature, invisible, highly penetrating, and capable of ionizing gases, of affecting photographic film, and of causing fluorescence. They are not deflected by an electric or magnetic field, but may be reflected, diffracted, or polarized. These effects will be discussed in subsequent chapters.

X-rays are produced when high-speed electrons from the cathode, accelerated by a high potential, strike a dense target. The cathode is cup-shaped to concentrate the stream of electrons on the target.

The Roentgen X-ray tube was constructed as a gas-filled tube. The Coolidge tube (Fig. 35.17) uses a low-voltage filament inside a focusing cup. An exceedingly high potential difference causes the electron stream to impinge

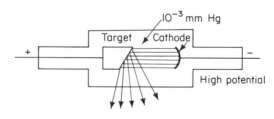

Fig. 35.17

upon a tungsten target. Some of the X-rays are emitted immediately; some of the electrons give up their energies to this additional energy, and eventually also give up X-rays. The acceleration and numbers of electrons may be controlled at the filament. In this tube, the glass envelope is evacuated and the electrons are accelerated and proceed to the target at full potential with minimum collisions.

The energy of impact is a function of accelerating voltage and the velocity acquired by the electrons. Thus

$$Ve = \tfrac{1}{2}mv^2$$

Also,

$$Ve = hF_{max} = \frac{hc}{\lambda_{max}}$$

Problems

1. A cathode-ray tube has a pair of deflecting plates which are 1 cm apart and 2 cm long. The distance from the center of the deflecting plates to the screen is 24 mm. The accelerating potential difference is 1200 volts and the deflecting voltage is 80 volts. Calculate (a) the horizontal velocity of the electrons; (b) the deflection of the beam at the screen.
2. In Prob. 1, change the accelerating voltage to 1000 volts and the deflecting voltage to 100 volts. Calculate the deflection of the electron beam at the screen.
3. The average velocity of the electrons in a cathode ray is 2×10^6 m/sec when a magnetic field, 3×10^{-4} wb/m², bends the ray into a circular path. Calculate (a) the radius of the beam; (b) the force on the beam as it enters the magnetic field; (c) the centripetal force on the electron.
4. An aluminum ion carries a $+3$ charge when entering a magnetic field, flux density 3×10^{-2} wb/m². The velocity of the ion is 2×10^5 m/sec. (a) What is the force on the ion? (b) What is the radius of curvature of the ion path?
5. The charge-to-mass ratio of a deuteron is 4.8×10^7 coul/kg. Assume that a potential difference of 1600 volts forces the particle into a circular path, radius 4 cm. Calculate the flux density of the magnetic field.
6. Calculate the charge-to-mass ratio of Cu^{++}.
7. Repeat Prob. 6 for the ion Ae^{+++}.
8. Repeat Prob. 6 for the ion O^{---}.
9. The intensity of the fields between the plates in a mass spectrograph is 3×10^4 nt/coul (electric) and 0.8 wb/m² (magnetic). Assuming the particles to be sodium, calculate (a) the velocity of the particles which pass between the plates; (b) the radius of curvature when the particles enter the second magnetic field, flux density 0.8 wb/m².
10. A cyclotron is operating with a flux density of 1.5 wb/m². The ion which enters the field is a proton having a mass of 1.67×10^{-27} kg. If the maximum orbit

of the particle is 0.5 m, find (a) the voltage necessary to orbit this proton; (b) the maximum velocity of the proton; (c) the kinetic energy of the particle; (d) the period and the frequency; (e) the energy in electron-volts.

11. A cyclotron operating with a flux density of 2 wb/m² is to accelerate alpha particles, mass 6.68×10^{-24} g, carrying a positive charge of 2. Find (a) the voltage required to put these particles into a maximum orbit of 40 cm; (b) the maximum velocity; (c) the frequency of oscillation; (d) the KE; (e) the angular velocity.

12. A retarding potential of 0.4 eV is required to block a wavelength of 5000 A in a photoelectric cell. Find (a) the frequency of this light; (b) the work function of this surface; (c) the threshold frequency; (d) the corresponding wavelength for this frequency.

13. The work function of a surface is 2.078 eV. What is the velocity of an electron after leaving the photoelectric surface when struck by light, wavelength 4000 A? What proof do you have that *any* electrons will leave the surface for this wavelength?

36

The Atom

36.1. Atomic Structure

The discovery of X-rays in 1897 and the discovery of the electron at about the same time led to Rutherford's nuclear theory of the atom in 1911. In 1913, Bohr presented his model of the hydrogen atom to the scientific world. Nineteen years after that, the discovery of the positron and the neutron gave Bohr the opportunity to expand his theory of the structure of the atom.

Essentially, the atom's heavy central core, called the *nucleus*, is positively charged. Spinning around this core are negatively charged particles, called *electrons* (e), of such a nature that the *net* charge on a *neutral* atom is zero. The electrons are very light. The core of the atom consists of a positively charged *proton* (p) and (except for hydrogen) the *neutron* (n), which has no charge. The neutron is approximately the same mass as the proton. The proton and neutron are called *nucleons* and carry a mass number of 1. Since the electron is about $\frac{1}{1840}$ lighter than the proton, it contributes little to the overall mass of the atom.

Thus the atomic-mass number A is the sum of the atomic number Z and the neutron number N; or

$$Z + N = A$$

The chemical symbols for the elements are used in atomic physics. In the symbol $_2\text{He}^4$, the superscript (upper right-hand) represents the *mass number;* the subscript (lower left-hand) represents the *atomic number*. Figure 36.1(a) shows the model of the neutral helium atom. This atom has 2 protons, 2 neutrons, and 2 electrons. The mass is 4, which indicates that there are 4 nucleons in the nucleus; or

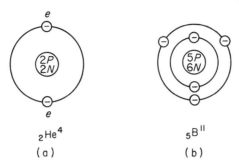

$_2\text{He}^4$
(a)

$_5\text{B}^{11}$
(b)

Fig. 36.1

$$N = A - Z = 4 - 2 = 2 \text{ neutrons} \qquad \left\{ \begin{array}{l} A = \text{mass number} = 4 \\ Z = \text{atomic number} = 2 \\ N = \text{number of neutrons} = 2 \end{array} \right\}$$

In Fig. 36.1(b), we see the symbol for neutral boron, $_5\text{B}^{11}$, which shows the mass number, $A = 11$, and the atomic number, $Z = 5$. Therefore, the number of neutrons is

$$N = A - Z = 11 - 5 = 6 \text{ neutrons}$$

The elements and several of their isotopes are shown in Appendix Table A.23.

Isotopes have the same atomic number, but differ in their mass and number of neutrons. Thus hydrogen has three isotopes: hydrogen, deuterium, and tritium. All are electrically neutral. Each hydrogen isotope has 1 electron and 1 proton, but different numbers of neutrons. These isotopes of hydrogen are so important that their nuclei have been assigned different names: the nucleus of the hydrogen atom is called the *proton;* of deuterium, the *deuteron;* of tritium, the *triton.*

Thus Fig. 36.2(a) shows the hydrogen atom; Fig. 36.2(b) shows one isotope of hydrogen, the deuterium atom; and Fig. 36.2(c) shows another isotope of hydrogen, the tritium atom.

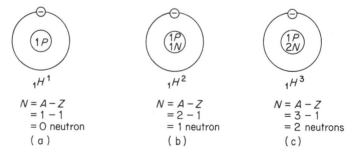

Fig. 36.2

36.2. The Hydrogen Series

The concept of electron shells developed as a result of the efforts of Balmer and Rydberg to find some logical order to the visible lines in the hydrogen spectrum, Fig. 36.3. The equation which resulted contains a constant known as the *Rydberg constant* ($R = 109{,}678/\text{cm}$) and yields the reciprocal of the wavelength of the light which causes the lines. The equation is

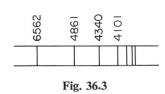

Fig. 36.3

$$w = \frac{1}{\lambda} = R\left(\frac{1}{2^2} - \frac{1}{n^2}\right) \quad \text{where } n = 3, 4, 5, \ldots \quad \begin{Bmatrix} w = \text{wave number} \\ R = \text{Rydberg constant} \\ \lambda = \text{wavelength} \end{Bmatrix}$$

As n approaches infinity, the limit of the *Balmer series* (visible spectrum) for hydrogen is approached. By substituting the respective values for n in the equation, the wavelengths for the various lines are found to check with the various lines in the spectrum.

Example 1

Find the wavelength for the first three lines of the visible spectrum of hydrogen.

Solution

1. The wavelength for $n = 3$:

$$\frac{1}{\lambda} = R\left[\frac{1}{2^2} - \frac{1}{n^2}\right] = 109{,}678\left[\frac{1}{2^2} - \frac{1}{3^2}\right]$$

$$\lambda = 6562 \text{ A}$$

2. For $n = 4$:

$$\frac{1}{\lambda} = R\left[\frac{1}{2^2} - \frac{1}{4^2}\right] \quad \begin{Bmatrix} R = 109{,}678/\text{cm} \\ \lambda = 3, 4, 5 \end{Bmatrix}$$

$$\lambda = 4861 \text{ A}$$

3. For $n = 5$:

$$\frac{1}{\lambda} = R\left[\frac{1}{2^2} - \frac{1}{5^2}\right]$$

$$\lambda = 4340 \text{ A}$$

Subsequently (Fig. 36.4 and Fig. 36.5) there were developed three additional series which apply to the infrared region of the hydrogen spectrum and one series which applies to the ultraviolet region of the hydrogen spectrum. The Balmer series was altered as follows:

The Paschen series:

$$\frac{1}{\lambda} = R\left[\frac{1}{3^2} - \frac{1}{n^2}\right]$$

where $n = 4, 5, 6, \ldots$ infrared.

The Brackett series:

$$\frac{1}{\lambda} = R\left[\frac{1}{4^2} - \frac{1}{n^2}\right]$$

where $n = 5, 6, 7, \ldots$ infrared.

The Pfund series:

$$\frac{1}{\lambda} = R\left[\frac{1}{5^2} - \frac{1}{n^2}\right]$$

where $n = 6, 7, 8, \ldots$ infrared.

The Lyman series:

$$\frac{1}{\lambda} = R\left[\frac{1}{1^2} - \frac{1}{n^2}\right]$$

where $n = 2, 3, 4, \ldots$ ultraviolet.

Example 2

Find the wavelengths for the lowest energy levels for each of the foregoing series of hydrogen.

Solution

1. The Paschen series—$n = 4$:

$$\frac{1}{\lambda} = R\left[\frac{1}{3^2} - \frac{1}{4^2}\right]$$

$$\lambda = 18{,}756 \text{ A}$$

2. The Brackett series—$n = 5$:

$$\frac{1}{\lambda} = R\left[\frac{1}{4^2} - \frac{1}{5^2}\right]$$

$$\lambda = 40{,}523 \text{ A}$$

3. The Pfund series—$n = 6$:

$$\frac{1}{\lambda} = R\left[\frac{1}{5^2} - \frac{1}{6^2}\right]$$

$$\lambda = 74,600 \text{ A}$$

4. The Lyman series—$n = 2$:

$$\frac{1}{\lambda} = R\left[\frac{1}{1^2} - \frac{1}{2^2}\right]$$

$$\lambda = 1216 \text{ A}$$

If the frequency is equal to

$$F = \frac{c}{\lambda}$$

the equation for frequency of the Balmer series becomes

$$F = Rc\left[\frac{1}{2^2} - \frac{1}{n^2}\right], \quad \text{etc.}$$

36.3. Bohr's Hydrogen Atom

According to classical theory, the energy of an electron that is orbiting about a nucleus is constantly decreasing because of the electron's centripetal acceleration. This presumes a constant radiation of energy. If the electron is constantly losing its energy, its orbit should constantly be getting smaller and the spectrum emitted should be continuous, so that eventually the electron would come to rest in the nucleus. This, of course, is not the case.

In an attempt to reconcile classical theory with these exceptions, Bohr proposed two postulates:

Postulate 1: There are certain permissible orbits which the electron may occupy; these orbits represent whole multiple-energy levels as a result of the angular momentum of the electron:

$$mvr = n\frac{h}{2\pi}$$

no energy is radiated when the electron is in one of these orbits, even though it may be in an excited state.

Postulate 2: Energy is emitted or absorbed by these electrons only when they jump from one energy level to another; this emission takes place as a quantum; so that

$$hF = \Delta E = E_i - E_f \quad \begin{cases} E_i = \text{initial energy} \\ E_f = \text{final energy} \\ F = \text{frequency} \\ h = \text{Planck's constant} \end{cases}$$

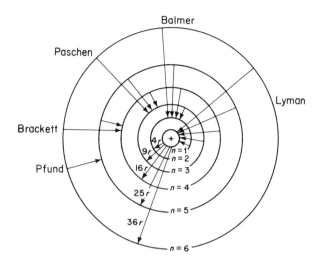

Fig. 36.4

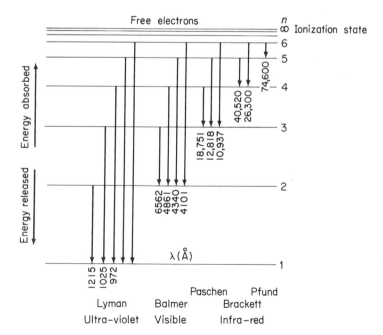

Fig. 36.5

Thus in Fig. 36.4, the electron in the hydrogen atom may find itself in any one of the orbits $n = 1, 2, 3, 4, \ldots$. The allowable number of electrons assigned to each shell is shown in Appendix Table A.24. An electron may be in a high-energy state as a result of energy which it has absorbed. As long as the electron is in one of the orbits, it will not radiate energy. If it does jump to one of the inner orbits, it will do so with the release of a characteristic energy hF. The lowest energy level is the inner orbit.

In Fig. 36.4, an electron may jump from $n = 6$ to $n = 5$ and then to $n = 1$; or from $n = 6$ to $n = 4$ to $n = 1$; or from $n = 6$ to $n = 3$ to $n = 2$ to $n = 1$; or any other decreasing order. Each time the jump occurs, it does so with a release of a packet of energy, characteristic of that jump.

Thus, if the Balmer (visible) series is considered, the electron may jump directly from $n = 6$ to $n = 2$, which is the lowest energy state for the *visible* spectrum. It may then jump to $n = 1$, the lowest energy state for the ultraviolet state of hydrogen. Of course, the jump may be from $n = 6$ directly to $n = 1$, which will yield a different wavelength than the ultraviolet, which is emitted if the jump is from $n = 2$ to $n = 1$.

Note that it is possible for an electron to jump from any outer orbit to any inner orbit provided that there is an opening, or hole, in that inner orbit, and provided the electron gives up a quantum of energy, according to the equation $hF = \Delta E$, characteristic of that jump. Since the outer orbits represent high-energy, or excited, states, the inner vacancies will be filled as soon as one of the inner electrons is forced out of its orbit by an absorption of a quantum of energy.

The normal distribution of electrons will be discussed later, but the probability is greater that the electron closest in energy to the electron removed will fill the "hole." Therefore, it is more probable that an electron from $n = 2$ will fill $n = 1$ than it is for an electron from $n = 3$, etc. These transitions take place only one at a time in each atom. The cumulative effect of this process in all the atoms in a given sample of an element yields the spectrum, with the various intensities for the spectrum lines depending on the number of like jumps in all the atoms.

The radius of the electron path and the velocity of the electron may be calculated by equating the force between two electron charges and the centripetal force. Thus

$$k\frac{e^2}{r^2} = \frac{mv^2}{r}$$

Solving for v^2,

$$v^2 = k\frac{e^2}{mr}$$

The centripetal angular momentum of the quantized electron may now be written

$$mvr = \frac{nh}{2\pi}$$

Solving for v,

$$v = \frac{nh}{2\pi mr}$$

Substituting for v^2 in $v^2 = ke^2/mr$,

$$\left(\frac{nh}{2\pi mr}\right)^2 = \frac{ke^2}{mr}$$

$$r = \frac{\varepsilon_0 n^2 h^2}{\pi me^2} \quad \text{where } n = 1, 2, 3, \ldots$$

$\left\{\begin{array}{l} h = \text{Planck's constant} \\ = 6.62 \times 10^{-34} \text{ joule/sec} \\ m = 9.11 \times 10^{-31} \text{ kg} \\ e = 1.6 \times 10^{-19} \text{ coul} \\ \varepsilon_0 = 8.85 \times 10^{-12} \text{ coul}^2 \text{ /nt-m}^2 \end{array}\right\}$

Example 3

Calculate the radius of the first quantum electron for the hydrogen atom.

Solution

The radius is

$$r = \frac{\varepsilon_0 n^2 h^2}{\pi me^2} = \frac{(8.85 \times 10^{-12}) \times 1^2 \times (6.62 \times 10^{-34})^2}{\pi (9.11 \times 10^{-31})(1.6 \times 10^{-19})^2}$$

$$= 5.3 \times 10^{-11} \text{ m}$$

The equation for the velocity of a quantized electron may be found from the centripetal equations:

$$k\frac{e^2}{r^2} = \frac{mv^2}{r} \quad \text{and} \quad mvr = \frac{nh}{2\pi}$$

Solving each for r,

$$r = \frac{ke^2}{mv^2} \quad \text{and} \quad r = \frac{nh}{2\pi mv}$$

Equating,

$$\frac{nh}{2\pi mv} = \frac{ke^2}{mv^2}$$

Solving for v,

$$v = \frac{k2\pi e^2}{nh}$$

Substituting $k = 1/4\pi\varepsilon_0$,

$$v = \frac{e^2}{2\varepsilon_0 nh}$$

Example 4

Calculate the velocity of the first-orbit electron for the hydrogen atom.

Solution

The velocity is

$$v = \frac{e^2}{2\varepsilon_0 nh} = \frac{(1.6 \times 10^{-19})^2}{2(8.85 \times 10^{-12}) \times 1 \times (6.62 \times 10^{-34})}$$
$$= 2.2 \times 10^6 \text{ m/sec}$$

The radii ratio for the excited orbits is

$$r_n = n^2 r$$

Therefore, the radii ratios for the excited orbits in Fig. 36.4 are

for $n = 2$,
$$r_2 = 2^2 r = 4r$$

for $n = 3$,
$$r_3 = 3^2 r = 9r$$

for $n = 4$,
$$r_4 = 4^2 r = 16r, \qquad \text{etc.}$$

The quantized velocities are

$$v_n = \frac{1}{n} v$$

An even better graphic diagram than the shell diagram is the energy-level diagram (Fig. 36.5). The orbital number n is shown on the right in the diagram. When $n = 1$, the electron is in its ground, or normal, state. As n increases, the electron finds itself in an excited state. The ultimate state of excitement is at $n = \infty$, which represents an ionization state. At this point, the electron has acquired enough energy to leave the atom.

Any *free* electron *may* jump in to fill the gap with a corresponding release of energy, depending upon the initial energy of the *free* electron. Since free electrons may have any amount of energy, the radiation emitted (as a result of the *excess* energy beyond that required to fill the gap) may range all the way from bright-line radiation to continuous spectra at the extreme end of the spectrum limit.

36.4. Electron Distribution

Bohr and Stoner combined their efforts and extended Bohr's model of the hydrogen atom to include all the elements on a building-block basis. They used the various *n*-orbits for the additional electrons needed in the circular shells.

Thus the hydrogen atom has 1 electron in the first shell; helium has 2 electrons in the first shell, which closes the first shell. To obtain lithium, which has 3 electrons, it becomes necessary to place 2 electrons in the first shell and 1 electron in the second shell. Neon, with its 10 electrons, has 2 in the first shell and 8 in the second shell. This closes the second shell, so that sodium, with 11 electrons, has 2 in the first shell, 8 in the second shell, and 1 in the third shell. The process continues for the rest of the elements.

If electrons were permitted to build up in circular shells about an ever-increasing nuclear mass and charge, the electrons in the inner shells would be drawn closer and closer to the nucleus as the nuclear charge increased. The size of the heavier elements, because of the additional attractive forces, would appear to become smaller. It is known that the atom diameter of all elements is approximately constant.

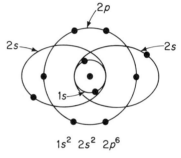

Fig. 36.6

To explain this discrepancy, Wolfgang Pauli retained the shell concept, but postulated the principle —known as the *Pauli exclusion principle*—which permits only 2 electrons for each orbit. Sommerfeld had extended the Bohr model to include elliptical orbits. Thus a shell, according to Sommerfeld, consists of a combination of circular and elliptical orbits. Instead of filling each circular orbit with the numbers of electrons permitted in that orbit as predicted by Bohr, Sommerfeld postulated a shell that is a combination of circular and elliptical orbits. He thus explained away the reduced size of the circular orbits for the heavier elements, so that the overall size of the atom remained approximately constant.

The first shell contains a maximum of 2 electrons—never any more. The second shell consists of 8 electrons, distributed 2 in one circular

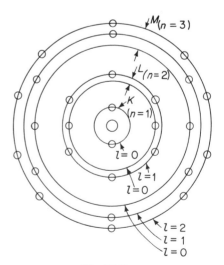

Fig. 36.7

Sec. 36.4   *Electron Distribution*   591

orbit and 2 in each of three elliptical orbits. For convenience in this chapter, the Bohr circular orbit (Fig. 36.7) will be used instead of the elliptical orbits shown in Fig. 36.6.

Appendix Table A.24 shows the elements, shells, and subshells. The shells are shown at the top of the table. Each shell should be closed before starting another; thus the operation is a building-block operation. These shells are designated by capital letters, starting with K, L, M, ... (these the X-ray notations).

The equation for closing a shell is

$$\text{X-ray notation} = 2n^2 \qquad \{ n = \text{the quantum number} \}$$

Example 5

Find the total number of electrons permitted in the K-, L-, and M-shells.

Solution

The K-shell:

$$2n^2 = 2(1)^2 = 2 \text{ electrons}$$

The L-shell:

$$2n^2 = 2(2)^2 = 8 \text{ electrons}$$

The M-shell:

$$2n^2 = 2(3)^2 = 18 \text{ electrons}$$

The subshell's quantum number, indicated by the letter l, may be found from the equation

$$\text{subshell} = 2(2l + 1) \qquad \text{where } l \leq n - 1$$

Therefore, in the K-shell, $n = 1$; $l = 0$. In the L-shell, $n = 2$; $l = 0$ and $l = 1$. In the M-shell, $n = 3$; $l = 0$, $l = 1$, $l = 2$, etc. These subshells carry the small letters s, p, d, f, g, h as shown in Appendix Table A.24. The sum of $s + p + d + \cdots = 2n^2$. The notation is $3p^6$, which shows the maximum electrons in the subshell as a superscript and the major electron shell as the coefficient.

Example 6

A zinc atom has an atomic number $Z = 30$. Find the number of electrons in the shells and subshells. (Use the circular instead of the elliptical diagram.) Draw the diagram showing the shells.

Solution

1. K-shell—$n = 1; l = 0$:
$$2n^2 = 2(1)^2 = 2 \text{ electrons in the first shell}$$
   The subshell distribution is
$$2(2l + 1) = 2[2(0) + 1] = 2 \text{ electrons in the 1s}^2 \text{ subshell}$$
2. L-shell—$n = 2; l = 0, l = 1$:
$$2n^2 = 2(2)^2 = 8 \text{ electrons in the second shell}$$
   The subshell distribution is
$$2(2l + 1) = 2[2(0) + 1] = 2 \text{ electrons in the 2s}^2 \text{ subshell}$$
$$2(2l + 1) = 2[2(1) + 1] = 6 \text{ electrons in the the 2p}^6 \text{ subshell}$$
3. M-shell—$n = 3; l = 0, l = 1, l = 2$:
$$2n^2 = 2(3)^2 = 18 \text{ electrons in the third shell}$$
   The subshell distribution is
$$2(2l + 1) = 2[2(0) + 1] = 2 \text{ electrons in the 2s}^2 \text{ subshell}$$
$$2(2l + 1) = 2[2(1) + 1] = 6 \text{ electrons in the 3p}^6 \text{ subshell}$$
$$2(2l + 1) = 2[2(2) + 1] = 10 \text{ electrons in the 3d}^{10} \text{ subshell}$$
4. The total electrons thus far are
$$2 \text{ (K-shell)} + 8 \text{ (L-shell)} + 18 \text{ (M-shell)} = 28 \text{ electrons}$$
Thus for the N-shell, since the s-subshell ($l = 0$) is 2, the 2 additional electrons needed for zinc must go into the subshell $l = 0$, which is the fourth shell, so that
$$28 + 2 = 30 \text{ electrons for zinc}$$
5. Figure 36.7 shows the diagram for zinc.

36.5. X-Rays

The theory just discussed applies to the production of X-rays, which results from the collision of a high-energy electron with one of the inner electrons of a heavy atom. This high-energy electron can penetrate the shells because of its high velocity. Once the inner K-electron is knocked out of the atom, a hole remains in the K-shell [Fig. 36.8(a)]. One of the high-energy L-electrons moves in to fill the hole [Fig. 36.8(b)], with a corresponding release in energy equal to hF_k. The hole left in the L-shell is filled by an electron from the M-shell with a release of another high-energy X-ray (hF_L but of a different frequency than the hF_k energy).

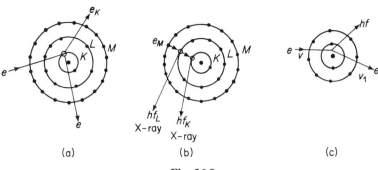

(a)   (b)   (c)

Fig. 36.8

Another phenomenon which yields apparent continuous spectrum (X-ray) is the *bremsstrahlung effect*. A bombarding electron may release its maximum energy on collision; or it may release a part of its energy when it passes through the atom, close to the nucleus. In the latter case, the electron deflects and is slowed down to a new velocity [Fig. 36.8(c)]. The deflection and new velocity indicate a new kinetic energy. The difference in kinetic energy released is an X-ray photon; or,

$$E_i - E_f = hF$$

36.6. Electron Spin

The Pauli exclusion principle was altered to accommodate the spinning of electrons on their own axes, as shown in Fig. 36.9(a). When an electron spinning on its own axis sets up a tiny charge—depending upon the direction of spin—it becomes, in effect, a small magnet.

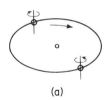

(a)

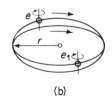

(b)

Fig. 36.9

Since two electrons in the same shell, rotating in opposite directions, alter each other's orbit slightly, the original model of a single orbit may now be regarded as a "paired" orbit having the same radius. Thus in Fig. 36.9(b), electrons e and e_1 rotate clockwise about the nucleus. Electron e rotates clockwise and electron e_1 rotates counterclockwise about its own axis. As a result, the orbits become slightly separated.

This further innovation eventually led to the theories of wave mechanics and the conclusion that in some instances, the electron has a "wave nature" and, in other cases, has a fixed "corpuscular nature." In wave mechanics, electrons are no longer thought of as occupying fixed orbits, but as residing anywhere within an orbital ring of a certain width. The orbit is a *region*, rather than a *line*. Remember that Bohr's orbit is an energy orbit, each element having a different K-orbit, etc. Wave mechanics does not abandon the energy relationships of the Bohr theory, but reinforces and extends them.

Problems

1. How many protons, neutrons, and electrons are there in (a) $_7N^{14}$? (b) $_6C^{12}$? (c) $_{19}K^{39}$? (c) $_{26}Fe^{56}$? (e) $_{46}Pd^{106}$? (f) $_{80}Hg^{202}$? (g) $_{82}Pb^{208}$? (h) $_{48}Ca^{112}$? (i) $_{74}W^{184}$? (j) $_{54}Xe^{129}$?
2. Repeat Prob. 1 for the following: (a) $_{80}Hg^{202}$; (b) $_{82}Pb^{208}$; (c) $_{48}Ca^{112}$; (d) $_{74}W^{184}$; (e) $_{54}Xe^{129}$.
3. Given the following data, write the symbol for each element (a) atomic number = 24, mass number = 52; (b) 81 nucleons, 35 electrons; (c) 232 nucleons, 90 protons; (d) 79 protons, 118 neutrons; (e) 227 nucleons, 89 protons.
4. Repeat Prob. 3 for the following data: (a) 27 nucleons, 13 electrons; (b) 52 nucleons, 24 neutrons; (c) 42 protons, 56 neutrons; (d) atomic number 53, mass number 127; (e) 134 protons, 87 electrons.
5. Find the wavelength of spectrum line, $n = 5$, for each of the five series of hydrogen—that is, when the quantized electron jumps from $n = 5$ to $n = 1$; from $n = 5$ to $n = 2$, etc. ($R = 109{,}678/\text{cm}$.)
6. Repeat Prob. 5 for $n = 6$.
7. Calculate (a) the radius of the quantized electron, $n = 6$; (b) the velocity of the electron.
8. What are the velocity and the radius of the quantized electron, $n = 5$?
9. A silicon atom has a value of $Z = 14$. Find the number of shells and subshells and draw the diagram showing the distribution of electrons. (Use the model of Fig. 36.7.)
10. Repeat Prob. 9 for $_{10}Ne^{20}$.

37

Transmutation

37.1. Radioactivity

Figure 37.1 shows the emission of alpha and beta particles and of gamma rays from a radioactive material influenced by a magnetic field. The direction of alpha deflection indicates that alpha particles carry a positive charge. After these particles had been identified as each having an atomic weight of 4 and a positive charge of 2, Rutherford and his co-workers then showed that these particles were helium particles with their outer electrons removed. The helium particle with its 2 electrons removed is called an *alpha particle*.

The alpha particle has a mass and velocity (comparatively constant for a particular material) which is quite high on being released from a radioactive material. This makes its kinetic energy high. In spite of this high kinetic energy, it has a range of only 6 to 7 cm before it is absorbed in air. It has the ability to ionize air and will cause fluorescent materials to "scintillate."

It was the ionization ability of these particles which led to the development of the Geiger counter. The *Geiger counter* is essentially a metal cylinder containing air or argon, with a wire inside the cylinder at a negative potential. If external particles ionize the gas inside the cylinder, these ions set up a current surge which may be amplified and measured.

Beta particles are emitted at very high velocities, approaching the speed of light. The emission of beta rays

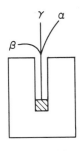

Fig. 37.1

covers a wide range of velocities, from very low to very high. Sharing in the total energy of the beta rays is the *neutrino*, which is assumed to have a negligible rest mass and a neutral charge. The charge on the beta particle is negative. Beta rays may be electrons from an atom's inner shells, released by the gamma rays emitted by the nucleus, or they may be emitted from the nucleus itself. In the former case, a sharpline spectrum is produced; in the latter case, a continuous spectrum is produced.

Gamma rays are high-energy waves, closely resembling X-rays except that the diffraction angles needed to detect gamma rays are much smaller than the diffraction angles needed to detect X-rays. This would indicate that gamma rays are of much higher frequency (shorter wavelength) than X-rays. A magnetic field will not deflect them, indicating no charge. Evidence indicates that a nucleus which emits an electron or an alpha particle is left in an excited state. Since this state is unstable, the nucleus will release a gamma ray as it returns to its ground state.

When an element emits an alpha particle, the *atomic number* is reduced by 2 and the *mass number* is reduced by 4. Thus,

$$_5B^{11} \rightarrow {}_3Li^7 + {}_2He^4 \quad (alpha\ particle)$$

Here the *atomic-mass number* becomes

$$11 - 4 = 7$$

and the *atomic number* becomes

$$5 - 2 = 3$$

Thus with an atomic number of 3, the new element is lithium and the other end product is the alpha particle.

When an element releases a negative beta particle ($_{-1}e^0$), the *atomic-mass number* is unaffected, but the *atomic number* is *increased* by 1. Thus

$$_{11}Na^{24} \rightarrow {}_{12}Mg^{24} + {}_{-1}e^0$$

Here the *atomic-mass number* becomes

$$24 - 0 = 24$$

and the *atomic number* becomes

$$11 - (-1) = 11 + 1 = 12$$

Thus with an atomic number of 12, the new element is magnesium and the other end product is a beta particle.

Note that the same procedure is followed for neutron ($_0n^1$) release, for proton ($_1H^1$) release, for positron ($_{+1}e^0$) release, etc.

37.2. The Radioactive Series

There are four known natural *radioactive series:* (1) uranium, (2) actinium, (3) neptunium, and (4) thorium. These isotopes disintegrate by alpha and beta emission. Figure 37.2 shows the uranium series. This disintegration takes place through the series of steps shown, until stable lead is reached. The atomic

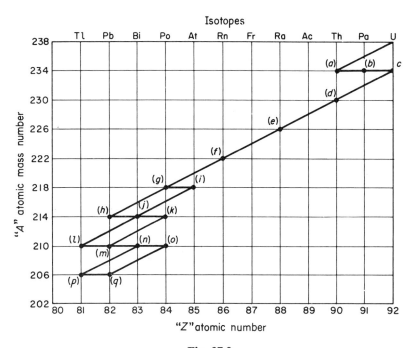

Fig. 37.2

numbers are plotted as the abscissa at the bottom of the graph, the corresponding isotopes are shown at the top of the graph, and the atomic-mass numbers are shown on the ordinate of the graph.

Example 1

Starting with radioactive uranium, trace its disintegration by alpha and beta emission to stable lead.

Solution

An atom of uranium-238 decays with an alpha-particle emission and leaves an atom of thorium-234 as shown in Fig. 37.2 at a:

$$_{92}U^{238} \rightarrow {_{90}Th^{234}} + {_2He^4}$$

Thorium-234 disintegrates with the emission of a beta particle and leaves protactinium-234, as shown (in Fig. 37.2) at b:

$$_{90}Th^{234} \rightarrow {_{91}Pa^{234}} + {_{-1}e^0}$$

Protactinium-234 disintegrates with the emission of another beta particle and leaves uranium-234, as shown in Fig. 37.2 at c:

$$_{91}Pa^{234} \rightarrow {_{92}U^{234}} + {_{-1}e^0}$$

The student should write the equations for the rest of the disintegrations in Fig. 37.2 and indicate the resulting elements.

A radioactive material disintegrates according to a concept known as its *half-life*. This half-life is equal to the time needed for half the nuclei of a given sample to decay. Thus, at the end of 1 half-life, one-half the nuclei will have decayed and half remain. At the end of a successive half-life, one-half of the remaining nuclei will have decayed and one-fourth of the *original* number of nuclei will be left. At the end of 3 half-life periods, one-eighth of the original number of nuclei will be left. We cannot determine which particular nuclei in a sample will disintegrate.

The time required for half-life decay ranges from a nondetectably small interval of time to a very long half-life decay. Elements which decay with long half-lives are said to be *stable*. The probability is that all elements have characteristic half-life decay.

Thus if T equals the half-life of a particular element, a radioactive nucleus will disintegrate according to the product of the number of half-lives involved. Letting n equal the number of half-lives of duration T, then

$$T = n \times \text{(original quantity of material)}$$

At the end of the first half-life interval T, the equation becomes

$$T_1 = \tfrac{1}{2} \times \text{(the material left)}$$

At the end of the second half-life period,

$$T_2 = \tfrac{1}{2} \times \tfrac{1}{2} = \tfrac{1}{4} \text{ (the material left)}$$

At the end of the third half-life period,

$$T_3 = \tfrac{1}{2} \times \tfrac{1}{2} \times \tfrac{1}{2} = \tfrac{1}{8} \times \text{(the material left)}, \quad \text{etc.}$$

Sec. 37.3   *The Cloud Chamber*   599

The general equation for the number of nuclei left at the end of a number of half-life periods is*

$$T_n = \frac{1}{2^n} \times \text{(original quantity of material)}$$

Example 2

Thirty-two grams of polonium disintegrates with a half-life of 3 min. (1) How much of the original material will be left at the end of 15 min of disintegration? (2) Set up a table to check your answer in part (1).

Solution

1. There are

$$n = \frac{15}{3} = 5 \text{ half-life intervals}$$

Therefore, there will be left

$$T_5 = \frac{1}{2^5} = \frac{1}{32} \times \text{(the original mass)}$$

The mass left is

$$\frac{1}{32} \times 32 = 1 \text{ g}$$

2. The material left at the end of each 3-min period is

at $T_0 = (0 \text{ min}) = 32$ g
at $T_1 = (3 \text{ min}) = \frac{1}{2} \times 32 = 16$ g
at $T_2 = (6 \text{ min}) = \frac{1}{2} \times 16 = 8$ g
at $T_3 = (9 \text{ min}) = \frac{1}{2} \times 8 = 4$ g
at $T_4 = (12 \text{ min}) = \frac{1}{2} \times 4 = 2$ g
at $T_5 = (15 \text{ min}) = \frac{1}{2} \times 2 = 1$ g   (*check*)

37.3. The Cloud Chamber

In his investigations, Rutherford used a chamber that was evacuated and then filled first with hydrogen and then with air. Alpha particles were permit-

---

*A more accurate solution would involve natural logarithms. The equation for the amount of material disintegrated equals the original amount times 2.718^{-kt}, where k is a constant for each material.

ted to travel through the gas and impinge upon an aluminum screen. This caused scintillation, which he was able to observe through a telescope. With hydrogen in the chamber, the tracks from the scintillation were much longer than the tracks left when alpha particles travel through air. Rutherford concluded that, since the hydrogen atom is much lighter than the alpha particle, the hydrogen would pick up kinetic energy which would give it more penetrating power after collision.

C. R. Wilson developed a *cloud chamber* used for tracking particles (Fig. 37.3). It consists of a piston inside a cylinder, capped with a glass top. A telescope is used to view the tracks inside the chamber.

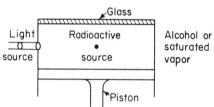

Fig. 37.3

Vapor particles form when the piston is pulled down: the lowered pressure causes vapor to become supersaturated; alpha particles, passing through the air in the chamber, ionize the air, and vapor particles form on the charged ions—or on any other charged particles present—to produce tracks. Any deviation from a straight-line path indicates collision.

Figure 37.4(a) shows the tracks, essentially straight, which result from an alpha particle's collision with another alpha particle.

Electrons leave a vapor trail which is not continuous, as shown in Fig. 37.4(b). These electrons have been subjected to a magnetic field and show a curved path. The momentum of a charged particle in a magnetic field may be determined in this manner.

Figure 37.4(c) shows alpha particles in nitrogen. The very heavy tracks are left by the nitrogen atoms; the medium-straight tracks are lighter

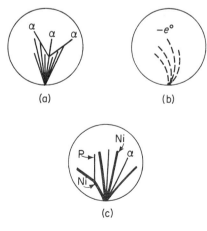

Fig. 37.4

Sec. 37.3   The Cloud Chamber   601

alpha particles; the very fine track is produced by the light proton. Again, any deviation from a straight line indicates collision.

If a collision takes place between an alpha particle and a helium atom, because the masses are almost identical and because we are considering collision as elastic, the angle between the alpha particle and the recoil helium atom is $90°$ (see Fig. 37.5a). If an alpha particle collides with a lighter hydrogen atom (Fig. 37.5b), the angle between the two tracks is less than $90°$. If it collides with a heavier particle, the angle between the tracks will be greater than $90°$, as shown in Fig. 37.5(c).

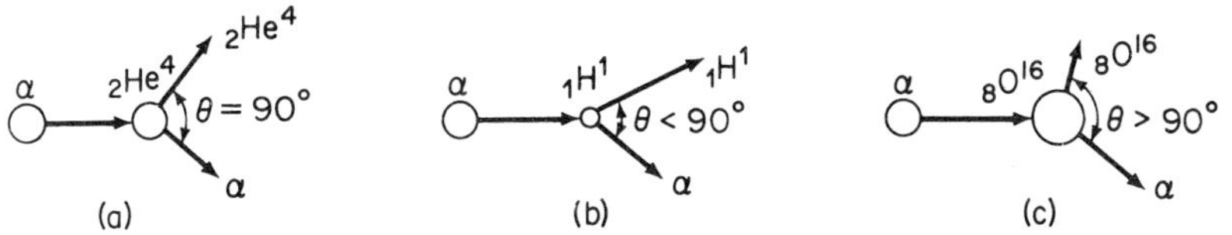

Fig. 37.5

The angle formation between both particles after recoil is demonstrated vividly when slow- or fast-moving electrons (beta rays) are used as projectiles. Figure 37.6(a) shows a fast-moving electron colliding with another, stationary electron. If the *fast*-moving electron exhibits a mass relativistic increase, an angle of less than $90°$ between the two tracks should be observed. This is exactly what is seen. If the bombarding electron is slow-moving, the mass of the moving and stationary electrons will be essentially the same and the angle after collision should be $90°$ [see Fig. 37.6(b)].

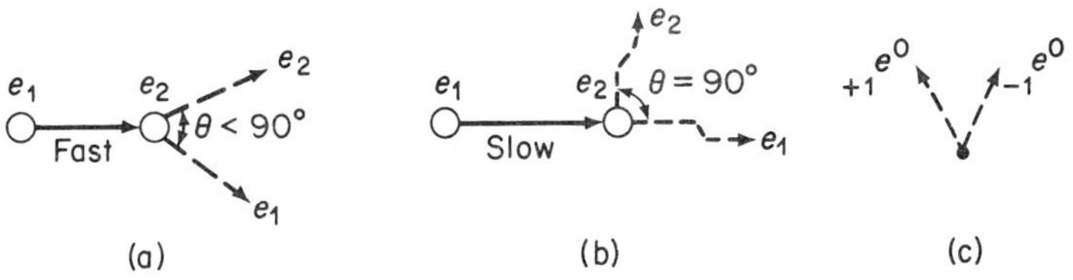

Fig. 37.6

Figure 37.6(c) shows a curved track to the right of an electron in a magnetic field. Anderson found that the curved track to the left has a positive charge but also has the same mass as an electron. This particle is called the *positron*.

Another method of detection is the *scintillation counter*. Alpha, beta, or gamma rays will cause fluorescence when they pass through certain materials. Part of this energy is reemitted, after absorption, as ultraviolet or visible blue light. If this energy is permitted to fall on a photoelectric-cathode surface *a*, it will emit electrons which may be accelerated by a potential difference to plate *b*. On striking plate *b*, each electron will cause 2 or more secondary electrons to leave plate *b*. If these electrons are now accelerated by a + potential toward plate *c*—where the process just described is repeated—a much larger quantity of electrons may be collected at the collector plate than was released at the first photoelectric plate (see Fig. 37.7).

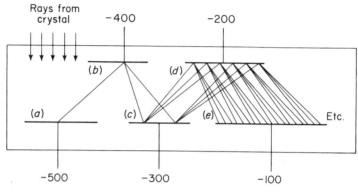

Fig. 37.7

37.4. Artificially Induced Transmutation

In 1936, Bohr first indicated that a nucleus bombarded by other particles may capture these particles and then be transformed (the process is called transmutation) into a new element with the ejection of another primary element and a quantity of energy (the "capture" stage of this process is shown in Fig. 37.8):

$$_2He^4 + {_7N^{14}} \rightarrow {_9F^{18}}$$
$$_9F^{18} \rightarrow {_8O^{17}} + {_1H^1} + E$$

Alpha particles ($_2He^4$) are used to bombard nitrogen ($_7N^{14}$). The alpha particle is captured, as shown in Fig. 37.8, by the nitrogen to form fluorine-18,

α   N¹⁴   F¹⁸   O¹⁷   P   **Fig. 37.8**

Sec. 37.4 Artificially Induced Transmutation 603

which is unstable. This fluorine breaks up immediately into oxygen ($_8O^{17}$), a proton ($_1H^1$), and a release of energy (E). The energy involved will be discussed in Chapter 38. It will be seen that if E is $+$, energy is released, if E is $-$, energy is absorbed.

Note that the *atomic-mass numbers* before bombardment are equal to the *atomic-mass numbers* after bombardment:

$$4 + 14 = 18 = 17 + 1$$

Also, the *atomic numbers* before bombardment are equal to the *atomic numbers* after bombardment:

$$2 + 7 = 9 = 8 + 1$$

It is because the sum of the *actual* masses after bombardment is not exactly equal to their sum before bombardment that we get the energy E.

The type of reaction just explained is called the α–p reaction because an alpha particle is used to bombard the nucleus, resulting in the release of a proton.

From our discussion thus far, it is evident that bombarding an element with an alpha particle will increase the atomic-mass number A by 4 and increase the atomic number Z by 2. Thus aluminum-27, when bombarded by alpha particles, results in phosphorus-31, which is unstable:

$$_{13}Al^{27} + {}_2He^4 \rightarrow {}_{13+2}P^{27+4} \rightarrow {}_{15}P^{31}$$

If the reaction is an α–p type, one of the end products must be a proton, or $_1H^1$. Therefore,

$$_{15}P^{31} \rightarrow {}_1H^1 + {}_{15-1}X^{31-1} \rightarrow {}_1H^1 + {}_{14}X^{30}$$

Since the new element must have an atomic-mass number of 30 and an atomic number of 14, the element can only be silicon; as

$$_{15}P^{31} \rightarrow {}_{14}Si^{30} + {}_1H^1$$

The reaction may be symbolized (omitting the atomic number) as

$$\underset{\text{(target)}}{\underset{\text{(projectile)}}{Al^{27}} (\alpha, p) \underset{\text{(end product)}}{Si^{30}}}$$

Using the methods of capture and disintegration, new elements and isotopes have been discovered and produced in the laboratory. A typical reaction would be americum-241 from uranium-238 in an α–n reaction with beta emission:

$$_{92}U^{238} + {}_2He^4 \rightarrow {}_{94}Pu^{241} + {}_0n^1$$
$$_{94}Pu^{241} \rightarrow {}_{95}Am^{241} + {}_{-1}e^0$$

Example 3

Write the reaction equation for K^{39} (α–p) Ca^{42}.

Solution

1. From $_{19}K^{39} + {_2}He^4$,

   atomic mass:
   $$39 + 4 = 43$$
   atomic number:
   $$19 + 2 = 21$$

2. From Appendix Table A.25, the element is
   $$_{21}Sc^{43}$$

3. Since this is unstable, it disintegrates into a new element and a proton:
   $$_{21}Sc^{43} \rightarrow {_1}H^1 + {_{20}}X^{42}$$

4. From Table A.25, the unknown element is calcium and the equation becomes
   $$_{19}K^{39} + {_2}He^4 \rightarrow {_{21}}Sc^{43} \rightarrow {_1}H^1 + {_{20}}Ca^{42}$$

Problems

1. Draw the graphs for the following alpha–beta-radioactive series:

   (a) Thorium $\alpha \rightarrow 2e \rightarrow 3\alpha \left\langle \begin{array}{c} \alpha \rightarrow e \\ e \rightarrow \alpha \end{array} \right\rangle \hspace-1ex \times \hspace-1ex \left\langle \begin{array}{c} \alpha \rightarrow e \\ e \rightarrow \alpha \end{array} \right\rangle Pd$

   (b) Neptunium $e \rightarrow 2\alpha \rightarrow e \rightarrow 2\alpha \rightarrow e \rightarrow 3\alpha \left\langle \begin{array}{c} \alpha \rightarrow e \\ e \rightarrow \alpha \end{array} \right\rangle \rightarrow e(Bi)$

   (c) Actinium $\alpha \rightarrow e \rightarrow \alpha \left\langle \begin{array}{c} e \rightarrow \alpha \\ \alpha \rightarrow e \end{array} \right\rangle \rightarrow 2\alpha \rightarrow \left\langle \begin{array}{c} e \rightarrow \alpha \\ \alpha \rightarrow e \end{array} \right\rangle \hspace-1ex \times \hspace-1ex \left\langle \begin{array}{c} e \rightarrow \alpha \\ \alpha \rightarrow e \end{array} \right\rangle Pd$

2. The element $_9F^{17}$ disintegrates with the emission of a positron. Its half-life is 1.2 min. (a) If 8 g of this unstable isotope are present, how much is left after 6 min? (b) How many atoms are left at the end of this time period?

3. How long will it take before 90% of the atoms disappear in Prob. 2.

4. Two grams of sodium-24 disintegrate with the emission of an electron. Sodium's half-life is 15 hr. How long will it take for the 1.8 g to disappear?

5. What element is needed to complete the following equations?
   (a) $_5B^{11} + {_1}H^1 \rightarrow {_6}C^{11} + (\quad)$
   (b) $_{24}Cr^{51} + (\quad) \rightarrow {_{23}}V^{51}$
   (c) $_2He^3 + {_1}H^1 \rightarrow {_2}He^4 + (\quad)$
   (d) $4(_1H^1) \rightarrow {_2}He^4 + 2(\quad)$
   (e) $_1H^3 + {_1}H^2 \rightarrow {_2}He^3 + (\quad)$

Problems

6. Write the reaction equation for the following disintegrations:
   (a) S^{32} (α,p)
   (b) Be^9 (α,γ)
   (c) Al^{27} (n,α) followed by beta emission
   (d) Li^6 (d,α)
   (e) Be^9 (γ,p)
7. Write the reaction equations for the following disintegrations:
   (a) K^{39} $(n,2n)$ followed by beta emission
   (b) C^{13} (d,t)
   (c) Be^9 (γ,n)
   (d) H^2 (d,n)
   (e) S^{32} (n,p) followed by beta emission
8. Complete the following reactions:
   (a) Na^{23} (p,α) (g) C^{12} (d,n)
   (b) Be^9 (p,d) (h) B^{10} (n,α)
   (c) Al^{27} (α,n) (i) N^{14} $(n,2\alpha)$
   (d) Cu^{65} (n,p) (j) Li^7 (p,γ)
   (e) C^{12} (d,p) (k) H^2 (γ,n)
   (f) H^1 (n,γ) (l) B^{11} (γ,t)
9. Write the equations for the following reactions:
   (a) N^{13} disintegrates into an element and a positron
   (b) V^{49} captures an electron
   (c) Mo^{95} (d,n)
   (d) U^{238} (n,γ)
   (e) Np^{239} disintegrates with an electron emission
10. The following reactions are classified as α–2n disintegrations. Write the equations and find the element resulting:
    (a) Bi^{209}
    (b) Pu^{239}
    (c) Am^{241}
    (d) Cm^{242}

38

Energy: The Elementary Particle

38.1. Nuclear Reactions

In Sec. 37.4, we said that if, instead of a whole number, the exact atomic mass is used to balance the transmutation equations, the sum of the masses before reaction will not equal the sum of the masses after reaction. The "mass defect" will be enough to balance the equations, if the energy released or absorbed is taken into account. Thus, the binding energy required to hold the nucleons together is equal to the energy released when the nucleus is altered. The reverse of the foregoing statement is also true: if the energy defect in the equation is positive, energy is released during reaction; if the energy defect is negative, then energy is absorbed by the reaction.

According to Einstein's equation for the conversion of rest mass to energy, or of energy to rest mass,

$$E = m_0 c^2 \quad \left\{ \begin{array}{l} E = \text{energy} \\ m_0 = \text{rest mass} \\ c = \text{speed of light} \end{array} \right\}$$

The unit of energy is the electron-volt (eV) and is calculated on the mass of the proton (1.66×10^{-24} g):

$$E = mc^2 = (1.66 \times 10^{-24} \text{ g})(2.998 \times 10^{10} \text{ cm/sec})^2$$
$$= 1.49 \times 10^{-3} \text{ erg}$$
$$= 1.49 = 10^{-10} \text{ joule}$$

Sec. 38.1   *Nuclear Reactions*   607

By definition, 1 joule is equal to 1 volt-coulomb. But 1 eV is equal to 1.6×10^{-19} joule. Therefore, 1 million electron-volts (MeV) is equal to

$$1 \text{ MeV} = 10^6 \text{ volts} \times (1.6 \times 10^{-19} \text{ coul})$$
$$= 1.6 \times 10^{-13} \text{ joule}$$

One *atomic-mass unit* (amu) is equal to

$$1 \text{ amu} = \frac{1.49 \times 10^{-10}}{1.6 \times 10^{-13}} = 931 \text{ MeV}$$

Since 1 amu is equal to the unit mass of the proton,

$$1 \text{ amu} = 1.66 \times 10^{-24} \text{ g}$$

Because the principle of conversation of energy must be adhered to, the sum of the masses (energy) before reaction must equal the sum of the masses (energy) after reaction.

Example 1

Find the energy needed to balance the equations $_5B^{10}$ (α,p) $_6C^{13}$. Is the energy released or absorbed?

Solution

1. The reaction equation is

$$_5B^{10} + {}_2He^4 \rightarrow {}_7N^{14} \rightarrow {}_6C^{13} + {}_1H^1 \pm E$$

2. From Appendix Table A.23, the atomic masses are

$$_5B^{10} = 10.012939$$
$$_2He^4 = 4.002604$$
$$_6C^{13} = 13.003354$$
$$_1H^1 = 1.007825$$

3. Mass defect—

   mass before reaction:

   $$10.12939 + 4.002604 = 14.015543 \text{ amu}$$

   mass after reaction:

   $$13.003354 + 1.007825 = 14.011179 \text{ amu}$$

   change in mass:

   $$\Delta m = 14.015543 - 14.011179 = 0.004364 \text{ amu}$$

4. Converting to MeV units,

$$E = 0.004364 \text{ amu} \times 931 \frac{\text{MeV}}{\text{amu}} = 4.063 \text{ MeV/atom}$$

5. The energy is *released* because the mass before reaction is greater than the mass after reaction.

Alternate Solution

1. Energy:

$$E = \Delta mc^2 = 0.004364 \, (1.66 \times 10^{-24})(2.998 \times 10^{10})^2$$
$$= 6.511 \times 10^{-6} \text{ erg}$$

2. Energy in MeV:

$$E = \frac{6.511 \times 10^{-6}}{1.6 \times 10^{-12}} = 4.07 \times 10^6 \text{ ev} = 4.063 \text{ MeV}$$

38.2. Binding Energy

If we consider the *change* in atomic mass—the difference between the sum of the component *nucleons* before reaction and the resulting atomic mass after reaction—a residual atomic mass will result. This residual atomic mass is converted to energy units and the resulting energy is called the *binding energy* of the nucleus.

It has been shown that

$$N = A - Z \qquad \left\{ \begin{array}{l} N = \text{number of neutrons} \\ Z = \text{atomic number} \\ A = \text{mass number} \end{array} \right\}$$

Therefore, from the preceding statement,

$$\text{binding energy} = Z_m + N_m - m \qquad \left\{ \begin{array}{l} m = \text{mass of the stable atom} \\ N_m = \text{mass of the neutrons} \\ Z_m = \text{mass of the hydrogen atom} \end{array} \right\}$$

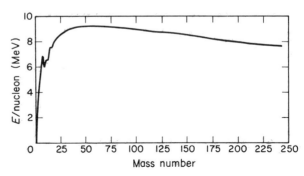

Fig. 38.1

Sec. 38.2 *Binding Energy* 609

Using the mass of the hydrogen atom takes into consideration the sum of the electron mass and the proton mass as a unit. High binding energies signify a stable atom.

Example 2

(1) What is the binding energy of $_6C^{12}$? (2) What is the binding energy per nucleon?

Solution

1. The binding energy of $_6C^{12}$ is

    (a) The isotope $_6C^{12}$ breaks down into

    $$6 \text{ neutrons; 6 protons; 6 electrons}$$
    or
    $$6 \text{ neutrons and 6 hydrogen atoms}$$

    (b) The atomic masses are

    $$\text{neutron} = 1.008665 \text{ amu}$$
    $$\text{hydrogen} = 1.007825 \text{ amu}$$

    (c) The total mass—

    total neutron mass:
    $$6 \times 1.008665 = 6.051990 \text{ amu}$$
    total hydrogen mass:
    $$6 \times 1.007825 = 6.046950 \text{ amu}$$
    total nucleon mass:
    $$6.053922 + 6.048870 = 12.102792 \text{ amu}$$

    (d) The mass of the stable $_6C^{12}$ atom = 12.000000 amu.

    (e) The mass difference is
    $$12.098940 - 12.000000 = 0.098940 \text{ amu}$$

    (f) The binding energy in MeV is
    $$0.098940 \times 931 = 92.31 \text{ MeV}$$

2. Since there are 12 nucleons, the binding energy per nucleon is
    $$\frac{92.31}{12} = 7.69 \text{ MeV nucleon}$$

38.3. Fusion

Fusion is the process by which light elements are combined to form a new light element with a release of energy: the two elements fuse into a heavier element.

1. The formation of helium may take place as follows:

    (a) A proton may capture a neutron to form a deuteron:
    $$_1H^1 + _0n^1 \rightarrow {_1H^2}$$

    (b) Two deuterons may react and produce either or both of the following reactions:

    i. Two protons form a triton and a proton:
    $$_1H^2 + _1H^2 \rightarrow {_1H^3} + {_1H^1}$$

    ii. Two protons form a helium-3 atom and a neutron:
    $$_1H^2 + _1H^2 \rightarrow {_1He^3} + {_0n^1}$$

    (c) The formation of helium in the foregoing reactions may then take place by either or both of the following reactions:

    i. Two triton atoms form helium-4 and 2 neutrons:
    $$_1H^3 + _1H^3 \rightarrow {_2He^4} + {_0n^1} + {_0n^1}$$

    ii. Or a triton and a deuteron form helium-4 and a neutron:
    $$_1H^3 + _1H^2 \rightarrow {_2He^4} + {_0n^1}$$

2. Helium may also be formed in the following manner (the proton-proton chain):

    (a) Two protons combine to form a deuteron and a positron:
    $$_1H^1 + _1H^1 \rightarrow {_1H^2} + {_{+1}e^0}$$

    (b) A deuteron and another proton combine to form the isotope He^3:
    $$_1H^2 + _1H^1 \rightarrow {_2He^3}$$

    (c) The helium-3 atom may combine with another proton to yield stable helium-4 and a positron:
    $$_2He^3 + _1H^1 \rightarrow {_2He^4} + {_{+1}e^0}$$

    (d) Or 2 helium-3 atoms may combine to form helium-4 and 2 protons:
    $$_2He^3 + _2He^3 \rightarrow {_2He^4} + {_1H^1} + {_1H^1}$$

3. Another fusion reaction of interest is the formation of helium-4 and carbon-12 from a proton and carbon-12:

    (a) Carbon-12 may combine with a proton to form carbon-13 with the emission of a gamma ray and a positron:
    $$_1H^1 + _6C^{12} \rightarrow {_7N^{13}} + \gamma \text{ ray} \rightarrow {_6C^{13}} + {_{+1}e^0}$$

(b) Carbon-13 then combines with another proton to form nitrogen-14 and a gamma ray:

$$_6C^{13} + {}_1H^1 \rightarrow {}_7N^{14} + \gamma \text{ ray}$$

(c) Nitrogen-14 combines with a proton to form nitrogen-15 and a positron:

$$_7N^{14} + {}_1H^1 \rightarrow {}_8O^{15} + \gamma \text{ ray} \rightarrow {}_7N^{15} + {}_{+1}e^0$$

(d) Finally, nitrogen-15 combines with still another proton to form helium and carbon-12:

$$_1H^1 + {}_7N^{15} \rightarrow {}_6C^{12} + {}_2He^4$$

4. Still another formation of helium-4 may take place if 4 protons are combined:

$$_1H^1 + {}_1H^1 + {}_1H^1 + {}_1H^1 \rightarrow {}_2He^4 + {}_{+1}e^0 + {}_{+1}e^0$$

38.4. Fission

We have just seen some examples of the release of energy when light nuclei are joined to form heavier nuclei. Hahn showed that heavy nuclei, when bombarded with neutrons, will break into two new elements with atomic numbers somewhere near the middle of the periodic table. The end result is several pairs of nearly equal elements and many fast-moving neutrons. This process is called *fission*.

If these fast neutrons are slowed down and are permitted to undergo "capture" by the additional heavy nuclei present, and if each neutron captured releases more than 1 neutron which is itself capable of being captured, a chain

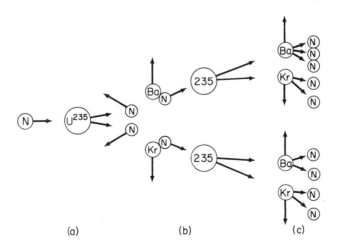

Fig. 38.2

builds up rapidly with a rapid release of energy. The *chain reaction* for uranium-235 bombarded by a neutron is shown in Fig. 38.2.

This chain can be supported if a characteristic mass, called the *critical mass*, is present. The size of the critical mass depends upon such conditions as the shape of the mass, the energy of the neutron used, the purity of the uranium, etc. If less than the critical mass (a *subcritical mass*) is present, there are not enough neutrons remaining to support a chain reaction—that is, too many neutrons are escaping from the surface of the mass.

The separation of two subcritical masses keeps them inactive. When two subcritical masses are merged so that the total mass is greater than a critical mass, fission occurs with the accompanying release of energy.

If the uranium isotope $_{92}U^{235}$ captures a neutron, the reaction equation becomes, for the moment,

$$_{92}U^{235} + {_0}n^1 \rightarrow {_{56}}Ba^{141} + {_{36}}Kr^{92}$$

Before capture, the nuclei contain

$$\begin{aligned} _{92}U^{235} &= 92 \text{ protons} \quad \text{and} \quad 143 \text{ neutrons} \\ _{0}n^1 &= \phantom{92 \text{ protons} \quad \text{and} \quad 14} 1 \text{ neutron} \\ \hline total &= 92 \text{ protons} \phantom{\quad \text{and} \quad 1} 144 \text{ neutrons} \end{aligned}$$

After capture, the nuclei contain

$$\begin{aligned} _{56}Ba^{141} &= 56 \text{ protons} \quad \text{and} \quad 85 \text{ neutrons} \\ _{36}Kr^{92} &= 36 \text{ protons} \quad \text{and} \quad 56 \text{ neutrons} \\ \hline total &= 92 \text{ protons} \phantom{\quad \text{and} \quad 1} 141 \text{ neutrons} \end{aligned}$$

Since the net neutrons are $144 - 141 = 3$ neutrons, which are not needed to form Ba and Kr, the equation should read

$$_{92}U^{235} + {_0}n^1 \rightarrow {_{56}}Ba^{141} + {_{36}}Kr^{92} + 3({_0}n^1) + 203 \text{ MeV}$$

The isotopes Ba^{141} and Kr^{92} are unstable, with the result that they emit beta rays until they have reached a stable state. The emission of a beta particle ($_{-1}e^0$) reduces the number of neutrons by 1 and increases the number of protons by 1. Thus reduction takes place until the stable praseodymium-141 and stable zirconium-92 are reached. Thus

$$_{92}U^{235} + {_0}n^1 \rightarrow [{_{56}}Ba^{141} + {_{36}}Kr^{92} + 3({_0}n^1)]$$
$$\rightarrow {_{59}}Pr^{141} + {_{40}}Zr^{92} + 7({_{-1}}e^0) + 3({_0}n^1)$$

The energy calculations are:

1. Energy before capture:

$$\begin{aligned} _{92}U^{235} &= 235.043943 \\ _{0}n^1 &= 1.008665 \\ \hline total &= 236.052608 \text{ amu} \end{aligned}$$

Sec. 38.4   Fission   613

2. Energy after capture:

$$\begin{aligned} _{59}Pr^{141} &= 140.899696 \\ _{40}Zr^{92} &= 91.905031 \\ 3(_0n^1) &= 3.025995 \\ 7(_{-1}e^0) &= 0.003843 \\ \hline total &= 235.834565 \text{ amu} \end{aligned}$$

3. Mass defect:

$$236.052608 - 235.834565 = 0.218043 \text{ amu}$$

4. Energy in MeV:

$$0.218043 \times 931 = 203 \text{ MeV}$$

Another fission reaction is that of the element $_{92}U^{238}$ when bombarded by slow-moving neutrons. These neutrons are captured and, by beta emission, reach relatively stable plutonium-239:

$$_{92}U^{238} + {_0n^1} \to {_{92}U^{239}} \quad (unstable) \to {_{93}Np^{239}} + {_{-1}e^0}$$

and

$$_{93}Np^{239} \quad (unstable) \to {_{94}Pu^{239}} + {_{-1}e^0}$$

The long-lived plutonium-239 is radioactive, emitting alpha particles, so that

$$_{94}Pu^{239} \to {_{92}U^{235}} + {_2He^4}$$

Another fission chain may be started by the production of uranium-233 from thorium-232 and neutrons. This reaction takes place in the following manner:

$$_{90}Th^{232} + {_0n^1} \to {_{90}Th^{233}} \to {_{91}Pa^{233}} + {_{-1}e^0} \to {_{92}U^{233}} + {_{-1}e^0}$$

Slow-moving neutrons have been found to effect the fission of U^{235} and Pu^{239}. Fast-moving neutrons have been used to effect the fission of U^{238}. In fact, the fission of elements with atomic masses larger than 231 will result from the capture of slow- or fast-moving neutrons.

It is interesting to note that the energy of 203 MeV from the U^{235} reaction may be converted into other energy units.

Example 3

Convert 203 MeV into (1) MeV per g; (2) watt-sec. per gram; (3) kilowatt-hours per gram; (4) foot-pounds per gram-second; (5) horsepower per gram; (6) calories per gram; (7) bituminous coal equivalent per gram.

Solution

1. MeV per gram

$$\frac{1 \text{ g} \times 6.02 \times 10^{23}}{235} = 2.56 \times 10^{21} \text{ atom/g}$$

$$2.56 \times 10^{21} \frac{\cancel{\text{atom}}}{\text{g}} \times 203 \frac{\text{MeV}}{\cancel{\text{atom}}} = 5.2 \times 10^{23} \text{ MeV/g}$$

2. Watt-seconds per gram

$$5.2 \times 10^{23} \frac{\cancel{\text{MeV}}}{\text{g}} \times 1.6 \times 10^{-13} \frac{\text{joule}}{\cancel{\text{MeV}}} = 8.32 \times 10^{10} \text{ joule/g}$$

$$= 8.32 \times 10^{10} \text{ watt-sec/g}$$

3. kilowatt-hours per gram

$$8.32 \times 10^{10} \frac{\cancel{\text{watt-sec}}}{\text{g}} \times \frac{1 \text{ kW-hr}}{3.6 \times 10^6 \cancel{\text{watt-sec}}} = 2.3 \times 10^4 \text{ kW-hr/g}$$

4. Foot-pounds per gram-second

$$8.32 \times 10^{10} \frac{\cancel{\text{watt-sec}}}{\text{g}} + 0.737 \frac{\text{ft-lb}}{\cancel{\text{watt-sec}}} = 6.13 \times 10^{10} \text{ ft-lb/sec-g}$$

5. Horsepower per gram

$$\frac{6.13 \times 10^{10} \cancel{\text{ft-lb-sec}}}{550 \cancel{\text{ft-lb-sec}} \cdot \text{g}} = 1.1 \times 10^8 \text{ hp/g}$$

6. Calories per gram

$$5.2 \times 10^{23} \frac{\text{MeV}}{\text{g}} \times 3.8 \times 10^{-14} \frac{\text{cal}}{\text{MeV}} = 1.976 \times 10^{10} \text{ cal/g}$$

7. Bituminous coal equivalent per gram of U^{235}

$$1.976 \times 10^{10} \frac{\cancel{\text{cal}}}{\text{g}} \times \frac{1 \text{ ton of coal}}{7.2 \times 10^9 \cancel{\text{cal}}} = 2.74 \frac{\text{tons of coal}}{\text{gram } U^{235}}$$

$$2.74 \frac{\text{tons of coal}}{\cancel{\text{gram}} \ U^{235}} \times 454 \frac{\cancel{\text{gram}}}{\text{lb}} = 1244 \frac{\text{tons of coal}}{\text{pound of } U^{235}}$$

38.5. Elementary Particles

Cosmic rays are composed of charged and uncharged particles originating in outer space. These rays are presently grouped into primary and secondary cosmic rays. The primary rays originate in outer space, have extremely high energies, and are mostly protons and other heavy particles. These particles apparently possess extremely high energies because, in spite of their charge, they are able to penetrate the earth's magnetic field and atmosphere. The secondary rays result from primary-particle collision with nuclei in the earth's atmosphere and represent a greater part of the cosmic particles which reach the earth.

The list of particles is growing (see Table 38.1). It seems that particles exist either as neutrally charged particles or in positive-negative pairs.

The process of pair-formation (materialization of energy) and the combining of a pair of particles into a single particle (annihilation of matter) is going on all the time. Some of the particles involved in these transformations will now be discussed briefly.

Table 38.1

Leptons	Symbol	Rest Mass*	Mesons	Symbol	Rest Mass	Baryons	Symbol	Rest Mass
electron–neutrino	ν_e	$< 1 m_e$	pion	π^0	265	nucleons:		
				$\pi^+, (\pi^-)$	275	neutron	n^0	1836
muon–neutrino	ν_μ	$< 1 m_e$	kaon	K^0	975	protons	P^+, P^-	
				$K^+, (K^-)$	966	hyperons:		
electron	$e^-, (e^+)$	$1 m_e$	eta	η	1000	lambda	Λ^0	2180
muon	$\mu^-, (\mu^+)$	200					Σ^+	2330
						sigma	Σ^0	2334
							Σ^-	2343
						xi	Ξ^0	2570
							$\Xi^- (\Xi^+)$	2585
photon	γ					omega	$\Omega^- (\Omega^+)$	3275

* m_e = mass related to mass of an electron.

In 1935, the Japanese physicist Yukawa postulated a field of force called the *meson field*. He credited this field with providing the force holding the nucleus of the atom together. In 1936, the first class of subatomic particles whose mass is between the electron and the proton was discovered. All known subatomic particles are now classified into three major classes of paired particles: the *leptons*, the *baryons*, and the *mesons* (see Table 38.1).

The *leptons* comprise the electron neutrino, the muon neutrino, the muon, and the electron. These are the electron-type particles. Except for photons, they are the lightest of all subatomic particles. Nuclear forces have no affect on muons. Electrical fields, however, do affect the muons and will deflect them. Thus a muon can penetrate deeply into matter and is eventually stopped by the electric fields—not by the nuclear forces. Neutrinos are affected by neither nuclear nor electric forces. Thus they penetrate all matter in space.

The *baryons* are proton-type particles. Protons and the neutrons-nucleons—belong to this class of particles. Baryons other than protons and neutrons are classed as *hyperons;* these are heavier than protons and neutrons, are short-lived, and eventually decay into protons and neutrons.

The *mesons* are the particles which generate the field forces credited with holding the nucleus together. The generation of these forces results from an exchange of mesons between protons and neutrons.

Some typical interactions that product elementary particles are shown below:

1. The unstable positron (unstable when alone) forms a gamma-ray photon:
$$_{+1}e^0 + {}_{-1}e^0 \rightarrow hF \quad \text{(annihilation of matter)}$$
2. The production of an electron–positron pair from a gamma-ray photon:
$$hF \rightarrow {}_{+1}e^0 + {}_{-1}e^0 \quad \text{(materialization of matter)}$$
3. Negative μ-mesons may interact with nuclear protons to yield a neutron and a neutrino:
$$_{-1}\mu + {}_1H^1 \rightarrow {}_0n^1 + \nu$$
4. One method for producing a positive π-meson is through the interaction of two protons:
$$_1H^1 + {}_1H^1 \rightarrow {}_1H^2 + {}_{+1}\pi$$
or
$$hF + {}_1H^1 \rightarrow {}_0n^1 + {}_{+1}\pi$$
5. The probable production of the negative π-meson is
$$hF + {}_1H^1 \rightarrow {}_0n^1 + {}_{+1}\pi \rightarrow {}_{-1}\pi + {}_1H^1$$
6. The probable production of the neutral meson is
$$hF + {}_1H^1 \rightarrow {}_1H^1 + {}_0\pi$$
or
$$hF + {}_0n^1 \rightarrow {}_0n^1 + {}_0\pi$$

Most elementary particles are unstable. The following list gives some of the products of decay:

1. When outside the nucleus, the unstable neutron decays with the production of a proton, a beta particle, and a neutrino:
$$_0n^1 \rightarrow {}_1H^1 + {}_{-1}e^1 + {}_0\nu$$
2. L-mesons decay as follows:
$$_{+1}\mu \rightarrow {}_{-1}e^1 + {}_0\nu + {}_0\nu$$
$$_{-1}\mu \rightarrow {}_{+1}e^1 + {}_0\nu + {}_0\nu$$
$$_0\pi \rightarrow {}_{+1}e^0 + {}_{-1}e^0 + {}_0\nu \quad \text{or} \quad \rightarrow hF + hF$$
$$_{+1}\pi \rightarrow {}_{+1}\mu + {}_0\nu$$
$$_{-1}\pi \rightarrow {}_{-1}\mu + {}_0\nu$$

3. K-mesons may decay as follows:

$$_0K \rightarrow {}_{+1}\pi + {}_{-1}\pi$$
$$_{+1}K \rightarrow {}_{+1}\pi + {}_{+1}\pi + {}_{-1}\pi$$
$$_{-1}K \rightarrow {}_{-1}\pi + {}_{-1}\pi + {}_{+1}\pi$$

4. Y-particles may decay as follows:

$$\Lambda \rightarrow {}_1H^1 + {}_{-1}\pi \quad \text{or} \quad \rightarrow {}_0n^1 + {}_0\pi$$
$$_0\Sigma \rightarrow {}_0\Lambda + hF$$
$$_{+1}\Sigma \rightarrow {}_1H^1 + {}_0\pi \quad \text{or} \quad \rightarrow {}_0n^1 + {}_{+1}\pi$$
$$_{-1}\Sigma \rightarrow {}_0n^1 + {}_{-1}\pi$$
$$_{-1}\Xi \rightarrow {}_0\Lambda + {}_{-1}\pi$$
$$_0\Xi \rightarrow {}_0\Lambda + {}_0\pi$$

Problems

1. What is meant by "critical mass"?
2. The isotope $_1H^2$ (α,n) reacts during a fusion process. What energy is released?
3. Solve Prob. 2 using $E = mc^2$.
4. Find the equivalent energy per gram in Prob. 2 in (a) kilowatt-hours; (b) foot-pounds.
5. The reaction P^{31} (p,α) releases 2.94 MeV of energy. If $P^{31} = 30.983613$ amu and $\alpha = 4.002775$ amu, what is the energy value of the proton in MeV and amu units?
6. Calculate the total energy release in the following sequence:

$$2(_1H^1) \rightarrow {}_1H^2 + {}_1e^0$$
$$_1H^1 + {}_1H^2 \rightarrow {}_1He^3$$
$$2(_2He^3) \rightarrow {}_2He^4 + 2(_1H^1)$$

7. Find the energy in the following sequence:

$$_{92}U^{238} + {}_0n^1 \rightarrow {}_{93}Np^{239} + {}_{-1}e^0$$
$$_{93}Np^{239} \rightarrow {}_{94}Pu^{239} + {}_{-1}e^0$$
$$_{94}Pu^{239} \rightarrow {}_{92}U^{235} + {}_2He^4$$

8. (a) What is the binding energy of $_{13}Al^{27}$? (b) What is the binding energy per nucleon?
9. (a) What is the binding energy for $_{18}Ar^{40}$ from the graph (Fig. 38.1)? (b) Check your results, assuming the stable element argon-40 can be built up from the required number of individual nucleons and electrons.

APPENDIX

Table A.1. Standard Prefixes—Multiple and Submultiple Units†

Prefix	Symbol	Multiplication Factor
atto	a	10^{-18}
femto	f	10^{-15}
pico* *or*	p	10^{-12}
micromicro	$\mu\mu$	
nano* *or*	n	
millimicro	$m\mu$	10^{-9}
micro	μ	10^{-6}
milli	m	10^{-3}
centi	c	10^{-2}
deci	d	10^{-1}
no prefix	——	1
deka	da	10
hecto	h	10^{2}
kilo	k	10^{3}
mega	M	10^{6}
giga* *or*	G	10^{9}
kilomega	kM	
tera* *or*	T	10^{12}
megamega	MM	

*Preferred
†As a result of the Eleventh General Conference on Weights and Measures (1960), as amended in 1962 by the Executive Board.

Table A.2. International System of Units (SI Units)*

Quantity	Unit	Symbol SI	Symbol Text
acceleration (linear)	meter per second^2	m/s^2	m/sec^2
acceleration (angular)	radian per second^2	rad/s^2	rad/sec^6
area	meter^2	m^2	m^2
charge, electric	coulomb	C	coul
density	kilogram per meter^3	kg/m^3	kg/m^3
density (magnetic flux)	tesla	T	wb/m^2
energy	joule	J	joule
field strength (electric)	volt per meter	V/m	volt/m
field strength^2 (magnetic)	ampere per meter	A/m	amp/m
flux (luminous)	lumen	lm	lum.
flux (magnetic)	weber	Wb	web
force	newton	N	nt
force (electromotive)	volt	V	volt
force (magnetomotive)	ampere	A	amp
frequency	hertz	Hz	cps
heat quantity	joule	J	joule
illumination	lux	lx	ft-candle
luminance	candles per meter^2	cd/m^2	candle/m^2
potential difference	volt	V	volt
power	watt	W	watt
pressure	newton per meter^2	N/m^2	nt/m^2
resistance (electric)	ohm	Ω	Ω
stress	newton per meter^2	N/m^2	nt/m^2
velocity (linear)	meter per second	m/s	m/sec
velocity (angular)	radian per second	rad/s	rad/sec
viscosity (kinematic)	meter^2 per second	m^2/s	m^2/sec
voltage	volt	V	volt
volume	meter^3	m^3	m^3
work	joule	J	joule

*From the Eleventh General Conference on Weights and Measures (1960), as amended in 1962 by the Executive Board. The French-language form for the system—"Système International d'Unitès"—accounts for the term "SI units."

Table A.3. Conversion Units

1 cm	$= 10^{-2}$ m	1 Btu	$= 252$ cal
	$= 0.3937$ in.		$= 778$ ft-lb
1 cm^2	$= 10^{-4}$ m^2		$= 1055$ watt-sec (joules)
1 cm^3	$= 10^{-6}$ m^3		$= 2.93 \times 10^{-4}$ kw-hr
1 in.	$= 2.54$ cm	1 cal	$= 4.18$ watt-sec (joules)
	$= 10^3$ mils	1 joule	$= 1$ watt-sec
1 in.^{2}	$= 6.452$ cm^2		$= 1$ nt-m
	$= 6.944 \times 10^{-3}$ ft^2		$= 10^7$ ergs
	$= 1.273 \times 10^6$ circular mils		$= 9.481 \times 10^{-4}$ Btu
1 in.^{3}	$= 16.39$ cm^2		$= 0.2389$ cal
	$= 5.787 \times 10^{-4}$ ft^3		$= 2.778 \times 10^{-7}$ kw-hr
1 ft^3	$= 7.48$ gal		$= 6.705 \times 10^9$ amu
1 gal	$= 231$ in.^{3}		$= 6.242 \times 10^{12}$ MeV
1 A (angstrom)	$= 10^{-8}$ cm		$= 1.113 \times 10^{-14}$ g
	$= 10^{-7}$ mm	1 kW	$= 3412$ Btu/hr
1 amp	$= 3 \times 10^9$ statamp		$= 2.389 \times 10^{-4}$ cal/sec
1 g	$= 10^{-3}$ kg		$= 1.34$ hp
	$= 2.205 \times 10^{-3}$ lb		$= 737.6$ ft-lb/sec
	$= 980.7$ dynes	1 hp	$= 2545$ Btu/hr
1 lb	$= 453.6$ g		$= 1.782 \times 10^{-4}$ cal/sec
1 slug	$= 1.459 \times 10^4$ g		$= 746$ watts
	$= 32.17$ lb		$= 33,000$ ft-lb/min
1 dyne	$= 1.02 \times 10^{-3}$ g		$= 550$ ft-lb/sec
	$= 10^{-5}$ nt	1 coul	$= 3 \times 10^9$ statcoul
1 nt	$= 10^5$ dynes	1 amp	$= 3 \times 10^9$ statamp
1 erg	$= 1$ dyne-cm	1 farad	$= 8.987 \times 10^{11}$ statfarads
	$= 10^{-7}$ volt-coul	1 statvolt	$= 300$ volts
	$= 10^{-7}$ watt-sec	1 ohm	$= 1.113 \times 10^{-12}$ statohm
1 atmosphere	$= 14.7$ psi	1 henry	$= 1.113 \times 10^{-1}$ stathenry
	$= 1.013 \times 10^5$ nt/m^2	1 maxwell	$= 10^{-8}$ wb $= 1$ line
	$= 1.013 \times 10^6$ dynes/cm	1 gauss	$= 10^{-4}$ wb/m^2
	$= 406.8$ in. water	1 oersted	$= 79.58$ amp-turns/m
	$= 29.92$ in. Hg		
	$= 76$ cm Hg		
	$= 34$ ft water		

Table A.4. Determination of Radius of Gyration

Lines	Areas	Solids
Circle line radius $= R$, length $= 2\pi R$ $K_o = \dfrac{R}{\sqrt{2}}$ $K_p = R$	**Circle area** radius $= R$, area $= \pi R^2$ $K_o = \dfrac{R}{2}$ $K_p = \dfrac{R}{\sqrt{2}}$	**Sphere** radius $= R$, volume $= \dfrac{4}{3}\pi R^3$ $K_o = R\sqrt{\dfrac{2}{5}}$ $K_p = K_o$
Semicircular arc c.g. to $B = \dfrac{2R}{\pi}$ radius $= R$, length $= \pi R$ $K_o = R\sqrt{\dfrac{1}{2} - \dfrac{4}{\pi^2}}$	**Semicircular area** c.g. to $B = \dfrac{4R}{3\pi}$ radius $= R$, area $= \dfrac{\pi R^2}{2}$ $K_o = R\sqrt{\dfrac{1}{4} - \dfrac{16}{9\pi^2}}$	**Hemisphere** c.g. to $B = \dfrac{3R}{8}$ radius $= R$, volume $= \dfrac{2}{3}\pi R^3$ $K_o = R\sqrt{\dfrac{2}{5} - \dfrac{9}{64}}$
Quadrant arc c.g. to $B = \dfrac{2R}{\pi}$ (both) radius $= R$, length $= \dfrac{\pi R}{2}$ $K_o = R\sqrt{\dfrac{1}{2} - \dfrac{4}{\pi^2}}$	**Quadrant area** c.g. to $B = \dfrac{4R}{3\pi}$ (both) radius $= R$, area $= \dfrac{\pi R^2}{4}$ $K_o = R\sqrt{\dfrac{1}{4} - \dfrac{16}{9\pi^2}}$	**Cylinder** radius $= R$, volume $= \pi R^2 h$ $K_o = \dfrac{1}{2}\sqrt{R^2 + \dfrac{h^2}{3}}$ $K_p = \dfrac{R}{\sqrt{2}}$
Rectangular loop length $= 2b + 2h$ $K_o = \dfrac{h}{2}\sqrt{\dfrac{h+3b}{3(h+b)}}$	**Rectangular area** area $= bh$ $K_o = \dfrac{h}{\sqrt{12}}$ $K_p = \sqrt{\dfrac{h^2 + b^2}{12}}$	**Parallelepiped** volume $= bh\tau$ $K_o = \sqrt{\dfrac{h^2 + \tau^2}{12}}$
Uniform rod length $= L$ $K_o = \dfrac{h}{\sqrt{12}}$	**Triangle area** area $= \dfrac{bh}{2}$ c.g. at $\dfrac{h}{3}$ $K_o = \dfrac{h}{\sqrt{18}}$	**Right cone** base radius $= R$ volume $= \dfrac{\pi R^2 h}{3}$ c.g. at $\dfrac{h}{4}$ $K_p = R\sqrt{\dfrac{3}{10}}$ $K_o = \sqrt{\dfrac{3}{20}\left(R^2 + \dfrac{h^2}{4}\right)}$
M of $I = LK^2$	M of $I = AK^2$	M of $I = VK^2$

Table A.5. Specific Weight

Material	Specific Weight (lb/ft³ at 68°F)	Material	Specific Weight (lb/ft³ at 68°F)
air (32°F)	0.8006	nickel	540
aluminum	168.5	oxygen (32°F)	0.0890
brass	535	platinum	1334
concrete	144	silver	655
copper	555	steel	488
cork	15	tin	455
glass (crown)	160	tungsten	1175
glass (flint)	247	water (68°F)	62.32
gold	1203	water (39.4°F)	62.43
ice (32°F)	57.2	wood (pine)	30
cast iron	440	wood (oak)	47
lead	708	zinc	445
mercury	845.6		

Table A.6. Coefficients of Friction

Material	Static Friction*	Sliding friction*	Rolling friction*
wood on wood	0.70	0.40	—
metal on metal	0.35	0.18	—
metal on leather	0.60	0.54	—
metal on wood	0.55	0.22	—
wood on masonite	0.50	0.20	—
leather on masonite	0.45	0.38	—
brass on masonite	0.63	0.26	—
steel on macadam	0.45	0.30	0.150
rubber tire on road (dry)	0.90	0.65	0.025
rubber tire on road (wet)	0.70	0.55	0.020

*Values are approximate and depend on the condition of the surfaces.

Table A.7. Elasticity

Material	Modulus of Elasticity*: Tension or Compression (psi $\times 10^6$)	Modulus of Rigidity: Shear (psi $\times 10^6$)	Bulk Modulus (psi $\times 10^6$)	Compressibility (psi $\times 10^{-6}$)
aluminum	10.3	3.7	10.0	0.1
brass	16.0	6.0	8.0	0.125
cast iron	18.2	7.0	14.0	0.070
mercury	—	—	3.8	0.260
steel	29.5	11.5	23.0	0.044
water (liquid)	—	—	0.3	3.300

*Additional values may be found in any standard engineering handbook. Multiply values by 6.9×10^4 to change to dynes/cm^2.

Table A.8. Density of Solids and Liquids

Material	Specific Weight (lb/ft³ at 68°F)	Density: g/cm³ Specific Gravity: in any unit system
alcohol	49.2	0.790
aluminum	168.5	2.700
brass	535.0	8.600
copper	555.0	8.900
cork	15.0	0.240
gold	1203.0	19.300
ice	57.2(32°F)	0.918
cast iron	440.0	7.050
lead	708.0	11.040
mercury	845.6	13.600
oil (lubricating)	54.2	0.870
platinum	1334.0	21.400
silver	655.0	10.500
steel	488.0	7.830
tin	455.0	7.300
tungsten	1175.0	18.900
water	62.4	1.000
wood (pine)	30.0	0.482
wood (oak)	47.0	0.755

Table A.9. Surface Tension

Liquids (in contact with air)	lb/ft^2 at 68°F
alcohol (ethyl)	0.00153
benzene	0.00198
carbon tetrachloride	0.00184
mercury	0.03190
soap solution	0.00192
water	0.00499

Table A.10. Capillarity

Liquid	Glass α	Air cos α
water (distilled)	0°	1.000
water (slightly impure)	25°	0.906
mercury (pure)	148°	−0.848
mercury (slightly impure)	140°	−0.766
turpentine	17°	0.956
petroleum	26°	0.898
paraffin	107°	−0.292

Table A.11. Expansion of Solids and Liquids

*Coefficients of Linear Expansion (a)**		
Material	Coefficient × 10^{-6}/F°	Coefficient × 10^{-6}/C°
aluminum	13.33	24.0
brass	10.55	19.0
glass (ordinary)	4.89	8.8
ice	2.10	3.6
cast iron	5.90	10.8
lead	15.70	28.3
platinum	4.90	8.8
quartz	6.60	11.8
steel	6.30	11.3
Coefficients of Cubical Expansion (δ)		
Material	Coefficient × 10^{-3}/F°	Coefficient × 10^{-3}/C°
alcohol (methyl)	0.630	1.130
carbon tetrachloride	0.700	1.240
mercury	0.100	0.182
water (0°C)	−0.035	−0.064
water (39.2°–100°F)	0.115	0.207

*Multiply the coefficient of linear expansion by 2 to obtain the coefficient of area expansion.

Table A.12. Specific Heat and Latent Heat of Solids

Material	Specific Heat (cal/g-C° or Btu/lb-F°)	Latent Heats	
		of Fusion (Btu/lb)	of Vaporization (Btu/lb)
aluminum	0.220	160.0	3506.0
brass (60 Cu, 40 Zn)	0.092	—	—
bronze	0.087	—	—
copper	0.093	78.0	3160.0
glass (thermometer)	0.199	—	—
cast iron	0.119	54.0	—
lead	0.032	11.3	401.0
mercury	0.033	5.0	127.0
nickel	0.110	130.0	2790.0
porcelain	0.260	—	—
steel	0.117	—	—
tin	0.055	26.0	690.0
zinc	0.096	43.0	765.0

Table A.13. Heat of Combustion

Material	Btu/lb	cal/g
coal (bituminous)	11,600	6,450
oil (crude)	19,500	10,840
oil (gas)	19,200	10,675
oil (fuel)	19,375 (145,000 Btu/gal)	10,762
oil (furnace)	19,000	10,555
kerosene	19,800	11,010
gasoline	20,750	11,535
natural gas	1,000 Btu/ft^3	8.89 cal/cm^3

Table A.14. Thermal Conductivity K

Material	$K_F =$ Btu-in./ft²-hr-F°	$K_C =$ cal-cm/cm²-sec-C°
Solids		
aluminum	1450.0	0.501
asbestos	0.6	0.00002
brass	625.0	0.216
copper	2670.0	0.918
cork	2.04	0.0007
gypsum	8.75	0.003
ice	14.5	0.005
iron (cast)	320.0	0.110
iron (wrought)	407.0	0.140
mica	0.524	0.00018
paper	0.87	0.0003
plaster of Paris	2.04	0.0007
porcelain	7.25	0.0025
steel	320.0	0.110
wood	0.262	0.00009
Liquid and Gases		
mercury	57.4	0.0197
oil (petroleum)	1.02	0.00034
water (0°C)	0.404	0.000139
glycerin	1.86	0.00064
air	0.165	0.0000568
hydrogen	0.96	0.00033

Table A.15. Specific Heat of Gases

Material (at 15 °C)	Specific Heat		Molecular Weight
	at Constant Pressure	at Constant Volume	
air (100 °C)	0.241	0.170	28.8
ammonia	0.52	0.395	17
argon	0.125	0.075	40
carbon dioxide	0.200	0.157	44
carbon monoxide	0.248	0.174	28
carbon tetrachloride	0.14	0.124	154
helium	1.25	0.745	4
hydrogen	3.4	2.41	2
nitrogen	0.248	0.176	28
oxygen	0.2175	0.155	32
sulphur dioxide	0.152	0.118	64
water vapor	0.48	0.36	——

Table A.16. Dialectric Coefficient k

Material	k	Material	k
vacuum	1	wood	5
glass	7	alcohol	28.5
ice	3	benzine	2.3
mica	5	glycerin	55
paper	2	oil	2–3
paraffin	2.3	water	81
rubber	2–30	air (1 atm)	1.0006
shellac	3	H_2 (1 atm)	1.000264
sulfur	4	water vapor (4 atm)	1.00705

Table A.17. Resistivity ρ and Temperature Coefficient of Resistance α

Material	ρ (ohm-cm/ft at 0°C)	α [(C°)$^{-1}$ at 0°C)]
amber	3.00×10^{22}	—
aluminum	15.70	4.23
brass	37.80	1.58
carbon	24.06×10^3	-0.30
constantan	295.00	0.01
copper	9.37	4.10
glass	1.00×10^{20}	—
gold	13.70	3.40
iron (cast)	588.00	5.3
iron (wrought)	70.00	6.00
manganin	249.00	0.01
mercury	566.00	0.86
mica	12.00×10^{23}	—
nichrome	587.00	0.45
nickel	59.00	3.90
platinum	66.00	3.67
quartz	3.00×10^{26}	—
silver	8.80	4.00
steel (soft)	72.00	4.23
steel (hard)	275.00	1.60
tungsten	32.50	4.50
wood	18.00×10^{16}	—

Table A.18. Velocity of Sound in Various Materials

Material	Temperature 0°C	Velocity ft/sec	Velocity m/sec
Solids			
aluminum	0°C	16,740	5,104
brass	0°C	11,480	3,500
copper	20°C	11,670	3,560
steel	0°C	16,410	5,000
glass	0°C	16,410	5,000
Liquids			
alcohol	20°C	3,890	1,213
benzene	17°C	3,826	1,166
turpentine	15°C	4,351	1,326
water	19°C	4,794	1,461
Gases			
air (dry, 1 atm)	0°C	1,087	331.36
carbon dioxide	0°C	846	258
carbon monoxide	0°C	1,106	337
hydrogen	0°C	4,165	1,170
oxygen	0°C	1,041	317

Table A.19. Luminous Efficiency

Source	Luminous Efficiency (lumens/watt)	lumens
tungsten lamp (watts):		
40	11	440
75	14	1,050
100	15	1,500
mercury vapor (watts):		
150 (low)	13	1,950
400 (High)	30	12,000
sodium vapor, 220 watts	50	11,000
fluorescent lamp (watts):		
10	40	400
30	50	1,500
40	60	2,400

Table A.20. Index of Refraction n (for wavelength of sodium D, 5893 Å)

Material	n
air	1.0029
alcohol (ethyl)	1.35
carbon dioxide	1.00045
carbon tetrachloride	1.46
calcite (ordinary)	1.66
calcite (extraordinary)	1.49
Canadian balsam	1.53
diamond	2.42
fluorite	1.43
glass (crown)	1.52
glass (flint)	1.63
glass (heavy flint)	1.65
glycerin	1.48
ice	1.30
quartz	1.46
water	1.33
water vapor	1.00025

Table A.21. Fraunhofer Lines

Color	Letter	Wavelength (Å)	Element
red	B	6867	O
red	C	6563	H
yellow	D_1	5893	Na
yellow	D_2	5890	Na
green	E	5270	Fe
blue	F	4861	H
blue	G'	4340	H
blue	G	4308	Fe
violet	H	3968	Ca
violet	K	3934	Ca

Table A.22. Index of Refraction for Types of Glass

Light	Ordinary Crown	Borosilicate Crown	Barium Flint	Light Flint	Medium Flint	Dense Flint
C	1.5146	1.5219	1.5648	1.5764	1.6224	1.6501
D_1	1.5171	1.5243	1.5682	1.5804	1.6273	1.6555
F	1.5233	1.5301	1.5765	1.5903	1.6394	1.6691
G′	1.5282	1.5347	1.5833	1.5986	1.6497	1.6808
H	1.5325	1.5387	1.5912	1.6092	1.6626	1.6940

Table A.23. Isotopes*

Element	Symbol	Atomic No. (Z)	Atomic-Mass No. (A)	Atomic Mass (amu)
electron	$_{-1}e^0$	1	0	0.000549
positron	$_{+1}e^0$	1	0	0.000549
neutron	$_0n^1$	0	1	1.008665
proton	p	1	1	1.007277
alpha particle	α	2	4	4.002775
hydrogen	H	1	1	1.007825
			2	2.014102
			3	3.016050
helium	He	2	3	3.016030
			4	4.002604
lithium	Li	3	6	6.015124
			7	7.016004
beryllium	Be	4	9	9.012186
boron	B	5	10	10.012939
			11	11.009305
carbon	C	6	12	12.000000
			13	13.003354
nitrogen	N	7	13	13.005738
			14	14.003074
			15	15.000108
oxygen	O	8	15	15.003070
			16	15.994915
fluorine	F	9	19	18.998405
sodium	Na	11	22	21.994437
aluminum	Al	13	27	26.981539
silicon	Si	14	28	27.976929
phosphorus	P	15	31	30.973765
chlorine	Cl	17	35	34.968851
argon	Ar	18	40	39.962384
copper	Cu	29	65	64.927786
zirconium	Zr	40	92	91.905031
antimony	Sb	51	121	120.903816
praseodymium	Pr	59	141	140.899696
thorium	Th	90	230	230.033159
			232	232.038079
			233	233.041604
protactinium	Pa	91	231	231.035903
			233	233.040268
uranium	U	92	233	233.039654
			234	234.040976
			235	235.043943
			238	238.050819
			239	239.054328
neptunium	Np	93	239	239.052951
plutonium	Pu	94	239	239.052175

*This is a partial table of isotopes. For a complete table, refer to John E. Munzer, *Radiological Health Handbook*, U.S. Dept. of Health, Education and Welfare (Washington, D.C.: U.S. Government Printing Office, Jan. 1970).

Table A.24. Shell Distribution of Electrons

n, l Element Z		1, 0 1s	2, 0 2s	2, 1 2p	3, 0 3s	3, 1 3p	3, 2 3d	4, 0 4s	4, 1 4p	4, 2 4d	4, 3 4f
Shell		K	L		M			N			
H	1	1									
He	2	2									
Li	3	2	1								
Be	4	2	2								
B	5	2	2	1							
C	6	2	2	2							
N	7	2	2	3							
O	8	2	2	4							
F	9	2	2	5							
Ne	10	2	2	6							
Na	11	\multicolumn{3}{l}{}	1						
Mg	12				2						
Al	13				2	1					
Si	14	\multicolumn{3}{l}{10-electron core}	2	2					
P	15	\multicolumn{3}{l}{re: neon}	2	3					
S	16				2	4					
Cl	17				2	5					
A	18				2	6					
K	19							1			
Ca	20							2			
Sc	21						1	2			
Ti	22						2	2			
V	23						3	2			
Cr	24						5	1			
Mn	25						5	2			
Fe	26						6	2			
Co	27	\multicolumn{5}{l}{18-electron core}	7	2			
Ni	28	\multicolumn{5}{l}{re: argon}	8	2			
Cu	29						10	1			
Zn	30						10	2			
Ga	31						10	2	1		
Ge	32						10	2	2		
As	33						10	2	3		
Se	34						10	2	4		
Br	35						10	2	5		
Kr	36						10	2	6		

Table A.24. (continued)

n, l Element Z	1	2	3	4,0 4s	4,1 4p	4,2 4d	4,3 4f	5,0 5s	5,1 5p	5,2 5d	5,3 5f	5,4 5g	6,0 6s	6,1 6p	6,2 6d	6,3 6f	6,4 6g	6,5 6h
Shell	K	L	M			N				O					P			
Rb 37								1										
Sr 38								2										
Y 39						1		2										
Zr 40						2		2										
Nb 41	\multicolumn{4}{}{36-electron core}		4		1										
Mo 42	\multicolumn{4}{}{re: krypton}		5		1										
Tc 43						6		1										
Ru 44						7		1										
Rh 45						8		1										
Pd 46						10												
Ag 47								1										
Cd 48								2										
In 49								2	1									
Sn 50	\multicolumn{7}{}{46-electron core}	2	2									
Sb 51	\multicolumn{7}{}{re: palladium}	2	3									
Te 52								2	4									
I 53								2	5									
Xe 54								2	6									
Cs 55	\multicolumn{7}{}{54-electron core}						1					
Ba 56	\multicolumn{7}{}{re: xenon}						2					
La 57								2	6	1			2					
Ce 58							1	2	6	1			2					
Pr 59							2	2	6	1			2					
Nd 60							3	2	6	1			2					
Pm 61							4	2	6	1			2					
Sm 62							5	2	6	1			2					
Eu 63							6	2	6	1			2					
Gd 64	\multicolumn{4}{}{46-electron core}			7	2	6	1			2					
Tb 65	\multicolumn{4}{}{re: 1s to 4d}			8	2	6	1			2					
Dy 66							9	2	6	1			2					
Ho 67							10	2	6	1			2					
Er 68							11	2	6	1			2					
Tm 69							13	2	6	0			2					
Yb 70							14	2	6	0			2					
Lu 71							14	2	6	1			2					
Hf 72										2			2					
Ta 73										3			2					
W 74										4			2					
Re 75	\multicolumn{7}{}{68-electron core}			5			2					
Os 76	\multicolumn{7}{}{re: 1s to 5p}			6			2					
Ir 77										7			2					
Pt 78										9			1					
Au 79										10			1					

Table A.24. (continued)

n, l Element Z	1	2	3	4	5,0 5s	5,1 5p	5,2 5d	5,3 5f	5,4 5g	6,0 6s	6,1 6p	6,2 6d	6,3 6f	6,4 6g	6,5 6h	7,0 7s	7,1 7p
Shell	K	L	M	N	O					P						Q	
Hg 80										2							
Tl 81										2	1						
Pb 82		78-electron core								2	2						
Bi 83		re: 1s to 5d								2	3						
Po 84										2	4						
At 85										2	5						
Rn 86										2	6						
Fr 87		86-electron core														1	
Ra 88		re: 1s to 6p														2	
Ac 89										2	6	1				2	
Th 90								1		2	6	1				2	
Pa 91								2		2	6	1				2	
U 92								3		2	6	1				2	
Np 93								4		2	6	1				2	
Pu 94								5		2	6	1				2	
Am 95								6		2	6	1				2	
Cm 96								7		2	6	1				2	
Bk 97								8		2	6	1				2	
Cf 98								9		2	6	1				2	

SOURCE: Henry Semat, *Introduction to Atomic and Nuclear Physics*, 3rd ed. (New York: Holt, Rinehart & Winston, Inc., 1958).

Table A.25. The Elements

Element	Symbol	Atomic Number (Z)	Atomic Mass (A)	Element	Symbol	Atomic Number (Z)	Atomic Mass (A)
actinium	Ac	89	227.000	lithium	Li	3	6.940
aluminum	Al	13	26.980	lutetium	Lu	71	174.990
americium	Am	95	243.000	magnesium	Mg	12	24.320
antimony	Sb	51	121.760	mendelevium	Mv	101	256.000
argon	Ar	18	39.944	mercury	Hg	80	200.610
arsenic	As	33	74.910	molybdenum	Mo	42	95.950
astatine	At	85	210.000	neodymium	Nd	60	144.270
barium	Ba	56	137.360	neon	Ne	10	20.183
berkelium	Bk	97	245.000	neptunium	Np	93	237.000
beryllium	Be	4	9.013	nickel	Ni	28	58.690
bismuth	Bi	83	209.000	niobium	Nb	41	92.910
boron	B	5	10.820	nitrogen	N	7	14.008
bromine	Br	35	79.916	nobelium	No	102	
cadmium	Cd	48	112.410	osmium	Os	76	190.200
calcium	Ca	20	40.080	oxygen	O	8	16.000
californium	Cf	98	246.000	palladium	Pd	46	106.700
carbon	C	6	12.010	phosphorus	P	15	30.975
cerium	Ce	58	140.130	platinum	Pt	78	195.230
cesium	Cs	55	132.910	plutonium	Pu	94	242.000
chlorine	Cl	17	35.457	polonium	Po	84	210.000
chromium	Cr	24	52.010	potassium	K	19	39.100
cobalt	Co	27	58.940	praseodymium	Pr	59	140.920
copper	Cu	29	63.540	promethium	Pm	61	145.000
curium	Cm	96	243.000	protactinium	Pa	91	231.000
dysprosium	Dy	66	162.460	radium	Ra	88	226.050
einsteinium	E	99	255.000	radon	Rn	86	222.000
erbium	Er	68	167.200	rhenium	Re	75	186.310
europium	Eu	63	152.000	rhodium	Rh	45	102.910
fermium	Fm	100	255.000	rubidium	Rb	37	85.480
fluorine	F	9	19.000	ruthenium	Ru	44	101.700
francium	Fr	87	223.000	samarium	Sm	62	150.430
gadolinium	Gd	64	156.900	scandium	Sc	21	44.960
gallium	Ga	31	69.720	selenium	Se	34	78.960
germanium	Ge	32	72.600	silicon	Si	14	28.090
gold	Au	79	197.200	silver	Ag	47	107.880
hafnium	Hf	72	178.600	sodium	Na	11	22.997
helium	He	2	4.003	strontium	Sr	38	87.630
holmium	Ho	67	164.940	sulfur	S	16	32.006
hydrogen	H	1	1.008	tantalum	Ta	73	180.880
indium	In	49	114.760	technetium	Tc	43	99.000
iodine	I	53	126.910	tellurium	Te	52	127.610
iridium	Ir	77	193.100	terbium	Tb	65	159.200
iron	Fe	26	55.850	thallium	Tl	81	204.390
krypton	Kr	36	83.800	thorium	Th	90	232.120
lanthanum	La	57	138.920	thulium	Tm	69	169.400
lead	Pb	82	207.210	tin	Sn	50	118.700

Table A.25. (continued)

Element	Symbol	Atomic Number (Z)	Atomic Mass (A)	Element	Symbol	Atomic Number (Z)	Atomic Mass (A)
titanium	Ti	22	47.900	ytterbium	Yb	70	173.040
tungsten	W	74	183.920	yttrium	Y	39	88.920
uranium	U	92	238.070	zinc	Zn	30	65.380
vanadium	V	23	50.950	zirconium	Zr	40	91.220
xenon	Xe	54	131.300				

SOURCE: C. H. Blanchard, C. R. Burnett, R. G. Stoner, and R. L. Weber, *Introduction to Modern Physics*, (Englewood Cliffs, N. J.: Prentice-Hall, Inc., 1958).

Table A.26. Physical Constants

Constant	Symbol	Quantity
charge on an electron	e	1.6×10^{-19} coul
		4.8×10^{-10} statcoul
1 coulomb		6.25×10^{18} charges
mass of an electron	m	9.108×10^{-28} g
ratio e/m (electron)		1.76×10^{8} coul/g
electron-volt	eV	1.6×10^{-19} joule
gravitational acceleration	g	980.7 cm/sec^2
		32.17 ft/sec^2
universal gravitation	G	6.67×10^{-11} nt-m^2/kg^2
		6.67×10^{-8} dyne-cm^2/g^2
universal gas constant	R	1543 ft-lb/lb-mol/°R
		8.314 joule/g-mol/°K
Faraday constant	F	96,520 coul
Planck's constant	h	6.624×10^{-34} joule-sec
Stefan-Boltzman constant	σ	5.67×10^{-8} watt/(K°)^4-m^2
velocity of light	c	2.998×10^{10} cm/sec
		186,000 mi/sec
velocity of sound		
(0°C, 14.7 psi)		33,170 cm/sec
		1,088 ft/sec
Wien's constant	C	2.899×10^{-3} mK°
atomic-mass unit	amu	1.66×10^{-24} g
(1 amu)		931 MeV
absolute zero		-273.2°C
		-459.7°F
Rydberg constant		109,678
Avogadro's number		6.023×10^{23} per g-mol^3
proportionality constant		
(Coulomb's law)	k (esu)	1 dyne-cm^2/statcoul^2
	k (mks)	9×10^9 nt$-$m^2/coul^2
	ε_0 (mks)	8.85×10^{-12} coul^2/nt-m^2

Table A.27. Trigonometric Functions

Angle (degrees)	sin	cos	tan	Angle (degrees)	sin	cos	tan
0	0.0000	1.0000	0.0000	45	0.7071	0.7071	1.0000
1	0.0175	0.9998	0.0175	46	0.7193	0.6947	1.0355
2	0.0349	0.9994	0.0349	47	0.7314	0.6820	1.0724
3	0.0523	0.9986	0.0524	48	0.7431	0.6691	1.1106
4	0.0698	0.9976	0.0699	49	0.7547	0.6561	1.1504
5	0.0872	0.9962	0.0875	50	0.7660	0.6428	1.1918
6	0.1045	0.9945	0.1051	51	0.7771	0.6293	1.2349
7	0.1219	0.9925	0.1228	52	0.7880	0.6157	1.2799
8	0.1392	0.9903	0.1405	53	0.7986	0.6018	1.3270
9	0.1564	0.9877	0.1584	54	0.8090	0.5878	1.3764
10	0.1736	0.9848	0.1763	55	0.8192	0.5736	1.4281
11	0.1908	0.9816	0.1944	56	0.8290	0.5592	1.4826
12	0.2079	0.9781	0.2126	57	0.8387	0.5446	1.5399
13	0.2250	0.9744	0.2309	58	0.8480	0.5299	1.6003
14	0.2419	0.9703	0.2493	59	0.8572	0.5150	1.6643
15	0.2588	0.9659	0.2679	60	0.8660	0.5000	1.7321
16	0.2756	0.9613	0.2867	61	0.8746	0.4848	1.8040
17	0.2924	0.9563	0.3057	62	0.8829	0.4695	1.8807
18	0.3090	0.9511	0.3249	63	0.8910	0.4540	1.9626
19	0.3256	0.9455	0.3443	64	0.8988	0.4384	2.0503
20	0.3420	0.9397	0.3640	65	0.9063	0.4226	2.1445
21	0.3584	0.9336	0.3839	66	0.9135	0.4067	2.2460
22	0.3746	0.9272	0.4040	67	0.9205	0.3907	2.3559
23	0.3907	0.9205	0.4245	68	0.9272	0.3746	2.4751
24	0.4067	0.9135	0.4452	69	0.9336	0.3584	2.6051
25	0.4226	0.9063	0.4663	70	0.9397	0.3420	2.7475
26	0.4384	0.8988	0.4877	71	0.9455	0.3256	2.9042
27	0.4540	0.8910	0.5095	72	0.9511	0.3090	3.0777
28	0.4695	0.8829	0.5317	73	0.9563	0.2924	3.2709
29	0.4848	0.8746	0.5543	74	0.9613	0.2756	3.4874
30	0.5000	0.8660	0.5774	75	0.9659	0.2588	3.7321
31	0.5150	0.8572	0.6009	76	0.9703	0.2419	4.0108
32	0.5299	0.8480	0.6249	77	0.9744	0.2250	4.3315
33	0.5446	0.8387	0.6494	78	0.9781	0.2079	4.7046
34	0.5592	0.8290	0.6745	79	0.9816	0.1908	5.1446
35	0.5736	0.8192	0.7002	80	0.9848	0.1736	5.6713
36	0.5878	0.8090	0.7265	81	0.9877	0.1564	6.3138
37	0.6018	0.7986	0.7536	82	0.9903	0.1392	7.1154
38	0.6157	0.7880	0.7813	83	0.9925	0.1219	8.1443
39	0.6293	0.7771	0.8098	84	0.9945	0.1045	9.5144
40	0.6428	0.7660	0.8391	85	0.9962	0.0875	11.4301
41	0.6561	0.7547	0.8693	86	0.9976	0.0699	14.3007
42	0.6691	0.7431	0.9004	87	0.9986	0.0524	19.0811
43	0.6820	0.7314	0.9325	88	0.9994	0.0349	28.6363
44	0.6947	0.7193	0.9657	89	0.9998	0.0175	57.2900
45	0.7071	0.7071	1.0000	90	1.0000	0.0000	

Appendix

Table A.28. Logarithms

N	0	1	2	3	4	5	6	7	8	9
0	0	0000	3010	4771	6021	6990	7782	8451	9031	9542
1	0000	0414	0792	1139	1461	1761	2041	2304	2553	2788
2	3010	3222	3424	3617	3802	3979	4150	4314	4472	4624
3	4771	4914	5051	5185	5315	5441	5563	5682	5798	5911
4	6990	7076	7160	7243	7324	7404	7482	7559	7634	7709
5	6990	7076	7160	7243	7324	7404	7482	7559	7634	7709
6	7782	7853	7924	7993	8062	8129	8195	8261	8325	8388
7	8451	8513	8573	8633	8692	8751	8808	8865	8921	8976
8	9031	9085	9138	9191	9243	9294	9345	9395	9445	9494
9	9542	9590	9638	9685	9731	9777	9823	9868	9912	9956
10	0000	0043	0086	0128	0170	0212	0253	0294	0334	0374
11	0414	0453	0492	0531	0569	0607	0645	0682	0719	0755
12	0792	0828	0864	0899	0934	0969	1004	1038	1072	1106
13	1139	1173	1206	1239	1271	1303	1355	1367	1399	1430
14	1461	1492	1523	1553	1584	1614	1644	1673	1703	1732
15	1761	1790	1818	1847	1875	1903	1931	1959	1987	2014
16	2041	2068	2095	2122	2148	2175	2201	2227	2253	2279
17	2304	2330	2355	2380	2405	2430	2455	2480	2504	2529
18	2553	2577	2601	2625	2648	2672	2695	2718	2742	2765
19	2788	2810	2833	2856	2878	2900	2923	2945	2967	2989
20	3010	3032	3054	3075	3096	3118	3139	3160	3181	3201
21	3322	3243	3263	3284	3304	3324	3345	3365	3385	3404
22	3424	3444	3464	3483	3502	3522	3541	3560	3579	3598
23	3617	3636	3655	3674	3692	3711	3729	3747	3766	3784
24	3802	3820	3838	3856	3874	3892	3909	3927	3945	3962
25	3979	3997	4014	4031	4048	4065	4082	4099	4116	4133
26	4150	4166	4183	4200	4216	4232	4249	4265	4281	4296
27	4314	4330	4346	4362	4378	4393	4409	4425	4440	4456
28	4472	4487	4502	4518	4533	4548	4564	4579	4594	4609
29	4624	4639	4654	4669	4683	4698	4713	4728	4742	4757
30	4771	4786	4800	4814	4829	4843	4857	4871	4886	4900
31	4914	4928	4942	4955	4969	4983	4997	5011	5024	5308
32	5051	5065	5079	5902	5105	5119	5132	5145	5159	5172
33	5185	5198	5211	5224	5237	5250	5263	5276	5289	5302
34	5315	5328	5340	5353	5366	5378	5391	5403	5416	5428
35	5441	5453	5465	5478	5490	5502	5514	5527	5539	5551
36	5563	5575	5587	5599	5611	5623	5635	5647	5658	5670
37	5682	5694	5705	5717	5729	5740	5752	5763	5775	5786
38	5798	5809	5821	5832	5843	5855	5866	5877	5888	5899
39	5911	5922	5933	5944	5955	5966	5977	5988	5999	6010
40	6021	6031	6042	6053	6064	6075	6085	6096	6107	6117
41	6128	6138	6149	6160	6170	6180	6191	6201	6212	6222
42	6232	6243	6253	6263	6274	6284	6294	6304	6314	6325
43	6335	6345	6355	6365	6375	6385	6395	6405	6415	6425
44	6435	6444	6454	6464	6474	6484	6493	6503	6513	6522
45	6532	6542	6551	6561	6571	6580	6590	6599	6609	6618
46	6628	6637	6647	6656	6665	6675	6684	6693	6702	6712
47	6721	6730	6739	6749	6758	6767	6776	6785	6794	6803
48	6812	6821	6830	6839	6848	6857	6866	6875	6884	6893
49	6902	6911	6920	6928	6937	6946	6955	6964	6972	6981

Table A.28. (continued)

N	0	1	2	3	4	5	6	7	8	9
50	6990	6998	7007	7016	7024	7033	7042	7050	7059	7067
51	7076	7084	7093	7101	7110	7118	7126	7135	7143	7152
52	7160	7168	7177	7185	7193	7202	7210	7218	7226	7235
53	7243	7251	7259	7267	7275	7284	7292	7300	7308	7316
54	7324	7332	7340	7348	7356	7364	7372	7380	7388	7396
55	7404	7412	7419	7427	7435	7443	7451	7459	7466	7474
56	7482	7490	7497	7505	7515	7520	7528	7536	7543	7551
57	7559	7566	7574	7582	7589	7597	7604	7612	7619	7627
58	7634	7642	7649	7657	7664	7672	7679	7686	7694	7701
59	7709	7716	7723	7731	7738	7745	7752	7760	7767	7774
60	7782	7789	7796	7805	7810	7818	7825	7832	7839	7846
61	7853	7860	7868	7875	7882	7889	7896	7903	7910	7917
62	7924	7931	7938	7945	7952	7959	7966	7973	7980	7987
63	7993	8000	8007	8014	8021	8028	8035	8041	8048	8055
64	8062	8069	8075	8082	8089	8096	8102	8109	8116	8122
65	8129	8136	8142	8149	8156	8162	8169	8176	8182	8189
66	8195	8202	8209	8215	8222	8228	8235	8241	8248	8254
67	8261	8267	8274	8280	8287	8293	8299	8306	8312	8319
68	8325	8331	8338	8344	8351	8357	8363	8370	8376	8382
69	8388	8395	8401	8407	8414	8420	8426	8432	8439	8445
70	8451	8457	8463	8470	8476	8482	8488	8494	8500	8506
71	8513	8519	8525	8531	8537	8543	8549	8555	8561	8567
72	8573	8579	8585	8591	8597	8603	8609	8615	8621	8627
73	8633	8639	8645	8651	8657	8663	8669	8675	8681	8686
74	8692	8698	8704	8710	8716	8722	8727	8733	8739	8745
75	8751	8756	8762	8768	8774	8779	8785	8791	8797	8802
76	8808	8814	8820	8825	8831	8837	8842	8848	8854	8859
77	8865	8871	8876	8882	8887	8893	8899	8904	8910	8915
78	8921	8927	8932	8938	8943	8949	8954	8960	8965	8971
79	8976	8982	8987	8993	8998	9004	9009	9015	9020	9025
80	9031	9036	9042	9047	9053	9058	9063	9069	9074	9079
81	9085	9090	9096	9101	9106	9112	9117	9122	9128	9133
82	9138	9143	9149	9154	9159	9165	9170	9175	9180	9186
83	9191	9196	9201	9206	9212	9217	9222	9227	9232	9238
84	9243	9248	9253	9258	9263	9269	9274	9279	9284	9289
85	9294	9299	9304	9309	9315	9320	9325	9330	9335	9340
86	9345	9350	9355	9360	9365	9370	9375	9380	9385	9390
87	9395	9440	9405	9410	9415	9420	9425	9430	9435	9440
88	9445	9450	9455	9460	9465	9469	9474	9479	9484	9489
89	9494	9499	9504	9509	9513	9518	9523	9528	9533	9538
90	9542	9547	9552	9557	9562	9566	9571	9576	9581	9586
91	9590	9595	9600	9605	9609	9614	9619	9624	9628	9633
92	9638	9643	9647	9652	9657	9661	9666	9671	9675	9680
93	9685	9689	9694	9699	9703	9708	9713	9717	9722	9727
94	9731	9736	9741	9745	9750	9754	9759	9763	9768	9773
95	9777	9782	9786	9791	9795	9800	9805	9809	9814	9818
96	9823	9827	9823	9836	9841	9845	9850	9854	9859	9863
97	9868	9872	9877	9881	9886	9890	9894	9899	9903	9908
98	9912	9917	9921	9926	9930	9934	9939	9943	9948	9952
99	9956	9961	9965	9969	9974	9978	9983	9987	9991	9996
100	0	0004	0009	0013	0017	0022	0026	0030	0035	0039

Answers to Even-Numbered Problems

Chapter 1
4. (a) 180 ft
   (b) 4.76 ft
   (c) 5.06×10^{-6} ft/sec²
   (d) 67.1 mph
   (e) 458×10^7 Btu/lb
6. 1.8×10^{10} cal-cm³ C°/sec
8. $B = 25.5°$
   $C = 139.5°$
   $c = 30.15$ in.

Chapter 2
2. (a) 240 lb at 16°
   (b) 135 lb at 57°
4. $5\frac{3}{8}$ lb at 60°24′
6. $5\frac{3}{8}$ lb at 60°24′
8. (a) $F_x = 4.70$ lb
       $F_y = 1.71$ lb
   (b) $F_x = 2.74$ lb
       $F_y = 7.52$ lb
   (c) $F_x = -3.86$ lb
       $F_y = 4.60$ lb
   (d) $F_x = 2.09$ lb
       $F_y = 11.82$ lb
   (e) $F_x = 13.59$ lb

   $F_y = -6.35$ lb
10. $R = 5.40$ lb at 60°24′
12. $R = 6.3$ lb at 30°

Chapter 3
4. $A = 44.4$ lb; $B = 62.4$ lb
6. $\theta = 49°27′$; $\alpha = -22°20′$
8. $B = 154$ lb; $C = 154$ lb
10. $B = 231$ lb; $T = 115.5$ lb
12. $AD = 319$ lb; $AB = 1320$ lb
14. $C = 2000$ lb; $B = 1732$ lb
    $D = 2261$ lb; $C = 1454$ lb
    $CN = 3000$ lb at 54°47′
16. $AB = 3464$ lb; $AC = 1732$ lb
    $BC = 3464$ lb; $BD = 3464$ lb
    $CD = 0$ lb; $CE = 4732$ lb
    $DF = BD = 3464$ lb
18. $AG = 3464$ lb; $AB = 4000$ lb
    $BC = 3500$ lb; $BG = 866$ lb
    $GC = 866$ lb; $GF = 2598$ lb
    $CF = 866$ lb; $CD = 3500$ lb
    $R_1 = R_2 = 3000$ lb
20. $\alpha = 41°50′$; $\beta = 16°36′$
    $C = 1044$ lb; $R = 299$ lb
    at F: $B = 1342$ lb; $F = 1194$ lb

Chapter 4
2. $M = 160$ lb
4. $f_1 = 62.5$ lb; $f_2 = 87.5$ lb
6. $M = 120$ ft-lb
8. $f = 88$ lb; $d = 1.27$ ft from 60 lb
10. $R_1 = 370$ lb; $R_2 = 1270$ lb
12. $P = 600$ lb; $f = 200$ lb
14. $B = 79.2$ lb; $A = 145.8$ lb
16. $R_1 = 317.5$ lb; $R_2 = 432.5$ lb
18. $f = 111$ lb
20. $P = 6.6$ lb; $f = 75.3$ lb at $84°58'$
22. $P = 101$ lb; $f = 226.7$ lb at $67°32'$
24. $W = 1385.6$ lb; $f = 1060$ lb at $19°6'$
26. $T = 804$ lb; $f = 644$ lb at $36°36'$
28. $x_e = 6.75$ in.; $y_e = 4.52$ in.
30. $X_e = 4.49$ in.; $Y_e = 4.7$ in.
32. $X_e = 3.45$ ft

Chapter 5
2. $\mu = 0.33$
4. (a) $f_r = 16.50$ lb
   (b) yes
   (c) accelerates
6. $f = 15.2$ lb; $N = 37.6$ lb
   $f_r = 13.16$ lb
8. (a) $w = 168$ lb;
   (b) $N = 81.2$ lb
   (c) $f_r = 33$ lb
10. $N = 410$ lb; $f_r = 82$ lb
    $f = 369$ lb
12. $f_r = 111$ lb; $\mu = 0.31$
14. topple
16. $f = 72.5$ lb (motion)
    $f_r = 70$ lb; $N = 218.8$ lb
    topple
18. $H = 12$ lb; $\mu = 0.202$
    $R_v = 55.9$ lb; $R_h = 11.3$ lb
20. $P = 550$ lb; $N' = 489$ lb
    $f = 486.8$ lb

Chapter 6
4. strain $= 0.3125$
6. $D = 0.683$ in.
8. $\Delta x = 2.61 \times 10^{-4}$ in.
10. $P = 294$ psi; $\beta = 6.5 \times 10^6$ psi
12. $A = \tfrac{5}{8}$ in^2; $f = 10,000$ lb

Chapter 7
2. $s = 4.75 \times 10^4$ ft
4. (a) 1050 mph
   (b) 250 mph
6. $x = 938$ ft
8. $t = 8.18$ sec
10. $l = 8$ ft
12. (a) $l_0 = 2,500$ m
    (b) $t = 10^{-5}$ sec
14. $m = 2.92 \times 10^{-30}$ kg
16. $x = 20$ mph; $y = 34.64$ mph
18. $s = 76 \times 10^5$ ft
20. (a) $v_{av} = 36.7$ ft/sec
    (b) $s = 66,000$ ft
22. (a) $v_f = 773.4$ ft/sec
    (b) $s = 12.8 \times 10^4$ ft
24. (a) 39 ft
    (b) $a = -8.67$ ft/sec^2
26. $a = -1.28 \times 10^6$ ft/sec^2
28. (a) $v_f = 85.4$ ft/sec
    (b) $t = 1.73$ sec
30. $s = 132$ ft
32. (a) $v_f = 80$ ft/sec; $v_h = 30$ ft/sec
    (b) $s_f = 100$ ft; $s_h = 75$ ft
34. (a) $v = 315$ ft/sec
    (b) $\theta = 37°30'$
36. $v = 283$ ft/sec at $1°13'$

Chapter 8
2. (a) $f = 570$ lb
   (b) $f = 287$ lb
4. $f = 168,000$ lb
6. $a = 8$ ft/sec^2; $T = 7.5$ lb
8. $a = 19.2$ ft/sec^2; $T = 8$ lb
10. $a = 10.72$ ft/sec^2; $T = 27.9$ lb
12. $T = 5625$ lb
14. $T = 4375$ lb
16. (a) $T = 212.5$ lb
    (b) $T = 3212.5$ lb
    (c) $T = 3187.8$ lb
18. $a = 32$ ft/sec^2
20. $v = 19.2$ ft/sec
22. incline: $v = 233.6$ ft/sec; $s = 3.23$ mi

Answers to Even-Numbered Problems

Chapter 9
2. (a) 144°
   (b) $157\frac{1}{2}°$
   (c) $231\frac{3}{7}°$
   (d) 1080°
   (e) 315°
4. (a) 76.4 rpm
   (b) 55.7 rpm
6. (a) $v = 22.7$ mph
   (b) $\omega = 26.7\pi$ rad/sec
8. (a) 191 rpm
   (b) 15.0 rad/sec; 20 rad/sec
10. $\alpha = -0.314$ rad/sec² (decel)
12. $\alpha = \frac{3}{4}$ rad/sec²; $\omega = 900$ rpm
14. (a) $v = 37.7$ rad/sec
    (b) $a_r = 2128$ ft/sec²
16. $v = 12.56$ ft/sec
    $a_c = 31.5$ ft/sec²
18. (a) $\omega = 8\pi$ rad/sec; $v_r = 0$ ft/sec
    (b) $a_c = 0$ ft/sec²; $a_c = 789$ ft/sec
20. $g = 981.5$ cm/sec²
22. (a) $a = 5.57$ ft/sec²
    (b) $s = 0.435$ ft
    (c) $T = 1.76$ sec
    (d) $F = 0.57$ vib/sec
    (e) tension $= 31.5$ lb

Chapter 10
2. $I = 62.5$ ft-lb-sec²
4. $I = 11{,}250$ ft-lb-sec²
6. $I = 14.5$ ft-lb-sec²
8. (a) $I_{p-p} = 402$ in.⁴
   (b) $I_{p-p} = 104$ in.⁴
10. (a) $I_{z-z} = 1005$ in.⁴
    (b) $I_{z-z} = 129.7$ in.⁴
    (c) $I_{z-z} = 288$ in.⁴
    (d) $I_{z-z} = 512$ in.⁴
12. (a) $I_{o-o} = 301$ in.⁴
    (b) $I_{x-x} = 6405$ in.⁴
    (c) $I_{z-z} = 979$ in.⁴
    (d) $I_{p-p} = 452$ in.⁴
14. $T = 1500$ ft-lb
16. (a) $I = 156$ ft-lb-sec²
    (b) $L = 375$ ft-lb
    (c) $\alpha = 2.4$ rad/sec²
    (d) $\omega = 108$ rad/sec

   (e) 0.29 rpm
18. $v = 20$ ft/sec
20. $u = 0.68$

Chapter 11
2. wk $= 1880$ ft-lb
4. wk $= 5.2 \times 10^6$ ft-lb
6. $w = 8.3 \times 10^4$ lb
8. $t = 10$ min
10. $\Delta PE = 4.92$ ft-lb
12. KE $= 402$ ft-lb
14. hp $= 4.7$
16. (a) hp $= 320$
    (b) $v = 46.9$ mph
18. (a) $E_o = 20.8$ ft-lb
    (b) $E_i = 22.5$ ft-lb
    (c) DR $= 72$
    (d) MA $= 66.7$
    (e) eff $= 92.6\%$
20. (a) in $= 2.67 \times 10^{-3}$ hp
    (b) out $= 2.42 \times 10^{-3}$ hp
22. (a) 12
    (b) 300 lb
24. (a) $S_i = 6.3$ ft
    (b) $S_o = 1.055$ ft
    (c) DR $= 6$
    (d) MA $= 5.83$
    (e) eff $= 97.2\%$

Chapter 12
2. (a) 1.8×10^4 lb-sec
   (b) KE $= 1.1 \times 10^7$ ft-lb
   (c) $f = 3.6 \times 10^4$ lb
4. (a) $ft = 9 \times 10^3$ lb-sec
   (b) $t = 3.3$ sec
   (c) $f = 2.7 \times 10^3$ lb
6. (a) $ft = 3.42$ lb-sec
   (b) $f = 107$ lb
8. $v_2 = 33.5$ ft/sec
10. $v_1 = 118$ ft/sec
    $v_2 = 88$ ft/sec
12. $v_2 = 57$ ft/sec
    $v_1 = -87$ ft/sec
14. $V = 57.2$ ft/sec
16. angular moment $= 65.4$ ft-lb-sec

Chapter 13
2. (a) $P_{ab} = 3960$ psfa
   (b) 27.5 psia
   (c) $h = 760$ in. water abs
   (d) $h = 55.9$ in. Hg abs
4. (a) 37.3 psia
   (b) 1033 in. water abs
   (c) $h = 75.9$ in. Hg abs
   (d) $h = 193$ cm Hg abs
6. (a) $p = 16.48$ psia
   (b) $P_{Hg} = 15.01$ psia
   (c) $P_{H_2O} = 14.826$ psia
8. (a) $P = 433.6$ psf
   (b) $f = 5446$ lb
   (c) 17,020 lb
10. (a) $h = 48.9$ ft (oil)
    (b) $h = 41.6$ ft water
    (c) $h = 36.7$ in. Hg
12. (a) $D_2 = 6$ in.
    (b) DR = 143
    (c) $S_2 = 0.028$ in.
    (d) $P = 382.2$ psi
    (e) $f = 10,800$ lb
14. $P = 2.2528$ psig
16. $P = 0.5776$ psig
18. $h = 25.8$ in.
20. $h_{0.06} = 44.288$ in.
    $h_{0.2} = 43.748$ in.
    $h_{0.3} = 43.673$ in.
22. $T = 0.0032$ lb/ft
24. $T = 0.0318$ lb/ft

Chapter 14
2. $T = 6.80$ lb
4. $T = 7.5$ lb sub
6. (a) 90.4% below surface
   (b) 5.424×10^5 lb below surface
   (c) 57,600 lb above surface
      volume = 1000 ft³
8. sp wt = 534 lb/ft³
10. (a) $h = 16$ in.
    (b) least weight = 997.6 lb
    (c) $h = 2$ ft, 1.6 in.
12. $V = 22.5$ ft³; $B = 1.8$ lb
14. (a) $B_s = 0.5$ lb
    (b) $B_b = 7$ lb
    (c) $B_b = 6.5$ lb
    (d) weight of water displaced = 6.5 lb
    (e) $V = 180$ in.³
    (f) $W = 0.0278$ lb/in.³
    (g) sp gr = 0.77
16. sp gr (metric) = 1.6
    sp wt (British) = 0.05776 lb/in.³
18. $\Delta L = 15.4$ in.
20. $\Delta L = 3.64$ in.

Chapter 15
2. (a) $v_2 = 13.3$ ft/sec
   (b) $v_3 = 19.2$ ft/sec
   (c) $Q_1 = Q_2 = Q_3 = 2.62$ ft³/sec
4. (a) 0 ft
   (b) 0 ft
   (c) 100 ft
6. $P_2 = 35$ psf
8. $v_1 = 3.7$ ft/sec; $v_2 = 22.9$ ft/sec
   $P_2 = 45.3$ psi
10. input head = -15.135 ft
    output head = 107.4 ft
    pump energy = 123.4 ft
    hp = 55.27
12. (a) $v_2 = 35.8$ ft/sec
    (b) $Q_a = 5.6$ ft³/sec
    (c) $t = 116.7$ min
14. $t = 85.4$ min
16. $Q_a = 0.422$ ft³/sec
18. (a) $v_2 = 80.3$ ft/sec
    (b) $P_1 = 6162.7$ psf
20. $Q_a = 56.8$ ft³/sec
22. $\Delta h = 1.85$ ft
24. $v = 21.9$ ft/sec

Chapter 16
2. (a) R = 510°R
   (b) K = 323°K
   (c) F = -410°F
   (d) C = -223°C
   (e) F = 45°C
   (f) C = 40°F
4. $\Delta L = 0.00378$ ft
6. $\Delta L = 0.180$ in.
8. $d_f = 1.0077$ in.
10. $d_i = 4.1282$ in.

Answers to Even-Numbered Problems 655

chapter 16 (cont.)

12. $T_f = 366.4\,°F$
14. overflow $= 33.108\ cm^3$
16. (a) $h = 2.39$ in.
    (b) $h = 2.038$ in.
18. reads $= 240.101$ in.
20. $V_f = 159.2\ ft^3$
22. $T_f = 166.5\,°C$

Chapter 17

2. water equivalent $= 0.702$ lb
4. $Q = 714$ Btu
6. $T_f = 593\,°F$
8. $T = 55.9\,°F$
10. $m_w = 64.5$ lb (approx. 8.9 gal)
12. $T = 25.8\,°C$
14. $T = 930.7\,°F$
16. $Q = 10,840$ cal/g
18. $T = 90.6\,°F$
20. $Q = 9$ lb of ice melted

Chapter 18

2. $V_2 = 12.1\ ft^3$
4. $V_2 = 2.5\ ft^3$
6. $P_2 = 300$ psia
8. $P_2 = 20.5$ psig
10. $P_2 = 15.7$ psig
12. $T_2 = 227.5\,°F$
14. $V_2 = 44.9\ ft^3;\ M = 33.5$ lb
16. hydrogen
18. $v_{av} = 1340$ ft/sec; KE $= 1.23 \times 10^6$ ft-lb
20. monatomic $c_v = 0.074$
    table 0.075
    $c_p = 0.124$
    table 0.125
    diatomic $c_v = 0.155$
    $c_p = 0.217$
    polyatomic $c_v = 0.093$
    $c_p = 0.125$
22. argon $= 1.7$
    oxygen $= 1.4$
    ammonia $= 1.34$
    helium $= 1.66$
    nitrogen $= 1.4$
    carbon dioxide $= 1.33$

24. $T = 1932\,°R$
    $v = 3206$ ft/sec

Chapter 19

2. (a) $W_i = 88.9$ Btu
   (b) $W_e = 26.4$ Btu
   (c) $Q = 115.3$ Btu
4. (a) $T_i = 142.6\,°F$
   (b) $V = 14.4\ ft^3$
   (c) $P_2 = 88.3$ psia
6. (a) $m = 1.057$ lb
   (b) $P_2 = 20$ psi
   (c) 1.66×10^5 ft-lb/lb
   (d) 225 Btu
8. (a) $T = 952\,°F$
   (b) $V_2 = 11.67\ ft^3$
   (c) $W_e = 189,000$ ft-lb
   (d) $Q = -243$ Btu; $W_i = 0$
10. (a) $m = 29.15$
    (b) $V_2 = 55.84\ ft^3;\ V_1 = 1.16\ ft^3$
    (c) $P_1 = 721$ psia
    (d) 4.67×10^5 ft-lb
12. (a) $Q = 8.414$ Btu
    (b) $Q = 8.414$ Btu
    (c) eff $= 15.6\%$

Chapter 20

2. $f = 30.7 \times 10^{-4}$ nt
4. $f = 16.6$ nt at $77\,°8'$
6. $f = 5.04 \times 10^{-4}$ nt at $83\,°10'$
8. $r = 0.236$ m
10. $E = 2.5 \times 10^4$ nt/coul
12. $E = 27.1 \times 10^4$ nt/coul at $47\,°9'$
14. $V = -2880$ volts
16. $V_{LM} = +1800$ volts (PD)
18. (a) $E = 0;\ V = 8000$ volts
    (b) $E = 6.4 \times 10^4$ nt/coul;
        $V = 4.8 \times 10^3$ volts
20. (a) $E = 1.09 \times 10^5$ volts
    (b) $v = 1.93 \times 10^{-3}$ m/sec

Chapter 21

2. $V = 71$ nt/coul
4. $N = 31$ plates
6. $t = 1.06$ cm
8. (a) $C = 5.8\ \mu f$
   (b) $q = 1276$ coul

Answers to Even-Numbered Problems

chapter 21 (cont.)

- (c) $V_1 = 106.3$ volts; $V_2 = 70.9$ volts; $V_3 = 42.5$ volts
10. (a) $C = 60 \ \mu f$
    (b) voltage $= 220$ volts
    (c) $q_1 = 2640 \ \mu\text{coul}$; $q_2 = 3960 \ \mu\text{coul}$; $q_3 = 6600 \ \mu\text{coul}$
12. $q = 989 \ \mu\text{coul}$
14. $W = 116 \times 10^{-6}$ joule
16. (a) $C = 185.85 \ \mu\mu f$
    (b) $W = 1.33 \times 10^{-6}$ joule
    (c) $E = 6 \times 10^4$ nt/coul
    (d) $Q = 0.02 \ \mu\text{coul}$

Chapter 22

2. $R = 5$ ohms
4. $R = 12$ ohms
6. (a) $I = 5.9$ amp
   (b) $I_{10} = 2.4$ amp; $I_{12} = 2.0$ amp; $I_{16} = 1.5$ amp
   (c) voltage $= 24$ volts
8. (a) $R = 5$ ohms
   (b) $I = 6$ amp
   (c) $I_{1.08} = I_3 = 6$ amp; $I_5 = I_7 = I_9 = 0.25$ amp $I_2 = 2.76$ amp; $I_4 = 1.38$ amp; $I_6 = 0.92$ amp; $I_8 = 0.69$ amp
   (d) $V_{1.08} = 6.48$ V
       $V_3 = 18$ V
       $V_5 = 1.25$ V
       $V_7 = 1.75$ V
       $V_9 = 2.25$ V
       $V_2 = V_4 = V_6 = V_8 = 5.52$ V
10. (a) $R = 4$ ohms;
    (b) $I = 6$ amp
    (c) parallel $= I = 3.6$ amp and 2.4 amp at second 4-ohm resistor $I = 1.2$ amp
12. (a) 10 volts
    (b) 6.7 ohms
    (c) 1.5 amp
    (d) 2 batteries $= 1.4$ volts
        2 batteries $= 1.7$ volts
        1 battery $= 1.25$ volts
14. (a) $R = 8.67$ ohms
    (b) $I = 1.38$ amp

   (c) $I_{12} = 0.46$ amp (each)
   (d) $V = 0.23$ volts
   (e) $V = 11.77$ volts
16. $L = 300$ m
18. $L = 12$ ft
20. $R = 205$ ohms
22. $R_{20} = 24.7$ ohms; $L = 775.6$ m
24. $t_0 = 250°C$; $R = 137$ ohms
26. $R = 300$ ohms
28. power consumed $= 1.645$ kw-hr

Chapter 23

2. $I = 3.64$ amp
4. (a) $I = 5$ amp
   (b) $V_{xy} = -45$ volts
   (c) $V_{zg} = 84$ volts
6. $I_1 = \frac{4}{21}$ amp;
   $I_2 = -\frac{3}{14}$ amp (wrong dir.)
   $I_3 = -\frac{1}{42}$ amp (wrong dir.)
8. $I_1 = 0.23$ amp
   $I_2 = 0.02$ amp
   $I_3 = 0.21$ amp
10. $I_2 = 1$ amp
    $E_2 = 7$ volts
12. (a) $K = 0.175 \times 10^{-6}$ amp/mm
    (b) $V = 7 \times 10^{-6}$ volt/mm
    (c) $I = 6.3 \times 10^{-6}$ amp (max)
14. (a) $R = 225 \times 10^3$ ohms
    (b) $K = 5001$ ohms/volt
    (c) $K_{40} = 600{,}070$ ohms
16. deflection 54.5 div; 65.5 div
18. 90 volts and 150 volts
20. $I_m = 0.124$ amp and $I_s = 19{,}876$ amp
22. (a) error $= 0.56\%$
    (b) error $= 0.004\%$
24. $R_x = 180.7$ ohms
26. $R_x = 18{,}000$ ohms
28. $R_x = 58.44$ ohms

Chapter 24

2. $f = 2.6 \times 10^{-4}$ newt
4. $H_r = 64.4$ amp-turns/m
6. $H_r = 172$ amp-turns/m at $60°44'$
8. $B = 8.09 \times 10^{-5}$ wb/m²
10. $B = 2.16 \times 10^{-6}$ wb/m²
12. $\tau = 3.8 \times 10^{-5}$ nt-m

Answers to Even-Numbered Problems

chapter 24 (cont.)

14. $\tau = 2.3 \times 10^{-3}$ nt-m
16. $I = 0.04$ μa

Chapter 25

2. (a) $H = 1.85 \times 10^4$ amp-turns/m (add)
   (b) $H = 0.65 \times 10^4$ amp-turns/m (subtract)
   (c) $B_1 = 2.324 \times 10^{-2}$ wb/m²
       $B_2 = 8.14 \times 10^{-3}$ wb/m²
4. $B = 9.5 \times 10^{-2}$ wb/m²
6. (a) $H = 2.4 \times 10^3$ amp-turns/m
   (b) 3.014×10^{-3} wb/m²
8. $I = 0.06$ amp
10. $I_2 = 3.3$ amp
12. $I = 15.5$ amp
14. (a) $I = 0.2$ amp
    (b) $\phi = 3.6 \times 10^{-4}$ wb
16. (a) $B = 0.48$ wb/m²
    (b) $H = 2667$ amp-turns/m
    (c) $\phi = 4.8 \times 10^{-4}$ wb
    (d) $u_r = 143.3$

Chapter 26

2. $E = 2.26 \times 10^{-3}$ volt
4. (a) $E_{inst} = 13.6$ volts
   (b) $E_{av} = 8.64$ volts
6. (a) $I = 1.3$ amp
   (b) $q = 0.53$ coul
8. $E_{av} = 8.6 \times 10^1$ volts
10. (a) $f = 0.18$ lb
    (b) $\tau = 1.06$ in.-lb
    (c) hp = 14.5
12. (a) $E = 2400$ volts
    (b) $\Delta\phi = 0.08$ wb
    (c) $E = 576$ joules
14. (a) $N_s = 1725$ turns
    (b) $I_p = 57.5$ amp
        $I_s = 230$ amp
        $R_s = 2.5$ ohms
16. $L_s = 4.5 \times 10^{-3}$ henry
18. $t = 0$ sec, $I = 0$ amp
    $t = 0.1$ sec, $I = 1.26$ amp
    $t = 0.2$ sec, $I = 1.72$ amp
    $t = 0.3$ sec, $I = 1.90$ amp
    $t = 0.4$ sec, $I = 1.963$ amp
20. $I = 0.2789$ amp
22. (a) $Q = 72 \times 10^{-6}$ coul
    (b) $V = 3$ volts

Chapter 27

2. (a) $\omega = 120\pi$
   (b) $E_{max} = 622$ volts
   (c) $I_{max} = 113$ amp
   (d) $E_{inst} = 621$ volts
4. (a) $X_L = 2.26$ ohms
   (b) $Z = 2.26$ ohms
   (c) $I = 53.1$ amp
6. (a) $X_c = 332$ ohms
   (b) $Z = 332$ ohms
   (c) $I = 0.36$ amp
8. (a) $X_L = 2.26$ ohms
   (b) $Z = 30.08$ ohms
   (c) $I = 3.99$ amp
   (d) $\theta = 4°26'$
10. (a) $X_L = 332$ ohms
    (b) $Z = 333$ ohms
    (c) $I = 0.36$ amp
    (d) $\theta = 84°47'$
    (e) I leads E
12. (a) $X_L = 251.2$ ohms
    (b) $X_C = 53.2$ ohms
    (c) $Z = 204.2$ ohms
    (d) $I = 1.08$ amp
    (e) $\theta = 75°50'$
14. (a) $z = 256$ ohms
    (b) $I = 0.86$ amp
    (c) $\theta = 76.5°$
16. (a) $\cos\theta = 0.469$
    (b) $P = 266$ watts
18. (a) $\cos\theta = 0.618$
    (b) $P = 115.6$ watt
20. $F_r = 25.3$ cps

Chapter 28

2. (a) $\lambda = 16.8$ ft
   (b) $F = 285.7$ vib/sec
   (c) $F = 285.7$ vib/sec (alternate method)
4. $v = 136$ ft/sec
6. $v = 92$ ft/sec
8. $v = 3.32 \times 10^4$ cm/sec
10. $v = 16.5 \times 10^3$ ft/sec

Answers to Even-Numbered Problems

chapter 29 (cont.)

12. $v = 3.49 \times 10^3$ ft/sec
14. (a) $F_1 = 17.9$ vib/sec
    (b) $\lambda = 20$ ft
    (c) $F_2 = 35.8$ vib/sec
16. (a) $T = 2940$ g
    (b) $T = 117.5$ g
18. (a) $f = 89$ lb
    (b) $\lambda = 1.175$ ft
    (c) $v = 517$ ft/sec
20. (a) $F = 1465$ vib/sec
    (b) $v_r = 11{,}720$ ft/sec
    (c) $v_2 = 1152$ ft/sec
22. $d_g = 3.18$ in.
24. (a) $F = 90.7$ vib/sec
    (b) $\lambda = 2.4$ ft
26. (a) $l = 1.168$ ft
    (b) $v_2 = 1264$ ft/sec; $\Delta l = 0.79$ in.
28. (a) $F_h = 842$ vib/sec
    (b) $\lambda = 4.95$ ft
    (c) $l = 1.24$ ft
30. (a) $F = 600$ vib/sec
    (b) $\lambda = 1.898$ ft
32. (a) $v = 1106$ ft/sec
    (b) $\lambda = 2.16$ ft
    (c) $\Delta\lambda = 1.85$ ft
34. (a) $F_{o \to s} = 1297.1$ vib/sec
    (b) $F_{-os} = 1102.9$ vib/sec
36. (a) $F = 5487.6$ vib/sec
    (b) $F = 696.6$ vib/sec

Chapter 29

2. $s = 23.3$ mi
4. (a) 445 lm/watt
   (b) 0.674 watt
   (c) 0.438 watt
6. (a) 45 watts
   (b) 10,994.25 lm
   (c) 244.3 lm/watt
8. $I = 720$ candles
10. $s = 2.59$; $\Delta s = 0.41$ ft
12. (a) 63 lamps
    (b) 21 fluorescent lamps
    (c) takes $\frac{1}{3}$ as many
14. (a) $I = 155$ candles
    (b) $E = 60.5$ lm/m²
    (c) $\omega = 0.44$
    (d) $F = 68.2$ lm
    (e) $F = 1947$ lm
    (f) eff $= 97.35$ lm/watt
16. (a) $I = 3200$ candles
    (b) $F = 40{,}192$ lm
    (c) 670 watts
    (d) $E = 12.2$ ft-candles
18. $E = 9.8$ ft-candles
20. $E = 9.21$ ft-candles
22. $I_x = 50.4$ candles

Chapter 30

2. (a) $q = -2.5$ in. (virtual)
   (b) $M = -0.75$ in. (erect)
   (c) virtual, erect
   (d) (drawing)
4. (a) $q = 30$ in.
   (b) (drawing)
6. (a) $q = -8.6$ in.
   (b) $M = 3.44$ in.
   (c) virtual, erect
   (d) (drawing)
8. (a) $v_1 = 2.26 \times 10^{10}$ cm/sec
   (b) $\lambda = 4.52 \times 10^{-5}$ cm
10. $q = -6.76$ in. below
12. (a) $\theta_c = 42°30'$
    (b) $x = 7.33$ in.
14. $q = 10$ in.; $M = 1\times$; inverted; real
16. $q = -24$ in.; $M = -2\times$; virtual; erect
18. (a) $q = -3.4$ in.
    (b) $M = -1.14$ in.
    (c) virtual, erect
    (d) (drawing)
20. $f = -5.3$ in.
22. (a) $q = -3.2$ in.
    (b) $M = -0.4\times$
    (c) virtual, erect
    (d) (drawing)
24. -2.96 diopters
26. (a) $q = -10.5$ in.
    (b) $M = -\frac{1}{8}$ in.
    (c) virtual, erect
28. (a) $q = -11$ in.
    (b) $M = -0.21$ in.

chapter 30 (cont.)

(c) virtual, erect
30. $q_2 = -0.976$ in.
32. $q_2 = -7.1$ in.
34. $q_2 = 5.22$ in.
36. $f_1 = 58.4$ cm; $n_1 = 1.154$

Chapter 31
2. $f = -60$ cm diverging
4. $q = -32.1$ cm left of lens
6. $f = -60$ diverging
8. (a) $q_2 = 7.8$ cm
   (b) $h_i = 1.12$ cm
   (c) real, erect
10. (a) $q_2 = -3.3$ cm
    (b) $h_o = -3.3$ cm
    (c) inverted, virtual
    (d) drawing
12. (a) $q_2 = -9.8$ cm
    (b) $h_i = -0.369$ cm
    (c) inverted, virtual
    (d) drawing
14. (a) $q_2 = -12$ cm (real)
    (b) $M = +0.021 \times$ (erect)
    (c) (drawing)
16. $f = 1.32$ cm
18. (a) $f = 2.3$ cm
    (b) $h_i = 72$ cm
    (c) $p = 2.04$ cm
20. (a) $f_2 = 18.8$ cm
    (b) $M = -7.14 \times$
22. (a) $M = -25 \times$
    (b) $l_a = 52$ cm
    (c) $l_t = 92$ cm
24. $l = 138.3$ cm
26. $M = 200 \times$
28. (a) $f/9$
    (b) $f/18$
    (c) $t_2 = 0.02$ sec
30. $f = 27.7$ in.

Chapter 32
2. (a) $\theta = 45°$
   (b) $t = 4.000$ in.
4. leaves the bottom at $20°$ normal
6. $D_m = 22°36'$

8. assume incident angle $\theta = 50°$ and $\theta = 40°$; then D_m lies between both angles
10. (a) $\omega = 0.024$
    (b) $v = 41.67$
12. (a) $A_1 = 6.34°$
    (b) $D_{F_1-H_1} - D_{F-H} = 0.0441$
14. (a) $A_{DF} = 6.51$
    (b) $\Delta_D - \Delta_D' = 0.9563°$

Chapter 33
2. $\lambda = 5330$ A
4. $l = 0.324$ cm
6. $m = 125$
8. $\lambda = 6720$ A
10. (a) $t = 1.35 \times 10^{-4}$ cm
    (b) $r = 0.0735$
        $d = 0.147$ cm
    (c) dark
12. (a) $t = 1.46 \times 10^{-4}$ cm
    (b) $m = 155$ dark bands
14. (a) $\lambda = 6349$ A
    (b) $\theta = 19$ sec
16. $l = 51.84$ cm
18. (a) $\lambda = 5085$ A
    (b) $m = 3$
20. (a) separation $= 4°26'$
    (b) separation $= 9°31'$
    (c) $L_R = 21.195$ cm; $\}$
        $L_Y = 17.63$ cm; $\}$ first order
        $L_B = 13.4$ cm. $\}$
        $L_R = 42.39$ cm; $\}$
        $L_Y = 35.26$ cm; $\}$ second order
        $L_B = 26.8$ cm. $\}$
22. $d = 3.0303 \times 10^{-8}$ cm
24. (a) $R_G = 21,000$
    (b) $\Delta\lambda = 0.01181$
    (c) $\Delta\lambda = 0.01181$

Chapter 34
2. (a) 44.15%
   (b) 29.35%
   (c) 20.65%
   (d) 33.5%
4. 30.8%
6. $\theta = 64°32'$
8. B–line $= \theta = 56°50'$

chapter 34 (cont.)

C–line $= \theta = 56°51'$
D–line $= \theta = 56°54'$
F–line $= \theta = 57°1'$
H–line $= \theta = 57°11'$

10. $\lambda_o = 3713$ A
    $\lambda_e = 4411$ A

Chapter 35

2. $y = 3.5 \times 10^{-3}$ m
4. (a) $f = 2.88 \times 10^{-15}$ nt
   (b) $r = 0.626$ m
6. 3.03×10^6 coul/kg
8. 1.21×10^7 coul/kg
10. (a) $V = 2.7 \times 10^7$ volts
    (b) $v_{max} = 7.185 \times 10^7$ m/sec
    (c) $KE = 4.31 \times 10^{-12}$ joule
    (d) $t = 2.185 \times 10^{-8}$ sec
        $F = 2.29 \times 10^7$ cps
    (e) $E = 2.7 \times 10^7$ ev
12. (a) $F = 6 \times 10^{14}$ vib/sec
    (b) 2.078 ev
    (c) 5.031×10^{14} vib/sec
    (d) $\lambda = 5963$ A

Chapter 36

2. (a) e = 80; p = 80; n = 122
   (b) e = 82; p = 82; n = 126
   (c) e = 48; p = 48; n = 64
   (d) e = 74; p = 74; n = 110
   (e) e = 54; p = 54; n = 75
4. (a) $_{13}Al^{27}$
   (b) $_{24}Cr^{52}$
   (c) $_{42}Mo^{98}$
   (d) $_{53}I^{127}$
   (e) $_{83}Fr^{221}$
6. (a) $\lambda = 937.8$ A
   (b) $\lambda = 4{,}103$ A
   (c) $\lambda = 10{,}941$ A
   (d) $\lambda = 26{,}260$ A
   (e) $\lambda = 74{,}599$ A
8. (a) $r_n = 1.325 \times 10^{-7}$ cm
   (b) $v_n = 4.38 \times 10^5$ cm/sec
10. (a) K-shell 2 in $1s^2$ subshell
    (b) L-shell 6 in $2P^6$ subshell
    (c) M-shell 2 in $3s^2$ subshell
    (d) none left for $3s^6$

Chapter 37

2. (a) $\frac{1}{4}$ gm remains
   (b) 8.85×10^{21} atoms
4. 50 hr
6. (a) $_{17}Cl^{35} + {_1}H^1$
   (b) $_6C^{13} + hF$
   (c) $_{-1}e^0 + {_{12}}Mg^{24} + {_2}He^4$
   (d) $_2He^4 + {_2}He^4$
   (e) $_3Li^8 + {_1}H^1$
8. (a) $_2He^4 + {_{10}}Ne^{20}$
   (b) $_4Be^8 + {_1}H^2$
   (c) $_{15}P^{30} + {_0}n^1$
   (d) $_{28}Ni^{65} + {_1}H^1$
   (e) $_6C^{13} + {_1}H^1$
   (f) $_1H^2 + hF$
   (g) $_7N^{13} + {_0}n^1$
   (h) $_3Li^7 + {_2}He^4$
   (i) $_3Li^7 + {_2}He^4 + {_2}He^4$
   (j) $_4Be^8 + hF$
   (k) $_1H^1 + {_0}n^1$
   (l) $_4Be^8 + {_1}H^3$
10. (a) $_{85}At^{211} + 2(_0n^1)$
    (b) $_{96}Cm^{241} + 2(_0n^1)$
    (c) $_{97}Bk^{243} + 2(_0n^1)$
    (d) $_{98}Cf^{244} + 2(_0n^1)$

Chapter 38

2. 3.27 Mev
4. (a) 8.76×10^4 kw-hr/g
   (b) 2.324×10^{11} ft-lb/g
6. $E = 19.28$ Mev
8. (a) $E = 224.845$ Mev
   (b) 8.33 Mev/nucleon

Index

A

Absolute temperature, 7
Absolute zero, 217
Absolute-zero scale, 217
Absorption by polarization, 551–552
Acceleration, 85–86, 555–580
   angular, 124
   centripetal, 115–117
   constant, 87–91
   gravitational, 91–98
   negative, 87
   particle, 571–575
   uniform, 86
Achromatization, 520–523
Action lines, 16
Addition:
   graphical, 19–21
   mathematical, 22–24
   vector, 17
Adhesion, 176–180
Adiabatic expansion, 264–266
Air wedge, 532–533
Algebraic sum, 282
Alpha particles, 595, 596
Alternating current, 408–428
   power in circuit, 425–426
Alternating-current generators, 392
Ammeters, 339–340
Ammeter-voltmeter method, 341–342

Amperes, 6, 237, 365–366
   defined, 307, 365
Ampere's theorem, 372
Ampere-turn per meter, 360
Amplitude, 433
Analyzers, 546
Angle:
   critical, 478
   deviation, 513
   incident, 469
   of reflection, 469
   of repose, 64
Angular separation, 520–523
Anodes, 556
Antinodes, 437
Appendix, 619–650
Aqueous humor, 494
Archimedes' principle, 184–186
Armatures, 394
Atmospheres:
   one, 164
   two, 164
Atomic numbers, 305, 581–582, 596, 603
Atomic-mass numbers, 596, 603
Atoms, 304–307, 581–594
   neutral, 304
   nucleus of, 304
   structure of, 581–583
Avogadro's law, 248–249

B

Balmer, Johann, 583
Balmer series, 583
Barometers, 174–176
    aneroid, 175–176
Baryons, 615
Batteries, 310–311
    multiple circuits, 329–331
    in parallel, 315–317
    in series, 315–317
Bernoulli's theorem, 198–199
Beta particles, 595–596
Betatrons, 574
Bimetallic strips, 225
Binding energy, 608–609
Blind spot, 494
Block and tackle, 145–146
Bohr, Niels, 581, 589–590, 602
Bohr's hydrogen atoms, 586–592
Boom, the, 30–31
Bourdon gages, 176
Box Wheatstone bridge, 344–346
Boyle's Law, 246–247
Bremsstrahlung effect, 593
Brewster's law, 548–549
British units, 3–7, 229
Buoyancy, 184–197
Bureau des Poids et Mesures, 4

C

Calories, 229
Calorimeters, 236
Calorimetry, 232–236
Cameras, 507–508
Candela, 6
Candle, 460
Capacitance, 294–303
    decay, 403–404
    growth, 403–404
    pure, 417–418
    resistance and, 420–424
    spherical, 296–297
    tubular, 296–297
Capacitive reactance, 418
Capacitors, 295
    energy of a charged, 300–302
    in parallel, 297–300
    in series, 297–300
Carnot cycle, 267–269
Cathode-ray tubes, 564–568
Cathode rays, 563–564
Cathodes, 456
Celsius scale, 216
Center of gravity, 46–51
Centigrade scale, 216

Centimeters, 7
Centroids, 46
Cesium resonators, 5
Chain hoists, 146–147
Chain reactions, 612
Charles, Jacques, 247
Charles' Law, 247–248
Circuits, 372–383
    magnetic, 378–381
    multiple-battery, 329–331
    simple, 304–328
Clocks, 4–5
Cloud chambers, 599–602
Coefficient of discharge, 203–204
Coercivity, 377
Cohesion, 176–180
Coils:
    force on, 368–370
    induced voltage in, 385–387
Collinear reactions, 15
Compressibility, 78
Condensation, 431–432
Conduction, 240–242
Conductors, 304–307, 559
Continuity principle, 199–201
Convection, 240
Coolidge tubes, 578
Coplanar forces:
    concurrent, 26–38
    nonconcurrent, 39–58
Cosmic rays, 615
Coulombs, 275
Coulomb's law, 275–276, 358–360
Couples, 40–41
Crookes tube, 562–563
Current, 6, 304–328 (*see also* Alternating current; Direct-current generators)
    eddy, 405–406
    effective value, 412
    Kirchhoff's law, 332–335
    measurement of, 348–349
    -voltage measurement, 340–341
Cycles, 408
Cyclotrons, 571–575

D

Da Vinci, Leonardo, 453
D'Alembert, Jean, 108
D'Alembert's principle, 108–109
Dalton's law of partial pressures, 250
D'Arsonval galvanometer, 336–338
Deceleration, 87
Deflection, 555–580
DeForest, Lee, 557
Density:
    defined, 163

Index

Density (*cont.*):
  flux, 362–363
  weight, 162
Deuterons, 583
Deviation, 513–516
Diamagnetic material, 376
Dielectric coefficient, 296
Dielectric strength, 300
Dielectrics, 295–296, 304–307
Diffraction, 525–544
  crystal, 539–540
  gratings, 537–538
  single-slit, 534–536
Diopters, 481
Dip, 382
Direct-current generators, 393
Dispersion, 518–520
Displacement, 85, 511–513
Displacement ratio, 143
Division, 12
Domain, 357
Doppler effect, 447–449
Double slit, 525–530
Dyne-centimeter, 136

E

Earth, 456–457
  geographic axis, 381
  magnetic axis, 381–382
  magnetic field, 381–382
Eddy currents, 405–406
Edison, Thomas, 556
Einstein, Albert, 82, 454–455
Elastic limit, 74
Elasticity, 74–79
  modulus of, 75–76
  volume, 77–78
Electric charge, 272–274
Electric field, 280–281, 287–291
Electric flux, 280–281
Electric lines of force, 280–281
Electromotive force, 311
  average, 410–411
  counter-, 399
  effective, 412–413
  instantaneous, 408–410
  measurement of, 346, 347
Electron spin, 593–594
Electrons, 304–308, 581
  distribution, 589–592
  free, 589
  in orbit, 304–305
  valence, 559
Electron-volt, 556
Electroscopes, 273–274
Electrostatics, 272–293

Energy, 7, 137–140, 405, 606–617
  binding, 608–609
  of a charged capacitor, 300–302
  heat, 236–239
  kinetic, 137–140
  mechanical, 142–143
  potential, 137–140, 281–287
  radiant, 460
Energy shells, 306
Energy waves, 307
Equation of state, 250
Equilibrium, 26–29, 41–45
  rotational, 41
  through mixing, 231–232
Equipotential surfaces, 281–287
Expansion, 215–228
  area, 220–221
  of gases, 224–225
  linear, 219–220
  temperature in, 215–219
  volume, 221–224
Exponents, 10
Eye, the, 494–496

F

Fahrenheit scale, 216
Farads, 295
Farsightedness, 495
Ferromagnetic material, 376
Ferromagnetism, 376–378
Fission, 611–614
Fizeau, Armand, 457
Fluoresce, 563
Fluids, *see* Liquids
Flux:
  electric, 280–281
  luminous, 459–461
  radiant, 459–461
Flux density, 362–363
Flux lines, 357
Flux loops, 378
Foot, the, 7
Foot-pound, the, 135
Forces, 7, 14–25, 356–358
  (*see also* Electromotive force)
  applied, 164
  between two current-carrying wires, 375–376
  buoyant, 184
  centrifugal, 132, 304
  centripetal, 131–132, 304
  on a charge in a magnetic field, 366–367
  on a coil placed into a magnetic field, 368–370
  completely defined, 16
  concurrent coplanar, 26–39

664 *Index*

Forces (*cont.*):
    contact, 26–27
    defined, 14–15, 16
    electric lines of, 280–281
    field, 26–27
    gravitational, 14–15, 60
    impulse of, 152
    line of action of, 39
    magnetomotive, 380
    moment of, 39–41
    nonconcurrent coplanar, 39–59
    nonconcurrent nonparallel, 44–45
    normal, 60
    not parallel to the incline, 67–69
    not parallel to the plane, 62–63
    parallel, 41–44
    parallel to the incline, 66–67
    parallel to the plane, 61–62
    resisting, 59
    resolution into components, 22
    sliding under weight only, 64–66
    on a straight wire placed into a magnetic field, 367–368
Force diagrams, 26–29
Foucault, Jean, 458–459
Frames of reference, 80–81
Fraunhofer, Joseph, 517
Fraunhofer lines, 517–518
Free-body diagrams, 26–29
Frequency, 408, 433–434
    threshold, 575
Friction, 59–73
    coefficient of, 59–61
    dynamic, 60
    limiting, 59
    in machines, 147–148
    rolling, 69–70
    sliding, 60, 61, 64–65
    static, 59, 61, 64–65
    vectors and, 61–69
Fusion, 610–611
    latent heat of, 233

G

Galileo, 456
Galvanometers:
    d'Arsonval, 336–338
    tangent, 382
Gamma rays, 596
Gases, 246–258
    diatomic, 256
    expansion of, 224–225
    monatomic, 255–256
    polyatomic, 256
    specific heat and, 251–253
Geiger counters, 595

Generators, 390–397
    alternating-current, 392
    direct-current, 393
Grams, 6
Graphic representation, 413–415
Gravitational acceleration, 91–98
Gravitational attraction, 46
Gravitational forces, 14–15, 60
Gravity:
    center of, 46–51
    specific, 163–164, 186–189
        determinations, 190–193
Gyration, radius of, 125–126

H

Hahn, Otto, 611
Half-life, 598
Half-wavelength shift, 530–532
Heat, 215
    of combustion, 236
    constant, 264–266
    energy, 236–239
    kinetic theory of, 253–257
    latent, of fusion, 233
    latent, of vaporization, 233
    sensible, 230, 232
    specific, 229–231
        gases and, 251–253
    transfer, 240–244
Helical springs, 119–122
Helium, 304–305
Henry, the, 397
Hooks, Robert, 75
Horsepower, 140, 202
Huygens, Christian, 453
Hydrogen atoms, 304–306, 586–592
Hydrogen series, 583–586
Hydrometry, 193–195
Hyperons, 615
Hysteresis curve, 377
Hysteresis loops, 376–378

I

Impact:
    elastic, 153–158
    inelastic, 153–158
Impedance, 417
Impulse:
    of a force, 152
    momentum and, 151–153
Inductance:
    decay, 401–403
    growth, 401–403
    mutual, 397–399
    pure, 415–417

Inductance (*cont.*):
  resistance and, 418–420, 422–424
  -self, 399–401
Induction, 384–407
Inductive reactance, 416
Inertia, 102
  defined, 14
  electrical, 397
  moment of, 7, 102–103, 124–125
    about any axis, 126–128
    of an area, 126
Insulators, 304–307
Intensity, luminous, 6
Interference, 525–544
  constructive, 437
  destructive, 437
  polarization and, 552–553
  in a thin film, 530–532
Interferometers, Michelson, 528–529
International Committee on Weights and Measures, 91, 307
International Committee on Weights and Standards, 460
"International System of Units," 3
Ionization potential, 562
Ions, 569
Isentropic change, 264–266
Isobaric change, 261–263
Isometric change, 260–261
Isothermal change, 263–264
Isotopes, 306, 569

J

Joule, the, 136
Jupiter (planet), 456

K

Kelvins, 6, 217
Kilograms, 6
Kinetic energy, 137–140
Kinetic theory of heat, 253–257
Kirchhoff's current law, 332–335
Kirchhoff's voltage law, 332–335
Kundt's tube, 442

L

Lane patterns, 539
Lawrence, Dr. Ernest, 571
Length, 6, 7
  British units of, 5
Lenses:
  concepts and conventions
    multiple, 494–510
    negative, 480–481

Lenses (*cont.*):
  positive, 480
  resolving power of, 540–542
  systems, 496–500
  thin, 479–490
    aberrations, 490
    in contact, 488–489
Lenz's law, 388–390
Leptons, 615
Light, 453–467
  development of theory of, 453–455
  point source of, 461–464
  polarization of, 545–554
  velocity of, 455–459
Light year, 5
Liquids, 162–214
  buoyancy, 184–197
  flow through an orifice, 204–206
  impact of flow, 202–203
  motion, 198–214
  properties of, 162–164
  quantity flow, 199–201
Livingston, Dr. Stanley, 571
Loads:
  dead, 33
  live, 33
Lumen, 460, 462
Luminosity curve, 460
Luminous flux, 459–461
Luminous intensity, 6

M

Machines:
  defined, 142
  efficiency of, 143
  friction in, 147–148
  mechanical advantage of, 142
  non-reversible, 147
  reversible, 147
Magnetic fields, 372–383
  of the earth, 381–382
  force on a charge in, 366–367
  force on a coil placed into, 368–370
  force on a straight wire placed into, 367–368
  intensity, 360–362
  torque on a coil placed into, 368–370
Magnetic moment, 364
Magnetic poles, 356–358
Magnetism, 356–371
Magnetite, 356
Magnetomotive force, 380
Magnets, 356–358
  bar, 356
  torque on, 364–365
Magnifying glasses, 500–502

Magnitude, 15–16
Manometers, 170–176
Mass, 5, 6, 7
　standard, 103
　subcritical, 612
Mass numbers, 305, 581, 596
Mass spectrographs, 569–570
Maxwell, James, 454
Mechanical advantage:
　actual, 143
　ideal, 143
Meson field, 615
Mesons, 615–616
Meter-kilogram-second-ampere system, 275
Meters, 6
　defined, 4
Metric system, 3–4
Michelson, Albert, 459
Michelson interferometers, 528–529
Microscopes, compound, 502–503
Mirrors:
　concave, 472–475
　concepts and conventions, 470–472
　convex, 472–475
　plane, 468–470
Modulus:
　bulk, 77–78
　of elasticity, 75–76
　of rigidity, 76–77
　shear, 76–77
　Young's, 75
Molecules, 306
Moment arm, 39
Moment of force, 39–41
Moment of inertia, 7, 102–103, 124–125
　about any axis, 126–128
　of an area, 126
Momentum, 7
　angular, 159
　impulse and, 151–153
Motion:
　angular, 112–134
　　laws of, 124–134
　　relationship to linear motion, 112–115
　compressional, 429
　frames of reference and, 80–81
　harmonic, 117–122, 128–130
　　the helical spring, 119–122
　　simple, 117
　　simple pendulum, 118–119
　laws of, 15
　linear, 80–101
　　relationship to angular motion, 112–115
　liquids, 198–214
　longitudinal, 429

Motion (*cont.*):
　Newton's laws of, 102–111
　　applications of, 105–108
　　First, 102–104
　　Second, 104–105
　　Third, 15, 105
　transverse, 429
　uniform, 102
　wave, 429–452
Motors, 390–397
Multiplication, 12

N

Nearsightedness, 495–496
Neutrino, 596
Neutrons, 305–306, 581
Newton, Sir Isaac, 102–111, 453, 455
　laws of motion, 102–111
　　applications of, 105–108
　　First Law of, 102–104
　　Second Law of, 104–105
　　Third Law of, 15, 105
Newton-meter, 136
Newtons, 275
Newton's law of cooling, 243
Newton's rings, 532–534
Nodes, 437
Nonflow process, 267–269
Nozzles, 206–207
N-type (negative) crystals, 560
Nuclear reactions, 606–608
Nucleons, 305, 581, 608

O

Office of Weights and Measures, 3–4
Ohm, George S., 308
Ohmmeters, 343
Ohms, 308
Ohm's law, 307–310
Optical axis, 549
Organ pipes, 443–445
Oscillation, period of, 120

P

Parallelogram method, 17–19
Paramagnetic material, 376
Particle acceleration, 571–575
Particle displacement, 433–437
Particles:
　alpha, 595–596
　beta, 595–596
　elementary, 614–617
Pascal's principle, 168–169
Pauli, Wolfgang, 590

Index

Pauli exclusion principle, 590
Pendulums, 118–119
  torsion, 128–130
Periods, 118, 408, 433
Permeability:
  constant, 376–377
  relative, 379
Permittivity for a material, 296
Phase, change of, 232
Phosphoresce, 563
Photoelectric cells, 575
Photoelectric effect, 575–578
Photometers:
  Bunsen, 464–465
  Lummer-Brodhun, 464
Photometry, 464–466
Photons, 575–576
Physics, defined, 1
Pitot tubes, 208–211
Planck, Max, 454–455, 575
Planck's constant, 575
Planes:
  forces not parallel to, 62–63
  forces parallel to, 61–62
  inclined, 64–69, 143–144
  level, 61–63
Planimeters, 259
Polarization, 545–554
  by absorption, 551–552
  interference and, 552–553
  by reflection, 548–549
  by refraction, 549–551
  by scattering, 551–552
Polarizers, 546
Polaroid, 546
Poles, pairs of, 408
Polygon method, 21
Positrons, 601
Potential, 281–287
Potential difference, 281–287
Potential energy, 137–140, 281–287
Potentiometers, 346–349
Pound, the, 5
Power, 140–141, 321–324
  in an ac circuit, 425–426
Pressure, 77, 164–167
  constant, 261–263
  Dalton's law, 250
  defined, 164
  gage, 166, 170
  Pitot tube for measuring, 208–211
  transmission of, 168–170
Principia (Newton), 102
Prisms, direct-vision, 520–523
Projectors, 508
Prony brake, the, 141–142
Protons, 304–306, 581, 582

P-type (positive) crystals, 560
P-V diagrams, 259–271
Pyrometers:
  optical, 226
  thermoelectric, 226–227
Pythagoras, 453

Q

Quadratic equations, 10–12

R

Radian, the, 112
Radiant energy, 460
Radiant flux, 459–461
Radiation, 242–244
Radioactive series, 597–599
Radiactivity, 595–596
Radius of gyration, 125–126
Rarefaction, 431–432
Reactance:
  capacitive, 418
  inductive, 416
Reactions, collinear, 15
Rectifiers, 556–559
  full-wave, 557
  half-wave, 556
Reflection, 242
  angle of, 469
  law of, 469
  by polarization, 548–549
Refraction, 475–479, 511–524
  double, 550
  index of, 475–479
  by polarization, 549–551
Relativity, special, 81–84
Resistance, 307–308
  capacitance and, 420–424
  decay, 401–404
  growth, 401–404
  inductance and, 418–420, 422-424
  internal, 311
  measurement, 341–349
  pure, 415
  temperature coefficient, 319–321
Resistivity, 317–319
Resistors, 307
Resonance, 426–427, 445–447
Resonators, cesium, 5
Resultant torque, 41
Retentivity, 377
Roemer, Olaf, 456–457
Roentgen, Wilhelm, 578
Roentgen X-ray tubes, 578
Rotation, center of, 39
Rotational equilibrium, 41

Rowlands, 380
Rutherford, Ernest, 581, 595, 599–600
Rydberg, 583
Rydberg constant, 583

S

Scalars, 3
Scattering, 551–552
Scintillation counters, 601
Screw threads, 144–145
Seconds, 4–5, 6, 7
    defined, 4
Sine waves, 408–410
Siren disks, 445
Skin effect, 177
Slide rules, 12–13
Slide-wire Wheatstone bridge, 343–344
Slug, 7
Snell's law, 478
Solenoids, 374
Solids, 186–189
    center of gravity, 49–51
Sommerfeld, Arnold, 590
Sound, 429–452
    Doppler effect for, 447–449
Space diagrams, 16
Spectra, 516–518, 537
    band, 517
    bright-line, 517
    continuous, 516–517
    dark-line, 517
    emission, 517
Spectrographs, mass, 569–570
Spectrometers, 515–516
Speed, 85–86
Stationary obstruction, 202–203
Steady state, 102
Stefan-Boltzmann law, 243
Steradians, 462
Stoner, 589–590
Strain:
    compressive, 74
    shear, 74
    stress and, 74–75
    tensile, 74
Stress:
    compressive, 75
    shear, 75
    strain and, 74–75
    tensile, 75
Studying, dynamics of, 2–3
Sun, the, 456–457
Surface tension, 176–180
Synchrotrons, 574

T

Telescopes, 503–507
    astronomical, 503–505
Temperature:
    absolute, 7
    coefficient of resistance, 319–321
    constant, 263–264
    in expansion, 215–219
    instruments used as indicators of, 225–227
    thermodynamic scale, 269–270
Tension, surface, 176–180
Thermionic emission, 555–556
Thermodynamic temperature scale, 269–270
Thermodynamics, second law of, 269
Thermometers, 215–216
    constant-volume gas, 225–226
    resistance, 227
Thermopiles, 227
Thomson, J. J., 568
Time, 6, 7
Toggle, the, 31–33
Toroids, 374
Torque, 39–41
    on a coil placed into a magnetic field, 368–370
    on a magnet, 364–365
    resultant, 41
Trajectories, 91–98
Transformers, 398
Transistors, 559–561
    junction type N-P-N, 561
    point-contact, 561
Transmissibility, principle of, 16
Transmutation, 595–605
    artificially induced, 602–604
Triangle method, 19
Trigonometry, 9–10
Triodes, 556–559
Tritons, 582
Truss, the, 33–35
Tyratrons, 561–562

U

Units:
    British, 3–7
    derived, 7
    fundamental, 6
    operation with, 5–9
    standardizing basic, 3–5
    systems of, 7
Universal gas law, 248–250

V

Valence electrons, 559
Valence numbers, 306
Vaporization, latent heat of, 233
Vector addition, 17
Vector diagrams, 16
Vectors, 15–16
    friction and, 61–69
    graphic summation of, 17–19
    graphical addition of more than two, 19–21
    mathematical addition of, 22–24
    quantities, 3
Velocity, 85–86, 433–437, 576
    average, 86
    constant, 86–87
    instantaneous, 85–86
    of light, 455–459
    magnitude of, 116
    variable, 87–91, 92
Vena contracta, 203–204
Venturi meters, 207–208
Vibrating rod, 440–443
Vibrating string, 438–440
Vibration, 438–443
    first harmonic, 439
    second harmonic, 439
Viscosity, 195
Vitreous humor, 494
Voltage, 308
    -current measurement, 340–341
    effective value, 412
    ideal, 311
    induced in a moving coil, 385–387
    induced in a straight wire, 384–385
    Kirchhoff's law, 332–335
Voltmeter-ammeter method, 341–342
Voltmeters, 338–339
Volts, 237
Volume:
    constant, 260–261
    elasticity, 77–78
    expansion, 221–224

W

Water equivalent, the, 230
Watts, 140, 237
Wave fronts, 469
Wave motion, 429–452
Wavelength, 433
    half shift, 530–532
Waves:
    compressional, 429
    energy, 307
    longitudinal, 429–433, 435
    sine, 408–410
    stationary, 437–438
    transverse, 429–433, 434
Weights, 5, 46, 49, 60, 103
    density, 162
    molecular, 249
    sliding under force of, 64–66
    specific, 162, 164, 186–189
        determinations, 190–193
Weightlessness, 5–6
Wein's constant, 243
Wein's law, 243
Wheatstone bridge:
    box, 344–346
    slide-wire, 343–344
Wires:
    force on, 367–368
    forces between two current-carrying, 375–376
    induced voltage in, 384–385
    long straight, 375
    loop of, 373
Work, 135–137, 307
    external, 259–260
Work function, 555

X

X-rays, 578–579, 592–593

Y

Young, Thomas, 453–454, 525
Young's modulus, 75
Yukawa, 615

Z

Zero, absolute, 217